Springer Series in Advanced Microelectronics

Volume 37

The Springer Series in Advanced Microelectronics provides systematic information on all the topics relevant for the design, processing, and manufacturing of microelectronic devices. The books, each prepared by leading researchers or engineers in their fields, cover the basic and advanced aspects of topics such as wafer processing, materials, device design, device technologies, circuit design, VLSI implementation, and subsystem technology. The series forms a bridge between physics and engineering and the volumes will appeal to practicing engineers as well as research scientists.

More information about this series at http://www.springer.com/series/4076

Rino Micheloni · Alessia Marelli
Kam Eshghi
Editors

Inside Solid State Drives (SSDs)

Second Edition

Editors
Rino Micheloni
Microsemi Corporation
Vimercate, MB
Italy

Kam Eshghi
Lightbits Labs
San Jose, CA
USA

Alessia Marelli
Microsemi Corporation
Vimercate, MB
Italy

ISSN 1437-0387 ISSN 2197-6643 (electronic)
Springer Series in Advanced Microelectronics
ISBN 978-981-13-0598-6 ISBN 978-981-13-0599-3 (eBook)
https://doi.org/10.1007/978-981-13-0599-3

Library of Congress Control Number: 2018942187

Printed on acid-free paper

This Springer imprint is published by the registered company Springer Nature Singapore Pte Ltd.
The registered company address is: 152 Beach Road, #21-01/04 Gateway East, Singapore 189721, Singapore

To my wife Sabrina, and my daughters Laura and Greta

Rino Micheloni

To my husband Michele and my daughter Elena for their unconditional love

Alessia Marelli

To my wife Nazila, and my daughters Elika and Vionna, who brighten up my life with their love

Kam Eshghi

Foreword

Error Correcting Coding for Solid State Disk Data Storage

Wireless communication had existed for half a century when Information Theory was expounded by Claude Shannon in the Bell System Technical Journal in 1948. Error correcting coding followed in primitive formulations which brought early digital communication systems only a short way toward the Shannon capacity limit. Various generations of algebraic codes: Hamming, BCH and Reed Solomon made gradual progress. With the advent of digital satellite transmission and soft-decision decoding of convolutional codes, the gap between uncoded performance and the Shannon limit was cut in half. Similar technology was used in second and third generation (2 G and 3 G) mobile phone voice modems. Finally turbo codes and low density parity check (LDPC) codes, which arrived about two decades ago, gradually were shown to greatly decrease the distance to the capacity limit. These technologies have entered predominant use for data transmission in 3 G and 4 G mobile modems.

High density data storage technology has followed a similar trajectory though with a more contracted time span. BCH and Reed Solomon codes were the norm until recently for hard disk drives (HDD). Recently though LDPC has taken root here too with major improvements in data density and reading and writing controller speeds. With the advent of the "smart phones" and tablets, solid state drives (SSD) became ever more important for their low latency and low power operation. For this use LDPC is becoming the norm as well. This book which covers all aspects of SSD technology also provides coverage of the important topic of ECC.

La Jolla, CA, USA

Andrew Viterbi
President
Viterbi Group, LLC

Preface to the Second Edition

We started writing the first edition of *Inside Solid State Drives (SSDs)* back in 2011, and the book was first published in 2013. At that time, SSDs were considered as the "new" technology in the storage space, but not really a "shining star" as they are seen today.

Over the past few years, we have collected a lot of feedback and questions about our book. Moreover, both SSD and Flash technologies have significantly changed along the way. Therefore, we thought it was the right time to refresh the content of "Inside Solid State Drives."

As editors, we have pushed all co-authors to refresh each chapter (Thank You ALL!), in terms of both the content and the bibliography.

But this second edition is much more than that.

As mentioned, SSD technologies have significantly changed in the last five years and we realized that there was the need to add three completely new chapters: Chaps. 5, 7, and 9.

In 2013, Flash manufacturers were still fighting against the challenges of shrinking the size of planar memory cells to keep up with the expectations of the market in terms of $/bit. Now, Flash technology is 3D (i.e., vertically integrated) and there is a new dimension to consider: the number of memory layers (100+ in the near future). 3D NAND Flash appeared in the market at the end of 2015, but there is still a plethora of alternatives around, based on different architectures and memory technologies (floating gate and charge trap). Chapter 5 covers 3D Flash array architectures with a lot of bird's-eye views, to help the reader understand better the new challenges that technologists and developers have to face.

In all SSDs, a Flash microcontroller sits between one or multiple hosts (i.e., CPUs) and NAND Flash memories, and on each side, there are a lot of challenges that designers need to overcome. Moreover, a single controller can have multiple cores, with all the complexity associated with developing a multi-threaded firmware. Chapter 7 is about how to make simulations of such a complex system, by providing insights into design trade-off and simulation strategies. As usual, simulation speed and precision do not go hand in hand, so it is important to understand when to simulate what. Of course, being able to simulate SSD's performances is necessary to meet time-to-market, as well as price and quality targets.

Nowadays, SSDs are electronic systems much more complex than in the past, especially because they have to manage a lot of 3D memories, by using several algorithms (wear leveling, Error Correction Code, soft decoding, randomization, read retry, etc.) at a very high speed (especially with PCIe/NVMe drives).

Chapter 9 is exactly designed to offer a comprehensive overview of the most recent Flash management techniques (aka Flash Signal Processing). We are sure that technologists, engineers, and scientist will appreciate the unbelievable level of know-how required by the management of electrons and holes inside nonvolatile memory cells.

We really placed our best effort in updating this book. Enjoy the reading!

Vimercate, Italy Rino Micheloni
Vimercate, Italy Alessia Marelli
San Jose, USA Kam Eshghi

Preface to the First Edition

Solid State Drives (SSDs) are gaining momentum in enterprise and client applications, replacing Hard Disk Drives (HDDs) by offering higher performance and lower power. In the enterprise, developers of data center server and storage systems have seen CPU performance growing exponentially for the past two decades, while HDD performance has improved linearly for the same period. Additionally, multi-core CPU designs and virtualization have increased randomness of storage I/Os. These trends have shifted performance bottlenecks to enterprise storage systems. Business critical applications such as online transaction processing, financial data processing and database mining are increasingly limited by storage performance.

In client applications, small mobile platforms are leaving little room for batteries while demanding long life out of them. Therefore, reducing both idle and active power consumption has become critical. Additionally, client storage systems are in need of significant performance improvement as well as supporting small robust form factors. Ultimately, client systems are optimizing for best performance/power ratio as well as performance/cost ratio.

SSDs promise to address both enterprise and client storage requirements by drastically improving performance while at the same time reducing power.

Inside Solid State Drives walks the reader through all the main topics related to SSDs.

A Solid State Drive is a very complex system: Chapter 1 contains an overview of the main blocks, including hardware and software.

Chapters 1 and 2 cover different SSD implementations with host interfaces ranging from SAS/SATA to PCI Express (PCIe). SAS/SATA offer compatibility with legacy storage infrastructure. However, for many applications, NAND Flash read and write speeds are exceeding the capabilities of these legacy interconnects. PCIe SSDs overcome this bottleneck and deliver unparalleled performance while, at the same time, reducing latency, power and cost by eliminating the traditional storage infrastructure and attaching directly to a platform's PCIe I/O interconnect.

SSDs and HDDs can also be combined together in various forms, as explained in Chapter 3 where "hybrid" storage is analyzed.

At the end of the day, a SSD is made up of NAND memories and a controller. Therefore, to understand SSDs it is important to understand all the basics of NAND Flash technology (Chapter 4) as well as design (Chapter 6).

When aiming to replace HDDs, particularly in enterprise applications, another key consideration is reliability. SSDs are complex electronic systems prone to wear-out and failure mechanisms mainly related to NAND. SSD reliability is analyzed at different levels in Chapter 8. The basic physical mechanisms affecting the traditional floating-gate cells and the possibility of anomalous erratic behavior is discussed, as well as disturbs arising because several cells share the same control lines. Solutions adopted to improve system reliability are presented, such as the use of RAID and protection against power loss during write operations. Test methods for endurance and retention verification are also described.

The physical constraints of Flash memory pose a lifetime limitation on these storage devices. Multilevel Flash technologies (MLC) further degrade endurance, as 2 bits are stored in the same physical cell. As a result, NAND devices may experience an unexpectedly short lifespan, especially when accessing these devices at high frequencies. In order to enhance the endurance, wear leveling algorithms are used to evenly erase blocks. Chapter 10 describes some existing wear leveling algorithms, highlighting their pros and cons.

Despite all the possible Flash management algorithms run by the memory controller, the residual BER needs to be properly managed in order to achieve a reliable system. That is why *Error Correction Codes* (ECCs) are so important in SSD design. Two main issues arise when an ECC is used inside an SSD. First, the ECC engine should not limit the performance of the drive. This requirement is addressed with a hardware ECC implementation that supports multiple devices (channels) in parallel. Second, ECC must avoid erroneous corrections when the error correction capability of the code is overcome; that is, it must have a high detection property.

Nowadays, the most popular ECC approach in commercial SSDs is BCH, which is covered in Chapter 11. As the NAND technology scales down, NAND raw BER becomes worse and a more powerful ECC is needed. Chapter 12 covers LDPC codes which are capable to get closer to the Shannon limit; in other words, they can handle higher BER at the expense of a higher complexity.

SSD security is another key requirement because sensitive data must be protected against external attacks. Unfortunately, existing methods in the HDD world cannot be applied to SSDs. These days encryption is the most popular method to secure SSDs. Chapter 13 covers encryption basics and their application to solid state drives.

We are in the midst of an exciting storage market transition, where Flash is expanding its reach to replace HDDs with dramatically faster and more efficient SSDs. After reading this book, the reader will get a comprehensive look at SSD applications and technologies. As you'll see, a Solid State Drive is a complex mix of digital and analog circuits working in concert with firmware and I/O software protocols. We hope you enjoy this tour inside Solid State Drives.

Rino Micheloni
Alessia Marelli
Kam Eshghi

Acknowledgements

After completing a book on a complex system like a solid state drive, we really wish to thank all the authors of the contributed chapters; their expertise in several different fields made this book possible.

We are especially grateful to Luca Crippa for his tremendous dedication to this project.

We also want to thank Springer for giving us the opportunity to refresh and update this work with a second edition.

Last but not least, we express our gratitude to all the people who reviewed the chapters.

Vimercate, Italy — Rino Micheloni
Vimercate, Italy — Alessia Marelli
San Jose, USA — Kam Eshghi

Contents

1 SSD Architecture and PCI Express Interface 1
Kam Eshghi and Rino Micheloni

2 SAS and SATA SSDs 29
S. Yasarapu

3 Hybrid Storage Systems 43
Rino Micheloni, Luca Crippa and M. Picca

4 2D NAND Flash Technology 61
M. F. Beug

5 3D NAND Flash Memories 105
Rino Micheloni, Seiichi Aritome and Luca Crippa

6 NAND Flash Design 135
Luca Crippa and Rino Micheloni

7 Memory Driven Design Methodologies for Optimal SSD Performance 181
L. Zuolo, C. Zambelli, Rino Micheloni and P. Olivo

8 SSD Reliability Assessment and Improvement 205
C. Zambelli and P. Olivo

9 Reliability Issues in Flash-Memory-Based Solid-State Drives: Experimental Analysis, Mitigation, Recovery 233
Yu Cai, Saugata Ghose, Erich F. Haratsch, Yixin Luo and Onur Mutlu

10 Efficient Wear Leveling in NAND Flash Memory 343
Yuan-Hao Chang and Li-Pin Chang

11 BCH Codes for Solid-State-Drives 369
Alessia Marelli and Rino Micheloni

12 Low-Density Parity-Check (LDPC) Codes 407
E. Paolini

13 Protecting SSD Data Against Attacks 455
Alessia Marelli and Rino Micheloni

Index 479

Editors and Contributors

About the Editors

Dr. Rino Micheloni (rino.micheloni@ieee.org) is Vice President and Fellow at Microsemi Corporation, where he currently runs the Flash Signal Processing Labs in Milan, Italy, with special focus on NAND Flash, Error Correction Codes, and Machine Learning. Prior to joining Microsemi, he was Fellow at PMC-Sierra, working on NAND Flash characterization, LDPC, and NAND signal processing as part of the team developing Flash controllers for PCIe SSDs. Before that, he was with Integrated Device Technology (IDT) as Lead Flash Technologist, driving the architecture and design of the BCH engine in the world's first PCIe NVMe SSD controller. Early in his career, he led NAND design teams at STMicroelectronics, Hynix, and Infineon/Qimonda; during this time, he developed the industry's first MLC NOR device with embedded ECC technology and the industry's first MLC NAND with embedded BCH.

He is IEEE Senior Member, he has co-authored more than 70 publications, and he holds 278 patents worldwide (including 131 US patents). He received the STMicroelectronics Exceptional Patent Award in 2003 and 2004 and the Infineon/Qimonda IP Award in 2007.

He has published the following books with Springer: *Solid-State-Drives (SSDs) Modeling* (2017), *3D Flash Memories* (2016), *Inside Solid State Drives* (2013), *Inside NAND Flash Memories* (2010), *Error Correction Codes for Non-Volatile Memories* (2008), *Memories in Wireless Systems* (2008), and *VLSI-Design of Non-Volatile Memories* (2005).

Alessia Marelli is Technical Leader at Microsemi Corporation, where she takes care of the Error Correction Code and Machine Learning algorithms. She joined Microsemi from PMC-Sierra where she was part of the NAND characterization team as senior engineer with a special focus on data analysis and Flash management algorithms. Before that, she was with IDT as senior designer working on ECC solutions for Flash controllers. Prior IDT, she worked in Qimonda/Infineon as digital designer and in STMicroelectronics defining the Error Correction Code for the industry's first MLC NAND with embedded BCH. She received her degree in mathematical science from "Università degli Studi di Milano—Bicocca," Italy, in 2003 with a thesis about ECC applied to Flash memories.

She holds more than 20 patents regarding Error Correction Codes and is co-author of *Inside Solid State Drives* (Springer, 2013), *Inside NAND Flash Memories* (Springer, 2010), and *Error Correction Codes for Non-Volatile Memories* (Springer, 2008).

Kam Eshghi is Vice President of Strategy and Business Development at Lightbits Labs, a stealth mode start-up developing innovative storage technologies for cloud infrastructure. He joined Lightbits Labs from Dell EMC, where he was Vice President of Strategic Alliances for the DSSD division. He developed and managed start-up DSSD's strategic partnership with EMC, ultimately leading to EMC's acquisition of DSSD.

Previously, as Sr. Director of Marketing and Business Development at Integrated Device Technology (IDT) he build IDT's NVMe controller business from start-up to industry leader. That business was then sold to PMC and is today a successful product line at Microsemi. Earlier in his career, he helped build product lines in storage, compute and networking markets at HP, Intel, Crosslayer Networks, and Synopsys.

He has a M.S. in electrical engineering and computer science and a Master of Business Administration, from Massachusetts Institute of Technology and U.C. Berkeley, respectively.

Contributors

Seiichi Aritome IPCC, Industrial Property Cooperation Center, Tokyo, Japan

M. F. Beug Physikalisch-Technische Bundesanstalt (PTB), Division 2 "Electricity", Braunschweig, Germany

Yu Cai Carnegie Mellon University, Pittsburgh, PA, USA

Li-Pin Chang Department of Computer Science, National Chiao-Tung University, Hsinchu, Taiwan

Yuan-Hao Chang Academia Sinica, Institute of Information Science, Taipei, Taiwan

Luca Crippa Storage Solutions, Microsemi Corporation, Vimercate, MB, Italy

Kam Eshghi Lightbits Labs, San Jose, CA, USA

Saugata Ghose Carnegie Mellon University, Pittsburgh, PA, USA

Erich F. Haratsch Seagate Technology, Fremont, CA, USA

Yixin Luo Carnegie Mellon University, Pittsburgh, PA, USA

Alessia Marelli Storage Solutions, Microsemi Corporation, Vimercate, MB, Italy

Rino Micheloni Storage Solutions, Microsemi Corporation, Vimercate, MB, Italy

Onur Mutlu ETH Zürich, Zürich, Switzerland; Carnegie Mellon University, Pittsburgh, PA, USA

P. Olivo Engineering Department, Università di Ferrara, Ferrara, Italy

E. Paolini DEI, University of Bologna, Bologna, Italy

M. Picca STMicroelectronics, Cornaredo, Italy

S. Yasarapu SSD Product Marketing, Western Digital Corporation, Irvine, CA, USA

C. Zambelli Engineering Department, Università di Ferrara, Ferrara, Italy

L. Zuolo Microsemi Corporation, Vimercate, MB, Italy

Chapter 1
SSD Architecture and PCI Express Interface

Kam Eshghi and Rino Micheloni

Abstract Flash-memory-based solid-state drives (SSDs) provide faster random access and data transfer rates than electromechanical drives and today can often serve as rotating-disk replacements, but the host interface to SSDs remains a performance bottleneck. PCI Express (PCIe)-based SSDs together with the standard called NVMe (Non-Volatile Memory express) solves this interface bottleneck. This chapter walks the reader through the SSD block diagram, from the NAND memory to the Flash controller (including wear leveling, bad block management, and garbage collection). PCIe basics and different PCIe SSD architectures are reviewed. Finally, an overview on the standardization effort around PCI Express is presented.

1.1 Introduction

> Creativity is just connecting things. When you ask creative people how they did something, they feel a little guilty because they didn't really do it, they just saw something. It seemed obvious to them after a while.
>
> —Steve Jobs

Solid-state drives are greatly enhancing enterprise and data center storage performance. While electromechanical disk drives have continuously ramped in capacity, the rotating-storage technology doesn't provide the access-time or transfer-rate performance required in demanding enterprise applications, including on-line transaction processing, data mining, and cloud computing. Client applications are also in need of an alternative to electromechanical disk drives that can deliver faster response times, use less power, and fit in smaller mobile form factors.

K. Eshghi (✉)
Lightbits Labs, San Jose, CA, USA
e-mail: kamyar.eshghi@alum.mit.edu

R. Micheloni
Microsemi Corporation, Vimercate, MB, Italy
e-mail: rino.micheloni@ieee.org

R. Micheloni et al. (eds.), *Inside Solid State Drives (SSDs)*,
Springer Series in Advanced Microelectronics 37,
https://doi.org/10.1007/978-981-13-0599-3_1

Flash-memory-based *Solid-State Drives* (SSDs) can offer much faster random access to data and faster transfer rates. Moreover, SSD capacity is now at the point that the drives can serve as rotating-disk replacements. But for many applications the host interface to SSDs remains a bottleneck to performance. PCI Express (PCIe)-based SSDs together with flash-optimized host control interface standards address this interface bottleneck. SSDs with legacy storage interfaces are proving useful, and PCIe SSDs will further increase performance and improve responsiveness by connecting directly to the host processor.

1.2 SSD Architecture

Flash cards, USB keys and Solid State Drives are definitely the most known examples of electronic systems based on non-volatile memories, especially of NAND type (Sect. 1.4).

Several types of memory cards (CF, SD, MMC, …) are available in the market [1–3], with different user interfaces and form factors, depending on the needs of the target application: e.g. mobile phones need very small-sized removable media like μSD.

SSDs are the emerging application for NAND. A SSD is a complete, small system where every component is soldered on a PCB and is independently packaged: NANDs are usually available both in TSOP and BGA packages.

A basic block diagram of a Solid State Drive is shown in Fig. 1.1. In addition to Flash memories and a microcontroller, there are usually other components. For instance, an external DC-DC converter can be added in order to derive the internal

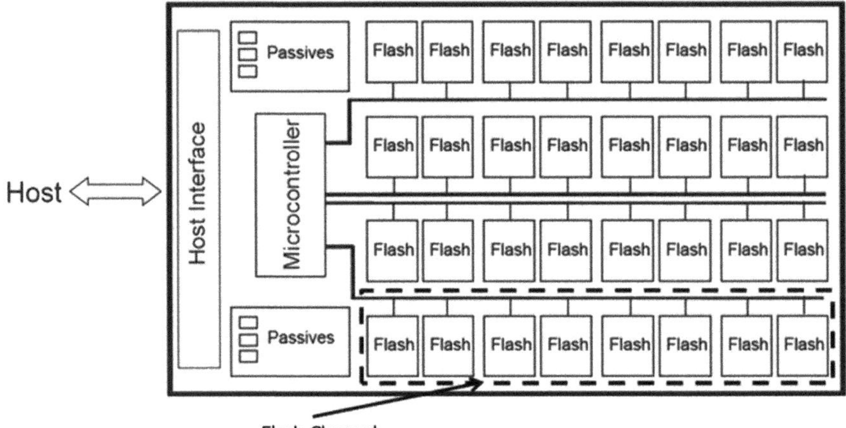

Fig. 1.1 Block diagram of a SSD

power supply, or a quartz can be used for a better clock precision. Of course, reasonable filter capacitors are inserted for stabilizing the power supply. It is also very common to have a temperature sensor for power management reasons. For data caching, a fast DDR memory is frequently added to the board: during a write access, the cache is used for storing data before transfer to the Flash. The benefit is that data updating, e.g. in routing tables, is faster and does not wear out the Flash.

In order to improve performances, NANDs are organized in different Flash channels, as shown in Fig. 1.1.

1.3 Non-volatile Memories

Semiconductor memories can be divided into two major categories: RAM (*Random Access Memories*) and ROM (*Read Only Memories*): RAMs lose their content when power supply is switched off, while ROMs virtually hold it forever. A third category lies in between, i.e. NVM (*Non-Volatile Memories*), whose content can be electrically altered but it is also preserved when the power supply is switched off. NVMs are more flexible than the original ROM, whose content is defined during manufacturing and cannot be changed by the user anymore.

NVM's history began in the 1970s, with the introduction of the first EPROM memory (*Erasable Programmable Read Only Memory*). In the early 1990s, Flash memories came into the game and they started being used in portable products, like mobile phones, USB keys, camcorders, and digital cameras. Solid State Drive (SSD) is the latest killer application for Flash memories. It is worth mentioning that, depending on how the memory cells are organized in the memory array, it is possible to distinguish between NAND and NOR Flash memories. In this book we focus on NAND memories as they are one of the basic elements of SSDs. NOR architecture is described in great details in [4].

NAND Flash cell is based on the Floating Gate (FG) technology, whose cross section is shown in Fig. 1.2. A MOS transistor is built with two overlapping gates rather than a single one: the first one is completely surrounded by oxide, while the second one is contacted to form the gate terminal. The isolated gate constitutes an excellent "trap" for electrons, which guarantees charge retention for years. The operations performed to inject and remove electrons from the isolated gate are called program and erase, respectively. These operations modify the threshold voltage V_{TH} of the memory cell, which is a special type of MOS transistor. Applying a fixed voltage to cell's terminals, it is then possible to discriminate two storage levels: when the gate voltage is higher than the cell's V_{TH}, the cell is on ("1"), otherwise it is off ("0").

It is worth mentioning that, due to floating gate scalability reasons, charge trap memories are gaining more and more attention and they are described in Chap. 5, together with their 3D evolution.

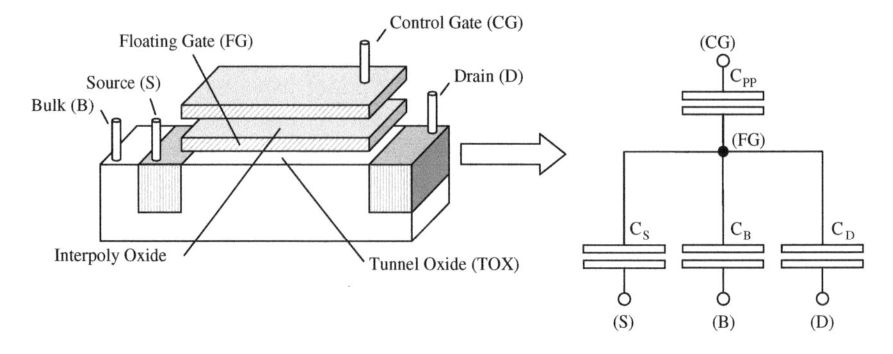

Fig. 1.2 Schematic representation of a floating gate memory cell (*left*) and the corresponding capacitive model (*right*)

1.4 NAND Flash

1.4.1 NAND Array

A Flash device contains an array of floating-gate transistors: each of them acts as memory cell. In Single Level Cell (SLC) devices, each memory cell stores one bit of information; Multi-Level Cell (MLC) devices store 2 bits per cell.

The basic element of a NAND Flash memory is the NAND string, as shown in Fig. 1.3a. Usually, a string is made up by 32 (M_{C0}–M_{C31}), 64 or 128 cells connected in series. Two selection transistors are placed at the edges of the string: M_{SSL} ensures the connection to the source line. M_{DSL} connects the string to the bitline BL. The cell's control gates are connected through the wordlines (WLs). Figure 1.3b shows how the matrix array is built starting from the basic string. In the WL direction, adjacent NAND strings share the same WL, DSL, BSL and SL. In the BL direction, two consecutive strings share the bitline contact. Figure 1.4 shows a section of the NAND array along the bitline direction.

All the NAND strings sharing the same group of WL's form a Block. In Fig. 1.3b there are three blocks:

- BLOCK0 is made up by WL_0 <31:0>;
- BLOCK1 is made up by WL_1 <31:0>;
- BLOCK2 is made up by WL_2 <31:0>.

Logical pages are made up of cells belonging to the same WL. The number of pages per WL is related to the storage capabilities of the memory cell. Depending on the number of storage levels, Flash memories are referred to in different ways:

- SLC memories stores 1 bit per cell;
- MLC memories stores 2 bits per cell;
- 8LC memories stores 3 bits per cell;
- 16LC memories stores 4 bits per cell.

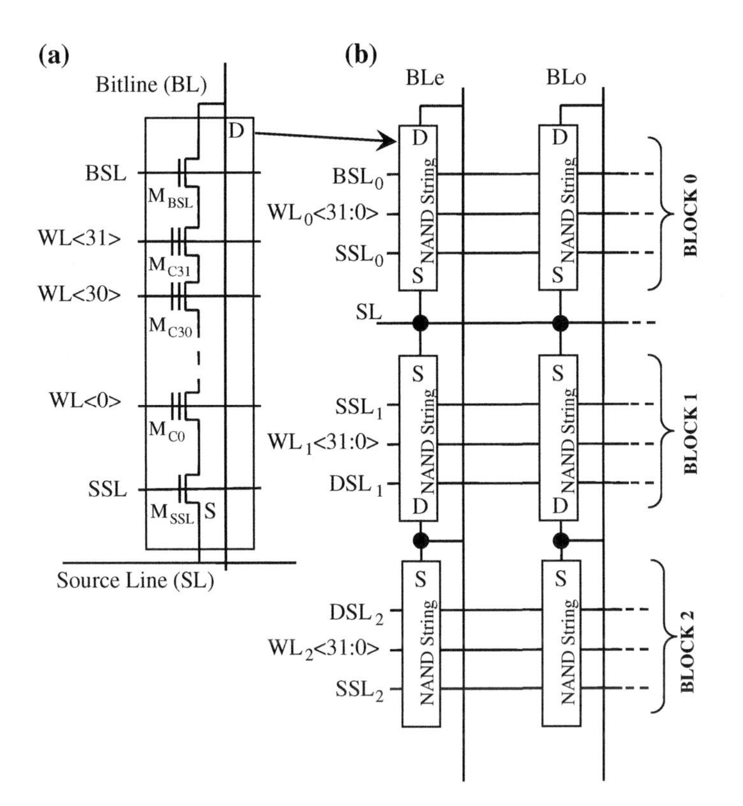

Fig. 1.3 NAND String (**a**) and NAND array (**b**)

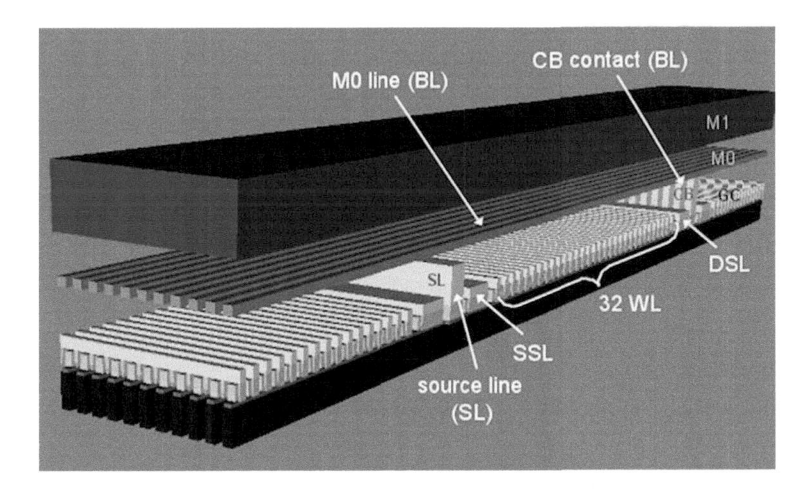

Fig. 1.4 NAND array section along the bitline direction

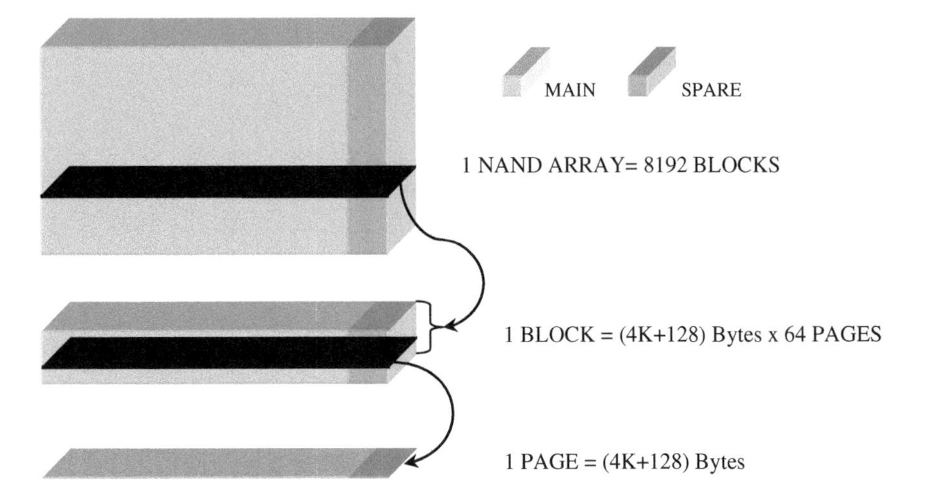

Fig. 1.5 32 Gbit memory logic organization

If we consider the SLC case with interleaved architecture (Chap. 6), even cells belong to the "even" page (BLe), while odd pages belong to the "odd" page (BL0). For example, a SLC device with 4 kB page has a WL of 32,768 + 32,768 = 65,536 cells. Of course, in the MLC case there are four pages as each cell stores one Least Significant Bit (LSB) and one\ Most Significant Bit (MSB). Therefore, we have MSB and LSB pages on even BL, and MSB and LSB pages on odd BL.

In NAND Flash memories, a logical page is the smallest addressable unit for reading and writing; a logical block is the smallest erasable unit (Fig. 1.5).

Each page is made up by main area (data) and spare area as shown in Fig. 1.5. Main area can be 4, 8 or 16 kB. Spare area can be used for ECC and is in the order of hundred of Bytes every 4 kB of main area.

Figure 1.5 shows the logic organization of a SLC device with a string of 32 cells, interleaving architecture, 4 kB page, and 128 Bytes of spare.

NAND basic operations, i.e. read, program, and erase are described in Chaps. 5 and 6 of this book.

1.4.2 NAND Interface

For many years, the asynchronous interface (Fig. 1.6) has been the only available option for NAND devices.

Asynchronous interface is described below.

- CE#: it is the Chip Enable signal. This input signal is "1" when the device is in stand-by mode, otherwise it is always "0".

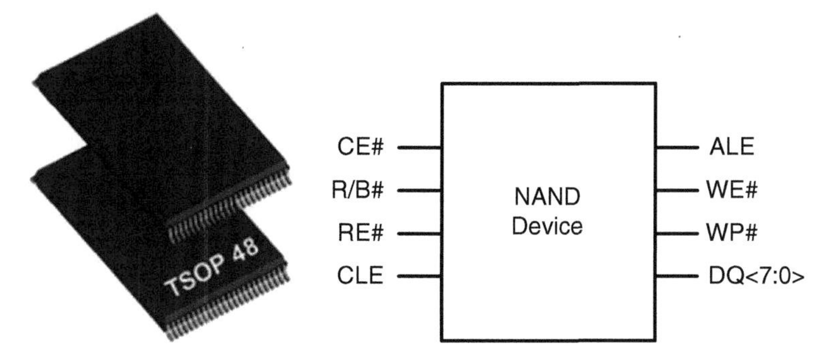

Fig. 1.6 TSOP package (*left*) and related pinout (*right*)

- R/B#: it is the Ready/Busy signal. This output signal is used to indicate the target status. When low, the target has an operation in progress.
- RE#: it is the Read Enable signal. This input signal is used to enable serial data output.
- CLE: it is the Command Latch Enable. This input is used by the host to indicate that the bus cycle is used to input the command.
- ALE: it is the Address Latch Enable. This input is used by the host to indicate that the bus cycle is used to input the addresses.
- WE#: it is the Write Enable. This input signal controls the latching of input data. Data, command and address are latched on the rising edge of WE#.
- WP#: it is the Write Protect. This input signal is used to disable Flash array program and erase operations.
- DQ <7:0>: these input/output signals represent the data bus.

As a matter of fact, this interface is a real bottleneck, especially looking at high performance systems like SSDs.

NAND read throughput is determined by array access time and data transfer across the DQ bus. The data transfer is limited to 40 MB/s by the asynchronous interface. As technology shrinks, page size increases and data transfer takes longer; as a consequence, NAND read throughput decreases, totally unbalancing the ratio between array access time and data transfer on the DQ bus. A DDR-like interface (Chap. 6) has been introduced to balance this ratio.

Nowadays two possible solutions are available on the market. ONFI (Open NAND Flash Interface) organization published the first standard at the end of 2006 [5]; other NAND vendors like Toshiba and Samsung use the Toggle-Mode interface. JEDEC [6] is now trying to combine these two approaches together.

Figure 1.7 shows ONFI pinout. Compared to the Asynchronous Interface, there are three main differences:

- RE# becomes W/R# which is the Write/Read direction pin;
- WE# becomes CLK which is the clock signal;

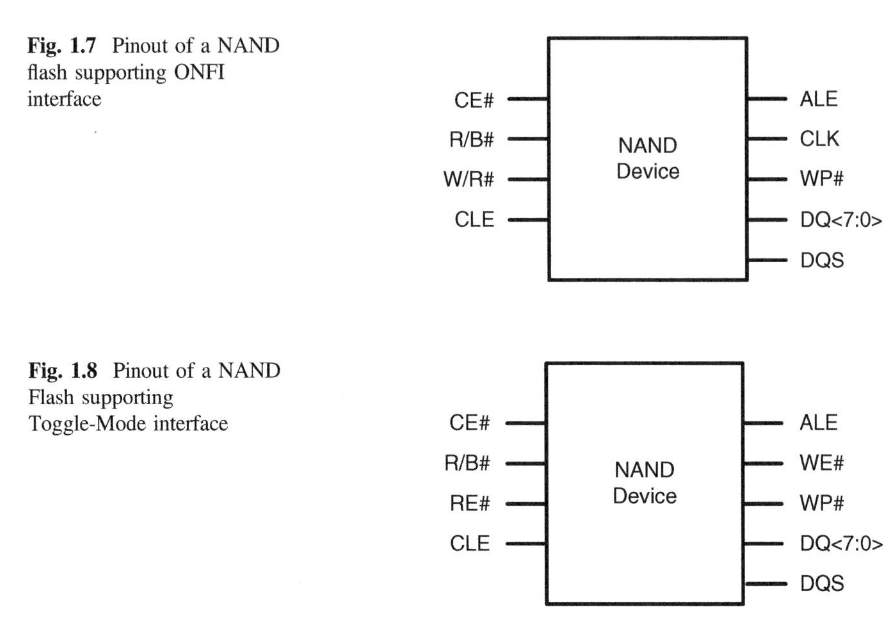

Fig. 1.7 Pinout of a NAND flash supporting ONFI interface

Fig. 1.8 Pinout of a NAND Flash supporting Toggle-Mode interface

- DQS is an additional pin acting as the data strobe,i.e. it indicates the data valid window.

Hence, the clock (CLK) is used to indicate where command and addresses should be latched, while a data strobe signal (DQS) is used to indicate where data should be latched. DQS is a bi-directional bus and is driven with the same frequency as the clock. Toggle-Mode DDR interface uses the pinout shown in Fig. 1.8.

It can be noted that only the DQS pin has been added to the asynchronous interface. In this case, higher speeds are achieved increasing the toggling frequency of RE#.

1.5 Memory Controller

A memory controller has two fundamental tasks:

1. to provide the most suitable interface and protocol towards both the host and the Flash memories;
2. to efficiently handle data, maximizing transfer speed, data integrity and information retention.

In order to carry out such tasks, an application specific device is designed, embedding a standard processor—usually 8–16 bits—together with dedicated hardware to handle timing-critical tasks.

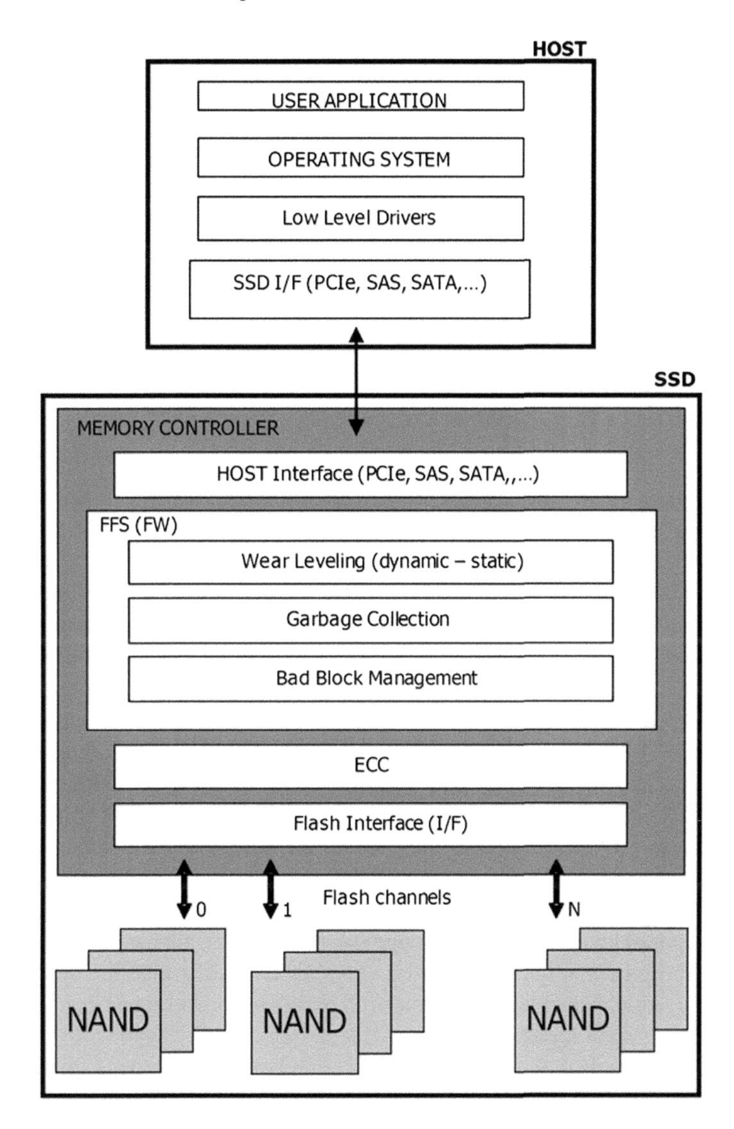

Fig. 1.9 High level view of a flash controller

Generally speaking, the memory controller can be divided into four parts, which are implemented either in hardware or in firmware (Fig. 1.9).

Proceeding from the host to the Flash, the first part is the host interface, which implements the required industry-standard protocol (PCIe, SAS, SATA, etc.), thus ensuring both logical and electrical interoperability between SSDs and hosts. This block is a mix of hardware—buffers, drivers, etc.—and firmware—command decoding performed by the embedded processor—which decodes the command

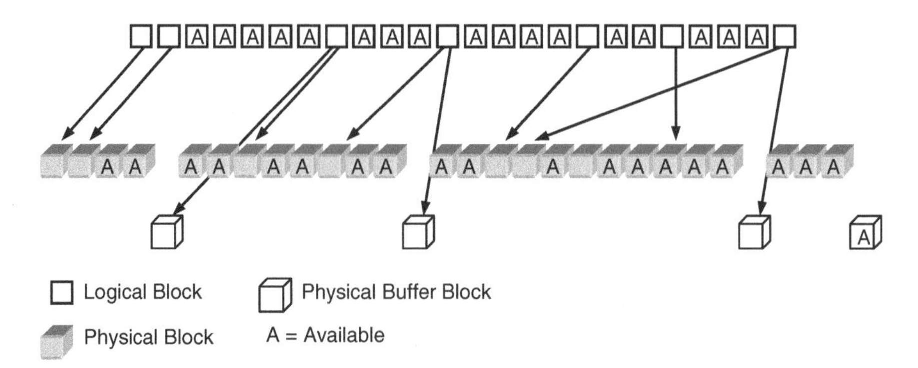

Fig. 1.10 Logical to physical block management

sequence invoked by the host and handles the data flow to/from the Flash memories.

The second part is the Flash File System (FFS) [7]: that is, the file system which enables the use of SSDs like magnetic disks. For instance, sequential memory access on a multitude of sub-sectors which constitute a file is organized by linked lists (stored on the SSD itself) which are used by the host to build the File Allocation Table (FAT). The FFS is usually implemented in form of firmware inside the controller, each sub-layer performing a specific function. The main functions are: *Wear leveling Management*, *Garbage Collection* and *Bad Block Management*. For all these functions, tables are widely used in order to map sectors and pages from logical to physical (Flash Translation Layer or FTL) [8, 9], as shown in Fig. 1.10. The upper block row is the logical view of the memory, while the lower row is the physical one. From the host perspective, data are transparently written and overwritten inside a given logical sector: due to Flash limitations, overwrite on the same page is not possible, therefore a new page (sector) must be allocated in the physical block and the previous one is marked as invalid. It is clear that, at some point in time, the current physical block becomes full and therefore a second one (Buffer) is assigned to the same logical block.

The required translation tables are always stored on the SSD itself, thus reducing the overall storage capacity.

1.5.1 Wear Leveling

Usually, not all the information stored within the same memory location change with the same frequency: some data are often updated while others remain always the same for a very long time—in the extreme case, for the whole life of the device. It's clear that the blocks containing frequently-updated information are stressed with a large number of write/erase cycles, while the blocks containing information updated very rarely are much less stressed.

In order to mitigate disturbs, it is important to keep the aging of each page/block as minimum and as uniform as possible: that is, the number of both read and program cycles applied to each page must be monitored. Furthermore, the maximum number of allowed program/erase cycles for a block (i.e. its endurance) should be considered: in case SLC NAND memories are used, this number is in the order of 100 k cycles, which is reduced to 10 k when MLC NAND memories are used.

Wear Leveling techniques rely on the concept of logical to physical translation: that is, each time the host application requires updates to the same (logical) sector, the memory controller dynamically maps the sector onto a different (physical) sector, keeping track of the mapping either in a specific table or with pointers. The out-of-date copy of the sector is tagged as both invalid and eligible for erase. In this way, all the physical blocks are evenly used, thus keeping the aging under a reasonable value.

Two kinds of approaches are possible: Dynamic Wear Leveling is normally used to follow up a user's request of update, writing to the first available erased block with the lowest erase count; Static Wear Leveling can also be implemented, where every block, even the least modified, is eligible for re-mapping as soon as its aging deviates from the average value.

1.5.2 *Garbage Collection*

Both wear leveling techniques rely on the availability of free sectors that can be filled up with the updates: as soon as the number of free sectors falls below a given threshold, sectors are "compacted" and multiple, obsolete copies are deleted. This operation is performed by the Garbage Collection module, which selects the blocks containing the invalid sectors, copies the latest valid copy into free sectors and erases such blocks (Fig. 1.11).

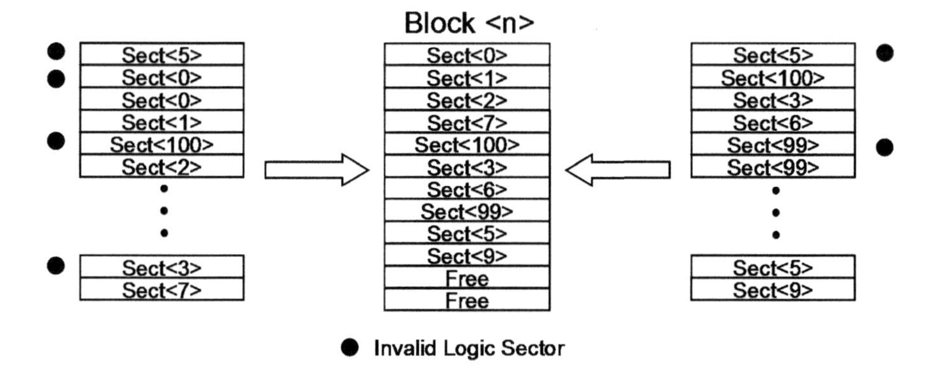

Fig. 1.11 Garbage collection

In order to minimize the impact on performance, garbage collection can be performed in background. The equilibrium generated by the wear leveling distributes wear out stress over the array rather than on single hot spots. Hence, the bigger the memory density, the lower the wear out per cell is.

1.5.3 Bad Block Management

No matter how smart the Wear Leveling algorithm is, an intrinsic limitation of NAND Flash memories is represented by the presence of so-called Bad Blocks (BB), i.e. blocks which contain one or more locations whose reliability is not guaranteed.

The Bad Block Management (BBM) module creates and maintains a map of bad blocks, as shown in Fig. 1.12: this map is created during factory initialization of the memory card, thus containing the list of the bad blocks already present during the factory testing of the NAND Flash memory modules. Then it is updated during device lifetime whenever a block becomes bad.

1.5.4 Error Correction Code (ECC)

This task is typically executed by a specific hardware inside the memory controller. Examples of memories with embedded ECC are also reported [10–12]. Most popular ECC codes, correcting more than one error, are Reed-Solomon and BCH [13]. Chapter 10 gives an overview of how BCH is used in the NAND world, including an analysis of its detection properties, which are essential for concatenated architectures. The last section of Chap. 10 covers the usage of BCH in high-end SSDs, where the ECC has to be shared among multiple Flash channels.

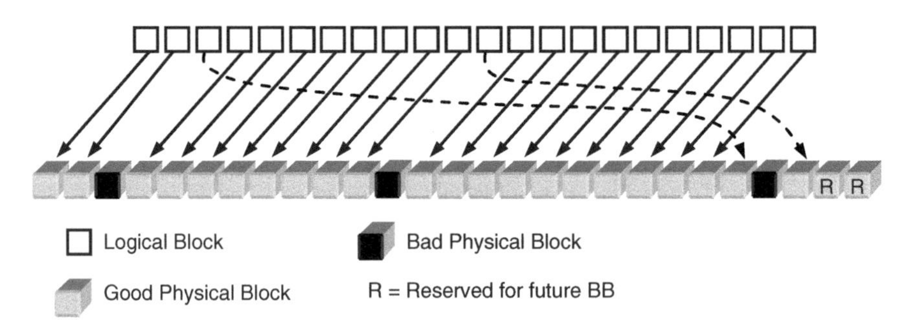

Fig. 1.12 Bad block management (BBM)

With the technology shrink, NAND raw BER gets worse, approaching the Shannon limit. As a consequence, correction techniques based on soft information processing are required: LDPC (Low Density Parity Check) codes are an example of this soft information approach and they are analyzed in Chap. 11.

1.6 Multi-channel Architecture

A typical memory system is composed by several NAND memories. Typically, an 8-bit bus, usually called channel, is used to connect different memories to the controller (Fig. 1.1). It is important to underline that multiple Flash memories in a system are both a means for increasing storage density and read/write performance [14].

Operations on a channel can be interleaved, which means that a second chip can be addressed while the first one is still busy. For instance, a sequence of multiple write operations can be directed to a channel, addressing different NANDs, as shown in Fig. 1.13: in this way, the channel utilization is maximized by pipelining the data load phase; in fact, while the program operation takes place within a memory chip, the corresponding Flash channel is free. The total number of Flash channel is a function of the target applications, but tens of channels are becoming quite common. Figure 1.14 shows the impact of interleaving. As the reader can notice, given the same Flash programming time, SSD's throughput greatly improves.

The memory controller is responsible for scheduling the distributed accesses at the memory channels. The controller uses dedicated engines for the low level communication protocol with the Flash.

Moreover, it is clear that the data load phase is not negligible compared to the program operation (the same comment is valid for data output): therefore, increasing I/O interface speed is another smart way to improve performances: DDR-like interfaces are discussed in more details in Chap. 6. Impact of DDR frequency on program throughput is reported in Fig. 1.15. As the speed increases, more NAND can be operated in parallel before saturating the channel. For instance,

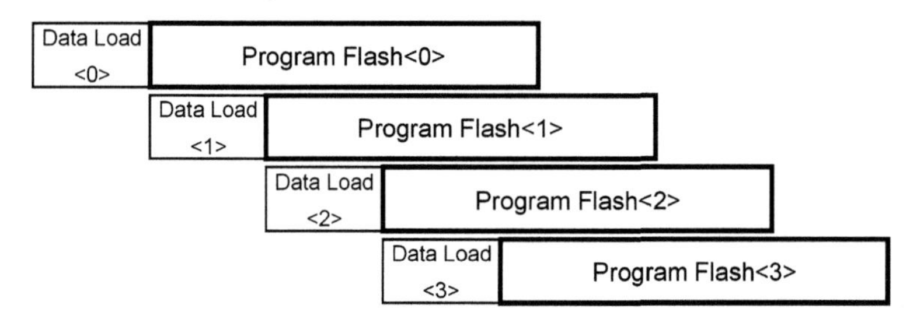

Fig. 1.13 Interleaved operations on one flash channel

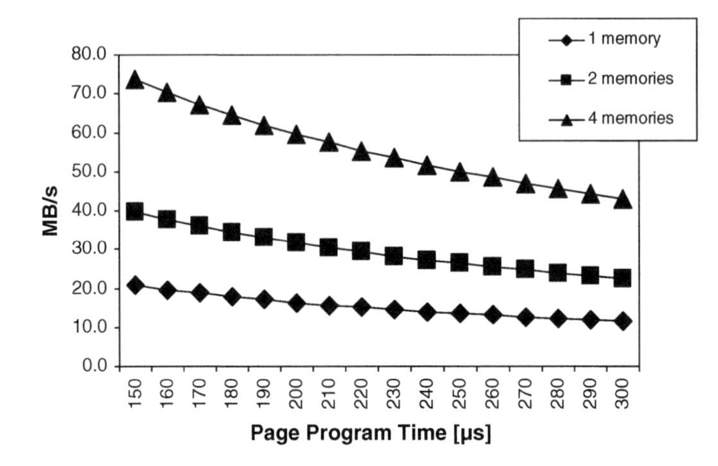

Fig. 1.14 Program throughput with an interleaved architecture as a function of the NAND page program time

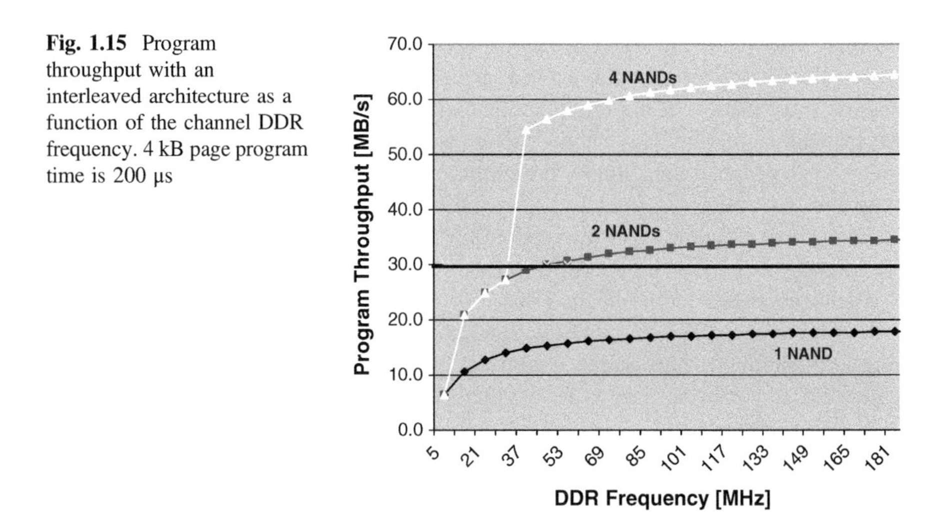

Fig. 1.15 Program throughput with an interleaved architecture as a function of the channel DDR frequency. 4 kB page program time is 200 μs

assuming a target of 30 MB/s, 2 NANDs are needed with a minimum DDR frequency of about 50 MHz. Given a page program time of 200 μs, at 50 MHz four NANDs can operate in interleaved mode, doubling the write throughput. Of course, power consumption has then to be considered.

After this high level overview of the SSD architecture, let's move to the interface towards the host. PCI Express (PCIe) is fast becoming the interface of choice for high performance SSDs.

1.7 What Is PCIe?

PCIe (Peripheral Component Interconnect Express) is a bus standard that replaced PCI and PCI-X. PCI-SIG (PCI Special Interest Group) creates and maintains the PCIe specification [15].

PCIe is used in all computer applications including enterprise servers, consumer personal computers (PC), communication systems, and industrial applications. Unlike older PCI bus topology, which uses shared parallel bus architecture, PCIe is based on point-to-point topology, with separate serial links connecting every device to the root complex (host). Additionally, a PCIe link supports full-duplex communication between two endpoints. Data can flow upstream (UP) and downstream (DP) simultaneously. Each pair of these dedicated unidirectional serial point-to-point connections is called a *lane*, as depicted in Fig. 1.16. The PCIe standard is constantly under improvement, with PCIe 5.0 being already announced (Table 1.1).

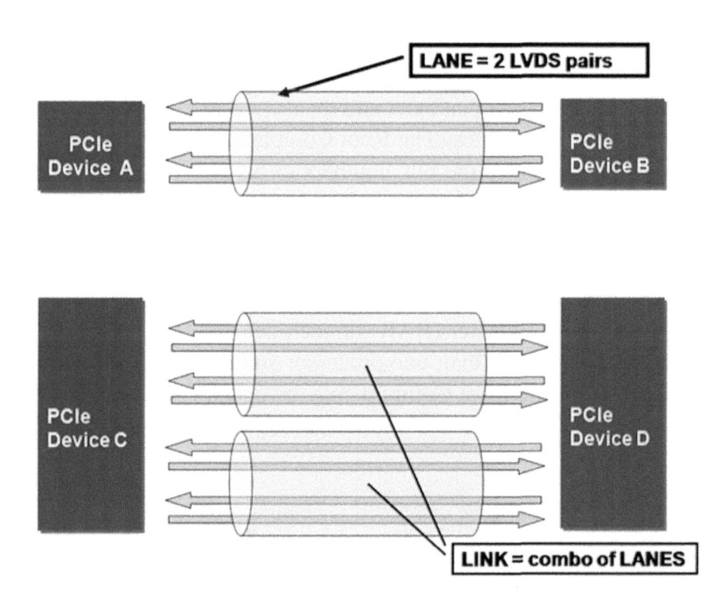

Fig. 1.16 PCI express lane and link. In Gen2, 1 lane runs at 5 Gbps/direction; a 2-lane link runs at 10 Gbps/direction

Table 1.1 Data rate of different PCIe generations

PCIe version	Year introduced	Data rate (GT/s)
PCIe 1.0 (Gen1)	2003	2.5
PCIe 2.0 (Gen2)	2007	5.0
PCIe 3.0 (Gen3)	2010	8.0
PCIe 4.0 (Gen4)	2018 (planned)	16.0
PCIe 4.0 (Gen5)	2020 (planned)	32.0

Other important features of PCIe include power management, hot-swappable devices, and the ability to handle peer-to-peer data transfers (sending data between two end points without routing through the host) [16]. Additionally, PCIe simplifies board design by utilizing serial technology, which eliminates wire count of parallel bus architectures.

The PCIe link between two devices can consist of 1–32 lanes. The packet data is striped across lanes, and the lane count is automatically negotiated during device initialization.

The PCIe standard defines slots and connectors for multiple widths: ×1, ×4, 8, ×16, ×32 (Fig. 1.17). This allows PCIe to serve lower throughput, cost-sensitive applications as well as performance-critical applications.

There are basically three different types of devices in a native PCIe system as shown in Fig. 1.18 [17]: *Root Complexes* (RCs), PCIe switches, and *EndPoints* (EPs). A Root Complex should be thought of as a single processor sub-system with a single PCIe port, even though it consists of one or more CPUs, plus their associated RAM and memory controller. PCIe routes data based on memory address or ID, depending on the transaction type. Therefore, every device must be uniquely identified within the PCI Express tree. This requires a process called enumeration. During system initialization, the Root Complex performs the enumeration process to determine the various buses that exist and the devices that reside on each bus, as well as the required address space. The Root Complex allocates bus numbers to all the PCIe buses and configures the bus numbers to be used by the PCIe switches.

A PCIe switch behaves as if it were multiple PCI-PCI Bridges, as shown in the inset of Fig. 1.18. Basically, a switch decouples every UP and DP ports so that each link can work as a point-to-point connection.

Within a PCIe tree, all devices share the same memory space. RC is in charge of setting the Base Address Register (BAR) of each device.

In multi-RC systems, more than one processor sub-system exists within a PCIe tree. For example, a second Root Complex may be added to the system via the DP of a PCIe switch, possibly to act as a warm stand-by to the primary RC. However,

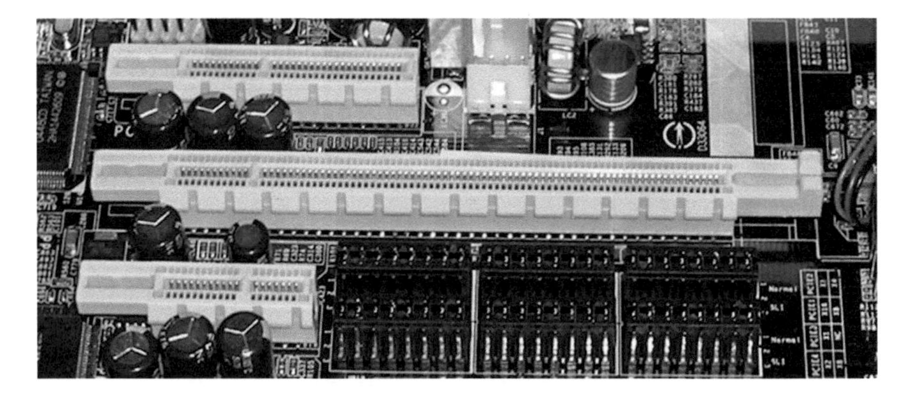

Fig. 1.17 Various PCIe slots. From top to bottom: PCIe × 4, PCIe × 16, PCIe × 1

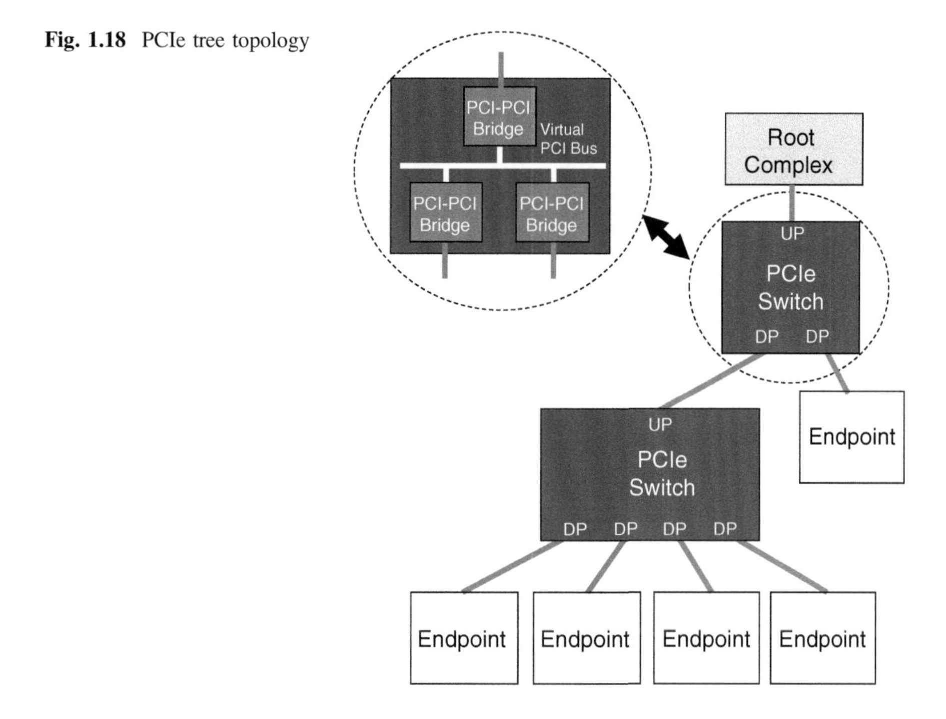

Fig. 1.18 PCIe tree topology

an issue arises when the second RC also attempts the enumeration process: it sends out Configuration Read Messages to discover other PCIe devices on the system. Unfortunately, configuration transactions can only move from UP to DP. A PCIe switch does not forward configuration messages that are received on its DP. Thus, the second RC is isolated from the rest of the PCIe tree and will not detect any PCIe devices in the system. So, simply adding processors to a DP of a PCIe switch will not provide a multi-Root Complex solution.

One method of supporting multiple RCs is to use a *Non-Transparent Bridging* (NTB) function to isolate the address domains of each of the Root Complexes [18]. NTB allows two Root Complexes or PCIe trees to be interconnected with one or more shared address windows between them.

In other words, NTB works like an address translator between two address domains. Of course, multiple NTBs can be used to develop multi-RC applications. An example of PCIe switch with embedded NTB functions is shown in Fig. 1.19: an additional bus, called NT Interconnect, is used for exchanging Transaction Layer Packets among RCs.

PCIe uses a packet-based layered protocol, consisting of a transaction layer, a data link layer, and a physical layer, as shown in Fig. 1.20.

The transaction layer handles packetizing and de-packetizing of data and status-message traffic. The data link layer sequences these *Transaction Layer Packets* (TLPs) and ensures they are reliably delivered between two endpoints (devices A and B in Fig. 1.5). If a transmitter device sends a TLP to a remote

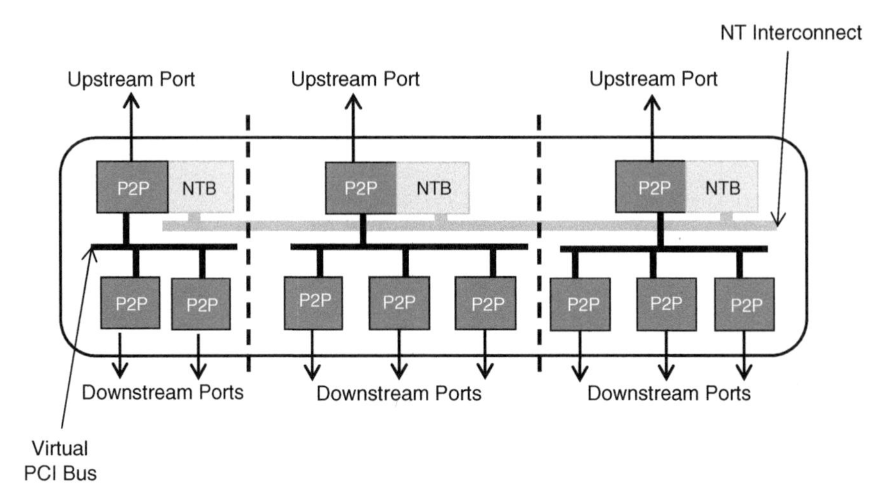

Fig. 1.19 PCIe switch with multiple NTB functions

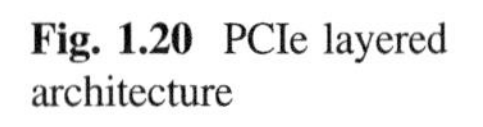
Fig. 1.20 PCIe layered architecture

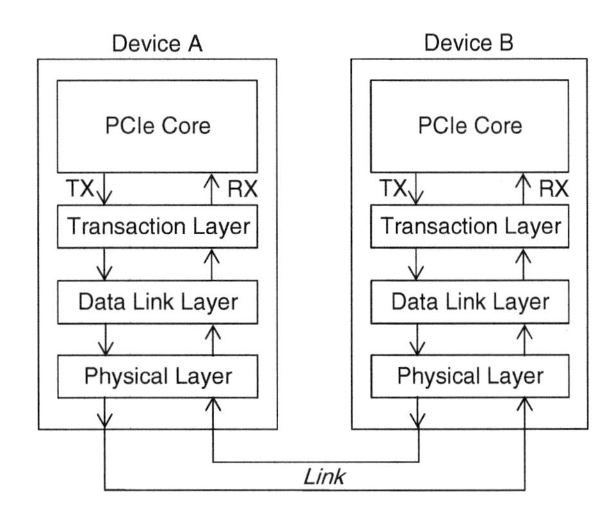

receiver device and a CRC error is detected, the transmitter device gets a notification back. The transmitter device automatically replays the TLP. With error checking and automatic replay of failed packets, PCIe ensures very low *Bit Error Rate* (BER).

The Physical Layer is split in two parts: the Logical Physical Layer and the Electrical Physical Layer. The Logical Physical Layer contains logic gates for processing packets before transmission on the Link, and processing packets from the Link to the Data Link Layer. The Electrical Physical Layer is the analog interface of the Physical Layer: it consists of differential drivers and receivers for each lane.

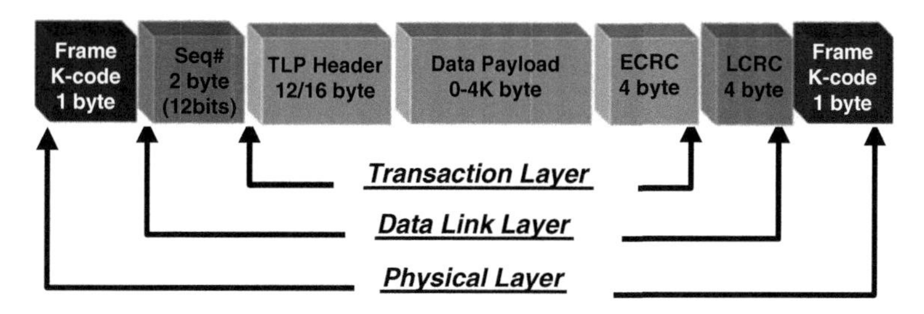

Fig. 1.21 Transaction layer packet (TLP) assembly

TLP assembly is shown in Fig. 1.21. Header and Data Payload are TLP's core information: Transaction Layer assembles this section based on the data received from the application software layer. An optional End-to-End CRC (ECRC) field is can be appended to the packet. ECRC is used by the ultimate targeted device of this packet to check for CRC errors inside Header and Data Payload. At this point, the Data Link Layer appends a sequence ID and local CRC (LCRC) field in order to protect the ID. The resultant TLP is forwarded to the Physical Layer which concatenates a Start and End framing character of 1 Byte each to the packet. Finally, the packet is encoded and differentially transmitted on the Link using the available number of Lanes.

Today, PCIe is a high volume commodity interconnect used in virtually all computers, from consumer laptops to enterprise servers, as the primary motherboard technology that interconnects the host CPU with on-board ICs and add-on peripheral expansion cards.

1.8 The Need for Storage Speed

The real issue at hand is the need for storage technology that can match the exponential ramp in processor performance over the past two decades. Processor vendors have continued to ramp the performance of individual processor cores, to combine multiple cores on one IC, and to develop technologies that can closely-couple multiple ICs in multi-processor systems. Ultimately, all of the cores in such a scenario need access to the same storage subsystem.

Enterprise IT managers are eager to utilize the multiprocessor systems because they have the potential of boosting the number of I/O operations per second (IOPS) that a system can process and also the number of IOPS per watt (IOPS/W) in power consumption. The ramping multi-processing computing capability offers better IOPS relative to cost and power consumption—assuming the processing elements can get access to the data in a timely fashion. Active processors waiting on data waste time and money.

There are of course multiple levels of storage technology in a system that ultimately feeds code and data to each processor core. Generally, each core includes local cache memory that operates at core speed. Multiple cores in a chip share a second-level and sometimes a third-level cache. And DRAM feeds the caches. The DRAM and cache access-time and data-transfer performance has scaled to match the processor performance.

The disconnect has come in the performance gap that exist between DRAM and rotating storage in terms of access time and data rate. Disk-drive vendors have done a great job of designing and manufacturing higher-capacity, lower-cost-per-GByte disk drives. But the drives inherently have physical limitations in terms of how fast they can access data and then how fast they can transfer that data into DRAM.

Access time depends on how fast a hard drive can move the read head over the required data track on a disk, and the rotational latency for the sector where the data is located to move under the head. The maximum transfer rate is dictated by the rotational speed of the disk and the data encoding scheme that together determine the number of Bytes per second read from the disk.

Hard drives perform relatively well in reading and transferring sequential data. But random seek operations add latency. And even sequential read operations can't match the data appetite of the latest processors.

Meanwhile, enterprise systems that perform on-line transaction processing such as financial transactions and that mine data in applications such as customer relationship management require highly random access to data. Cloud computing also requires random access to data, whether it's for unstructured databases or analytic workloads. This random access requirement is escalating with technologies such as virtualization, which expand the scope of different applications that a single system has active at any one time. Every microsecond of latency relates directly to money lost and less efficient use of the processors and the power dissipated by the system.

Fortunately Flash memory offers the potential to close the performance gap between DRAM and rotating storage. Flash is slower than DRAM but offers a lower cost per GByte of storage. That cost is more expensive than hard disk drive storage, but enterprises will gladly pay the premium because Flash also offers much better throughput in terms of MB/s and faster access to random data, resulting in better cost-per-IOPS compared to rotating storage.

Ramping Flash capacity and reasonable cost has led to a growing trend of SSDs that package Flash in disk-drive-like form factors. Moreover, the SSDs have most often utilized disk-drive interfaces such as SATA (serial ATA) or SAS (serial attached SCSI).

1.9 Why PCIe for SSD Interface?

The disk-drive form factor and interface allows IT vendors to substitute an SSD for a magnetic disk drive seamlessly. There is no change required in system hardware or driver software. An IT manager can simply swap to an SSD and realize significantly better access times and somewhat faster data-transfer rates.

Neither the legacy disk-drive form factor nor the interface is ideal for Flash-based storage. SSD manufacturers can pack enough Flash devices in a 2.5-in. form factor to easily exceed the power profile developed for disk drives. And Flash can support higher data transfer rates than even the latest generation of disk interfaces.

Let's examine the disk interfaces more closely (Fig. 1.22). Most mainstream systems have migrated to third-generation SATA and SAS that support 600 MB/s throughput, and drives based on those interfaces have found usage in enterprise systems. While those data rates support the fastest electromechanical drives, new NAND Flash architectures and multi-die Flash packaging deliver aggregate Flash bandwidth that exceeds the throughput capabilities of SATA and SAS interconnects. In short, the SSD performance bottleneck has shifted from the storage media to the host interface. Therefore, many applications need a faster host interconnect to take full advantage of Flash storage.

The PCIe host interface can overcome this storage performance bottleneck and deliver unparalleled performance by attaching the SSD directly to the PCIe host bus. For example, a 4-lane (×4) PCIe Generation 3 (Gen3) link can deliver 4 GB/s data rates. Simply put, PCIe affords the needed storage bandwidth. Moreover, the direct PCIe connection can reduce system power and slash the latency that's attributable to the legacy storage infrastructure.

Clearly an interface such as PCIe could handle the bandwidth of a multi-channel Flash storage subsystem and can offer additional performance advantages. SSDs that use a disk interface also suffer latency added by a storage-controller IC that handles disk I/O. PCIe devices connect directly to the host bus eliminating the architectural layer associated with the legacy storage infrastructure. The compelling performance of PCIe SSDs has resulted in system manufacturers placing PCIe SSDs in servers as well as in storage arrays to build tiered storage systems (Fig. 1.23) that accelerate applications while improving cost-per-IOPS (*Input/Output Operations per Second*).

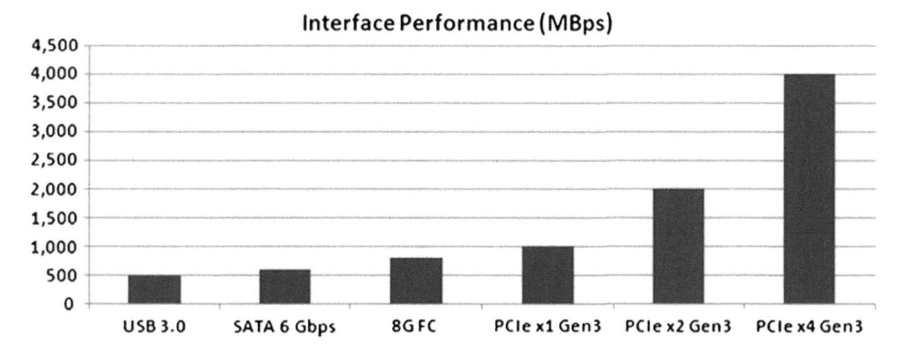

Fig. 1.22 Interface performance. PCIe improves overall system performance by reducing latency and increasing throughput

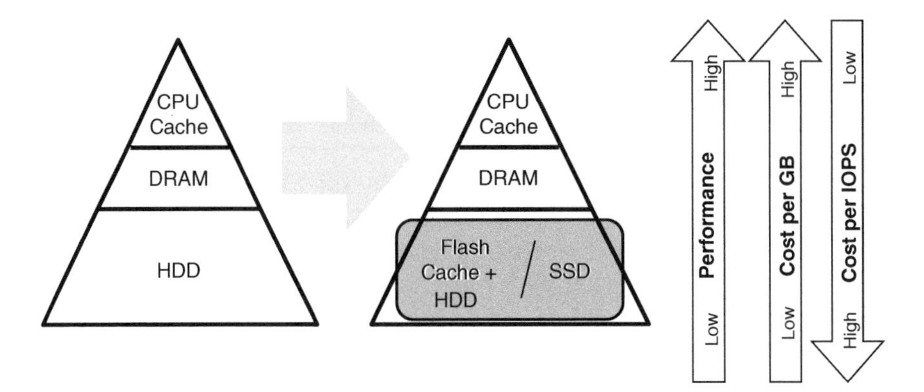

Fig. 1.23 Enterprise memory/storage hierarchy paradigm shift

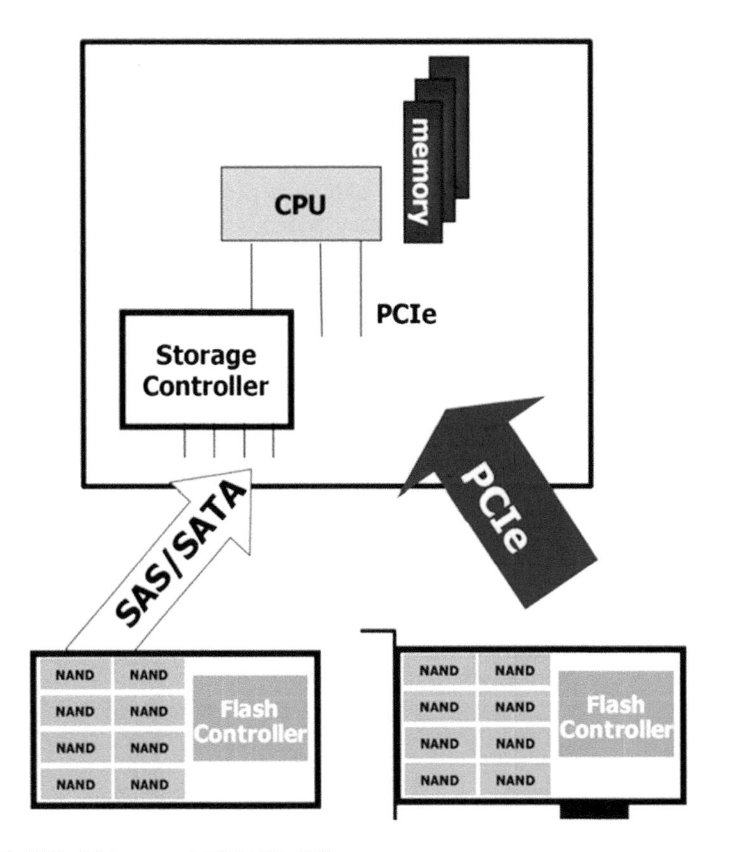

Fig. 1.24 PCIe SSD versus SAS/SATA SSD

Moving storage to a PCIe link brings additional challenges to the system designer. As mentioned earlier, the SATA- and SAS-based SSD products have maintained software compatibility and some system designers are reluctant to give up that advantage. Any PCIe storage implementation will create the need for some new driver software.

Despite the software issue, the move to PCIe storage in enterprises is well underway. Performance demands in the enterprise are mandating this transition. There is no other apparent way to deliver improving IOPS, IOPS/W, and IOPS per dollar characteristics that IT managers are demanding.

The benefits of using PCIe as a storage interconnect are clear. Already at Gen3, you can achieve over 6× the data throughput relative to SATA or SAS. You can eliminate components such as host bus adapters and SERDES ICs on the SATA and SAS interfaces—saving money and power at the system level. And PCIe moves the storage closer to the host CPU reducing latency, as shown in Fig. 1.24.

Let's now take a deeper look at PCIe-based SSD architectures.

1.10 PCIe SSD Implementations

The simplest PCIe SSD implementations can utilize legacy Flash memory controller ICs that while capable of controlling memory read and write operations, have no support for the notion of system I/O. Such Flash controllers would typically work behind a disk interface IC in existing SATA- or SAS-based SSD products (Fig. 1.25).

Alternatively, it is possible to run Flash-management software on the host processor to enable a simple Flash controller to function across a PCIe interconnect (Fig. 1.26).

That approach has several drawbacks. First it consumes host processing and memory resources that ideally would be available for application software. Second

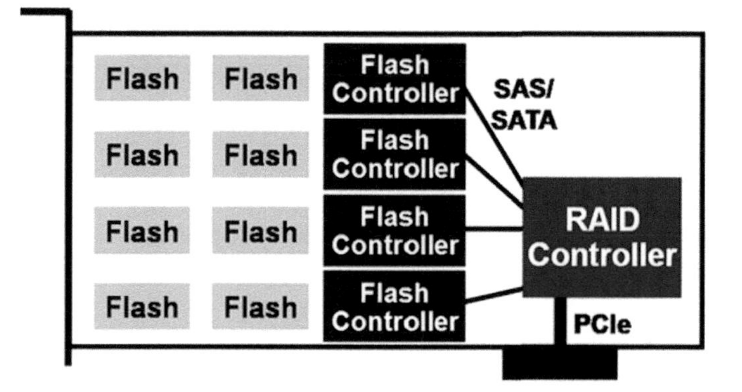

Fig. 1.25 RAID-based PCIe SSDs not optimized for performance/power

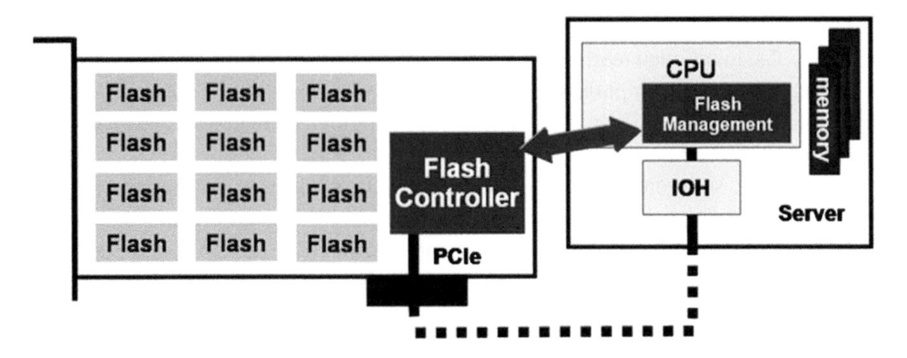

Fig. 1.26 Running flash management algorithms on the host drains the host CPU/RAM resources

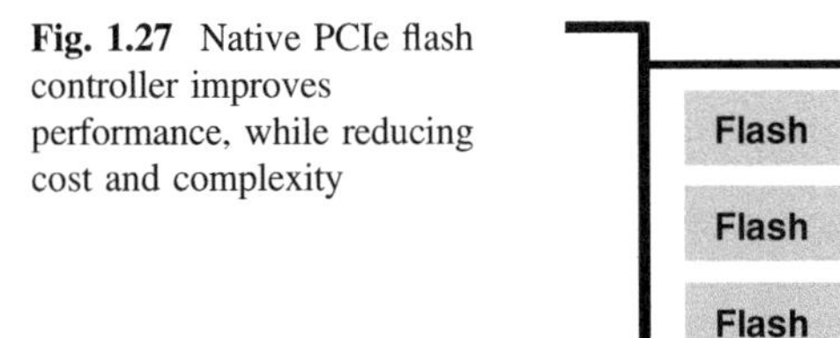

Fig. 1.27 Native PCIe flash controller improves performance, while reducing cost and complexity

it requires proprietary drivers and raises OEM qualification issues. And third it doesn't deliver a bootable drive because the system must be booted for the Flash-management software to execute and enable the storage scheme.

Clearly, these designs have found niche success. These products are used by early adopters as caches for hard disk drives rather than mainstream replacements of high-performance disk drives.

More robust and efficient PCIe SSD designs rely on a complex SoC that natively supports PCIe, integrates Flash controller functionality, and that completely implements the storage-device concept (Fig. 1.27). Such a product offloads the host CPU of handling Flash management, and ultimately enables standard OS drivers that support plug-and-play operations just as with SATA and SAS.

1.11 NVM Express Driving Broader Adoption of PCIe SSDs

The NVM Express (NVMe) 1.0 specification, developed cooperatively by more than 80 companies from across the industry, was released in March, 2011, by the NVMHCI Work Group—more commonly known as the NVMe Work Group.

The specification defines an optimized register interface, command set, and feature set for PCIe SSDs. The goal of the standard is to help enable the broad adoption of PCIe-based SSDs, and to provide a scalable interface that realizes the performance potential of SSD technology now and into the future. By maximizing parallelism and eliminating complexity of legacy storage architectures, NVMe supports future memory developments that will drive latency overhead below one microsecond and SSD IOPS to over one million. The NVMe specification may be downloaded from www.nvmexpress.org.

The NVMe specification is specifically optimized for multi-core system designs that run many threads concurrently with each thread capable of instigating I/O operations. Indeed it's optimized for just the scenario that IT managers are hoping to leverage to boost IOPS. NVMe specification can support up to 64 k I/O queues with up to 64 k commands per queue. Each processor core can implement its own queue.

In June, 2011, the NVMe Promoter Group was formed to enable the broad adoption of the NVMe Standard for PCIe SSDs. NVMe supporters include Cloud service providers, IC manufactures, Flash-memory manufacturers, operating-system vendors, server manufacturers, storage-subsystem manufacturers, and network-equipment manufacturers.

The original NVMe specification was focused on direct-attached PCIe SSD usage model. More recently, with the growth of scaleout cloud infrastructure, many data centers are moving from inefficient direct-attached storage model where compute and storage are deployed in fixed ratios, to a hyper-scale shared SSD model where compute and storage are scaled independently to achieve maximum resource utilization and drive down cost. This trend has triggered disaggregation of NVMe SSDs from compute servers and driven the need for an extension of NVMe outside of the box over networking fabrics.

The NVMe over Fabrics standard was born to define a common architecture that supports a range of storage networking fabrics for NVMe block storage protocol. This includes enabling a front-end interface into storage systems, scaling out to a large numbers of NVMe devices and extending the distance within a data center over which these devices can be accessed. The goal of NVMe over Fabrics is to provide remote connectivity to NVMe devices with minimal additional latency over a direct-attached NVMe device inside a compute server. The NVMe over Fabrics specification was published in June 2016.

The most recent addition to the list of fabric transports for NVMe is TCP (Fig. 1.28). NVMe over TCP block storage interface enables disaggregation of NVMe SSDs from compute servers without compromising latency and without requiring changes to networking infrastructure. The storage network in this case is standard TCP/IP over Ethernet, a high-performance ubiquitous networking architecture that is both scalable and reliable. The NVMe Work Group is standardizing TCP/IP transport binding, adding this to the NVMe Fabrics specification alongside RDMA and Fibre Channel.

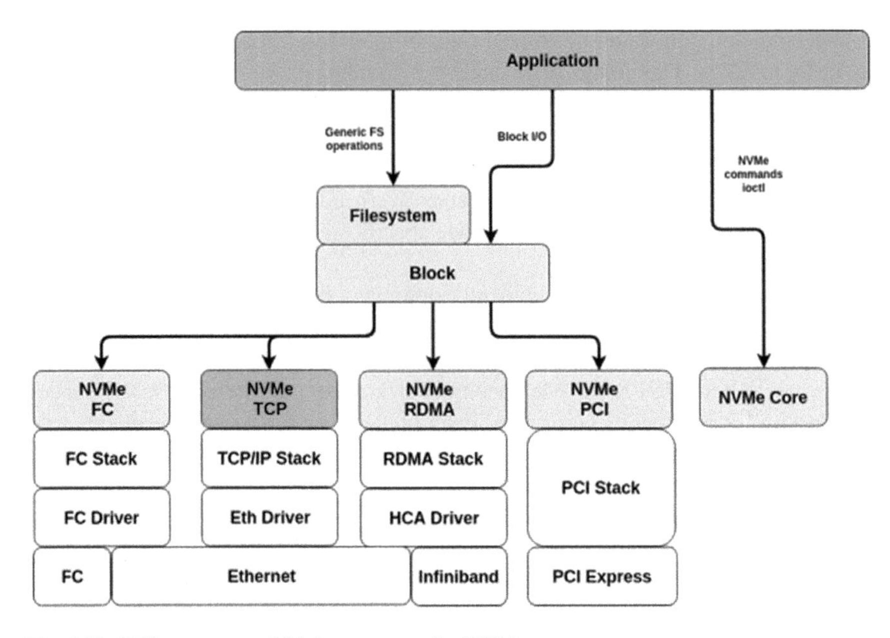

Fig. 1.28 Different types of fabric transports for NVMe

The building blocks are all falling into place for broader usage of PCIe-connected SSDs and deliverance of the performance improvements that the technology will bring to enterprise applications. And while the focus in the past has been more on the enterprise, the NVMe standard has already trickled down to client systems, offering a performance boost in notebook PCs while reducing cost and system power. The NVMe standard will continue to drive more widespread use of PCIe SSD technology as new compatible ICs and drivers come to market.

References

1. www.mmca.org
2. www.compactflash.org
3. www.sdcard.com
4. G. Campardo, R. Micheloni, D. Novosel, *VLSI-Design of Non-Volatile Memories* (Springer, Berlin, 2005)
5. www.onfi.org
6. www.jedec.org
7. A. Kawaguchi, S. Nishioka, H. Motoda, A flash-memory based file system, in *Proceedings of the USENIX Winter Technical Conference* (1995), pp. 155–164
8. J. Kim, J.M. Kim, S. Noh, S.L. Min, Y. Cho, A space-efficient flash translation layer for compact flash systems. IEEE Trans. Consum. Electron. **48**(2), 366–375 (2002)
9. S.-W. Lee, D.-J. Park, T.-S. Chung, D.-H. Lee, S.-W. Park, H.-J. Songe, FAST: A log-buffer based FTL scheme with fully associative sector translation, in *2005 US-Korea Conference on Science, Technology, & Entrepreneurship* (Seoul, Aug 2005)

10. T. Tanzawa, T. Tanaka, K. Takekuchi, R. Shirota, S. Aritome, H. Watanabe, G. Hemink, K. Shimizu, S. Sato, Y. Takekuchi, K. Ohuchi, A compact on-chip ECC for low cost flash memories. IEEE J. Solid-State Circuits **32**(May), 662–669 (1997)
11. G. Campardo, R. Micheloni et al., 40-mm^2 3-V-only 50-MHz 64-Mb 2-b/cell CHE NOR flash memory. IEEE J. Solid-State Circuits **35**(11), 1655–1667 (2000)
12. R. Micheloni et al., A 4 Gb 2b/cell NAND flash memory with embedded 5b BCH ECC for 36 MB/s system read throughput, in *IEEE International Solid-State Circuits Conference Dig. Tech. Papers* (Feb 2006), pp. 142–143
13. R. Micheloni, A. Marelli, R. Ravasio, *Error Correction Codes for Non-Volatile Memories* (Springer, Dordrecht, 2008)
14. C. Park et al., A high performance controller for NAND flash-based Solid State Disk (NSSD), in *IEEE Non-Volatile Semiconductor Memory Workshop NVSMW* (Feb 2006), pp. 17–20
15. www.pcisig.com
16. R. Budruk, D. Anderson, T. Shanley, Mindshare, *PCI Express System Architecture* (Addison-Wesley, Boston, 2003)
17. K. Kong, *Enabling Multi-peer Support with a Standard-Based PCI Express Multi-ported Switch*, White Paper (Jan 2006), www.idt.com
18. K. Kong, Non-Transparent Bridging with IDT 89HPES32NT24G2 PCI Express NTB Switch, AN-724 (Sept 2009), www.idt.com
19. www.ssdformfactor.org

Chapter 2
SAS and SATA SSDs

S. Yasarapu

Abstract This chapter focuses on the different types of solid state drives. The chapter details the differences between consumer and enterprise solid state drives and also details the differences between SAS and SATA solid state drive and what lies ahead for SATA and SAS protocols for SSDs.

2.1 Introduction

Data centers today require fast and reliable storage to provide end-users with high quality of service. Data centers operators are continuously challenged to improve performance to keep up with the demands of high performance applications. Space, power and cooling limitations require data centers to find the most cost-, space-, and energy efficient products. Solid state drives increase the performance and reliability of the enterprise while reducing the overall space, power, energy footprint of the data centers. However, not all data center and enterprise environments are created equal. Depending on the size, number of users, serviceability requirements and applications running in the data center, the need for performance and storage capacity varies and so do the solid state devices used within these environments.

In fact, not all SSDs are created the same. Some are designed for the enterprise and some are designed for consumer applications. Even in the enterprise segment, some are intended for direct attach to servers and some are designed for shared storage enclosures. Understanding the differences between the various solid state drives helps consumers, as well as, enterprises to select the right solution for their intended applications.

S. Yasarapu (✉)
SSD Product Marketing, Western Digital Corporation, Irvine, CA, USA
e-mail: swapna.yasarapu@wdc.com

R. Micheloni et al. (eds.), *Inside Solid State Drives (SSDs)*,
Springer Series in Advanced Microelectronics 37,
https://doi.org/10.1007/978-981-13-0599-3_2

2.2 Enterprise Versus Consumer SSDs

Let's first start by understanding the difference between enterprise and consumer solid state devices [1]. To really understand the differences between consumer and enterprise solid state drives, let's start by first observing where these devices are used. This will highlight the fundamental assumptions made by designers of consumer and enterprise solid state drives.

Consumer solid state devices are used in laptops, desktops, and mobile devices where conserving power is the most important criteria to ensure long battery life of the device. Now, let's think about the typical usage pattern of a laptop user. Laptop user, either a business or a home user, generally turns on the laptop at the beginning of the day. Typical applications running on the laptop are email, internet explorer, Microsoft Word, Excel, PowerPoint. For the majority of time, user reads information—reading emails, browsing the web etc. The laptop is perhaps left idle during meetings; is left idle during lunch time. Laptop is turned off at the end of the day. Let's take another example—a desktop user's typical day. In addition to everything the laptop user does, desktop users may also play video games, listen to music and access other digital content. Again, this involves fetching of relatively large amounts of data from the storage → fast reads. So, what makes the laptop and desktop users happy? Laptops should turn on as soon as they are powered on to minimize the wait for system boot up → fast boot times; user would like to use the laptop on battery for as long as possible → low power footprint and email and browser applications should load up fast → fast reads. Now let's contrast this with typical usage patterns in an enterprise data center.

Enterprise Solid State Drives are used in corporate and cloud data centers where uninterrupted operations and high reliability are the most important criteria. Data center of an enterprise is the information technology hub that holds the most important intellectual property of any business/enterprise—DATA. The data stored in the data center is made available via different applications such as Oracle databases, email applications, customer relation management systems and is used by multiple users—R&D, finance, sales, operations, customer service etc. Data is accessed from different locations, at different times. Loss of data is not an acceptable event because of its disastrous consequences to the business. Let's take the example of a financial institution where customers make deposits/withdrawals of money from their accounts. If a withdrawal transaction is lost due to loss of data in the institution's data center then the financial institution loses money. Now this may not seem like a big deal but if it happens systematically, then this could add up to millions of dollars in losses. Or worse, if a deposit amount is not posted to a customer's account, then customer loses money which could be even more disastrous because the bank loses its credibility and hence customers → loss of revenues. So what makes the *Chief Information Officer* (CIO) happy? All systems in the data center should run uninterrupted → 24 h/day—7 days/week—365 days/year operations with minimal maintenance; there should never be a case leading to

Table 2.1 Application level usage pattern

Criteria	Consumer SSD	Enterprise solid state drive
Hours of operation	Interrupted	24/7/365 Uninterrupted operation consistent and low latency and quality of service
	Fast boot time for frequent power up	
Performance	Fast large block reads only	Fast small block random reads and writes
Access pattern	Single threaded accesses	Multi user accesses
Power consumption	Low power to improve battery life	Reduce total data center power and energy footprint
High availability	Not required	High availability
Reliability	Ease of replacement	High reliability
	Loss of data is managed	Loss of data is catastrophic

data loss → high reliability; ability to service multiple users at any time → high performance.

As summarized in Table 2.1, we can conclude that the usage pattern of a consumer solid state drive is dramatically different from that of an enterprise solid state drive. This primarily drives completely different design criteria.

Let's see how the consumer and enterprise solid state drives differ in their construction. This will highlight the fundamental assumptions made by the testers and integrators of consumer Flash and enterprise solid state drives. To do this, let's first understand the composition of consumer and enterprise solid state drive. The basic composition of an SSD is a controller and a Flash as shown in Fig. 2.1. But that is where the similarity ends.

What really separate enterprise solid state drive from consumer SSD is the design of the controller hardware and more importantly the controller firmware features and the rigors of testing and qualification process the enterprise solid state drive is put through before it makes it to the market in a product form.

Controller hardware and firmware running on the Enterprise SSDs are the brains of the device. Their primary functions are to respond to host commands, to transfer data between the host and Flash media and to manage the Flash media to achieve high reliability and endurance throughout the operational lifetime of the drive. How well a controller handles Flash management and host data transfers simultaneously is what differentiates it from a consumer SSD. In addition, enterprise SSDs have additional built-in features to improve the reliability and endurance of the Flash and hence the enterprise SSD. Enterprise solutions require 24/7/365 uninterrupted operation. Therefore, controllers in enterprise SSDs are designed to maintain consistent performance behavior while transferring data irrespective of the amount of Flash capacity in use and also the traffic generated to the drive. Wear leveling operations and background media error correction algorithms are designed such that data transfer performance to the host is unchanged while these operations run in the background to the Flash.

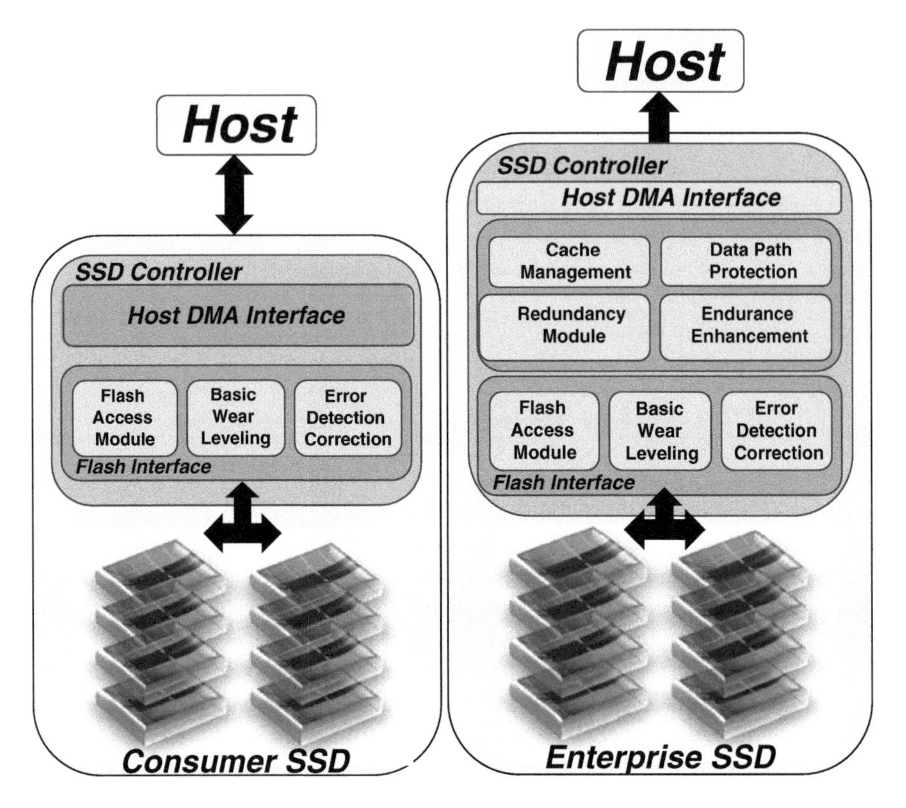

Fig. 2.1 Consumer (*left*) and enterprise (*right*) solid state drives

Enterprise solutions are required to support a large number of users, i.e., multiple initiators running different types of traffic patterns independent of one another resulting in random traffic. Therefore, the controller hardware and firmware is designed to support multi-threaded access where up to hundreds of threads of data per drive can be pushed between host and the device while maintaining the performance as well as integrity of data. Therefore, enterprise SSDs are designed to perform extremely well even for small transfers of varying sizes and for simultaneous reads and writes.

Data integrity and availability is of the highest importance in enterprise solutions. Therefore, enterprise SSDs are designed to provide full data path protection with ECC and CRC coverage and power fail protection against unscheduled power loss.

Reliability and endurance are extremely important for enterprise application because solutions deployed into enterprise have a longer working life. Unlike consumer deployments, enterprise deployments have a long service life. Therefore, enterprise SSDs are designed to survive in mission critical storage area networks under 24/7/365 workloads for over up to 5 years. To this effect, enterprise SSDs have built-in redundancy to ensure that even if Flash die fails, the SSD can successfully recover data by using the redundancy built into the data stored on the Flash.

Enterprise SSDs are also built with features to improve the endurance of the Flash. This is an extremely important capability required to counter the deterioration in Flash endurance as technology nodes shrink.

Enterprise SSDs have the characteristics of drives designed for use in all environments (like the ones on Mars): this allows for drive to operate in environments that do not require human presence and can handle unknown conditions as they arise.

The above mentioned design capabilities are driven by the application use cases where enterprise solid state drives are used. Consumer SSD, unlike enterprise SSD is not designed with these assumptions and is therefore unsuitable for enterprise applications.

Consumer SSDs are designed for cost, which may or may not include robust controller/Flash management technology. Consumer SSD doesn't have power fail protection and do not have the same stringent data protection capabilities of enterprise SSDs. Consumer SSDs are not designed to endure under enterprise workloads; they are designed for laptops and desktops not expected to work beyond a few years.

Since consumer SSD is focused on providing faster boot time, and application load time, they are optimized to provide fast large block read transfers. Given that consumer SSD is left idle for long durations, consumer SSD depends on host side to manage the SSD media. This in turn leads to short lifetime of the consumer SSD. In addition, consumer SSD is designed for single operation management, for data loading, installing, saving, etc.

Consumer SSDs have been designed for single user usage and are only designed for read focused operation, where only a small amount of data is written with many hours of idle time. Therefore, consumer SSDs though suitable for low end applications where the devices are not challenged to work at high performance levels, are not suitable for high performance, high reliability enterprise deployments. However, it is known that consumer SSDs are sometimes uses for boot use cases in enterprise and data center deployments. Typically consumer SSD uses SATA interface to connect to host systems with PCIe NVMe interface connected SSDs on the horizon.

Enterprise SSDs come in different form factors with different interfaces. There are 3 main interface protocols used to connect SSDs into server and/or storage infrastructure: *Serial Attached SCSI* (SAS), *Serial ATA* (SATA) and PCIe NVMe. SAS SSDs deliver high levels of performance and are used in both high end server and midrange—high end storage enclosures. SATA based SSDs are used mainly in client applications and in entry and midrange server and storage enclosures. PCIe based SSDs are newest of the three (3) types of SSDs. There are generally two classes of SSDs—those delivering the highest performance are mainly used in server based deployments with storage deployments in the near horizon and those delivering good enough performance to replace SATA SSDs in data centers. SAS and SATA SSDs combined continue to hold the lion share of the enterprise SSD market with PCIe SSDs showing the highest growth and adoption rate.

In this chapter, let's focus on the SAS and SATA SSD—protocol differences, key feature highlights, similarities and differences and where they are used.

2.3 SAS Versus SATA Protocol

Serial Attached SCSI (SAS) is a communication protocol traditionally used to move data between storage devices (target) and host (initiator). SAS defines how 1 or more initiators can connect to 1 or more SAS device targets. It uses a standard SCSI command set to drive device communications. Today, SAS based devices most commonly run at 12 Gbps. There is ongoing development of a faster 24 Gbps SAS interface speed which may be brought to market sometime in the future. On the other side, SAS interface can also be run at slower speeds—1.5, 3 Gbps and/or 6 Gbps to support legacy systems.

S also offers backwards-compatibility with second-generation SATA drives at the physical layer. The T10 technical committee of the International Committee for Information Technology Standards (INCITS) develops and maintains the SAS protocol; the SCSI Trade Association (SCSITA) promotes the technology.

Serial ATA (SATA or Serial Advanced Technology Attachment) is another interface protocol used for connecting host bus adapters to mass storage devices such as hard disk drives and solid state drives. Serial ATA was designed to replace the older parallel ATA/IDE protocol. SATA is also a point to point connection using a serial physical connection. It uses ATA and ATAPI command set to drive device communications. Today, SATA based devices most commonly run at 6 Gbps.

Serial ATA industry compatibility specifications originate from The Serial ATA International Organization [2] (aka. SATA-IO).

2.3.1 Connectivity and High Availability

A typical SAS eco-system consists of SAS SSDs plugged into a SAS backplane or a host bus adapter via a point to point connection, which in turn is connected to the host microprocessor either via an expander or directly, as shown in Fig. 2.2.

Each expander can support 255 connections to enable a total of 65,535 (64 K) SAS connections. Therefore, SAS based deployments enable use of a large number of SAS SSDs in a shared storage environment.

SAS SSDs are built with two ports. This dual port functionality allows host systems to have redundant connections to SAS SSDs. In case one of the connections to the SSD is either broken or malfunctions, host systems still have the second port that can be used to maintain continuous access to the SAS SSD. In enterprise

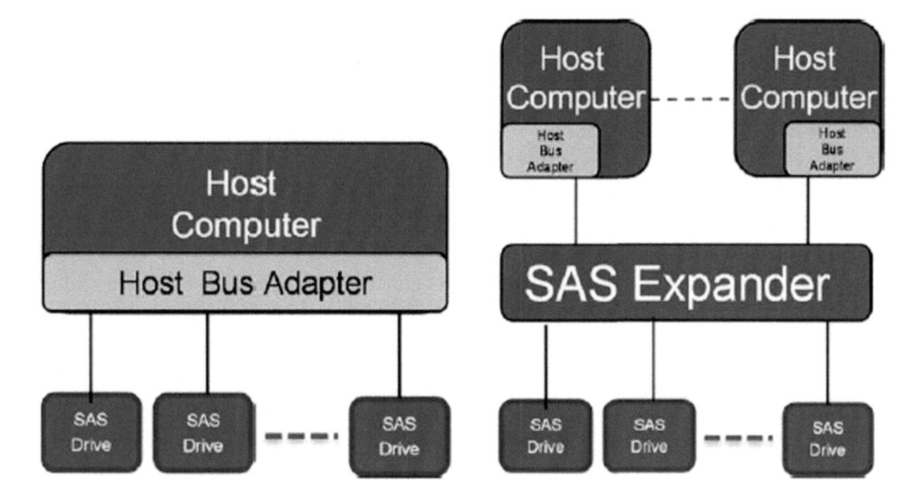

Fig. 2.2 SAS connectivity

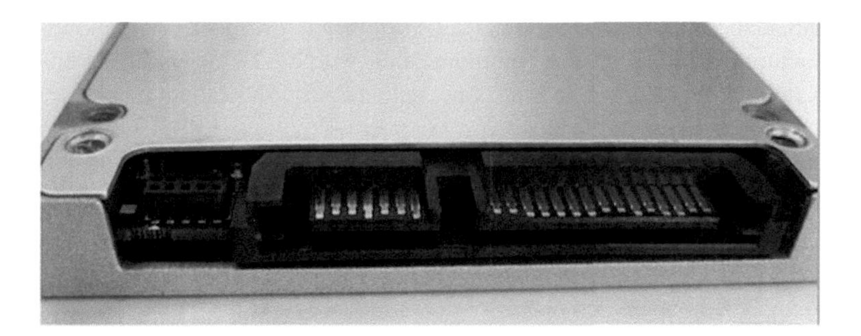

Fig. 2.3 Dual port SAS connector

applications where high availability is an absolute requirement, this feature, unique to SAS SSDs, makes it the SSD of choice for enterprise applications. Figure 2.3 below shows the dual port connector used with SAS SSDs.

SAS SSDs also support hot plug. Hot plug feature enables SAS SSDs to be dynamically removed or inserted while the system is running. This feature allows for automatic detection of newly inserted SAS SSDs. While a server or storage system is running, newly inserted SAS SSDs can be dynamically configured and put to use. Even more importantly, even if SAS SSDs are pulled out of a running system, all the in-flight data that is committed by the host system is properly stored inside a SAS SSD and can be accessed once the SSD is powered back on.

As opposed to SAS, a typical SATA eco-system consists of SATA SSDs connected to host bus adapter via a point to point connection, which in turn is connected to the host microprocessor. In addition, SATA SSDs are built with one port unlike SAS SSDs. These two main differences make SATA based SSDs more suited for entry or mid-range deployments and consumer applications.

SATA SSDs also support hot plug which enables SSDs to be dynamically removed or inserted while the system is running. While a server or storage system is running, newly inserted SATA SSDs can be dynamically configured and put to use. However, not all SATA SSDs are designed to withstand hot plug functionality and to ensure that if pulled out of a running system, all the inflight data that is committed by the host system is properly stored inside a SATA SSD. This capability, also commonly known as surprise removal, is an extremely important feature and is generally only supported by selected enterprise grade SATA SSD vendors.

SATA drives may be connected to SAS backplanes, but SAS drives may not be connected to SATA backplanes.

This is an important feature, in that physically SAS infrastructure is designed to accommodate SATA SSDs. Connector on SATA SSDs is designed such that they can be plugged into SAS receptacles though the reverse is not true. This enables SATA SSDs to be plugged into SAS based storage system making the SATA SSD more ubiquitous for use.

In addition, even though SAS uses SCSI as the primary communication protocol, SAS also supports STP (*Serial ATA Tunneled Protocol*) that allows SAS infrastructure is built to ensure communication with SATA SSDs hence enabling interoperability. Again, reverse is not true, in that SAS SSDs cannot be plugged into SATA based deployments.

Similarities between SAS and SATA technologies are summarized in Fig. 2.4; differences between the two are in Fig. 2.5.

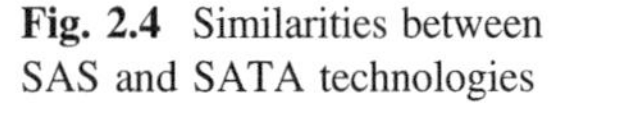
Fig. 2.4 Similarities between SAS and SATA technologies

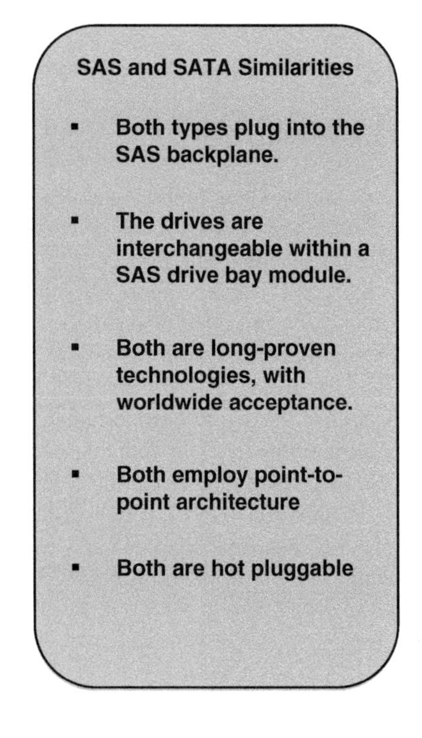

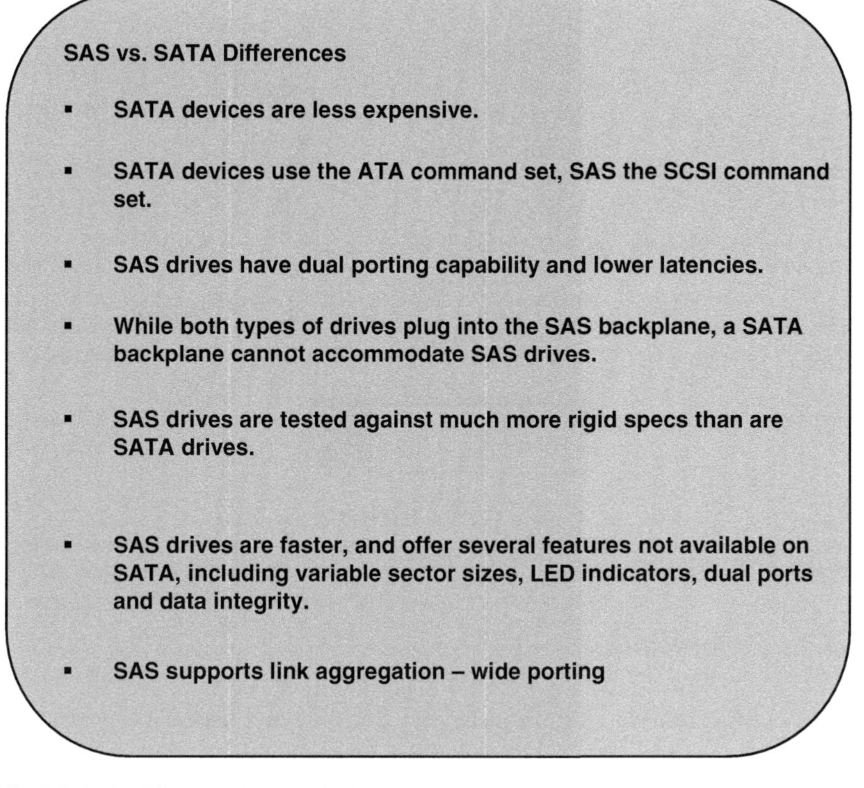

Fig. 2.5 Main differences between SAS and SATA

2.3.2 Form Factor and Capacity

SAS and SATA SSDs come in a variety of capacities and form factors.

SAS SSDs are designed in primarily to fit into 2.5″ form factor. This form factor is primarily defined and driven by the small form factor working group and the T-10 organization. Since SAS SSDs are designed for both server and storage applications, the capacity of SAS SSDs varies from 200 GB up to 30 TB in capacity for use depending on deployment and application requirement.

SATA SSDs are designed in a variety of form factors—2.5″, 1.8″ as well as smaller M.2 form factors (Fig. 2.6). Typical enterprise applications use either 2.5″ SATA SSDs or M.2 form factor SATA SSDs. For example, M.2 SATA SSDs are popularly used as boot devices. In addition, the smaller form factors enable SATA to be used in space constrained embedded applications. SATA SSDs in capacity vary anywhere between 32 GB and 8 TB and are generally used either in consumer, boot or entry and mid-range data center applications.

Fig. 2.6 SATA form factors

2.3.3 Performance

SAS uses SCSI command set to transfer data. SCSI is a more efficient command set with features such as command queuing that enable higher performance of SAS SSDs. Therefore, SAS SSDs are used where higher performance is required. Unlike SAS, SATA SSDs using the ATA protocol have lower performance compared to SAS and therefore are more widely for mid-range and entry level system.

However, a point to note is that both SATA and SAS SSDs are orders of magnitude faster than hard disk drives (HDDs). To better understand the performance characteristics of SSDs first, it is important to know what is inside an SSD compared to HDD.

Hard disk drives are electro-mechanical devices which inherently is limited by the mechanical element utilized to build them, i.e., rotating magnetic disk. In order to retrieve data that is stored on the magnetic disk, one must rotate the disk to place it under the media head (rotational latency), moving the head to the right track (seek latency) and then using a combination of electronics and mechanics to transfer the data to/from the host devices (transfer time). The only way to hide rotational and seek latencies is by transferring large sequential data from the disk once the right track on the disk is located. Therefore, hard drives are inherently sequential devices and limited in random performance. Sequential performance is generally measured in MBps or GBps, whereas, random performance is measured in IO per seconds (IOPs). The fastest hard drives on the market provide at best 350 IOPS under random workloads. However, real world applications are random by nature.

In contrast to hard drives, solid state drives are electronic devices. There are no mechanical elements on a solid state drive. Data is stored in NAND Flash devices, and is retrieved from the NAND Flash by on board controller. All blocks of data on the NAND Flash are equally accessible by the controller, i.e., there are no rotational and/or seek latencies to get to the right block of data.

Performance, reliability, and endurance of SSDs are highly dependent on the design of the SSD controllers as discussed in earlier sections.

How efficiently HW (Hardware) and FW (Firmware) of the SSD controller handle data streaming while also performing Flash management determines the performance of the SSD. Controllers in enterprise SAS and SATA SSDs are designed to maintain consistent performance behavior while transferring data, regardless of the amount of Flash capacity in use, and irrespective of the volume of traffic being generated to the drive at any point in time. Wear-leveling operations and background media error correction algorithms are designed so that data transfer performance to the host is unchanged while these operations run in the background. An enterprise-class SSD is designed to handle these heavy workloads 24/7/365 for 5 years or more.

To be of real value, SSD performance needs to be measured after the SSD reaches steady. Performance measured on a fresh out of the box SSD—SAS or SATA will not truly represent the performance of the drive in a real deployment. Therefore, before measuring SSD performance, one must precondition the SSD under test. This is accomplished by writing random data patterns to completely fill all NAND blocks and engage the drive's wear-leveling and Flash management routines. Properly managing data flow and internal NAND will make the measurement a more useful gauge of SSD performance under real-world conditions. Figure 2.7 illustrates the higher performance of fresh out of box SSDs that reach steady state after pre-conditioning the SSD.

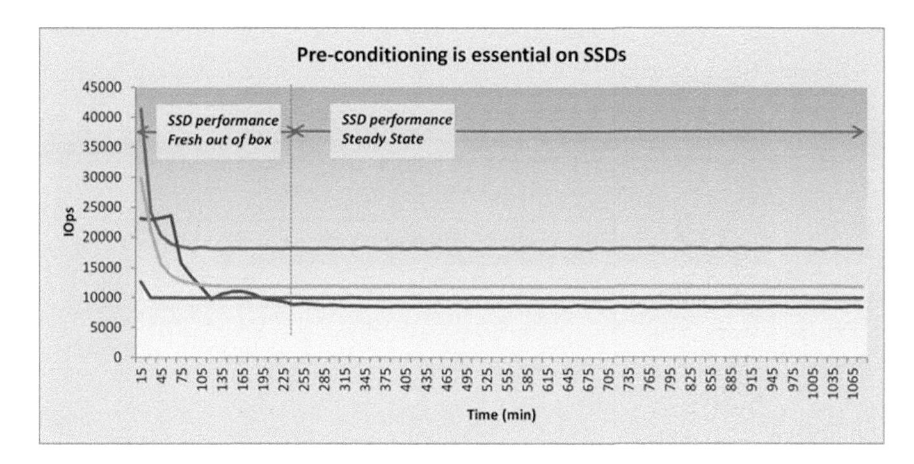

Fig. 2.7 Effect of preconditioning on performances

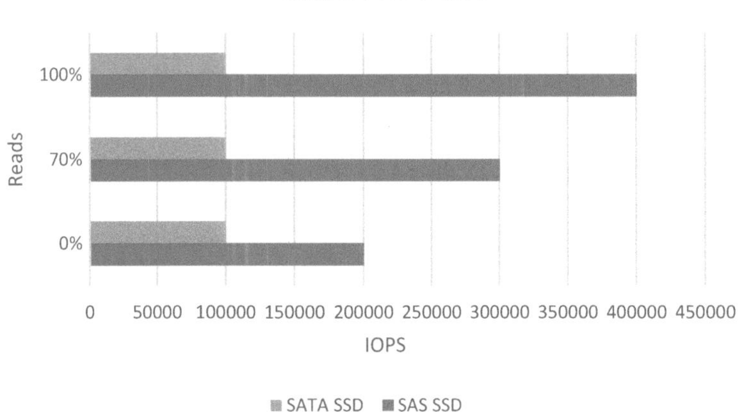

Fig. 2.8 SAS versus SATA under random workloads

To understand the real world benefits of SAS and SATA SSDs, performance is usually measured for large block 128 KB or larger sequential and small block 4 KB or 8 KB random read, write and mixed workloads.

Figures 2.8 and 2.9 show a real world comparison between SAS and SATA SSDs. As seen in these charts, SAS SSDs deliver almost 4× higher performance compared to SATA SSDs.

As seen from the charts above, both type of enterprise SSDs—SAS or SATA, have place in the data center. SAS SSDs are used for high end performance critical enterprise systems and SATA SSDs are used with mid-range or entry level systems.

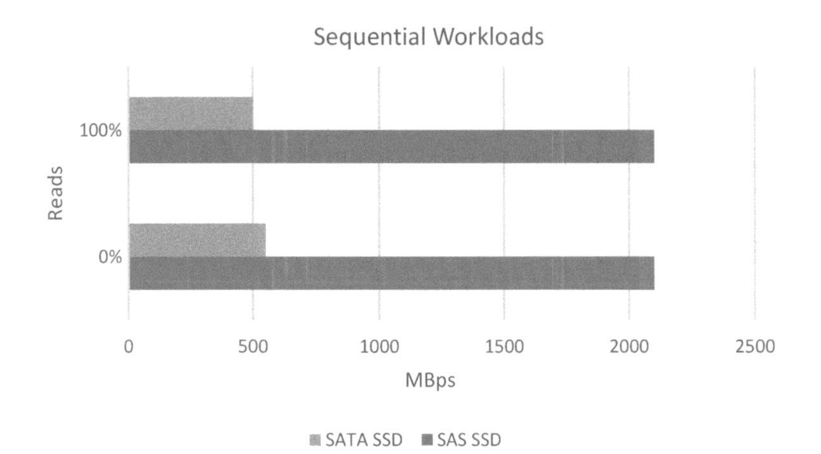

Fig. 2.9 SAS versus SATA under sequential workloads

2.4 What's Ahead

SATA and SAS based SSDs have been adopted in consumer and enterprise applications. This adoption is expected to continue and expand in the coming years. Data center and enterprise applications are using increasingly large amounts of SSDs and also SSDs of higher capacities to deliver on the need for ever increasing demands for data storage. As the NAND Flash geometries shrink, the capacity of SAS SSDs is expected to increase to address this need for higher capacity SSDs.

On the SATA front, the SATA protocol is expected to continue to deliver 6 Gbps interface speeds for the near future. However, for faster SSDs, industry is expected to adopt PCIe SSDs in future generations of server and storage environments.

On the SAS front, the protocol is expected to continue to deliver 12 Gbps interface speeds for the foreseeable future. There are development efforts ongoing to enable a higher speed SAS interface 24 Gbps. However, it is unclear to what extent the industry will widely adopt a faster SAS 24 Gbps interface or if enterprise will continue to use 12 Gbps SAS SSDs for vast majority of mainstream appplications.

As discussed above, enterprise solid state drives increase the performance and reliability of the enterprise while reducing the overall space, power, and energy footprint of the data centers. Key features for the enterprise are long service life, high endurance, consistent and high performance, high reliability which are delivered by SAS and/or SATA SSDs.

Choosing the right SSD—SATA or SAS, depends on the end user application. Use of SSDs leads to improved performance, higher reliability and reduced power space and energy consumption which reduces capital and operating expenses of next generation data centers. This is what makes SSDs a great product to enable highest levels of performance and fast access to data in consumer as well as enterprise applications.

References

1. www.hgst.com
2. http://www.t10.org

Chapter 3
Hybrid Storage Systems

Rino Micheloni, Luca Crippa and M. Picca

Abstract In recent years, both industry and academia have increased their research effort in the hybrid memory management space, developing a wide variety of systems. It is worth mentioning that "hybrid" is a generic term and it can have different meanings depending on the context. For instance, a storage system can be hybrid because it combines HDD and SSD; an SSD can be hybrid because it combines SLC, MLC and TLC Flash memories, or it combines NAND with *Storage Class Memories* (SCMs), which are non-volatile memories like ReRAM, PCM or MRAM. In this chapter we look at all these different meanings. The last section covers over-provisioning and the *Write Amplification Factor* (WAF): these parameters have a great impact on SSD performances and reliability, as well as on the available storage capacity.

3.1 NAND Flash Memory and HDD

If we look at the DRAM history [1], DRAM data access speeds have increased at a faster pace than *Hard Disk Drives* (HDDs), leaving a gap in the memory hierarchy as shown in Fig. 3.1. The gap in read and write performances between DRAM and HDD has widened over the last decade, thus leaving an opportunity for a new intermediate memory/storage technology between HDDs and DRAM: NAND Flash memories and SCMs can fill this performance gap.

While HDDs are common secondary storage devices, their high power consumption and low shock resistance limit them as an ideal mobile storage solution

R. Micheloni (✉) · L. Crippa
Storage Solutions, Microsemi Corporation, Vimercate, MB, Italy
e-mail: rino.micheloni@ieee.org

L. Crippa
e-mail: luca.crippa@ieee.org

M. Picca
STMicroelectronics, Cornaredo, Italy
e-mail: massimiliano.picca@st.com

R. Micheloni et al. (eds.), *Inside Solid State Drives (SSDs)*,
Springer Series in Advanced Microelectronics 37,
https://doi.org/10.1007/978-981-13-0599-3_3

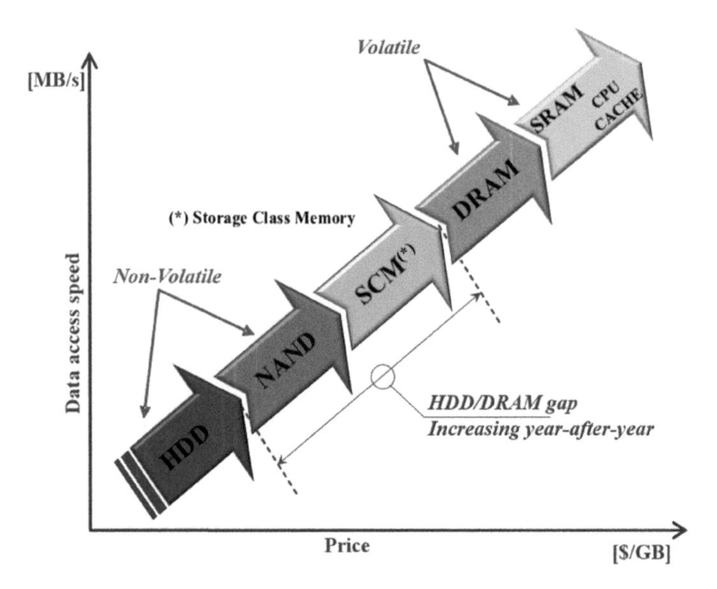

Fig. 3.1 Memory hierarchy

[2]. On the other hand, Flash memories (especially of NAND type) overcome the main problems of HDDs, but they are still more expensive and can only support a limited number of program/erase cycles [3].

Researchers generally agree that disk-storage performance is subject to the handling of small files and filesystem metadata. Unlike traditional disk storage, Flash memory has no seek penalty, but is subject to garbage collection and wear leveling.

To avoid excessive wear-out of Flash memories, and to mitigate their low write throughput, it is a good approach to migrate frequently-read data to the Flash and frequently-written data to HDD, as sketched in Fig. 3.2. In other words, there should be a caching software that dynamically manages the use of the entire drive capacity for superior overall storage performance, where the most frequently/recently used "hot" data are cached for ultra-fast access, while the "cold" data remains on the primary storage partition.

The trade-offs associated with HDDs and Flash memories motivate lots of storage system designs [4–8]. Many applications use Flash memory as a non-volatile cache storing data blocks which are likely to be accessed in the near future, and thus allowing the disk to spin down for longer periods.

However, these schemes treat flash memory as complement of DRAM buffer cache, and only a subset of data blocks are cached in flash memory; as a result, the disk is used quite frequently due to cache misses or flushing. As flash memory's capacity increases, a real hybrid secondary storage solution is expected to be more effective [9]. Different from data block level cache, Flash memory stores files and can be accessed independently in hybrid secondary storage system.

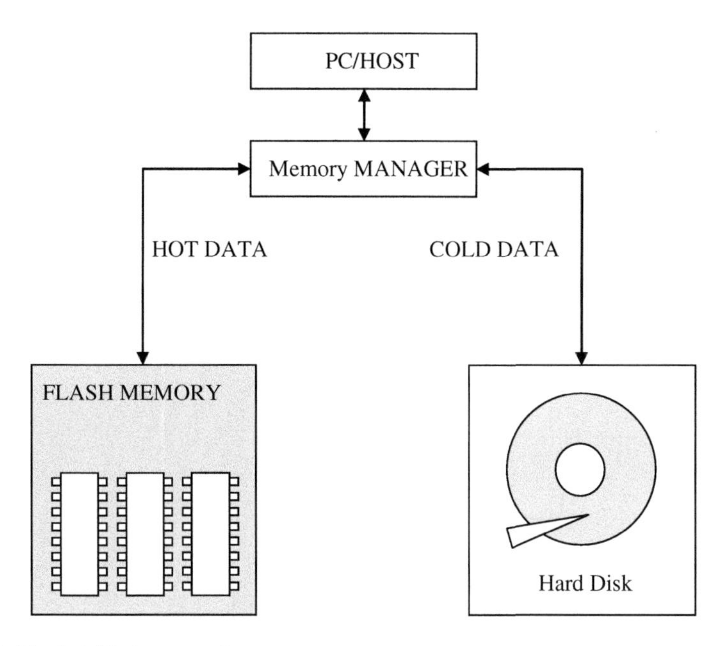

Fig. 3.2 The hybrid storage system

In recent years, both industry and academia have increased their research effort in the hybrid memory management space, developing a wide variety of systems [10–12]. At this point it is worth mentioning that “hybrid” is a generic term and it can have different meanings depending on the context. Figure 3.3 is a summary of what a hybrid storage could be.

We will look at various combinations of Flash memory and HDDs in the following sections.

3.2 External NAND + HDD

One of the first examples of NAND used as an external memory was ReadyBoost [13–15]. It works by using flash memory, a USB flash drive, SD card, CompactFlash or any kind of portable flash mass storage system as a cache, as shown in Fig. 3.4.

The core idea of ReadyBoost is that a flash drive has a much faster seek time than HDD, allowing it to satisfy requests faster than reading files from the hard disk.

When an *EXternal Memory* (EXM) is plugged into the computing device, the system populates EXM with disk sectors and/or memory sectors. The system routes I/O read requests directed to the sector to the EXM cache instead of the actual

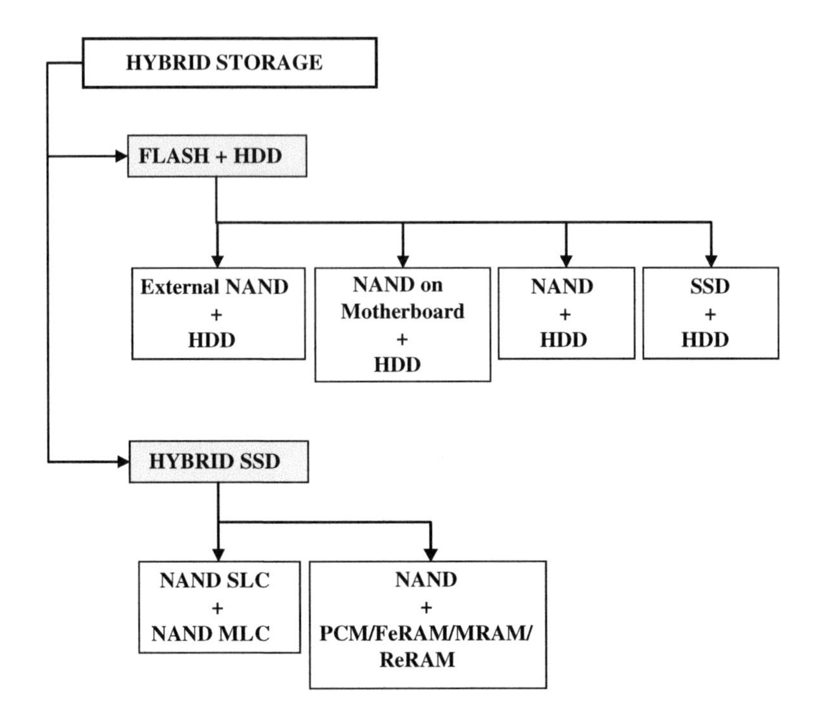

Fig. 3.3 Hybrid storage overview

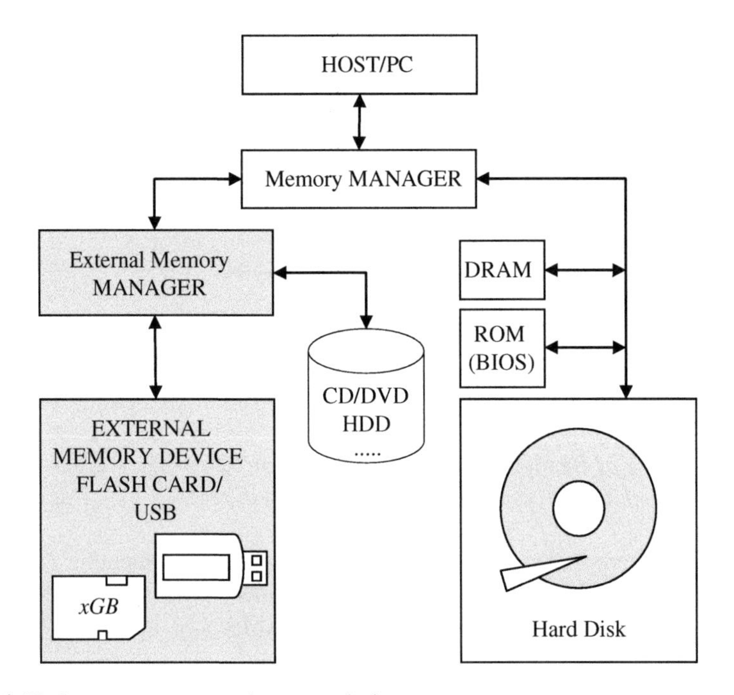

Fig. 3.4 Flash memory as external memory device

sector. The use of EXMs increases performance and productivity on the computing device systems for a fraction of the cost of adding memory to the computing device.

The system detects when an EXM is used for the first time. Once the type of EXM is discovered, a driver is installed and it is used to cache disk sectors on the external memory. Sectors from any disk and/or slower memory device on the system can be cached to EXM. Without a prior knowledge of which sectors are valuable in terms of frequent access, the system may use data on the computing machine to determine which sectors are used to populate the EXM cache. Alternatively, the system populates the EXM cache with a particular sector when that particular sector is accessed during operation. The next time that particular sector is to be accessed for a read operation, the system directs the read operation to access the copy from the EXM. The system may track usage patterns and determine which disk sectors are most frequently accessed. On subsequent uses of the EXM, the system caches those sectors that are most frequently accessed onto the EXM. If the EXM is present when the computing device is powered up, the EXM can be pre-populated with data during start-up of the operating system [13].

3.3 NAND on Motherboard + HDD

Computer motherboards contain the processor chip and some high performance SRAM and DRAM memories. In the last few years there have been proposals to add Flash memory to the computer motherboard for a non-volatile memory layer to the motherboard memory/storage architecture. The motherboard Flash memory could be inserted into the motherboard with an ONFI module or DIMMs similar to those currently used for DRAM, allowing memory replacement when faster or larger memory becomes available.

Intel introduced a motherboard Flash memory technology in 2007, known as "Robson Technology" or "Turbo Memory" [16, 17]. This early implementation ran into issues due to lack of support for management of Flash/HDD partition in main operating systems. In fact, central to the operation of any hybrid storage computer architecture is management to determine which data is to be kept on the HDD and which data will be kept on the Flash memory.

Figure 3.5 shows a storage management controller that determines what data should be stored on each memory device. This storage management function must balance the needs of data access, power savings opportunities, and data security.

As with any NAND based memory product solution, the NAND flash memory controller is also key in executing the NAND wear leveling algorithm, managing the reads, writes, erases, and performing the ECC (Error Correction Code) as needed [16].

With NAND moving into the demanding computing environment, the wear leveling algorithm must comprehend not only the usage statistics of the NAND flash but also track the key reliability statistics. In other words, the controller must

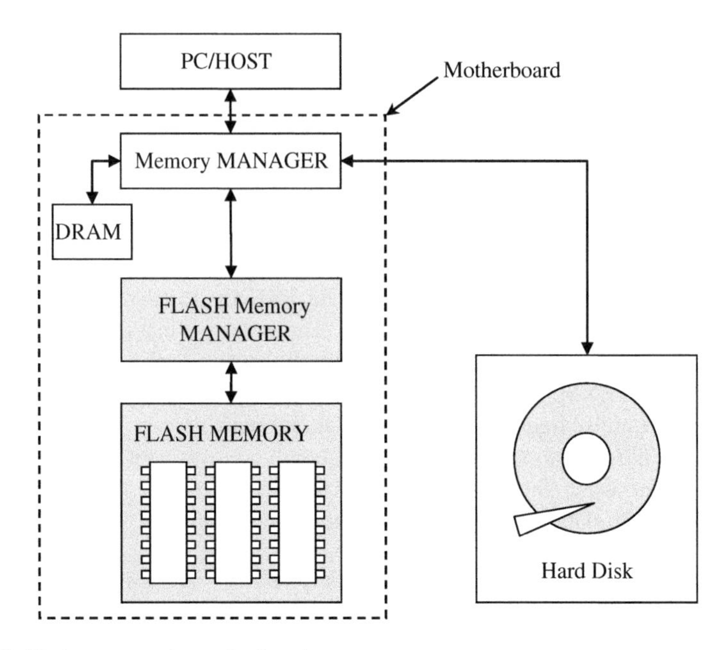

Fig. 3.5 Flash on computer motherboard

track all the failure mechanisms known in the NAND Flash industry (Chap. 9): program disturb, read disturb, program/erase cycles, data retention, etc.

In the next section, HDD is combined with another drive, a *Solid State Drive* (SSD).

3.4 NAND/SSD + HDD

A block diagram of the monolithic HDD + SSD solution, usually referred to as hybrid drive, is shown in Fig. 3.6 [18–25]. A Solid State Drive is made up of several NAND chips plus a controller: therefore, all the considerations of this section also apply to a storage system composed by HDD and a single NAND device. Unlike standard HDDs, the hybrid drive in its normal state has its platters at rest, without consuming power or generating heat. When reading data from the platters, extra data are read and stored in buffer memory in the hope of anticipating future requirements as in any disk cache. For example, data required for the next boot-up can be stored in the non-volatile buffer before shutting down the computer.

In 2010 Seagate released the Momentus XT [20, 21], which uses so-called "adaptive memory" for its SSD portion, which does not rely on driver support from the operating system. This removes the need for a special operating system, and the speed benefits can be used by any OS.

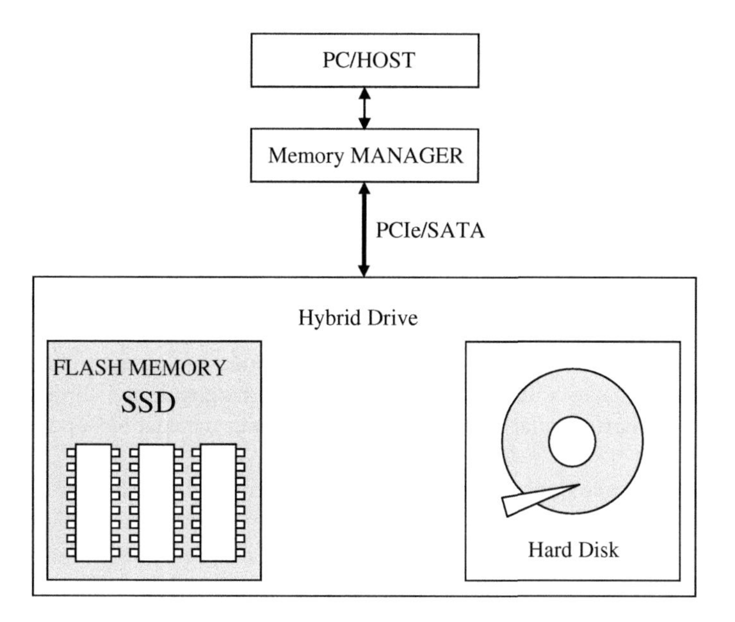

Fig. 3.6 Monolithic hybrid drive

The Flash memory is used to store frequently accessed content using an adaptive memory algorithm. This algorithm monitors data access transactions and maintains frequently accessed data on the Flash memory. The drive includes software that tracks a person's use trends and then uses the SSD component of the drive to optimize performance, and it can adjust that performance over time with changes in user behavior. Up to 50% performance improvement is seen between the first and second iteration of data access [18].

Manufacturers claim several benefits of the hybrid drive over standard hard drives, especially for use in notebook computers: among them, speed of data access and consequent faster computer boot process, decreased power consumption, and improved reliability.

There are some drawbacks too, especially when accessing non-cached data. In fact, if the data being accessed is not in the cache and the drive has spun down, access time will be greatly increased since the platters will need to spin up.

Another concern is the lower performance for small disk writes. NAND is significantly slower when writing small data; an effect that is amplified when the file system is using journaling techniques.

Anyhow, hybrid drives have a great potential and the industry is actively working in this field. As a matter of fact, Windows Vista and Windows 7 natively support the use of hybrid drives (ReadyDrive) [22].

As mentioned, a NAND device can experience a limited number of program/erase cycles. With the hybrid drive, a simple solution to mitigate this wear-out effect would be to place all the data that is accessed by read operations on the Flash memory device, and the remaining data on the HDD. This placement would save a substantial amount of the energy consumption while a longer lifetime for the Flash memory device is expected [12].

However, in practice, we cannot know in advance whether data should be placed on the Flash memory device or the hard disk.

We now review an existing method of skewing frequently accessed data, called *Popular Data Concentration* (PDC): it was proposed by Pinheiro and Bianchini [23] to deal with the highly skewed file access frequencies exhibited by the workloads of network servers. The idea of PDC is to concentrate the most popular (i.e. most frequently accessed) disk data by migrating it to a subset of the disks, so that the other disks can be sent to a low-power mode to conserve energy. PDC redistributes data across the disk array according to its popularity, so that the first disk stores the most popular data, the second disk stores the next most popular data, and so on.

However, if the frequency of file access varies significantly with time, PDC may cause a lot of file migrations, which will increase energy use, in particular by disturbing idle disks. This also happens when new files are created, because they will be stored on the disk with the least popular data, which has to be woken up.

PDC concentrates on popular data without considering whether I/O accesses are reads or writes. If we split I/O transactions into reads and writes and move only the data corresponding to one sort of access, we can reduce the amount of migrations. For instance, if the total amount of data associated with reads is less than that associated with writes, then transferring the data that is being read will be more profitable. This scheme is called PB-PDC (pattern-based PDC): it improves the PDC technique by moving frequently-accessed read and write data to separate sets of disks [9].

Thus, while the disks containing data which are accessed in one way (read or write) are being accessed frequently, the disks storing data accessed in the other way can be sent to a low power mode to conserve energy.

We can apply PB-PDC to a hybrid drive. Because a Flash memory device has low write throughput and limited erasure cycles, PB-PDC moves the popular write data to the hard disk and the popular read data to the Flash memory device.

Another possible approach when looking at data partitioning within a hybrid drive is to employ cache device organization where a subset of disks are treated in the storage system as cache disks to absorb I/O traffic [24].

Summarizing, PDC does not ask for file duplication while, in the caching approach, files in Flash memory are a copy of that on disk.

The cached file selection algorithm decides files to be cached in Flash. Usually, both static and dynamic types of selections can be used. The static approach is more suitable for files frequently accessed by users: for example, the operating system, compiler and some C libraries.

When the remaining capacity of Flash memory cache device reaches a threshold value, replacement is needed. The main guideline for replacement algorithm is that files accessed less frequently and files that will not be accessed in near future should

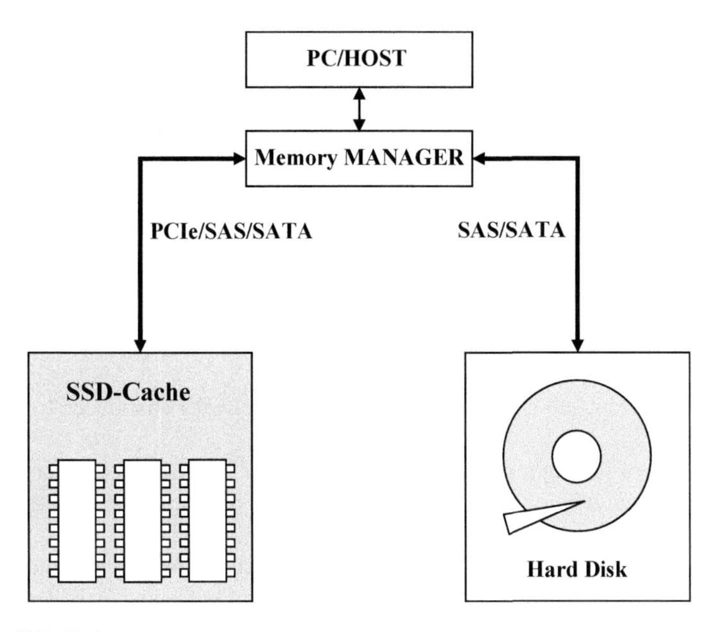

Fig. 3.7 SSD-Cache

be removed from Flash memory cache. The oldest and yet still widely used algorithm in cache management is LRU [12].

The above mentioned algorithms are just a small part of what is available in the open literature: it is clear that in order to really exploit all the benefits of hybrid storage, it is fundamental to decide where it is the right place to store data, depending on their characteristics. Of course, workloads are application and user specific: therefore, the storage management algorithm should be able to adapt to different needs.

At the end of this section it is worth mentioning that another term is becoming very popular in the hybrid storage world: SSD-Cache [26].

SSD-Cache is a discrete, separate memory component, as sketched in Fig. 3.7: in other words, HDD and SSD are housed separately. While all the hot/cold topics mentioned above remain valid, discrete cache SSDs and HDDs are easier to scale, with a broad selection of drive manufacturers [27–32].

3.5 Hybrid SSD

NAND Flash memories fall into different categories, depending on the number of bits stored inside the same physical cell [33], as shown in Fig. 3.8. SLC and MLC store 1 and 2 bits per cell, respectively. *Triple-Level Cell* (TLC) stores 3 bits within

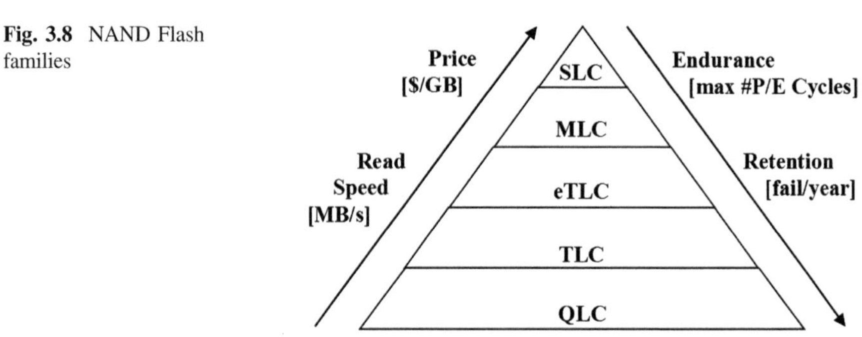

Fig. 3.8 NAND Flash families

a memory cell; 4 bit/cell is called QLC and has been already announced by all Flash manufacturers.

Downsides of storing more bits per cell are slower speeds, higher error rates and lower endurance/retention [34, 35]. The advantage is clearly the reduced silicon area, and therefore cost [30, 36, 37].

eTLC ("e" stands for enterprise) offers a higher number of erase/program cycles. For instance, if standard TLC runs for 3 k, eTLC can withstand 7–10 k [38].

Table 3.1 compares typical SLC, MLC, TLC and QLC specifications [38]: SLC is much faster than all the others during both read and write.

Performances of an individual Flash device are still insufficient to meet the bandwidth requirements of the interface (SAS/SATA/PCIe) and, therefore, interleaving is very common in most high-performance SSDs. The interleaving technique is also useful to extend the endurance because write operations can be distributed over multiple devices [39].

Because of the cost benefit, there have been many attempts to address performance and endurance problems in TLC-based storage systems. One possible approach is to combine SLC and TLC Flash memories inside a single SSD, which is then called "hybrid" [40–46]. A basic block diagram is shown in Fig. 3.9. The goal of this hybrid-SSD design is to achieve the response time of SLC, while having the cost structure of TLC. In other words, SLC capacity must be small. It is worthwhile to highlight that most of the modern TLC devices allow users to configure some or all the blocks in SLC mode. Therefore, in this case, the NAND itself can be viewed as a hybrid device.

The basic idea is to use SLC for storing small random (hot) data and TLC for large sequential (cold) data [47–54]. In fact, SLC has better endurance and small random data tend to be updated more frequently. However, TLC is still the limiting factor when long sequential data writes frequently occur to the storage.

Figure 3.10 shows a possible data flow during write. Every write request enters in the "Data Sensor": cold data directly go to TLC. Hot data move to another block called "Utilization Limiter". If the SLC NAND blocks wear out too fast, this limiter has the task to reduce the write traffic to SLC blcoks. In other words, a second level

Table 3.1 SLC, MLC, TLC and QLC specifications

NAND type	SLC	MLC	TLC	QLC
Page read (μs)	25	50–60	80–90	300
Page write (μs)	200	800–1,200	3,000	15,000
Block erase (ms)	10	10	10	10
Endurance (k)	50	20	3–7	1

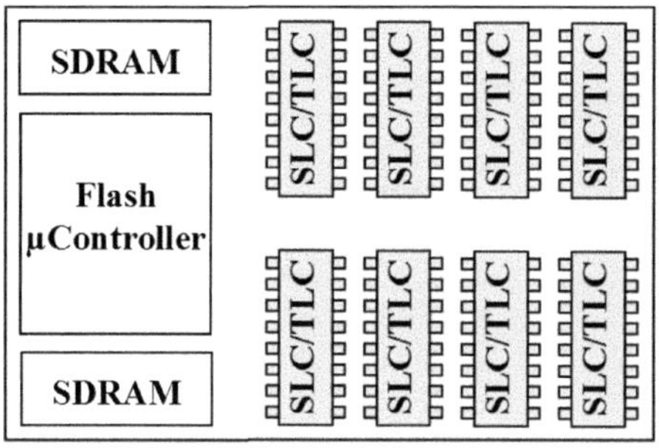

Fig. 3.9 SLC + MLC hybrid SSD

of data classification is adopted: hot-data go to SLC and quasi-hot-data are switched to TLC.

As mentioned, SLC capacity has to be small; therefore, when data become cold, they should be removed from SLC in order to maximize the space for hot data.

At this point it is clear that the foundation of this approach is the ability of classifying data. A lot of methods to identify hot data have proposed, including LRU, LRU-k [55], hash-table-based approaches [48, 49, 56]. The reader can refer to this extensive literature for more details.

Chang [49] showed that, by adding a 256 MB SLC Flash to a 20 GB MLC-Flash array, the hybrid SSD improves over a conventional SSD by 4.85 times in terms of average response. The average throughput and energy consumption are improved by 17% and 14%, respectively. The hybrid SSD is only 2% more expensive than a purely MLC-Flash-based SSD.

Of course, the hybrid concept can be extended to a Solid State Drive made up by different types of NAND memories, as shown in Fig. 3.11 [57, 58].

3.6 Over-Provisioning

When looking at the overall capacity of a solid state drive, over-provisioning must be taken into account. Over-provisioning is the difference between the physical capacity of the Flash memory and the logical capacity available for the user. Of course, this is also true for hybrid SSDs [59].

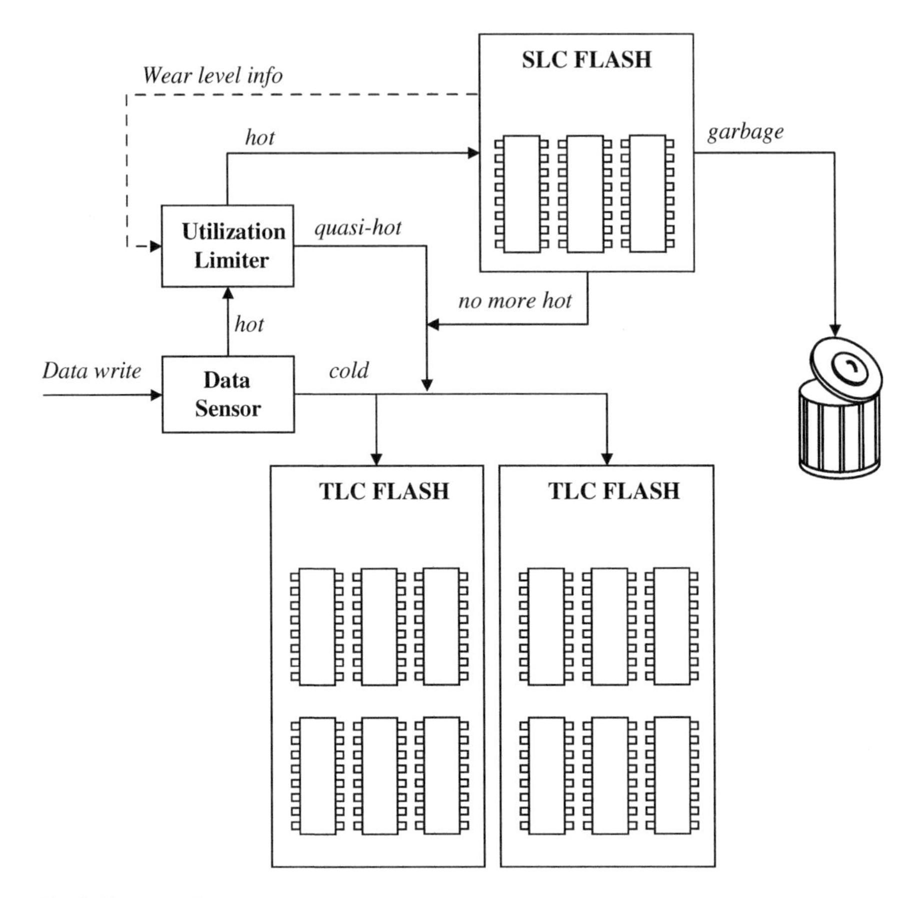

Fig. 3.10 Write flow

The idea behind over-provisioning is to have a "reserve" of spare blocks that can be used by the controller.

Let's assume an application that wants to randomly write data to the SSD drive. The drive controller writes these data to some erased pages in a particular block. After a while, the application decides to update the content: given the nature of Flash memories, this would imply erasing the block. In order to improve performances, the drive controller just marks those pages as unavailable and writes the new content to different physical pages: actually, no electrical erase takes place. When the entire block has been used and another write comes in, a real erase operation is needed. At this point, the controller needs to go through the following process:

- copy the entire content of the block to a temporary location (likely cache);
- remove the unused data from the cache;

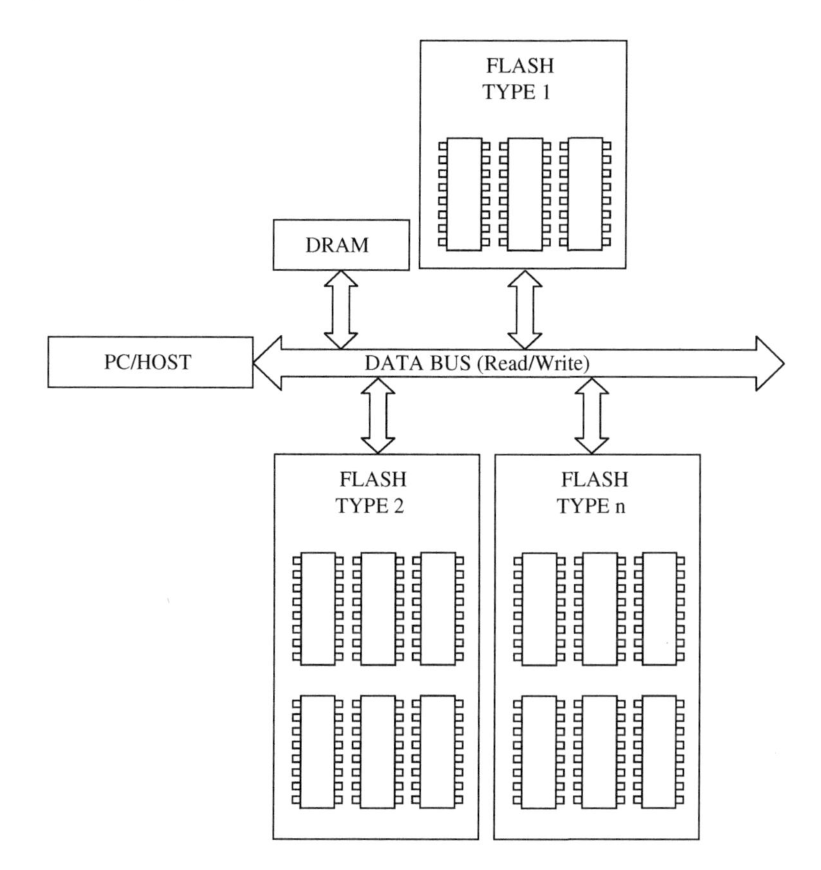

Fig. 3.11 Hybrid SSD including different types of NAND Flash memories

- add the new data to the block in cache;
- erase the addressed block on the SSD drive;
- copy the entire block from the cache;
- empty the cache.

This sequence is very time consuming and kills write throughput performances [59, 60]. When over-provisioning is used, the flow can be different. Instead of having to erase the unavailable portion of the block to accommodate new data, the controller can use some of the spare space instead. This means that the sequence of reading the entire block, merging the new data, erasing the block, and writing the entire new block back, can be avoided. The controller just maps spare space to be part of the drive capacity (so it is seen by the OS) and moves the unused pages to the spare capacity portion of the drive.

Anyhow, at some point the unavailable pages will have to be erased forcing the erase/write sequence mentioned above. In real world applications, 100% random writes are unlikely and the Flash controller does the erase/write sequence in

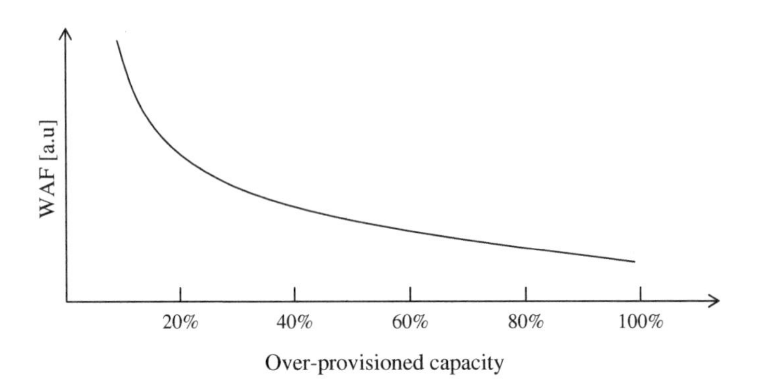

Fig. 3.12 Write amplification factor (*WAF*) versus over-provisioned capacity

background or when the drive is not in use. To get to the worst case, the host has to randomly write across all the drive's capacity without stopping to read.

Some controllers may not actively defragment the space to save costs, so the worst case performance becomes typical after the drive has been written few times.

Spare capacity can also be used when "bad" areas develop in the drive. For example, if a certain set of pages/blocks has much fewer remaining erase/write cycles than most of the drive, then the controller can remap them to spare pages/blocks. Moreover, the controller can watch for bad writes and use the spare capacity as a "backup" (similar to extra blocks on hard drives). The controller can check for bad writes by doing read-after-write (reads are much faster than writes).

During the garbage collection, wear-leveling, and bad block mapping operations inside the SSD, the additional space from over-provisioning helps lowering the *Write Amplification Factor* (WAF) [60–63]; this factor corresponds to the additional writes caused by garbage collection (see flow above) and wear leveling (Chap. 9). Jedec defines WAF as the data written to the Flash divided by data written by the host to the SSD [64].

Figure 3.12 sketches a typical behavior of WAF vs. over-provisioned capacity. In commercial products over-provisioned capacity is usually around 30%. On one side, with a very small over-provisioning percent, the amount of data "moves" that have to take place can be very high, lowering the achievable write IOPS. On the other side, still looking at Fig. 3.12, 30% looks a good trade-off between performances and area (cost): in fact, beyond 30% WAF reduces at a lower rate [60].

In summary, reducing the amount of over-provisioned capacity can lower the cost per GigaByte, but then WAF can become a real problem. Please bear in mind that the over-provisioned space shrinks over time as it is also intended to countermeasure wear out of Flash blocks.

References

1. The DRAM story, with articles by Dennard, Itoh, Koyanagi, Sunami, Foss and Isaac. IEEE SSCS News. **13**(1) (Winter 2008), www.ieee.org/sscs-news
2. D. Baral, *Life Cycle Power Consumption HDD Vs. SSD*, Flash Memory Summit, Session 101 (Storage Labs Samsung Information Systems America, San Jose, 2009)
3. V. Kasavajhala, *Solid State Drive vs. Hard Disk Drive Price and Performance Study* (Dell Technical White Paper, Dell Power Vault Storage Systems, May 2011), http://www.dell.com/downloads/global/products/pvaul/en/ssd_vs_hdd_price_and_performance_study.pdf
4. B. Marsh, F. Douglis, P. Krishnan, Flash memory file caching for mobile computers, in *Proceedings of the Twenty-Seventh Hawaii International Conference on System Sciences*, Wailea, HI (1994), pp. 451–460
5. T. Bisson, S.A. Brandt, D.D.E. Long, NVCache: increasing the effectiveness of disk spin-down algorithms with caching, in *MASCOTS 2006*, Monterey (2006), pp. 422–432
6. T. Bission, S. Brandt, Reducing energy consumption with a non-volatile storage cache, in *Proceedings of International Workshop on Software Support for Portable Storage*, San Francisco, CA (2005)
7. F. Chen, S. Jiang, X. Zhang, SmartSaver: turning flash drive into a disk energy saver for mobile computers, in *Proceedings of the 2006 International Symposium on Low Power Electronics and Design*, Tegernsee, Germany (2006), pp. 412–417
8. R. Panabaker, Hybrid hard disk and ReadyDrive™ technology, improving performance and power for windows vista mobile PCs, in *Proceedings of MicrosoftWinHEC*, Los Angeles, CA (2006)
9. Y.-J. Kim, K.-T. Kwon, J. Kim, Energy-efficient file placement techniques for heterogeneous mobile storage systems, in *Proceedings of the 6th ACM & IEEE International Conference on Embedded software*, Seoul, Korea (2006), pp. 171–177
10. T. Kgil, T. Mudge, FlashCache: a NAND Flash memory file cache for low power web servers, in *Proceedings of the International Conference on Compilers, Architecture and Synthesis for Embedded Systems*, Seoul, Korea (2006)
11. T. Kgil, D. Roberts, T. Mudge, Improving NAND Flash based disk caches, in *ISCA'08 Proceedings of the 35th Annual International Symposium on Computer Architecture*, Beijing, China
12. S. Liu, X. Cheng, X. Guan, D. Tong, in *Energy Efficient Management Scheme for Heterogeneous Secondary Storage System in Mobile Computers SAC'10*, Sierre, Switzerland, 22–26 March 2010
13. A. Kirshenbaum et al., Using external memory devices to improve system performance, U.S. Patent No. 7,805,571 and U.S. Patent application No. 20100217929, Assignee: Microsoft Corporation
14. Microsoft Windows, Windows 7 features—ReadyBoost—Microsoft Windows, http://windows.microsoft.com/en-US/windows7/products/features/readyboost
15. W.R. Stanek, *Windows 7: The Definitive Guide* (O'Reilly Media, 2010), Sebastopol, CA 95472, pp. 105–109
16. White Paper Intel® Flash Memory Intel® NAND Flash Memory for Intel® Turbo Memory (2007), http://download.intel.com/design/flash/nand/turbomemory/whitepaper.pdf
17. Intel® Turbo Memory—Overview and Support, http://www.intel.com/cd/channel/reseller/apac/eng/products/mobile/mprod/turbo_memory/396715.htm
18. T. Coughlin, J. Handy, *Two May Be Better Than One: Why Hard Disk Drives and Flash Belong Together* (White Paper SNIA, Feb 2011), http://www.snia.org/sites/default/files/Storage%20Pairing%20WP%20FEB%202011.pdf
19. T. Coughlin, J. Handy, *HDDs and Flash Memory: A Marriage of Convenience* (SNIA, Feb 2011), http://www.snia.org/sites/default/files2/SDC2011/presentations/Monday/TomCoughlin_and_Handy_HDDS_Flash_Memory.pdf

20. Seagate MomentusXT Datasheet, http://www.seagate.com/files/staticfiles/docs/pdf/datasheet/disc/momentus-xt-data-sheet-ds1704-4-1205-us.pdf
21. Seagate MomentusXT: Overview features and specs, http://www.seagate.com/internal-hard-drives/laptop-hard-drives/momentus-xt-hybrid/
22. R. Panabaker, Hybrid hard disk and ReadyDrive™ technology: improving performance and power for windows vista mobile PCs, in *Proceedings of Microsoft WinHEC* (2006)
23. E. Pinheiro, R. Bianchini, Energy conservation techniques for disk array-based servers, in *Proceedings of the 18th International Conference on Supercomputing (ICS'04)*, June 2004
24. D. Colarelli, D. Grunwald, Massive arrays of idle disks for storage archives, in *Proceedings of the 2002 ACM/IEEE Conference on Supercomputing*, Baltimore, MD (2002), pp. 1–11
25. G. Symons, *Hybrid SSD/HDD Storage: A New Tier?* Flash Memory Summit (Xiotech Corporation, Colorado Springs, 2011)
26. Intel® RAID SSD Cache 2.0, http://www.intelraid.com/uploads/Intel_RAID_SSD_Cache2_PB_080911.pdf
27. Intel® Solid-State Drive 313 Series, http://www.intel.com/content/www/us/en/solid-state-drives/solid-state-drives-313-series.html
28. Adaptec maxCache 2.0 Series, http://www.adaptec.com/en-us/_common/maxcache/
29. OCZ Synapse Cache SATA III 2.5″ SSD, http://www.ocztechnology.com/ocz-synapse-cache-sata-iii-2-5-ssd.html
30. Corsair Accelerator Series SSD Cache, http://www.corsair.com/ssd/accelerator-series-ssd-cache-drives.html
31. LSI Nytro MegaRAID Application, http://www.lsi.com/products/storagecomponents/Pages/NytroMegaRaid.aspx
32. Crucial Adrenaline Solid State Cache (Windows 7 PCs), http://www.crucial.com/store/ssc.aspx
33. R. Micheloni, L. Crippa, A. Marelli, *Inside NAND Flash Memories* (Springer, New York, 2010)
34. N. Duann, *SLC & MLC Hybrid*, Flash Memory Summit (Silicon Motion, Inc., 2011)
35. B. Chang, *SSD with Hybrid NAND* Novachips, Flash Memory Summit, 2011
36. Y. Koh, *NAND Flash Scaling beyond 20 nm*, in *IMW '09, IEEE International Memory Workshop* (2009)
37. White paper, *Engineering MLC Flash-Based SSDs to Reduce Total Cost of Ownership in Enterprise SSD Deployments*, STEC's CellCare™ Technology, http://www.stec-inc.com/downloads/MLC_flash_based_SSDs_Reduce_TCO.pdf
38. C.C. Wu, Quality comparison of SLC, MLC and eMLC., in *InnoDisk International Memory Workshop IMW*, San Diego, CA (2011)
39. E. Bek, A. Klein, *The Future of SSD Architectures, International Memory Workshop IMW*, SanDisk (2011)
40. W.H. Radke et al., Hybrid memory management, U.S. Patent No. 8,060,719, Assigned: Micron Technology, Inc., 28 May 2008
41. C. Lee et al., Hybrid SSD using a combination of SLC and MLC flash memory arrays, U.S. Patent No. 8078794, Assignee: Super Talent Electronics, Inc., San Jose, 29 Oct 2007
42. Y.S. Kim, Semiconductor memory device, and multi-chip package and method of operating the same, U.S. Patent No. 8085569, Assignee: Hynix Semiconductor Inc., 14 Dec 2010
43. H. Tan et al., Portable data storage using SLC and MLC flash memory, U.S. Patent App. No. 20080215801, Assignee: Trek 2000 International Ltd., 28 Sept 2005
44. M. Moshayedi, Enhanced MLC solid state device, U.S. Patent App. No. 20090327590, Assignee: STEC, Inc., 24 June 2009
45. M. Moshayedi, SLC-MLC combination flash storage device, U.S. Patent App. No. 20090327591, Assignee: STEC, INC., 24 June 2009
46. L.E. Aszmann et al., Solid state drive data storage system and method, U.S. Patent App. No. 20110010488 (12 Jul 2009)
47. T.-W. Kuo et al., Configurability of performance and overheads in Flash Management, in *11th Asia and South Pacific Design Automation Conference (ASP-DAC)* (2006)

48. L.-P. Chang, Hybrid solid-state disks: combining heterogeneous NAND flash in large SSDs, in *13th IEEE/ACM Asia and South Pacific Design Automation Conference (ASP-DAC)* (2008)
49. L.-P. Chang, A hybrid approach to NAND-flash-based solid-state disks. IEEE Trans. Comput. **59**(10), 1337–1349 (2010)
50. L.-P. Chang, Y.-C. Su, Plugging versus logging: a new approach to write buffer management for solid-state disks, in *The 48-th Design Automation Conference (DAC)*, Monterey, CA (2011)
51. S. Hong, D. Shin, NAND flash-based disk cache using SLC/MLC combined flash memory, in *2010 International Workshop on Storage Network Architecture and Parallel I/Os*
52. S. Jung, Y.H. Song, Hierarchical use of heterogeneous flash memories for high performance and durability. IEEE Trans. Consum. Electron. **55**(3), 1383–1391 (2009)
53. M. Murugan, D.H.C. Du, Hybrot: towards improved performance in hybrid SLC-MLC devices, in *20th IEEE International Symposium on Modeling, Analysis and Simulation of Computer and Telecommunication Systems (MASCOTS)* (Short Paper) (Aug 2012)
54. B.-W. Nam, A hybrid flash memory SSD Scheme for Enterprise Database applications, in *12th International Asia-Pacific Web Conference*, Busan, Korea, 2010
55. E.J. O'Neil, P.E. O'Neil, G. Weikum, The LRU-k page replacement algorithm for database disk buffering. ACM SIGMOD Rec. **22**(2), 297–306 (1993)
56. J.W. Hsieh, T.W. Kuo, L.P. Chang, Efficient identification of hot data for Flash memory storage systems. ACM Trans. Storage **2**(1), 22–40 (2006)
57. J. Niu, J. Xu, L. Xie, Hybrid storage systems: a survey of architectures and algorithms. IEEE Access (99) (2018)
58. C. Matsui, C. Sun, K. Takeuchi, Design of hybrid SSDs with storage class memory and NAND Flash memory. Proc. IEEE **105**(9), 1812–1821 (2017)
59. D.A. Heger, *SSD Write Performance—IOPS Confusion Due to Poor Benchmarking Techniques* (Aug 2011), http://www.cmg.org/measureit/issues/mit82/m_82_4.pdf
60. X.-Y. Hu, Write amplification analysis in Flash-based solid state drives, in *SYSTOR'09* (IBM Zurich Research Laboratory, Haifa, Israel)
61. K. Smith, *Benchmarking SSDs: The Devil is in the Preconditioning Details*, Flash Memory Summit (2009)
62. White Paper, *Intel High-Performance SATA Solid-State Drive: Over-Provisioning an Intel SSD*, http://www.matrix44.net/cms/wp-content/uploads/2011/07/intel_over_provisioning.pdf
63. T. Frankie, *SSD Trim Commands Considerably Improve Overprovisioning*, Flash Memory Summit (2011)
64. JEDEC STANDARD, Solid-State Drive (SSD) Requirements and Endurance Test Method, JESD218 (Sept 2010), http://www.jedec.org/sites/default/files/docs/JESD218A.pdf

Chapter 4
2D NAND Flash Technology

M. F. Beug

Abstract This chapter describes the basic operating principle and presents the major reliability and scaling limitations of floating gate NAND non-volatile memory as used in SSD applications. It further discusses charge trapping memory cells as a potential replacement for floating gate cells in the NAND array and evaluates the potential of both memory cell types with regard to 3D NAND applications as will be described in the next chapter.

4.1 Flash for SSD Application

Flash memory for non-volatile data storage was introduced commercially in the mid-1980s. Since then, common ground NOR and NAND architecture have become the most common memory array architectures. Traditionally, NOR Flash is used for code storage due to faster memory cell access. NAND Flash is used for mass data storage because of its higher memory density, enabling higher storage capacities.

The memory cell area difference can already be seen from the schematic NOR and NAND array images in Fig. 4.1. In the NOR array, two memory cells each share one contact to ground and one contact to the bit line (see Fig. 4.1a). This results in an effective memory cell area of about 10 F^2 (where F is the minimum feature size). The effective memory cell area of NAND cells is only slightly more than $4F^2$. Figure 4.1b shows the so-called NAND string with up to 64 memory cells connected in a row. To operate the NAND string two additional select transistor devices (GSL: "Ground Select Line" and SSL: "String Select Line") and contacts to ground (SL: "Source Line") and the bit line (BL) need to be added. These additional structures cause the effective cell area consumption to be slightly higher than

M. F. Beug (✉)
Physikalisch-Technische Bundesanstalt (PTB), Division 2 "Electricity", Bundesallee 100, 38116 Braunschweig, Germany
e-mail: Florian.Beug@ptb.de

R. Micheloni et al. (eds.), *Inside Solid State Drives (SSDs)*,
Springer Series in Advanced Microelectronics 37,
https://doi.org/10.1007/978-981-13-0599-3_4

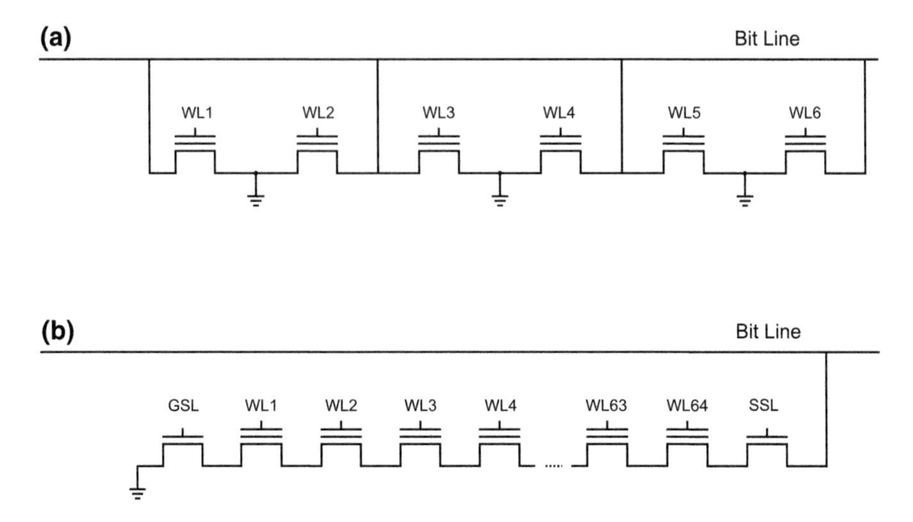

Fig. 4.1 Schematic memory cell organization of the NOR array (**a**) and the NAND array (**b**). The word lines (WL) run perpendicular to the bit lines (BL)

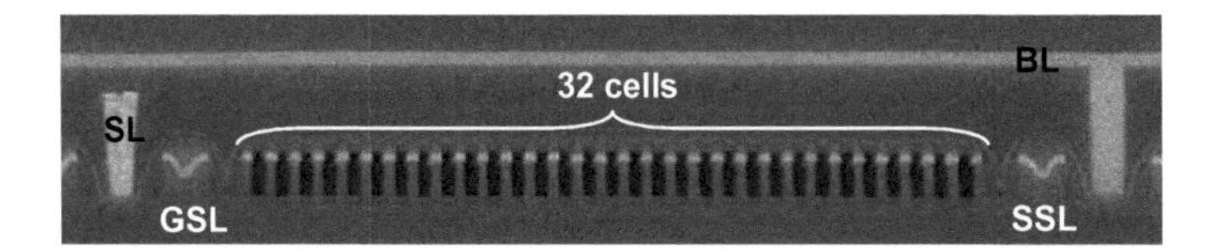

Fig. 4.2 SEM picture of a NAND string with 32 cells per string in a 48 nm floating gate NAND technology [2]

$4F^2$—the theoretically smallest effective cell size. The cross section of a 48 nm NAND technology with 32 cells per string is shown in Fig. 4.2.

For SSD application, only NAND Flash is a viable option due to the required high memory capacity and bit cost structure. Therefore, the following sections will focus on operation, reliability, and scaling topics of NAND Flash.

4.2 Introduction to Floating Gate NAND Operation

A floating gate memory cell stores information in terms of charge in an isolated gate electrode (floating gate: FG). The FG is located between the memory transistor channel and the active gate electrode (control gate: CG). This data storage principle was proposed by Kang and Sze in 1967 [1] and enables data to be stored without the connection of a supply voltage over time periods of several years.

4.2.1 The Floating Gate NAND Memory Structure

The schematic structure of floating gate NAND cells is shown in Fig. 4.3a, b. Figure 4.3c, d shows the cross sections of a 48 nm floating gate NAND technology [2]. The FG and the CG are typically made of polysilicon. For all operations of the floating gate cell, the active control gate electrode capacitive couples to the floating gate. The dielectric between the FG and the CG is referred to as inter-poly dielectric (IPD) and is typically made of a silicon oxide/silicon nitride/silicon oxide triple layer (ONO). The alterable threshold voltage of a floating gate cell, which represents the bit information, consequently depends on the coupling strength between the FG and the CG, and the amount of charge on the FG.

The FG NAND structure in word line direction is shown in Fig. 4.3a, c.

The CG is wrapped around the FG to improve the capacitive coupling from the CG to the FG. This reduces the operating voltages of the floating gate cells and

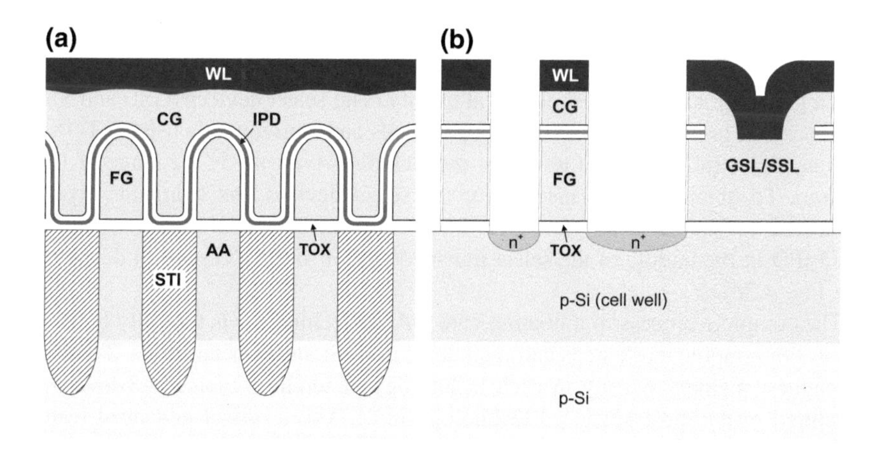

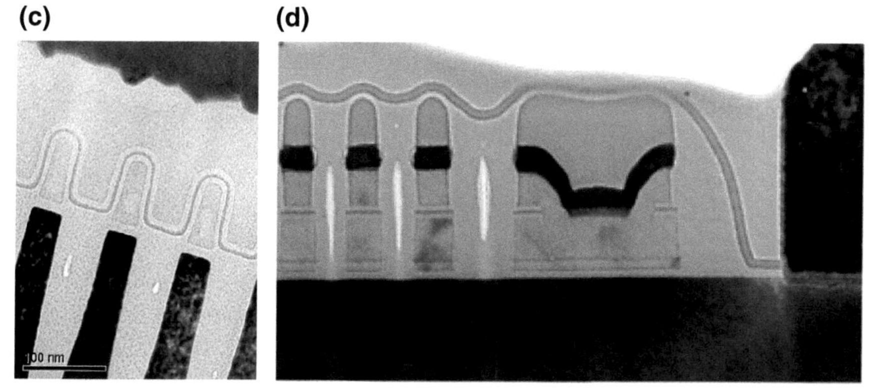

Fig. 4.3 Schematic structure of a floating gate NAND array in word line (WL) (**a**) and bit line (BL) direction (**b**). Corresponding TEM pictures of a 48 nm floating gate NAND technology [2] in WL direction (**c**) and BL direction (**d**)

ensures a reliable operation as will be described in the next section. The active areas (AA) of two neighboring NAND strings are separated by shallow trench insulation (STI) and are about 200 nm deep in current generations. The memory cell transistor gate oxide is denoted as tunnel oxide (TOX) because the charge for bit information storage is transferred through this SiO_2 dielectric by quantum mechanical tunneling.

Generally, it is a very crucial point for reliable floating gate cell operation that charge during program and erase operations is only transferred through the TOX. Every charge transfer through the IPD (between FG and CG) needs to be urgently avoided to prevent severe reliability issues.

In BL direction, the cell strings run as shown in Figs. 4.1a and 4.3c, d. The floating gate cells are patterned by a vertical WL etch step. In the etched spaces between the floating gate cells, shallow n^+ junctions are implanted in order to define the memory cell transistors and reduce the string resistance. To improve the charge retention of the memory cells, the side wall of the floating gate is passivated by a thermal oxidation process.

The generated high quality thermal side wall oxide (SWOX) forms an effective tunnel barrier against charge loss from the FG. Subsequently, the space between the FG cells is filled with a deposited silicon oxide (inter-word line dielectric: IWD) which generally has a reduced electrical quality. The select devices (GSL and SSL) are processed together with the floating gate cells and consequently use the TOX as the gate dielectric. The select transistor gate length is typically in the range of 150–200 nm. To obtain a real transistor for the select devices, the word line layer is connected to the floating gate layer. This contact is made by removing the ONO IPD in the middle of the select transistors prior to the CG poly-Si deposition (see Fig. 4.3d).

The complete process of a floating gate NAND technology is typically based on 30–40 lithographic mask steps and includes 2 poly-Si and 3 metal levels. To obtain the highest memory density in each technology generation, typically 3 levels are structured in the most advanced technology node. The levels of advanced feature size are active area/STI, word line and bit line. The bit line is either done in the first or second metal layer. There are some more process steps with stringent lithographic requirements, such as the contacts to the bit line, but also the source contacts, the CG to FG contacts in the select devices, and others.

4.2.2 The Floating Gate Cell Capacitive Coupling Model

It was described that floating gate NAND cells are arranged in strings with up to 64 memory cells in actual NAND technologies. However for the basic understanding of the floating gate cell functionality it is necessary to look at a single FG cell first.

Since the floating gate is isolated from the active control gate, all voltages for operation of the memory cell need to be capacitively coupled to the floating gate. In

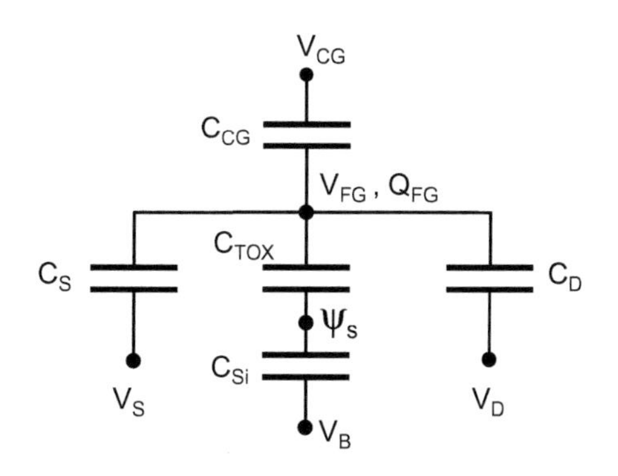

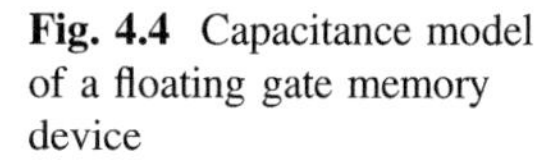
Fig. 4.4 Capacitance model of a floating gate memory device

principle, the floating gate cell forms a capacitive voltage divider which is typically described with the aid of the FG cell capacitive coupling model [3] as shown in Fig. 4.4.

It describes the voltage of the floating gate as a function of the other terminals of a FG cell. These terminals are typically source (V_S), drain (V_D), the bulk terminal (V_B), the control gate (V_{CG}), and a number of other (parasitic) terminals. All these terminal voltages are capacitive coupled to the floating gate. The floating gate voltage can be written as

$$V_{FG} = \alpha_G \cdot V_{CG} + \alpha_S \cdot V_S + \alpha_D \cdot V_D + \frac{C_{TOX}}{C_T} \cdot \psi_S \frac{Q_{FG}}{C_T} + \sum \alpha_{other} \cdot V_{other}. \quad (4.1)$$

The gate coupling ratio α_G in (4.1) is a key factor and is defined as

$$\alpha_G = \frac{C_{CG}}{C_T}. \quad (4.2)$$

C_T is the total capacitance and is given by

$$C_T = C_{CG} + C_{TOX} + C_S + C_D + \sum C_{other}. \quad (4.3)$$

The sum of C_{other} contains all other terminals which couple to a specific floating gate and represent neighboring bit and word lines or neighboring floating gates. The capacitive components in the sum are traditionally small compared to the other terms, but gain significantly in importance when floating gate cells are scaled to feature sizes below 50 nm [4].

The gate coupling ratio α_G describes the portion of the voltage applied between the CG and the channel that drops across the TOX. For grounded source, drain, bulk, and other terminals during program operation, the floating gate voltage is given by

$$V_{FG} = \alpha_G \cdot V_{CG}. \tag{4.4}$$

A control gate voltage V_{CG} = 20 V in combination with a gate coupling ratio of $\alpha_G = 0.6$ results in a voltage drop of V_{FG} = 12 V across the tunnel oxide. Consequently, the CG voltage is concentrated on the tunnel oxide, when a high C_{CG} to C_T ratio and therefore a high α_G can be realized.

Under such coupling conditions, the requested floating gate cell operation can be obtained, where charge is only transferred between the channel region and the floating gate.

The FG voltage formulation (4.1) and α_G formula in (4.2) were described in [5] and only take into account the voltage drop across the tunnel dielectric (across C_{OX}) and consequently include the channel surface potential (ψ_s). It does not consider the voltage drop in the Si substrate (across C_{Si}) [6].

The source and drain coupling ratios have the same form as the α_G expression (4.2) and are given by $\alpha_S = C_S/C_T$ and $\alpha_D = C_D/C_T$.

The capacitive coupling model and (4.1) also yield the formula for the floating gate cell threshold voltage shift ΔV_{th} caused by charge stored on the floating gate. The threshold voltage shift is in principle the voltage increase which is necessary at the control gate to compensate the floating gate charge induced field effect. Therefore, it is the additional CG voltage for resuming the floating gate voltage that would be present without the FG charge and results in a defined TOX field which is necessary to invert the memory cell channel. For constant potentials at source and drain during the read operation, (4.1) can be rearranged to

$$\Delta V_{th} = \Delta V_{CG}|_{\Delta V_{FG}=0} = -\frac{\Delta Q_{FG}}{\alpha_G \cdot C_T} = -\frac{\Delta Q_{FG}}{C_{CG}}. \tag{4.5}$$

This means that for an optimized high gate coupling ratio value and a given threshold voltage shift, the number of stored electrons is increased (which is beneficial for charge retention). The required high C_{CG} value can be either obtained by a large coupling area between the CG and the FG, (the previously described CG wrapped around the FG), or a reduction in the electrical IPD thickness.

The effect of the latter option on the ability to program and erase floating gate cells will be discussed in the following section.

4.2.3 Program and Erase of a Single Floating Gate Cell

Floating gate cells in NAND applications are programmed and erased by the Fowler-Nordheim (FN) tunneling mechanism [7]. This quantum mechanical tunneling mechanism is based on a strong electric field across the tunneling barrier of the TOX. The electric field across the typically 8 nm thick tunnel oxide causes a band distortion. The induced FN tunneling current has a strong electric tunnel oxide field (E_{TOX}) dependency. The FN current density changes over several orders of

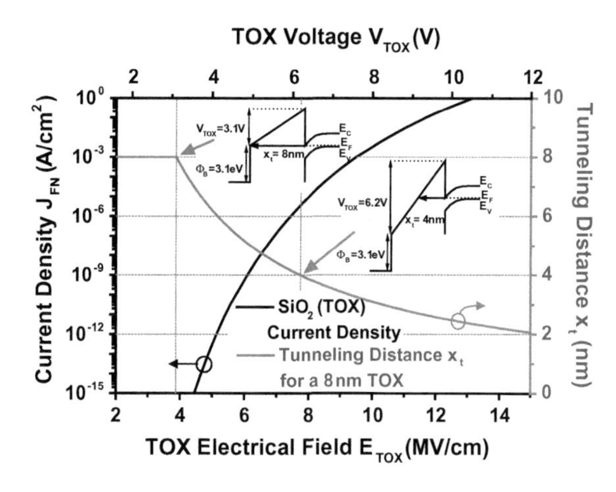

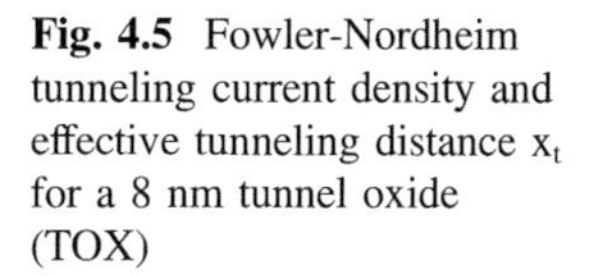
Fig. 4.5 Fowler-Nordheim tunneling current density and effective tunneling distance x_t for a 8 nm tunnel oxide (TOX)

magnitude and is the result of a significant reduction in the effective tunneling distance x_t, as shown in Fig. 4.5 and its inset.

The Fowler-Nordheim tunneling current density is given by

$$J_{FN} = A_t \cdot E_{ox}^2 \cdot \exp\left(-\frac{B_t}{E_{ox}}\right), \tag{4.6}$$

with the two tunneling constants A_t and B_t which are given by

$$A_t = \frac{q^3 m_e}{8\pi h m^* \Phi_B}; \quad B_t = \frac{8\pi\sqrt{2m^*\Phi_B^3}}{3qh}. \tag{4.7}$$

In (4.7), q is the electron charge, m_e and m* the mass of the electron and the effective electron mass in the SiO_2, h is Planck's quantum and Φ_B the tunnel barrier height between Si and SiO_2. The Fowler-Nordheim tunneling current density for a 8 nm thick $Si0_2$ tunnel dielectric with an exponential dependence on the electric oxide field E_{ox} is shown in Fig. 4.5.

Significant amounts of charge are transferred during a program pulse typically shorter than 1 ms, where the TOX electric field is in the strong Fowler-Nordheim tunneling regime above 10 MV/cm. Such strong oxide fields reduce the effective tunnel distance x_t of the triangular barrier to values below 3 nm as shown in Fig. 4.5.

When a floating gate cell is intended to be programmed to a certain V_{th} state, this is typically accomplished by the so-called "incremental step pulse programming" (ISPP) scheme [8]. To reach a targeted cell threshold voltage, programming pulses with durations in the range of $t_{pp} = 100$ µs are applied with increasing pulse amplitude. Each programming step is followed by a sense operation to evaluate whether the target V_{th} has already been reached. The increment of program pulse voltage steps depends on the required accuracy of the programmed V_{th} value.

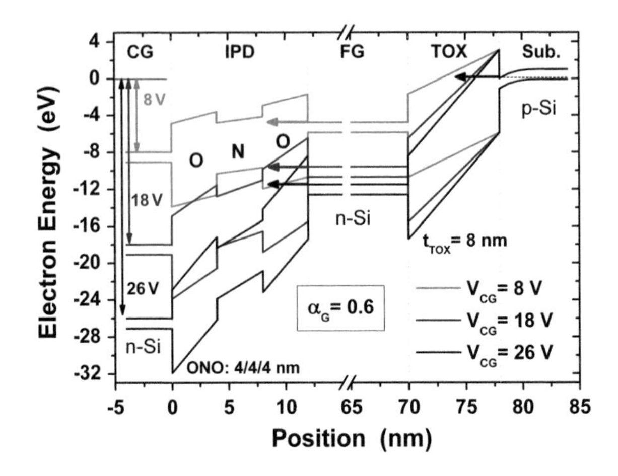

Fig. 4.6 Band diagram of a floating gate cell with t_{TOX} = 8 nm, an ONO IPD of 4/4/4 nm and a gate coupling ratio α_G = 0.6 for the program voltages V_{CG} = 8 V, V_{CG} = 18 V, and V_{CG} = 26 V after the program charge transfer, if applicable (compare Fig. 4.7). For V_{CG} = 8 V, the tunnel oxide field E_{TOX} is too low for electron injection through the TOX. For V_{CG} = 18 V, charge is injected into the FG until E_{TOX} is reduced to 12 MV/cm (shown here), the threshold program field. For V_{CG} = 26 V in the assumed simplified model, the FG charge increases until the electric fields in the TOX and the IPD suboxide equal each other. The FG charge remains constant in principle, but a strong tunneling current continuously passes through the hole FG stack and would in reality cause significant damage

Therefore, the program step voltage directly affects the cell V_{th} distribution width in a memory array with large numbers of cells [9].

For a relatively low programming voltage of only V_{CG} = 8 V at the beginning of the ISPP sequence, this voltage is divided between the tunnel oxide and the IPD according to the gate coupling ratio α_G. The band diagram of a floating gate cell for such a small voltage is shown in Fig. 4.6. However, for the assumed values α_G = 0.6 and IPD layer thicknesses of O/N/O = 4 nm/4 nm/4 nm, no significant amount of charge is transferred to the floating gate, since the TOX field is only 6 MV/cm (see Fig. 4.7). The assumed ONO layer thicknesses of 4 nm for each layer are already very small values as similarly used in state-of-the-art floating gate NAND Flash technologies in the range of 25 nm [10, 11]. Due to the exponential field dependency of Fowler-Nordheim tunneling, programming starts at a certain program threshold voltage which is equivalent to a fixed threshold electric TOX field. For the threshold field conditions, a significant amount of charge can be injected into the FG within the short program pulse time of typically t_{pp} = 100 μs. A typical value for the program start or threshold field is in the range of 12–13 MV/cm and depends on the process of the tunnel oxide formation which can influence the oxide barrier height. In addition, factors like the TOX thickness profile and the STI edge shape can affect this value. Due to this programming threshold field (which will be assumed to be 12 MV/cm in the following), it can be assumed that the same field strength will be present at the end of programming. This assumption

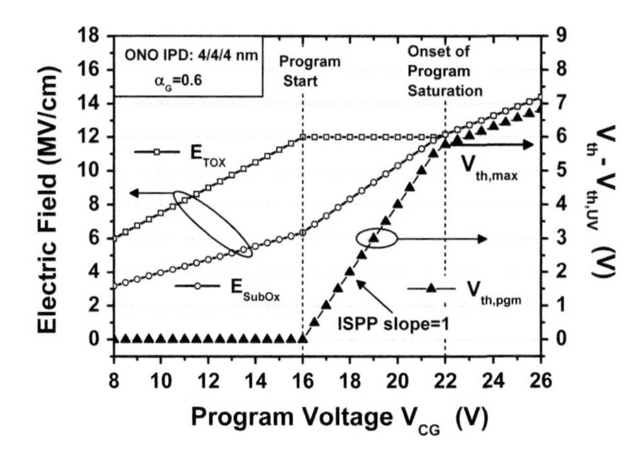

Fig. 4.7 Electric field condition in the tunnel oxide (E_{TOX}) and the IPD suboxide (E_{SubOx}) during ISPP programming of a floating gate cell with $t_{TOX} = 8$ nm, $\alpha G = 0.6$, and ONO IPD layer thicknesses of 4 nm each for the suboxide, the silicon nitride and the top oxide. Programming with an ideal ISPP slope = 1 takes place until E_{SubOx} at the end of programming equals the TOX electric threshold field of 12 MV/cm

is realistic because at a constant programming voltage, negative charge (electrons) is transferred to the floating gate as long as the additional charge has reduced the electric TOX field (4.1) to such an extent, that no more significant charge transfer can take place.

For the described exemplary FG cell configuration used for Figs. 4.6 and 4.7, programming with no significant IPD current takes place in the CG voltage range between $V_{CG} = 16$ V and $V_{CG} = 22$ V. The ISPP slope in this V_{CG} range is essentially at unity [2]. At around $V_{CG} = 22$ V and beyond this CG voltage value it can be observed that the TOX and the IPD suboxide electric fields equal each other. This results in an electron tunneling to the FG and at the same time an electron tunneling out of the FG towards the CG. For an IPD purely consisting of SiO_2, the same fields in TOX and IPD would result in the same currents tunneling into and out of the floating gate, which results in program saturation.

For an ONO IPD with additional SiN layer, charge can be injected into the SiN layer and will be stored in this layer as in a charge trapping memory cell storage layer. The charge injected and trapped in the ONO increases the effective barrier height [12] (compare Fig. 4.16b) and is therefore able to block weak and leaky spots of the ONO IPD by this means. This is one reason why an ONO IPD is generally used.

However, the electrons injected and finally stored in the ONO IPD beyond the program saturation starting point cause a permanent FG memory cell threshold voltage shift [10]. In addition to the stored charges, a large current is transferred through the whole FG cell stack from the channel towards the control gate which will substantially damage the memory cell. These large permanent currents become clear when looking at the strongly reduced TOX and IPD suboxide x_t for $V_{CG} =$ 26 V in Fig. 4.6.

By equating the electric fields in the TOX and the IPD suboxide, a simple model for the onset of program saturation can be derived [13].

Finally, an expression for the maximum reachable programmed threshold voltage (program saturation point) can be obtained, which is given by

$$V_{th,\,max} = 12\frac{\mathrm{MV}}{\mathrm{cm}} \cdot \left(t_{TOX} + t_{IPD-EOT} - \frac{t_{TOX}}{\alpha_G} \right) \tag{4.8}$$

It can be seen from (4.8) that in principle a thick tunnel oxide and a large equivalent oxide thickness of the IPD ($t_{IPD\text{-}EOT}$) are beneficial for good programmability of floating gate cells. Also a large gate coupling ratio improves $V_{th,max}$. However, due to the middle term in (4.7) the increase of the control gate to floating gate area is preferred over a reduction of $t_{IPD\text{-}EOT}$ to obtain a large α_G.

Figure 4.8 examines the effect of an increased α_G due to cell geometry means while keeping the TOX and IPD thicknesses unchanged.

It can be observed that for increasing the gate coupling ratio the initial (uncharged FG) field difference between the TOX and IPD electric fields increases. Consequently, FG cells with a higher gate coupling ratio can be programmed to higher V_{th} levels before program saturation occurs. The program saturation point ($V_{th,max}$) can be found in the V_{th} ISPP curves in Fig. 4.8, where the ISPP slope changes from unity to a value significantly lower than one. ISPP slopes lower than unity [14] generally show that the combination of cell geometry and IPD current blocking ability is not sufficient to avoid an IPD electron tunneling current during program operation.

The floating gate memory cell erase works principally in the same way, but with control gate voltages negative with respect to the cell channel region. Consequently, the electric field direction is reversed and the erase is mainly due to electron tunneling from the floating gate towards the channel. Again, as described for program saturation, the TOX erase field is reduced for decreasing erase cell V_{th}

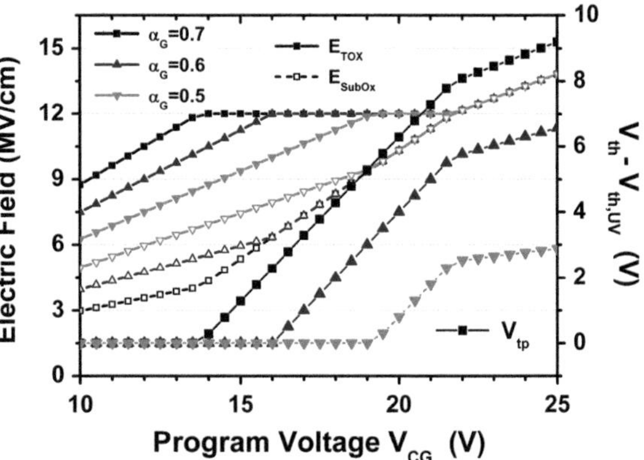

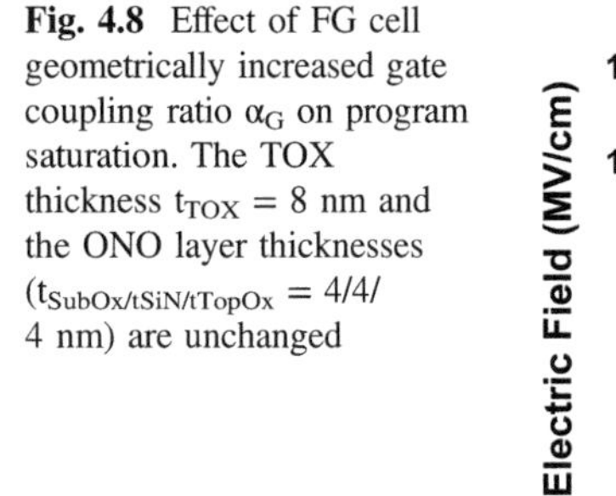
Fig. 4.8 Effect of FG cell geometrically increased gate coupling ratio α_G on program saturation. The TOX thickness t_{TOX} = 8 nm and the ONO layer thicknesses ($t_{SubOx}/t_{SiN}/t_{TopOx}$ = 4/4/4 nm) are unchanged

values while the IPD field increases. In practice, erase saturation can in principle also become a problem, e.g. for bi-layer high-k dielectric containing IPD options. However, for NAND FG Flash only one single erase V_{th} distribution needs to be placed in the negative V_{th} range which generally does not require erasing the cells to large negative threshold voltages. For the positive V_{th} range the situation is different, because for a multi-level cell (MLC cell (TLC), eight different V_{th} distributions need to be placed in the positive V_{TH} range, which requires at least that a V_{th} = +4 V can be programmed.

Consequently, program saturation is usually a more severe issue than erase saturation.

4.2.4 *Program, Erase, and Read of FG Cells in the NAND String*

When a large number of a floating gate cells need to be operated in the NAND array it has to be taken into account that one floating gate cell is located at every crossing point of bit lines and word lines. Therefore, the memory cells in the NAND array cannot be operated independently of each other anymore. In the word line direction (depending on the page size), a couple of thousand FG cells are controlled by the same word line. In bit line direction, the string size (64–66 cells in latest NAND generations) defines the number of cells that cannot be operated independently. Consequently, it is very important to bear in mind what is happening with all neighboring cells when one cell is treated. This is even more important since the threshold voltage of each memory cell needs to be carefully adjusted as shown for SLC and MLC cells in Fig. 4.9.

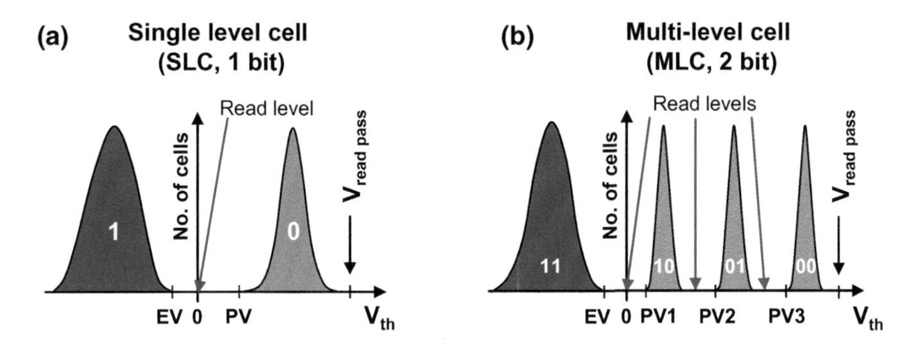

Fig. 4.9 Memory cell threshold voltage distributions for one bit per cell (SLC) data storage (**a**) and two bit per cell (MLC) data storage in a NAND flash array

The erased V_{th} cell distribution is placed at negative V_{th} values. In an ISPP-like sequence the erase voltage is increased until all cells are erased below the erase verify (EV) level. The programmed V_{th} distributions are placed in the positive Vth range. For a single level cell (SLC) the ISPP programming is continued until all cells designated for programming are above the program verify (PV) level. In the case of multi-level cells (MLC), there are consequently three program verify levels (PV1, PV2, and PV3). In addition, it has to be guaranteed that the margins between the different programmed V_{th} distributions are large enough to place the read levels and have sufficient margin for charge/retention loss-caused V_{th} reductions (see Sect. 4.3). To obtain these kinds of narrow cell V_{th} distributions it is necessary to apply a specific distribution shaping algorithm with a small program step increase in certain stages of ISPP programming [9].

4.2.4.1 NAND Cell Programming and Self-boosted Program Inhibit (SBPI)

Figure 4.10 shows the voltage condition in the NAND array when the FG cell at WL3 in BL2 is programmed. For this purpose, a program pulse with the pulse amplitude of $V_{pp} = 20$ V is applied to WL3. To conduct a successful program, it is also required to transfer 0 V to the channel region of the programmed cell as shown in Fig. 4.10 (i). Consequently, the 0 V potential is applied to BL2 and then needs to be transferred to the whole string including the programmed cell at WL3. This is done by applying the pass voltage (e.g. $V_{pass} = 10$ V) to all other word lines.

In principle, all cells addressed by WL3 could be programmed by this means at the same time. However, the programming of arbitrary information requires that specific memory cells at WL3 are excluded from programming. The cell at the crossing point of BL1 and WL3 represents, in this example, the cells which should be prevented from programming (program-inhibited cell in Fig. 4.10. In former FG NAND generations, programming in certain NAND strings was avoided by actively applying a positive voltage to the corresponding bit lines. As a result, the voltage difference between the channel and the control gate was not high enough for programming in these strings. This procedure was complicated and the voltage pumps used for this purpose required additional power and chip area. Therefore, in later generations the so-called "Self-Boosted Program Inhibit" (SBPI) scheme was introduced [8]. The principle of the SBPI scheme is that the channel potential in the inhibited strings is not actively raised by applying a voltage, but capacitively raised, as will be seen in the following.

The voltages applied to different word lines, bit lines and select devices in the SBPI sequence are shown in Fig. 4.10. The corresponding detailed timing of the signals at different signal lines is shown in Fig. 4.11. For a successful program inhibit at the programmed word line an inhibit channel potential in the range of typically 6–8 V is required. The exactly required channel potential further depends on the maximal used programming voltages.

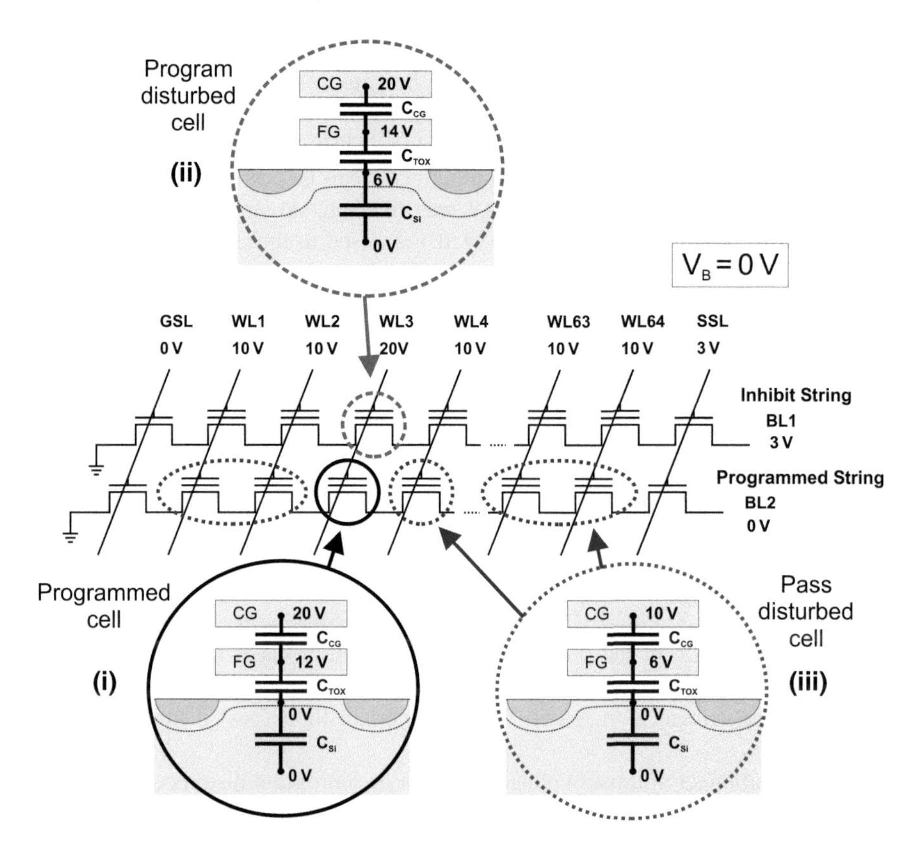

Fig. 4.10 Voltage conditions during program operation in the NAND array. The memory cell at the crossing point of WL3 and BL2 is programmed; several other cells are disturbed by either program disturb or pass disturb

Fig. 4.11 Signal timing for the self-boosted program inhibit (SBPI) scheme

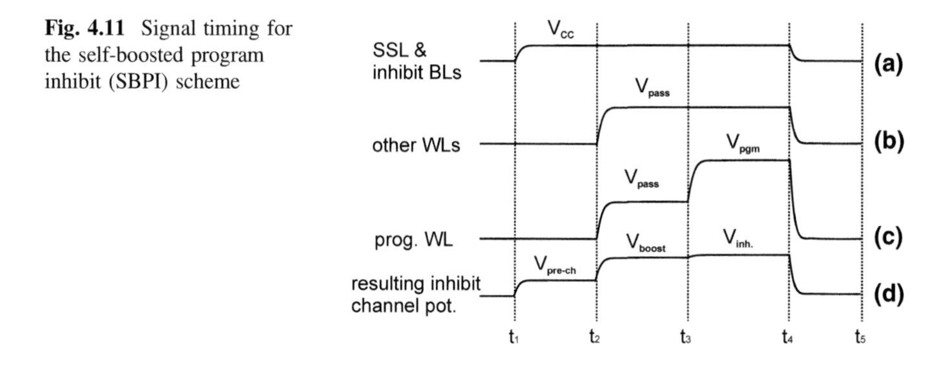

In the first step (t_1), V_{CC} (e.g. 3 V) is connected to the SSL and the inhibit strings at the same time (Fig. 4.11a). This results in a pre-charge of the inhibit string to a channel potential of $V_{pre\text{-}ch} = V_{CC} - V_{th,SSL}$ as shown in Fig. 4.11d. During this

pre-charge of the string the channel side of the select transistor acts as the source. Accordingly, a charging current flows until the gate-to-source voltage equals the threshold voltage of the select transistor. In the second time step t_2, all word lines are raised to the program pass voltage V_{pass} (Fig. 4.11b, c) and the channel inhibit potential is increased by capacitive coupling. This can be done because the select transistor is closed since the pre-charge was finished. At time t_3 the word line selected for programming (WL3 in Fig. 4.10) is raised to the full program voltage in the ISPP sequence which further increases the channel potential to its full inhibit voltage V_{inh}.

In this last step, only a small channel voltage increase is achieved which results from the CG to channel capacitance ratio of one cell in relation to the whole cell string. Therefore, a larger channel voltage increase can be obtained when not the whole string is boosted, but only a few cells in the vicinity of the programmed word line. Such an approach is called the "local self-boosted program inhibit" (local SBPI) scheme [15, 16].

It is clear that a major part of the inhibit channel potential depends on the pass voltage, since V_{inh} is partly generated by the capacitive channel boosting.

On the one hand, the ability to prevent programming at the "program disturbed cell" (WL3 of BL1 in Fig. 4.10(ii)) improves with increasing pass voltage V_{pass} as shown in Fig. 4.12. On the other hand, the pass cells located in a string with a memory cell dedicated for programming (BL2) experience a soft programming when the pass voltage is increased beyond a certain limit (pass disturbed cell in Fig. 4.10(iii)).

The general effect of a pass voltage variation on a program disturbed and a pass disturbed cell in a 48 nm FG NAND technology is shown in Fig. 4.12. Since both effects, program and pass disturb, result in a threshold voltage increase and are more severe on erased cells, the memory cells in Fig. 4.12 were first erased to a threshold voltage below $V_{th} = -4$ V before the program and pass disturbs could be measured. In addition to the pass voltage pulse amplitude value, the number of

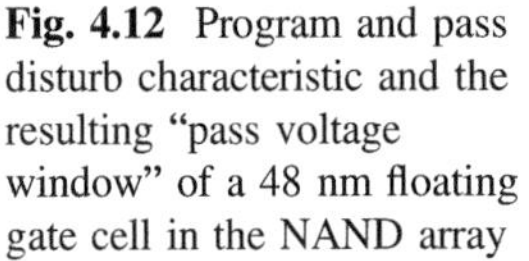

Fig. 4.12 Program and pass disturb characteristic and the resulting "pass voltage window" of a 48 nm floating gate cell in the NAND array

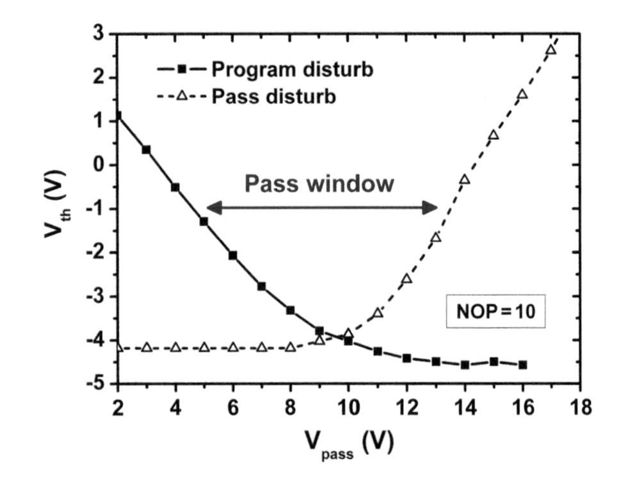

disturbing pulses is very important for the disturb strength. The determining factor here is the number of program operations (NOP) carried out at each word line [17]. In the example given in Fig. 4.12, the operation of a FG memory cell used in MLC mode was chosen which results, e.g., in NOP = 10. This is because every word line is logically divided into different pages which need to be separately programmed. Finally, a NOP = 10 results in approximately 100 program pulses with the highest program voltage assumed for the slowest cell in programming and about 5000 pass voltage pulses, because each of the 64 cells in the string needs to be programmed.

It can be observed that the selection of the pass voltage results in a trade-off between program and pass disturb. Generally it needs to be guaranteed that the V_{th} of all erased cells remains (with a certain margin) below $V_{th} = 0$ V.

Therefore, a "pass window" with suitable pass voltages could be determined at the level $V_{th} = -1$ V. The optimum for the trade-off between program and pass disturb can be found in Fig. 4.12 slightly below $V_{pass} = 10$ V.

4.2.4.2 Erase and Read of FG Cells in the NAND String

The advantage of the NAND Flash erase operation is that a whole erase block is erased at once. The voltage conditions during erase are shown in Fig. 4.13. All word lines are at ground potential ($V_{CG} = 0$ V) and the erase voltage is applied to the well of the erase block. Very important during erase is that the select transistors as well as the bit line and the source line are left floating. For this purpose, the usually grounded source line needs to be disconnected from the ground potential. By this means, the source line and the bit line, and to a certain extend the select transistors, can follow the bulk potential, and large currents into the source line and the bit line are avoided. Due to the improved coupling when the same voltage is applied to all cells, the voltage difference between the control gate and the channel required for erase (e.g. $V_B = 18$ V) is lower than the programming voltage. The erase operation is successful when all cells in the erase block are erased below the EV level as described above.

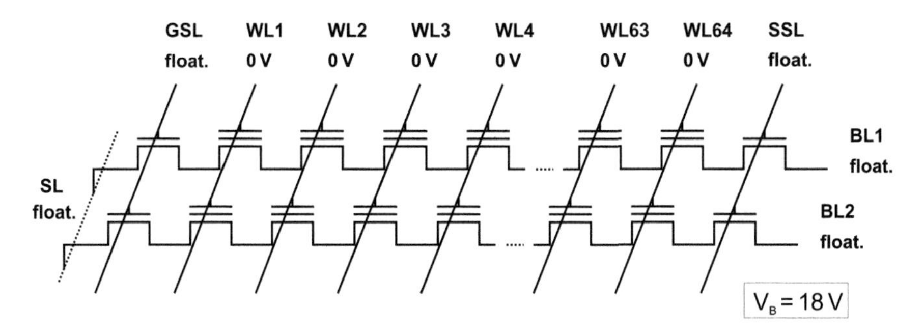

Fig. 4.13 The erase of floating gate cells in the NAND array is carried out in electrically separated erase sectors. By applying a positive voltage (e.g. $V_B = 18$ V) to the well of the erase sector, all cells are erased at the same time

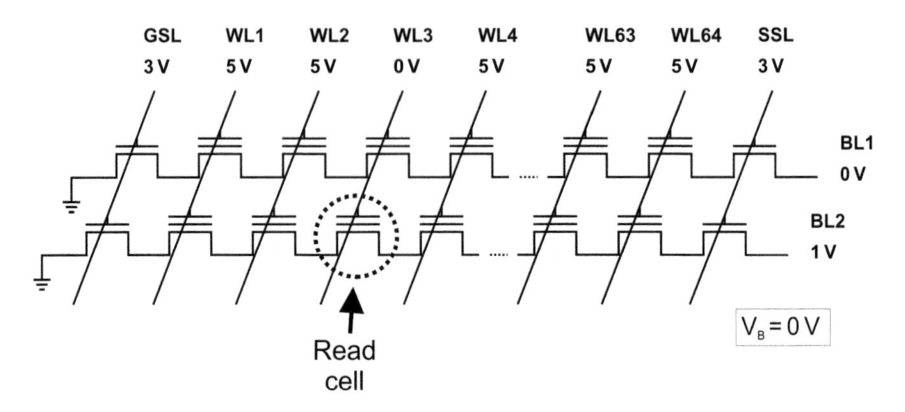

Fig. 4.14 Read operation in the NAND Flash array

The read operation in the NAND array is carried out word line by word line. For a current sensing read scheme [18] the bit line which is selected for read operation (BL2 in Fig. 4.14) can be set to the read voltage (e.g. $V_{BL2} = 1$ V). For a SLC read operation the word line at the read cell is set to 0 V, while typically 5 V are applied as read pass voltage for all other word lines.

By this means it can be detected if the cell at WL3 in the string of BL2 is in the programmed or erased cell. It is clear that for reading one cell, the read current needs to flow through all cells in the 64 cell string and that only one cell in the string can be read at a time.

It needs to be mentioned that also the read pass voltage of only $V_{rpass} = 5$ V can result in a change of the threshold voltage (read disturb [19]) when only the number of read operations is high enough. For SLC FG NAND cells it is assumed that 10^6 read operations with 15 μs durations need to be guaranteed without read fails. This results in a total disturb time of about 15 s. Again, erased cells are most susceptible to read disturb as described before for program and pass disturb.

4.3 Reliability of Floating Gate NAND Memory Cells

The reliability of FG NAND Flash memory is one of the most important criteria, since typically 10 years of charge retention and 1–100 k program/erase cycles need to be guaranteed for a NAND Flash product chip.

In Fig. 4.15, a typical charge retention requirement is shown. It needs to be guaranteed for a successful read-out of the stored information that the programmed V_{th} (above the PV level) is not decreased more than 10% over the product relevant time period of 10 years.

In principle, there are multiple leakage paths which can lead to a loss of the programmed floating gate electron charges as shown in Fig. 4.16a. The electrons can be lost through the IPD towards the control gate ($I_{IPD\text{-}leak}$) or leak through the

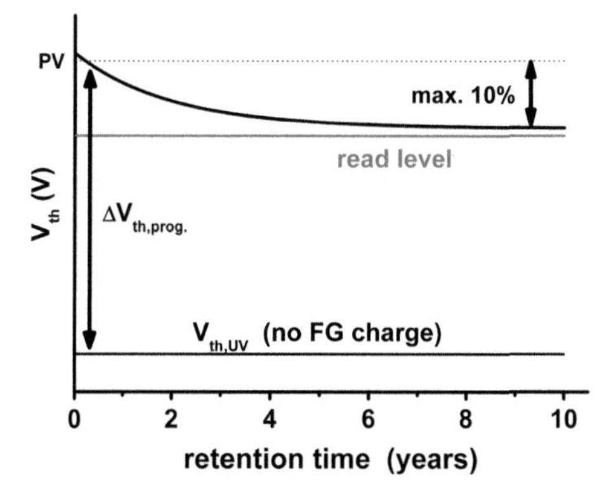

Fig. 4.15 Charge retention of an FG cell. A certain amount of charge loss needs to be tolerated (e.g. 10% V_{th} loss over the time period of 10 years)

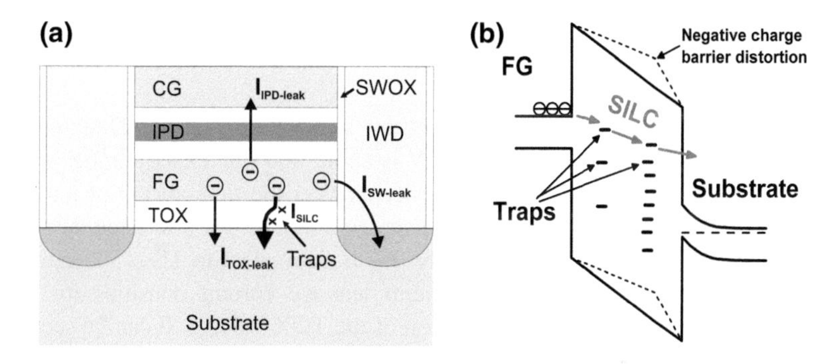

Fig. 4.16 Possible leakage path for charge loss from the floating gate (**a**). Tunnel oxide damage due to program/erase cycling and the resulting stress-induced leakage current (SILC) are usually the main reasons for retention loss (**a**, **b**). Negative trap charge built up over cycling additionally induces a barrier distortion which results in an increased tunnel barrier (**b**) [12]

cell side wall oxide (SWOX → $I_{SW\text{-leak}}$) and the inter-word line oxide (IWD → $I_{IWD\text{-leak}}$) to the cell junction area.

However, the most severe charge loss component of an optimized floating gate cell process is the leakage through the TOX ($I_{TOX\text{-leak}}$). This is not only because the TOX is physically the thinnest dielectric layer which holds the electrons on the floating gate, but there are additional processes which cause wear of the FG cells. As shown in Fig. 4.16a, b, the charge transfer during program and erase generates electric states in the TOX (and the TOX should be the only dielectric where charge is transferred, as previously discussed) which are called oxide traps. These traps are broken bonds of the atoms in the oxide matrix due to the electron tunneling processes [20]. The density of traps in the tunnel oxide consequently increases with the number of program/erase cycles which cause so-called oxide stress. The traps in the TOX barrier can act as stepping stones when floating gate electrons leak via a trap-assisted tunneling process towards the cell channel region. The probability of this trap-to-trap tunneling (called stress-induced leakage current, SILC) [21] is much higher than a direct tunneling process through the whole TOX thickness. The reason is that the effective tunnel distance of each tunneling step is significantly reduced for the SILC.

The TOX trap generation during the product lifetime and the corresponding SILC is the reason for a general TOX thickness scaling limitation in floating gate cells [22]. Therefore, the TOX cannot be scaled below 8.0–7.5 nm. To understand this TOX thickness limitation in more detail we need to determine the oxide electric field, or alternatively, the oxide voltage during retention conditions, which is given by

$$V_{FG,\,Ret.} = \alpha_G \cdot \Delta\, V_{th,\,prog.}. \tag{4.9}$$

where α_g is again the gate coupling ratio and $\Delta V_{th,prog.}$ is the programmed threshold voltage shift as shown in Fig. 4.15. For assumed values of $\Delta V_{th,prog.} = 4$–5 V and $\alpha_g = 0.6$, the TOX voltage under retention conditions is about 3 V. The second criterion of interest is the acceptable leakage current for the 10-year charge retention.

The number of stored floating gate electrons in a 50 nm FG NAND technology for a threshold voltage shift of $\Delta V_{th} = 4$ V is about 600 (the exact number will be discussed in Sect. 4.4.4). The 10% loss criterion over the time period of ten years results in a tolerable loss of one electron every two months (or a leakage current of 3E−26 A). Converted to a current density this is equivalent to 1E−15 A/cm^2.

Figure 4.17 shows the Fowler-Nordheim leakage current densities for TOX thicknesses of 6, 8, and 10 nm as a function of the TOX voltage. It can be seen that for an unstressed TOX and the estimated TOX retention voltage $V_{FG,Ret} = 3$ V and current criterion, a tunnel oxide thickness of 6 nm would be sufficient. However, 2 nm additional TOX thickness is required to fulfill the retention criterion for a damaged TOX with trap-to-trap SILC leakage as discussed above.

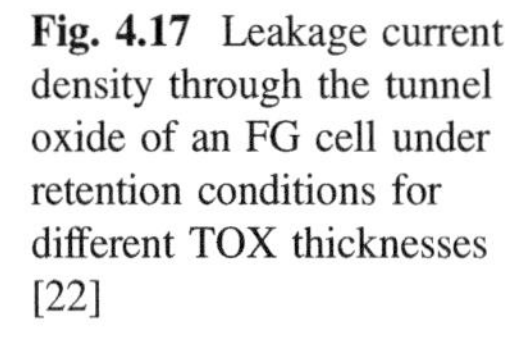

Fig. 4.17 Leakage current density through the tunnel oxide of an FG cell under retention conditions for different TOX thicknesses [22]

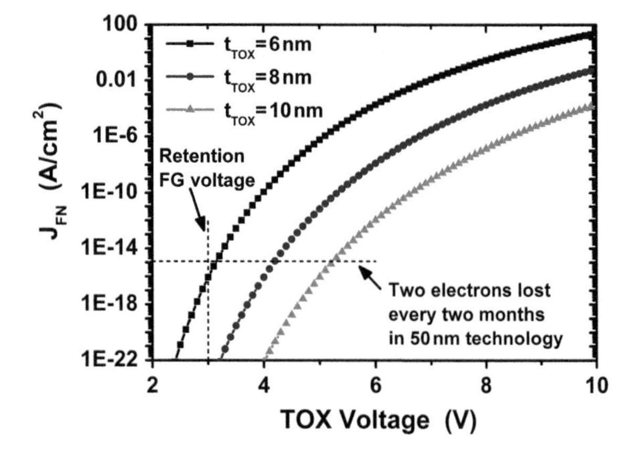

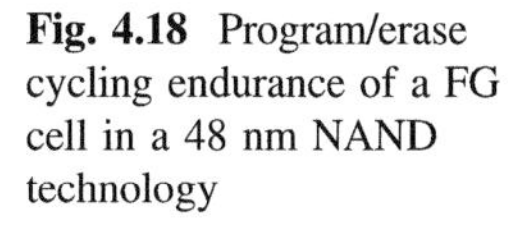

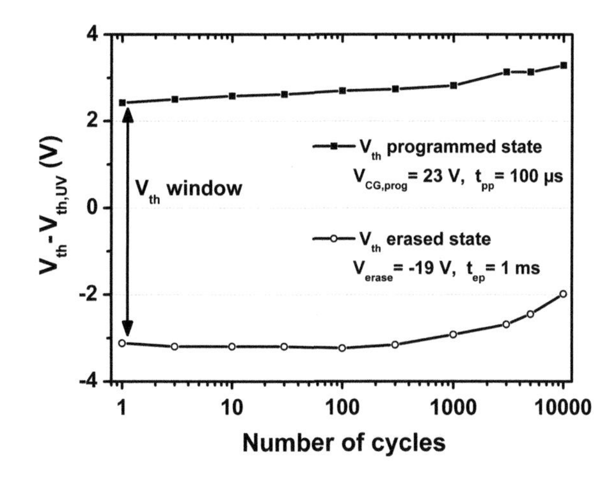

Fig. 4.18 Program/erase cycling endurance of a FG cell in a 48 nm NAND technology

Figure 4.18 shows the endurance of FG cells in a 48 nm NAND technology. All program and erase cycles were carried out with unchanged program and erase cycle voltages of $V_{CG,prog} = 23$ V and $V_{erase} = -19$ V for the indicated pulse times. For low cycle numbers, the V_{th} window is slightly increases, whereas for higher cycle number above 300 cycles the V_{th} window closes. Furthermore, a general V_{th} upward shift is visible.

This behavior can be explained with positive charge trapping at low cycle counts which leads to a reduced TOX barrier and negative charge trapping which results in an increased barrier height (see Fig. 4.16b) at higher cycle numbers.

For a reduced tunneling barrier, more electrons can be transferred through the TOX for unchanged program and erase voltages, whereas for an increased barrier this number of transferred electrons is reduced. Additionally, the fixed negative charges which are generated in the TOX for higher cycle counts generally increase the cell V_{th}. In the case shown in Fig. 4.18, the erased cell V_{th} is shifted by one volt after 10 k program/erase cycles. Besides the increased retention problem for higher cycle numbers due to trap generation, the window closing and the general V_{th} upward shift will result in increased pulse voltages, especially for erase.

4.4 Scaling of Floating Gate NAND Memory Cells

The NAND Flash memory scaling of the last 15 years was accomplished by reducing the cell dimensions, whereas the cell construction principle was unchanged. The effective cell size of NAND Flash in 1995 was in the range of 1 μm^2 which resulted in a product chip memory capacity of 32 Mb [8]. In 2010, the cell size was reduced to 0.0028 μm^2 [10] with a chip capacity of 64 Gb. This strong reduction of the cell geometry leads to scaling issues which are discussed in the following.

4.4.1 Scaling of the Floating Gate Cell Geometry

As described in Sect. 4.2.3, it is very important for a programmability of floating gate cells to have an enhanced control gate to floating gate area by a control gate which is wrapped around the floating gate. However, this requires a certain space between adjacent floating gates, since this space needs to fit two times the IPD thickness plus the poly plug. Depending on the FG NAND ground rule (or half pitch F), this has some implications for the remaining control gate plug width as shown in Fig. 4.19a.

Figure 4.19b shows the remaining control gate plug width as a function of the bit line half pitch F. To obtain more space for the control gate plug, the width of the floating gate can be reduced with respect to the space between the floating gates as done in the latest FG NAND generations [23, 24]. The space between adjacent floating gates consequently becomes wider, as indicated in Fig. 4.19a. Additionally, the physical IPD thickness can be reduced. These two options are combined in Fig. 4.19b with the result that for an FG width of 0.6 F and a physical IPD thickness of only 8 nm a control gate plug width of 10 nm can be realized down to a bit line half pitch of 20 nm. Due to this bit line pitch scaling limitation it can be observed in the latest FG NAND technology generations that the bit line pitch is less aggressively scaled than the word line pitch [10, 11, 23].

In case of very narrow control gate plugs, it may be that the poly-Si doping level in the CG plug cannot be maintained sufficiently high. This would result in poly-Si depletion and consequently in an electrically inactive CG plug. An alternative could be a metal control gate material as presented in [2].

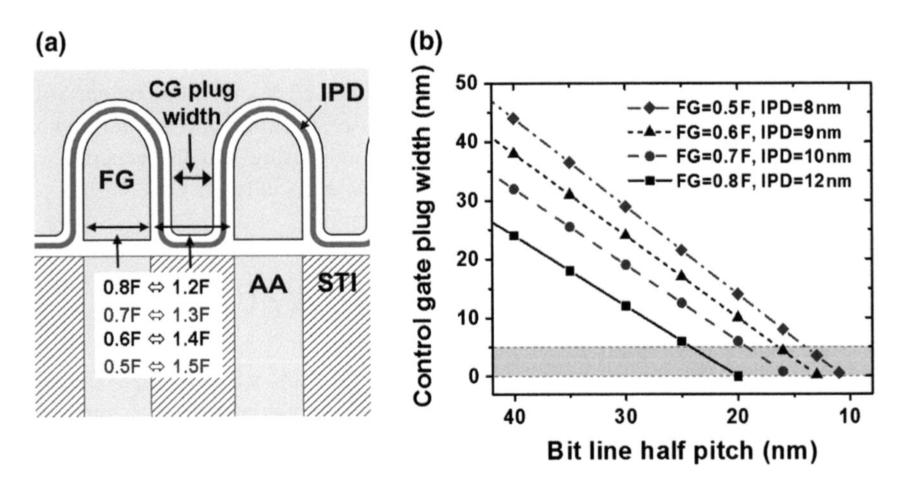

Fig. 4.19 Bit line pitch scaling limitation for the typical control gate to floating gate enhanced coupling area FG NAND cell. To fit two times the IPD thickness plus the poly plug (**a**) with an assumed minimum width of 10 nm, the active area (AA) width can be reduced below the half pitch F to clear a space for the CG plug (**b**)

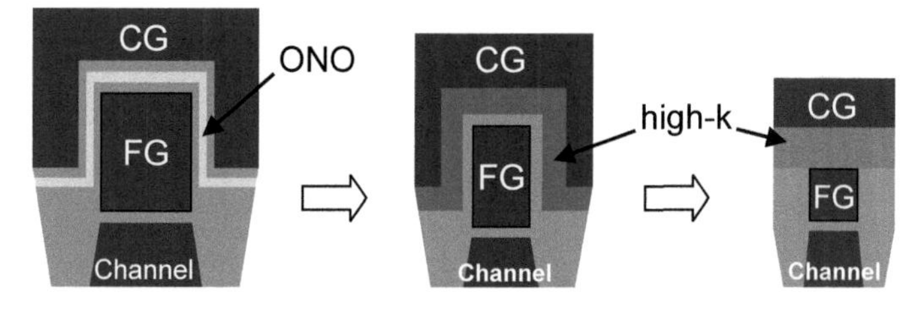

Fig. 4.20 Floating gate NAND cell scaling: The requirement for a continued reduction in the floating gate cell dimensions in combination with a high gate coupling ratio leads from the typical ONO IPD cell with a control gate wrapped around the floating gate to a high-k containing IPD, and finally due to the lack of space for the control gate plug to a planar floating gate cell

Continued scaling of floating gate NAND cells (see Fig. 4.20) in combination with a sufficiently high gate coupling ratio requires efforts to reduce the electrical IPD thickness (EOT). One option to do so is the introduction of high-k dielectrics in the IPD stack. However, at a certain floating gate NAND technology node there won't be sufficient space for the control gate plug, which automatically leads to a planar floating gate cell as shown in Fig. 4.20.

It was discussed in Sect. 4.2.3 that for insufficiently high gate coupling ratios together with an electrically thin IPD, tunnel currents can in principle flow through the IPD during the program and erase conditions. An IPD leakage can result in a degraded program and erase behavior, visible in reduced ISPP and erase slopes [25]. Consequently, a fully planar floating gate cell with ONO IPD cannot be programmed and erased in the traditional manner where charge is transferred through the tunnel oxide only. Even an IPD layer combination of SiO_2 and high-k or a pure high-k IPD layer is problematic with respect to program/erase saturation [13].

One possibility to improve the planar floating gate cell was the usage of a dual layer floating gate as proposed in [26]. Figure 4.21 illustrates the advantages of a dual layer floating gate with an n-doped poly-Si bottom part (adjacent to the tunnel oxide) and a high work function metal layer on top (adjacent to the high-k IPD) with respect to program and erase saturation.

Figure 4.21a, b shows the conditions during program operation. The n-poly-Si floating gate in Fig. 4.21a has the problem of the insufficient effective IPD barrier which does not provide sufficient current blocking margin to program the cells to high V_{th} levels. The situation is improved by the introduction of the high work function metal gate layer, as shown in Fig. 4.21b, where the barrier height and the effective electron tunneling barrier (shadowed area) is significantly larger. The advantage of the dual layer floating gate under erase conditions and why simply a single layer high work function metal FG cannot replace the poly FG are illustrated in Fig. 4.21c, d respectively. The single layer metal floating gate has a larger barrier between the FG and TOX which would hinder the erase when electrons are tunneling out of the FG towards the channel region (Fig. 4.21c). Consequently, a higher erase voltage would be necessary with the even more problematic effect that

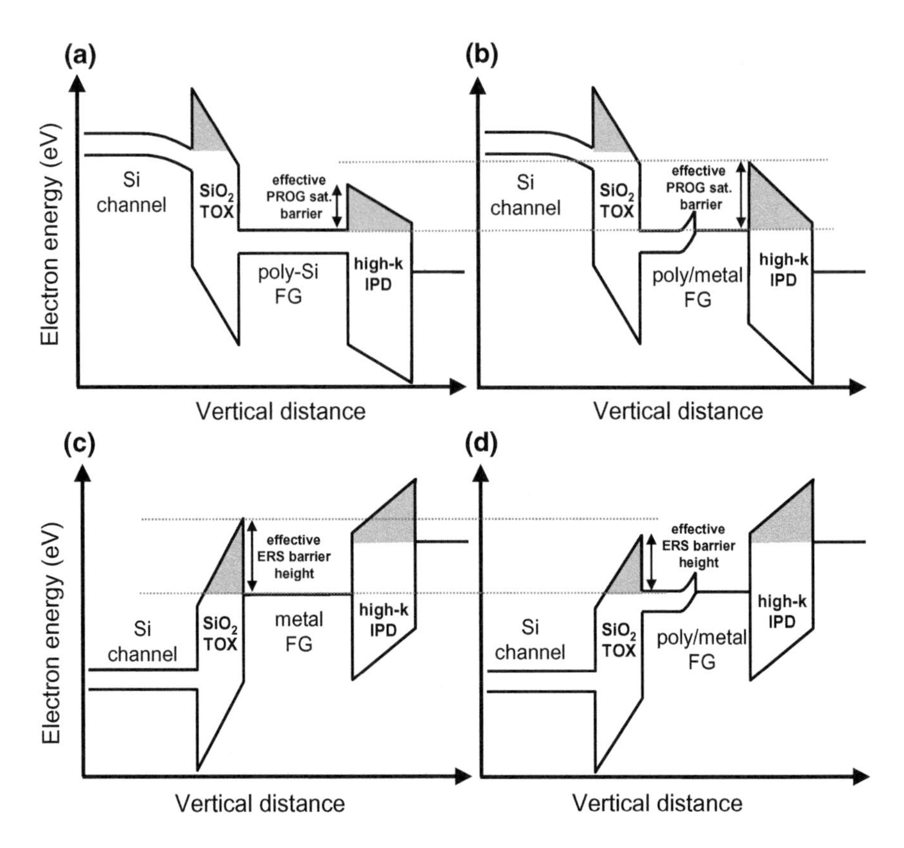

Fig. 4.21 Field improvement in planar floating gate cells and how program and erase saturation can be avoided by the usage of a dual layer FG structure [26]

at the same time electrons tunnel from the control gate to the floating gate (electron back tunneling) and cause erase saturation. This electron back tunneling will be seen in Sect. 4.6 to be one of the major issues of charge trapping memory cells, but is less problematic for the dual layer FG as seen in Fig. 4.21d.

4.4.2 *Floating Gate Cell Cross-Coupling*

Another general problem for floating gate NAND cells in technology generations below 50 nm is the cell-to-cell cross-coupling. This effect is the direct coupling from one floating gate to the nearest neighboring floating gates as shown in Fig. 4.22. It is clear that this direct coupling increases for reduced dimensions since the cells move closer together and therefore the relative coupling capacitance increases. Most significant is the FG to FG coupling in the direction along the bit lines (y-direction in Fig. 4.22). This is because the floating gates are directly face

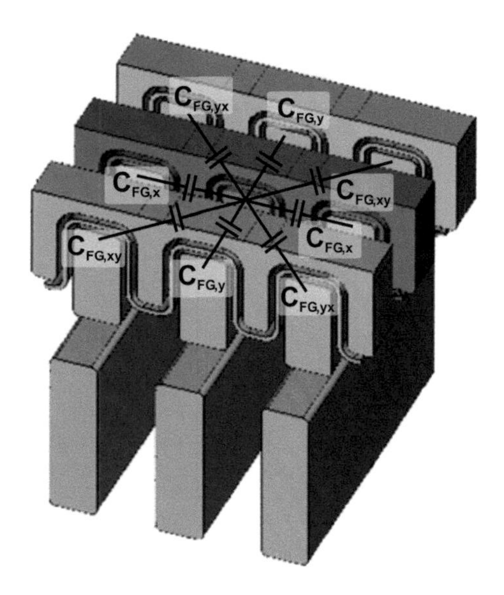

Fig. 4.22 Floating gate cross-coupling in scaled NAND Flash technologies [27]

each other with the full FG height and full FG width in this direction. Consequently, $C_{FG,y}$ is the largest of the FG to FG coupling capacitance terms. In the direction along the word lines (x-direction), parts of the FG to FG coupling are screened by the control gate plug and therefore $C_{FG,x}$ is typically smaller than $C_{FG,y}$. To minimize the coupling capacitance in x-direction it would be beneficial to have a very deep position of the CG plug, ideally down to the STI level, which would mean a complete screening in x-direction. However, the full programming voltage drop between the control gate plug and the channel limits the minimum CG plug to channel distance. The diagonal coupling components $C_{FG,xy}$ and $C_{FG,yx}$ are typically the smallest ones.

In cell programming schemes, where even and odd bit lines are programmed separately (because they belong to different logical pages), the programming of a cell can change the threshold voltage of a directly neighboring cell which was already programmed. This effect is called floating gate cross-coupling or floating gate interference [27].

The cell-to-cell coupling potentially leads to a decreased gate coupling ratio since all increased capacitance terms from FG cross-coupling are added in the denominator of the gate coupling ratio (4.3). Therefore, the gate coupling ratio decreases at least in the case where the floating gate cell dimensions are scaled proportional to the technology node, while TOX and IPD thicknesses are kept constant.

This behavior can be seen in the lowermost curve of Fig. 4.23a obtained from 3D simulations with a commercial field solver [28]. It can be seen that for a constant IPD EOT of 11 nm in combination with a floating gate whose height is two times the width (width = F, height = 2 F in points A, B, and C), the gate coupling ratio decreases from 0.63 in the 50 nm technology node to only 0.52 in the

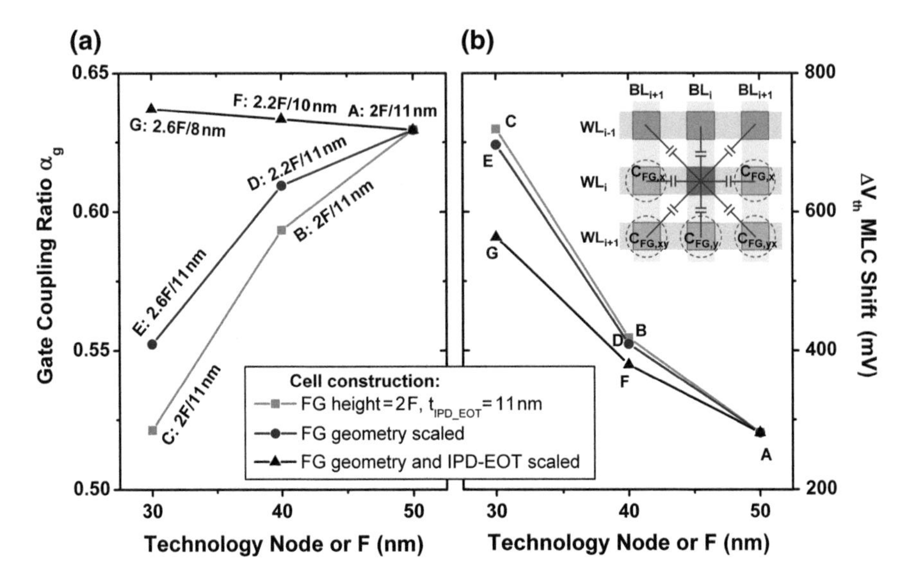

Fig. 4.23 Gate coupling ratio (**a**) and threshold voltage shift (MLC shift) due to the programming of five directly neighboring cells (**b**) by a $\Delta V_{th,prog} = 5$ V as a function of cell technology generation. For each point, the floating gate height and the IPD EOT value (e.g. 2 F/11 nm) are given. The floating gate width is 1 F for each technology node and the TOX thickness is always 8.5 nm

30 nm technology. A slight gate coupling ratio improvement can be seen for an increased floating gate height to width ratio with decreasing half pitch in the middle curve (points A, D, and E) of Fig. 4.23a. A slightly increasing α_g for smaller dimensions is only obtained here for an increased FG height to width ratio in combination with a decreased IPD effective thickness (points A, F, and G).

However, in Fig. 4.23b, it is apparent that all efforts to keep the gate coupling ratio value high do not significantly improve the V_{th} shift due to neighboring cell programming in conventional cell programming schemes. For the simulation of the depicted ΔV_{th} MLC shift it is assumed that five neighboring cells influence the V_{th} of each ready programmed cell in worst case, as indicated in the inset of Fig. 4.23b. In detail, these five cells consist of two neighboring cells in word line direction, two diagonal cells, and one directly neighboring cell in bit line direction, resulting from an assumed conventional word line by word line programming scheme for serial even and odd bit line addressing. In the NAND chip layout belonging to the serial WL programming of even and odd bit lines the serial treatment is necessarily performed, since two neighboring bit lines share one single sense amplifier for reading the V_{th} state during ISPP programming. The cross-coupling capacitance terms were again taken from the 3D field simulations, and for the depicted MLC shift it is assumed that all five cells are programmed by a $\Delta V_{th} = 5$ V. This would be the threshold voltage shift for erased FG cells which are programmed to $V_{th} = 4$ V.

The fact that for conventional programming schemes the simulated MLC shift at 30 nm cannot be reduced below 500 mV leads to the conclusion that at a certain point in shrinking the FG NAND Flash dimensions the program algorithm needs to take care of the floating gate cross-coupling issue. The strategy is simply to reduce the number of neighboring cells that are programmed after reaching the final programming target V_{th} of each cell, in combination with a reduction of the amount these neighboring cells increase their V_{th}.

One component for reducing the unwanted FG cross-coupling is the all bit line (ABL) architecture, where each bit line has a separate sense amplifier and therefore all bit lines can be programmed at the same time.

Together with the improved program algorithm with respect to the order in which the cells are programmed, it was possible to master FG cross-coupling even for three bits per cell (TLC) and four bit per cell (XLC) technologies [11, 29, 30].

4.4.3 Word Line to Word Line Leakage Current

The reduced cell–to-cell distances with scaled dimensions also cause strongly increased electric fields between neighboring word lines during program operation.

The WL-to-WL voltages during erase are uncritical because all cells are erased at the same time and therefore all word lines are at the same potential.

High WL voltage differences during program operation are even more critical since the programming voltage does not scale or rather increase slightly, as described above. As a result of the strong electric fields between word lines, electrons can tunnel from a programmed floating gate to the control gate that is on the high program voltage V_{pgm} [31] or generally introduce WL-to-WL leakage currents as shown in Fig. 4.24. The electric field strength in an assumed SiO_2 IWD is shown for different WL-to-WL distances as a function of the WL difference voltage in Table 4.1.

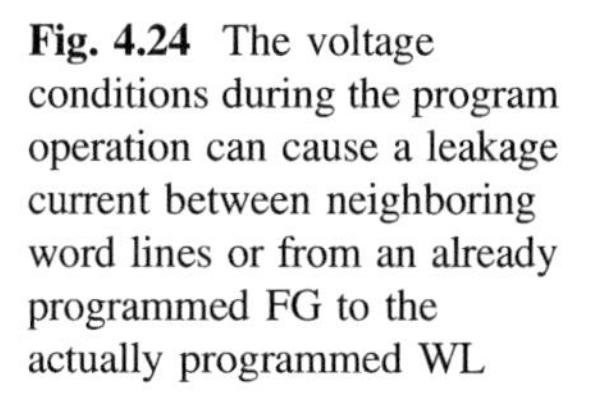

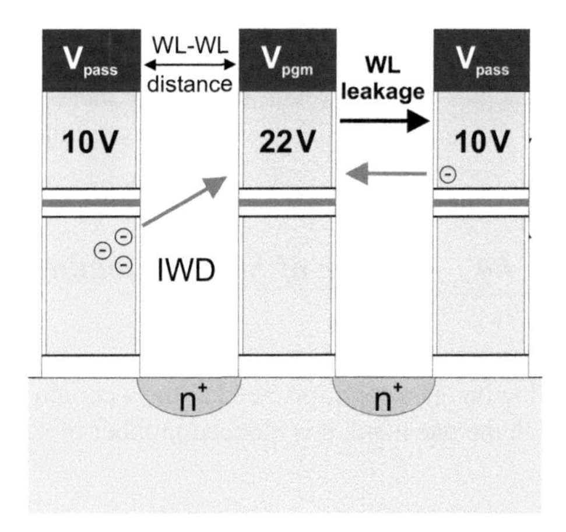

Fig. 4.24 The voltage conditions during the program operation can cause a leakage current between neighboring word lines or from an already programmed FG to the actually programmed WL

Table 4.1 WL-to-WL IWD (SiO_2) electric field in MV/cm as a function of the voltage and the distance between different word lines. The light grey shaded WL-WL distance and voltage combinations represent electric IWD fields above the usual 4 MV/cm operation conditions. The dark grey shaded electric IWD field range above 8 MV/cm represent very high values in the Fowler-Nordheim tunnelling regime (see Fig. 4.5)

WL-to-WL Distance

WL-to-WL Difference Voltage	10 nm	15 nm	20 nm	25 nm	30 nm	35 nm	40 nm	45 nm	50 nm
20 V	20.0	13.3	10.0	8.0	6.7	5.7	5.0	4.4	4.0
19 V	19.0	12.7	9.5	7.6	6.3	5.4	4.8	4.2	3.8
18 V	18.0	12.0	9.0	7.2	6.0	5.1	4.5	4.0	3.6
17 V	17.0	11.3	8.5	6.8	5.7	4.9	4.3	3.8	3.4
16 V	16.0	10.7	8.0	6.4	5.3	4.6	4.0	3.6	3.2
15 V	15.0	10.0	7.5	6.0	5.0	4.3	3.8	3.3	3.0
14 V	14.0	9.3	7.0	5.6	4.7	4.0	3.5	3.1	2.8
13 V	13.0	8.7	6.5	5.2	4.3	3.7	3.3	2.9	2.6
12 V	12.0	8.0	6.0	4.8	4.0	3.4	3.0	2.7	2.4
11 V	11.0	7.3	5.5	4.4	3.7	3.1	2.8	2.4	2.2
10 V	10.0	6.7	5.0	4.0	3.3	2.9	2.5	2.2	2.0
9 V	9.0	6.0	4.5	3.6	3.0	2.6	2.3	2.0	1.8
8 V	8.0	5.3	4.0	3.2	2.7	2.3	2.0	1.8	1.6
7 V	7.0	4.7	3.5	2.8	2.3	2.0	1.8	1.6	1.4
6 V	6.0	4.0	3.0	2.4	2.0	1.7	1.5	1.3	1.2
5 V	5.0	3.3	2.5	2.0	1.7	1.4	1.3	1.1	1.0

IWD Electric Field in MV/cm

Generally speaking, electric fields up to 4 MV/cm can be handled with deposited oxides as the IWD with sufficient reliability. The field range above 4 MV/cm becomes critical, but the range of 8 MV/cm and above is already in the Fowler-Nordheim tunneling regime for a thermally grown oxide which would not allow a reliable operation anymore.

Options to reduce WL-to-WL leakage by use of a special program algorithm would include limiting the difference voltage between adjacent word lines. This could be accomplished with a specific handling of the word lines close to the program word, similar to the individual word line treatment in local program inhibit schemes [16]. However, effectively increasing the pass voltage at the cells adjacent to the programmed cell will adversely affect the pass disturb.

4.4.4 Number of Stored Floating Gate Electrons

When the dimensions of floating gate cells are scaled down, also the number of floating gate electrons needed for a certain threshold voltage shift ΔV_{th} is reduced. On the one hand, this reduced number of stored floating gate electrons is critical for reliability and charge retention because the loss of one electron has increasing impact on the cell V_{th} loss. On the other hand, the charge granularity of single electrons affects, at a certain stage, the ability to program narrow V_{th} distributions. The effect is most critical in TLC or XLC NAND technologies with very narrow V_{th} distributions in case one electron causes a significant threshold voltage shift.

The approximated number of floating gate electron can be derived from (4.5) and is given as a function of the feature size F for different NAND technology nodes by

$$N = \frac{C_{CG}}{e} \cdot \Delta V_{th} = \frac{\varepsilon_0 \varepsilon_r}{e} \frac{A_{CG-FG}}{t_{IPD-EOT}} \cdot \Delta V_{th} = \frac{\varepsilon_0 \varepsilon_r}{e} \frac{A_{IPD}/A_{TOX}}{t_{IPD-EOT}} \cdot F^2 \cdot \Delta V_{th} \quad (4.10)$$

where e is the electron charge and A_{IPD}/A_{TOX} is the CG-FG area to TOX area ratio.

As shown in Fig. 4.25 and discussed beforehand, this area ratio needs to be increased in combination with a reduction of the IPD EOT value to have the programming voltages remain the same. The shown values for the A_{IPD}/A_{TOX} ratio and IPD EOT are similar to the values used by major NAND Flash manufacturers in recent generation.

The simple planar plate capacitor approximation of (4.10) results in the estimate of about 200 stored electrons, in case a 25 nm FG NAND cell is programmed to $\Delta V_{th} = 4$ V above the UV level, as depicted in Fig. 4.25. The tolerable electron loss per year for this technology node is already less than ten, if a relaxed retention criterion compared to Sect. 4.3 with 20% tolerable V_{th} loss after 5 years is assumed.

However, the general trend of the number of stored electrons as a function of the FG cell technology node in Fig. 4.25 shows a strong reduction with reduced dimensions.

A similar consideration based on TCAD simulations was carried out and presented in [10]. The result of the number of electrons stored in different FG cell locations (see Fig. 4.26) that cause a threshold voltage shift of $\Delta V_{th} = 100$ mV is shown in Table 4.2 for 50, 35, and 25 nm technology generations. The number of

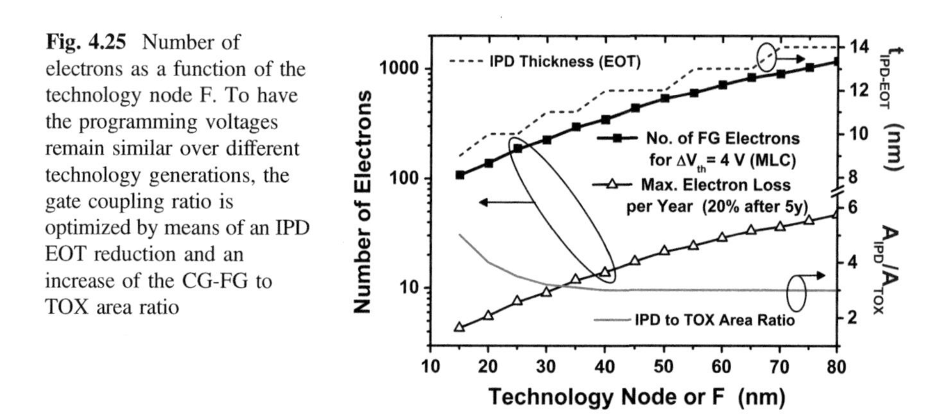

Fig. 4.25 Number of electrons as a function of the technology node F. To have the programming voltages remain similar over different technology generations, the gate coupling ratio is optimized by means of an IPD EOT reduction and an increase of the CG-FG to TOX area ratio

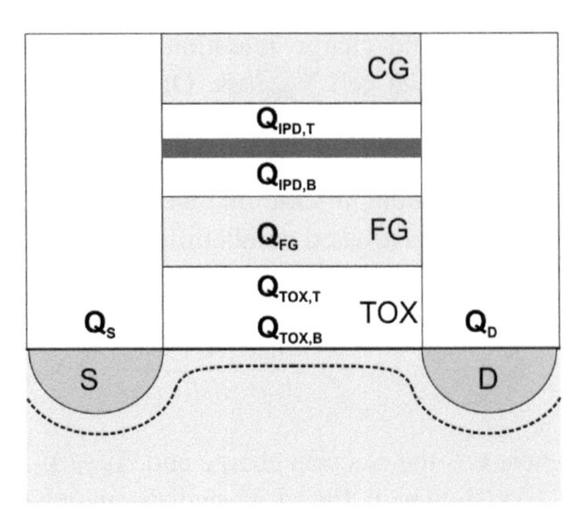

Fig. 4.26 Locations of trapped charges in an FG NAND memory cell which cause a threshold voltage shift

Table 4.2 Electron sensitivity of different FG NAND Flash technology generations. The table indicates the number of electrons required at different locations in an FG cell for a 100 mV threshold voltage shift as determined by TCAT simulations in [10]

Technology	50 nm	35 nm	25 nm
$Q_{TOX,B}/e$	4	2	1
$Q_{TOX,T}/e$	9	7	4
Q_{FG}/e	18	12	10
$Q_{IPD,B}/e$	22	17	11
$Q_{IPD,T}/e$	149	103	100
Q_S/e	33	9	5
Q_D/e	61	16	10

electrons required for a $\Delta V_{th} = 4$ V shift in a 25 nm technology taken from these values is 400 and therefore two times higher than the estimate of (4.10), but the trend over different technology generations is the same.

Table 4.2 indicates that especially electrons stored in tunnel oxide traps, which are generated during program and erase operations, cause higher V_{th} shifts per electron than electrons in the FG. Therefore, uncontrolled electron storage in the TOX can be a significant issue as discussed in the following section.

4.4.5 Random Telegraph Noise

Random telegraph noise (RTN) can be observed in different types of field effect devices and can be explained by electron capture and emission processes in oxide traps close to the channel of a MOSFET device [32]. As mentioned previously, the same process can take place in the TOX of a floating gate NAND cell [33, 34].

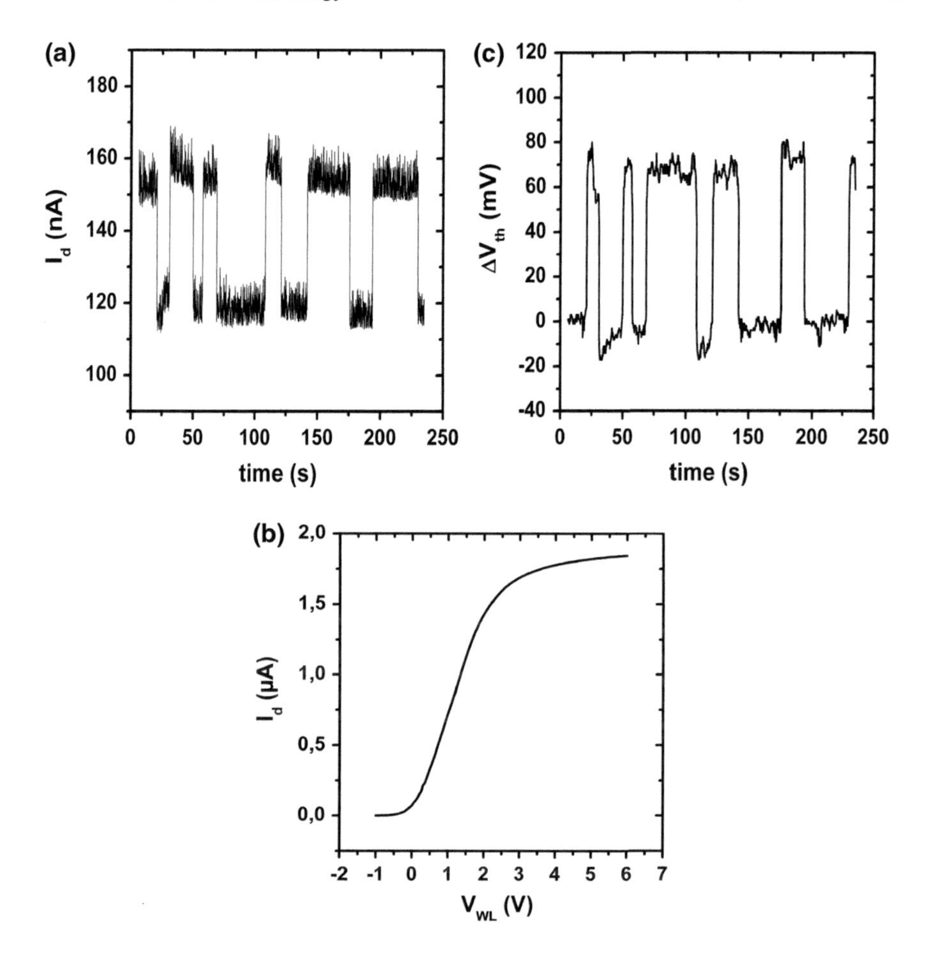

Fig. 4.27 Random telegraph noise of 48 nm FG cells in a NAND string configuration. The variation in the string current (**a**) due to charging and discharging of one oxide trap in the channel region can be converted by the string transfer curve (**b**) into a V_{th} variation (**c**)

Figure 4.27 shows RTN measurements in a 32 cell string of a 48 nm FG NAND technology. Operated in the sub-threshold region, the drain current of the investigated cell (or the string current) shows a characteristic two level I_d signature as shown in Fig. 4.27a. The two level signature and the time constants for capture and emission in the second range indicate that a single tunnel oxide trap about 1–2 nm from the channel/TOX interface [35] is charged and discharged by direct tunneling.

With the aid of the string $I_d - V_{WL}$ transfer curve in Fig. 4.27b, the current signal can be converted into a threshold voltage shift ΔV_{th} as depicted in Fig. 4.27c. The resulting RTN amplitude is about 70 mV and in this case higher than expected from the TCAD simulations [10] in Table 4.2.

However, for scaled dimensions the RTN threshold voltage shifts can cause read fails, which is even more significant for MLC and TLC functionality with small distances between V_{th} distributions.

4.5 Shrinking the Floating Gate NAND Technology Beyond the Direct Optical Lithography Limitation

The effects of scaled dimension on the functionality of floating gate NAND cells as described in the last section are one aspect of the shrinking issues. Another aspect is the generation of the extremely small structures in NAND Flash memory cells which currently arrived in the sub-20 nm range [23].

This development of the feature size or critical dimension (CD) is even more impressive, because the size of actual cell structures is one order of magnitude smaller than that of the 193 nm wavelength of the ArF laser which is used for illumination.

To understand the challenge to generate such small structures, Fig. 4.28 shows the CD development of the NAND Flash technology half pitch and the used lithography wavelength since 1996.

At the end of the 1990s, the NAND Flash CD in the cell array was close to the lithography wavelength. However, since the 193 nm was the last reduction of the wavelength used as a light source for lithography, the gap between the NAND Flash technology node and the lithographic wavelength has been increasing since then.

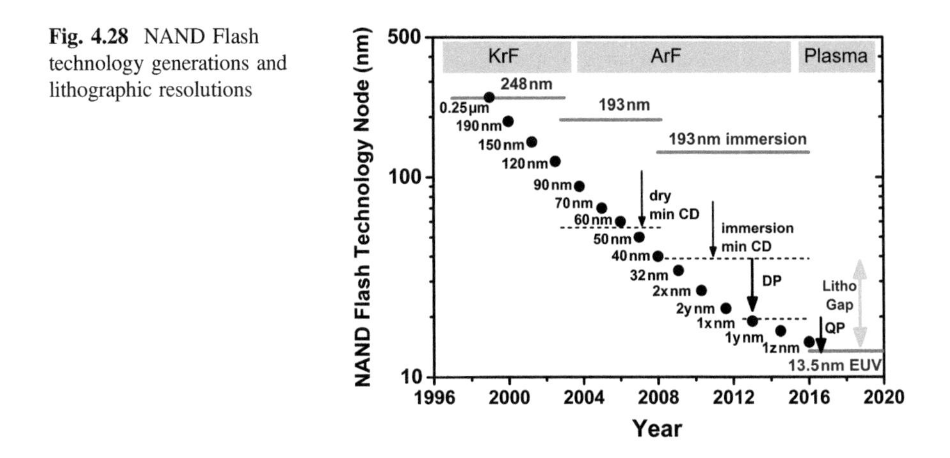

Fig. 4.28 NAND Flash technology generations and lithographic resolutions

The ability of a lithographic system to generate a minimum CD is described by

$$CD = k_1 \frac{\lambda}{NA} \tag{4.11}$$

where k_1 is a constant, λ is the wavelength, and NA is the numerical aperture of the optical illumination system. For a single exposure, dry 193 nm lithography with optimized illumination conditions with, e.g., $k_1 = 0.28$ in combination with a numerical aperture in the range of NA = 0.93, the minimum CD is limited to values slightly below 60 nm [36].

With the introduction of immersion lithography with a liquid on top of the wafer during illumination, the NA could be improved to 1.35, which is also the reason why the 193 nm immersion lithography wavelength is shown in Fig. 4.28 "virtually" reduced by this factor. The smallest achievable half pitch for single exposure 193 nm immersion lithography is therefore about 38 nm [37].

To bridge the gap to extreme UV (EUV) lithography (see litho gap in Fig. 4.28) , which was not available for industrial volume production in time, the semiconductor industry introduced (around 2009) special process sequences to generate small structures that cannot be obtained by single exposure direct printing.

For logic circuits, such as microprocessors, it is usually sufficient to generate the required small gate length by a trimming of larger lithographically generated structures. The required short gate length in logic circuits can therefore be obtained by tapered trim etch processes.

In memory products such as DRAM or NAND Flash it is not the small memory cell structure itself that is important, but the high memory cell density. Besides, the memory cell arrays have the great advantage that the basic structure consists of a very regular line and space pattern, which can be printed more easily than complex state-of-the-art SRAM structures.

Consequently, it is necessary to generate additional features that cannot be directly printed by lithography.

Most common for NAND Flash memory are process sequences which generate two smaller lines with a corresponding space out of one larger line that can be printed lithographically. These kinds of process sequences which basically make two lines out of one are known as self-aligned double patterning (SADP) [38, 39], or sometimes pitch fragmentation [28]. The typical SADP approach is schematically shown in Fig. 4.29.

The starting point is a multiple layer stack of CVD-deposited materials like a-Si, Si_3N_4, SiO_2, and carbon hard masks which can be selectively etched to each other. Double patterning starts with a directly printed equal line and space pattern which has two times the half pitch of the final structures (Fig. 4.29a). For a 20 nm target half pitch, the initial line and space half pitch consequently would be 40 nm. With the aid of the tapered trim etch process, this pattern is transferred to the underlying layer with a line width half of the initial line. Subsequently, a conformal liner is deposited (Fig. 4.29b) to generate a spacer with the width of the target half pitch as shown in (Fig. 4.29c). Proceeding from this processing stage, two different SADP

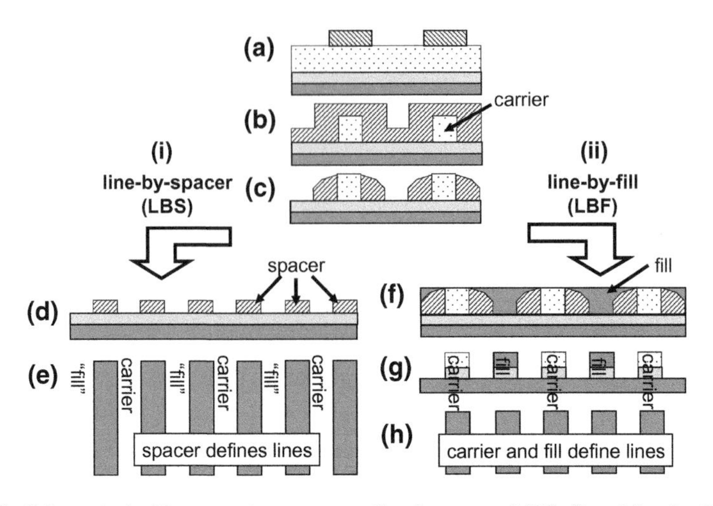

Fig. 4.29 Schematic double patterning sequences line-by-spacer (LBS) (i) and line-by-fill (LBF) (ii) [28]. The line width of an equal line and space pattern (**a**) is reduced by a trim etch process and a conformal liner is deposited (**b**) in order to generate spacer (**c**) of the same width as the trimmed lines. In the line-by-spacer sequence the spacers are used after line removal (**d**) to generate the final pattern (**e**), in contrast to the line-by-fill sequence where additional "fill" lines are generated in between the spacers (**f**) and the carrier and fill lines are used after spacer removal (**g**) to generate the target pattern (**h**)

final sequences can be principally chosen. Option (i) is the so-called line-by-spacer (LBS) sequence because it uses the generated spacer (Fig. 4.29d) to transfer the obtained pattern into the underlying hard mask. Prior to this, the carrier needs to be removed. The resulting hard mask structure is the equivalent of a single exposure lithographically generated pattern at larger half pitches, which is, in turn, used for patterning of the active chip structure as shown in Fig. 4.29e. Processing images of a LBS SADP sequence is shown in Fig. 4.30.

Figure 4.30a shows the situation after the trim etch step with a line width one quarter of the initial pitch. Figure 4.30b, c illustrates the process after the spacer etch and the carrier recess etch, where the trimmed initial line is removed. In Fig. 4.30d the spacer pattern is transferred into the hard mask and the spacer is removed in e. When the small SADP-generated structures in the memory array are generated, the close connection of every two neighboring lines needs to be etched away. This cut etch process can be carried out together with the patterning of periphery structures or, e.g., the select transistors as shown in Fig. 4.30f.

The second SADP processing option (ii) in Fig. 4.29 is the line-by-fill (LBF) sequence. Subsequent to the spacer formation in Fig. 4.29c, a material that can be as selectively etched to the spacer (e.g. the same material as the carrier) is filled in between the spacers. Therefore, the material is called "fill" as shown in Fig. 4.29f. Before the spacer material in between the carrier and fill lines can be removed as depicted in Fig. 4.29g, a chemical-mechanical planarization (CMP) process step is needed to have a better exposure of the spacer material to the etch chemistry. In the

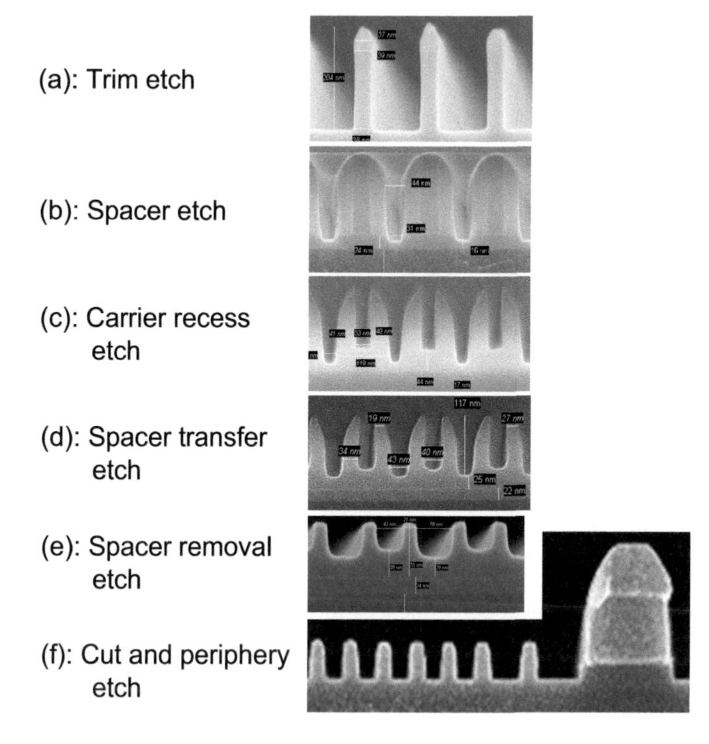

Fig. 4.30 Exemplary line-by-spacer process sequence [28, 39]

final step, the pattern can be transferred into the hard mask which is shown in Fig. 4.29h.

With respect to CD variations, it should be mentioned that generally the spacer width in SADP schemes can be better controlled than the carrier and fill width. The spacer width variations mostly depend on thickness conformity of the deposited spacer liner. In contrast, the carrier and fill line widths essentially depend on two critical processes, which are the carrier trim etch and the spacer formation.

The knowledge of this different CD control can be used to guarantee a reliable operation of FG NAND cells. It was described that the control gate plug is essential for the gate coupling ratio and consequently for the FG cell performance.

Based on this, it is beneficial to use the LBF sequence for the one-step patterning of the active area and floating gate width in a self-aligned STI (SA-STI) cell approach [40] as shown in Fig. 4.31a.

This choice has the major advantage that the space for the critical control gate plug has a good controllability [28]. For the patterning of the word line level which defines the length of the FG cells it could be beneficial to use the LBS sequence. The consequential spacer-defined good control of the FG cell length can help to reduce cell-to-cell V_{th} variations since the latest NAND cell generations are definitely in the short channel regime which increases cell length effects.

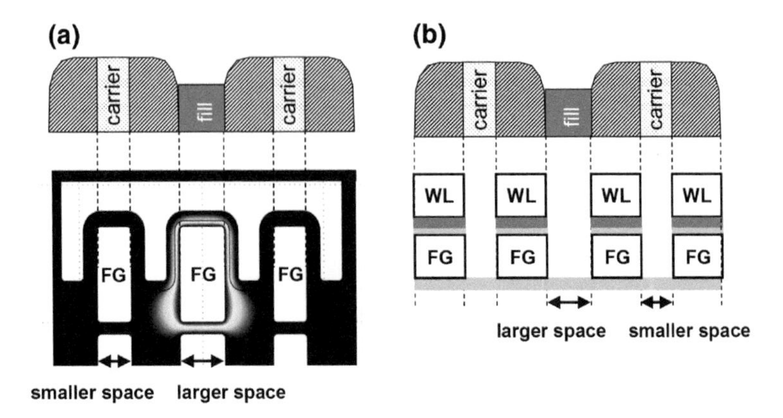

Fig. 4.31 Major variations in LBF (**a**) and LBS (**b**) pitch fragmentation sequences [28]

As shown in Fig. 4.28, it is required for FG NAND technologies beyond 20 nm half pitch to use quadruple patterning (QP) techniques [23, 37] to generate such small structures. Quadruple patterning is essentially two times the consecutive usage SADP with its logical consequences for the CD control of lines and spaces.

4.6 Planar NAND Memory Cells as Conventional Floating Gate Cell Replacement

In most cases charge trapping (CT) cells were discussed as a planar memory cell replacement of the conventional floating gate cell in 2D memory arrays.

However, the planar FG cell as shown in Fig. 4.20 is also an alternative when it is possible to overcome the program saturation issue.

The construction of CT memory cells for NAND application is at first glance not very different from the floating gate NAND cell construction. The major difference is that charge is stored in a non-conducting dielectric layer with high trap density instead of the conducting floating gate. This non-conducting charge storage layer has two major consequences:

(i) The surface of the dielectric charge storage layer is not an equipotential surface as the floating gate. The stored charge can be inhomogeneously distributed when the injection is locally enhanced.

(ii) In a planar cell structure, no capacitive voltage divider can be formed to concentrate the voltage drop and, therefore, the electric field to the tunnel oxide as in floating gate cells (with optimized gate coupling ratio α_g).

The typical layout of CT memory cells is shown in Fig. 4.32. The traditional SONOS (poly-Si/SiO_2/Si_3N_4/SiO_2/Si) cell, as shown in Fig. 4.32a, stores the charge in a Si_3N_4 (SiN) layer. SiN is widely used as the charge trapping layer (CTL) due to its high trap density of a few times 10^{19} cm^{-3} and its good process

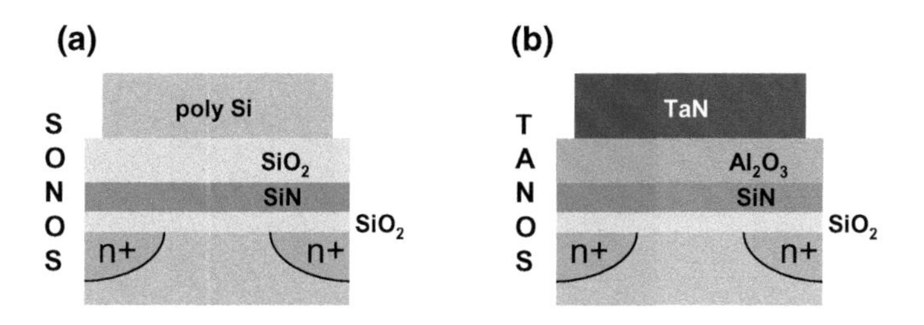

Fig. 4.32 Charge trapping stacks in SONOS (**a**) and TANOS [48] (**b**) memory cells

compatibility with Si and SiO_2. Sometimes other dielectrics are used for charge storage, such as Al_2O_3 [41].

CT memory cells typically have a planar cell layout and therefore resemble planar FG cells, layout-wise. Due to the lack of an increased gate coupling ratio it cannot be realized that charge is only transferred through the tunnel oxide during program and erase operation. Under the Fowler-Nordheim program condition in the CT cell the injected electron current tunnels through the whole CT stack. Only a certain part of this tunneling current is trapped in trap states and cause a V_{th} increase. The rest of the injected electron current leaves the charge trapping layer towards the gate electrode. Consequently, the ISPP slope for CT memory cells is not at unity, but rather in the range between 0.6 and 0.8 [42]. This tunneling current passing the whole memory cell stack resembles FG cells in the program saturation regime as described in Sect. 4.2.3.

However, the program operation is generally not the problem of CT cells, since usually high V_{th} levels (even suitable for MLC) can be reached.

One of the major issues of SONOS memory cells is the erase. It can be observed that the erasability of SONOS cells significantly deteriorates when the tunnel oxide thickness is increased above 2 nm [43]. In the TOX thickness range up to 2 nm the erase mechanism is based on direct tunneling of holes from the channel region to the SiN CTL. For thicker tunnel dielectric layers, the direct tunneling probability is significantly reduced and for an efficient erase operation the electric field strength needs to be increased up to the Fowler-Nordheim tunneling regime. The problem that occurs in SONOS cells with thick tunnel oxide under FN erase conditions is the so-called erase saturation which is illustrated in Fig. 4.33a. Under FN tunneling conditions for holes from the cell channel, the electric field in the top SiO_2 (blocking oxide: BLOX) layer is already high enough to inject electrons from the gate towards the storage SiN (back tunneling). These injected gate electrons compensate the positive charge of the injected holes and stop the V_{th} decrease (erase saturation). Other erase mechanisms which do not suffer from erase saturation, such as hot hole injection (HHI) [44], are limited to the NOR array structure where NROM-like cells [45] are commercially available, but cannot be implemented in the NAND array.

Erase saturation in planar CT cells can be improved when a gate material with high work function and/or a high-k blocking oxide is used, as shown in Fig. 4.33a.

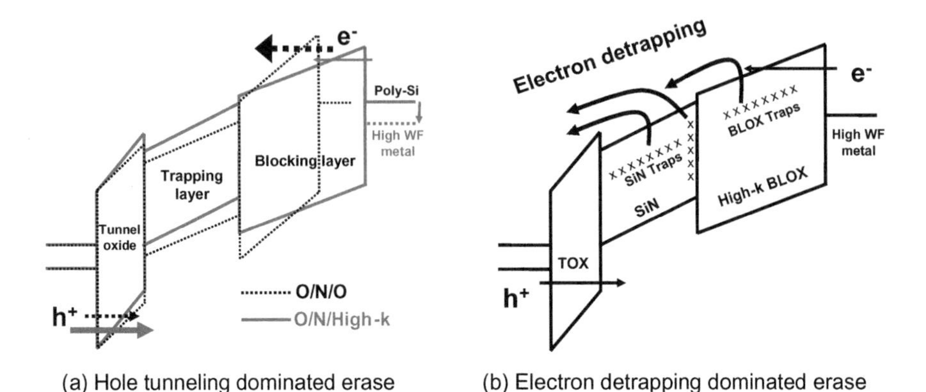

Fig. 4.33 TANOS erase due to reduced electron back tunneling [48]

A higher work function can be obtained by a p-doped poly-Si layer instead of the n-doped poly-Si gate [46], or by the use of a high work function metal gate [47]. The combination of both program saturation improvement approaches was the reason for the introduction of so-called TANOS (TaN/Al_2O_3/Si_3N_4/Si) CT memory cells [48]. In the ideal TANOS image, the erase mechanism is solely due to hole tunneling from the channel, the charge is only stored in the SiN CTL, and the Al_2O_3 blocking oxide is assumed to be trap free.

However, there are several indications that the ideal TANOS image is not fully true. Other investigations of the TANOS erase even describe that electron detrapping from SiN traps is the predominant effect [49], as illustrated in Fig. 4.33b.

It was additionally found that the Al_2O_3 BLOX of the TANOS stack is not trap-free and acts as a charge trapping layer as well [41, 42]. Consequently, detrapping from Al_2O_3 traps could be another contribution to the improved erase performance of TANOS memory cells.

The major reason why CT Flash memory cell containing NAND product chips are to date not commercially available is the observation of a general trade-off between erasability and retention of CT memory cells.

Assuming that detrapping is an important component for CT cell erase, this could be principally understood since energetically deep trap levels would be beneficial for a good retention, but hinder the erase, and vice versa.

Compared to FG NAND cells, the retention of TANOS memory cells is generally not sufficient for MLC application. This can be seen for TANOS cells in a 48 nm NAND Flash technology in Fig. 4.34. The TANOS cell (without sealing oxide) shows a good erase level for $V_{ers} = -23$ V with a long $t_{ers} = 300$ ms erase pulse, but the retention loss of nearly 550 mV after a 2 h retention bake at 200 °C is not suitable for MLC. This high retention loss is most likely due to a combination of electrons lost from the storage SiN due to hopping conduction over Al_2O_3 traps and a direct charge loss of electrons stored in Al_2O_3 BLOX traps. Figure 4.34 shows the retention improvements at the expense of erase performance when parts

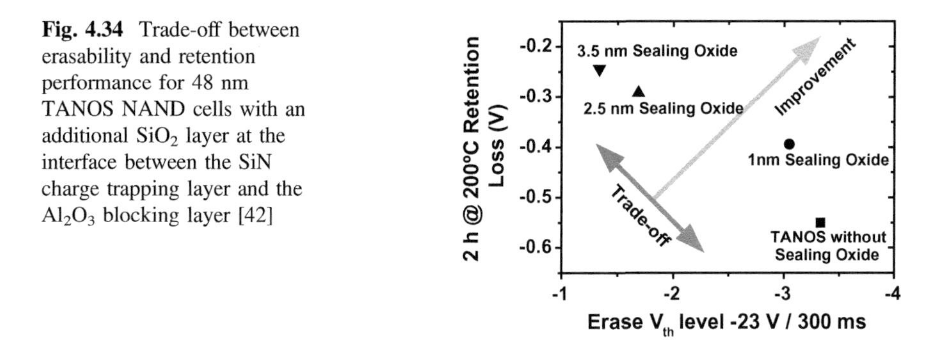

Fig. 4.34 Trade-off between erasability and retention performance for 48 nm TANOS NAND cells with an additional SiO_2 layer at the interface between the SiN charge trapping layer and the Al_2O_3 blocking layer [42]

of the Al_2O_3 BLOX adjacent to the SiN charge trap layer are replaced by an SiO_2 layer (sealing oxide) with identical electrical thickness (EOT). The reduction of the retention loss to 250 mV for the 3.5 nm sealing oxide results in CT TAONOS ($TaN/Al_2O_3/SiO_2/Si_3N_4/Si$) cells that can hardly be erased below $V_{th} = -1$ V (both values are critical for MLC).

A similar trade-off between erase performance and retention was obtained from large area CT memory cells in the μm range, where the SiN CT composition was varied with respect to the Si content [50] (see Fig. 4.35a), or with an additional high-k BLOX layer, introduced on top of the Al_2O_3 to reduce gate back tunneling during erase [51] (see Fig. 4.35b). In all cases shown in Fig. 4.35a, b, the standard TANOS cell behavior is among the best performing CT cells, or only the described trade-off between retention and erase performance is seen.

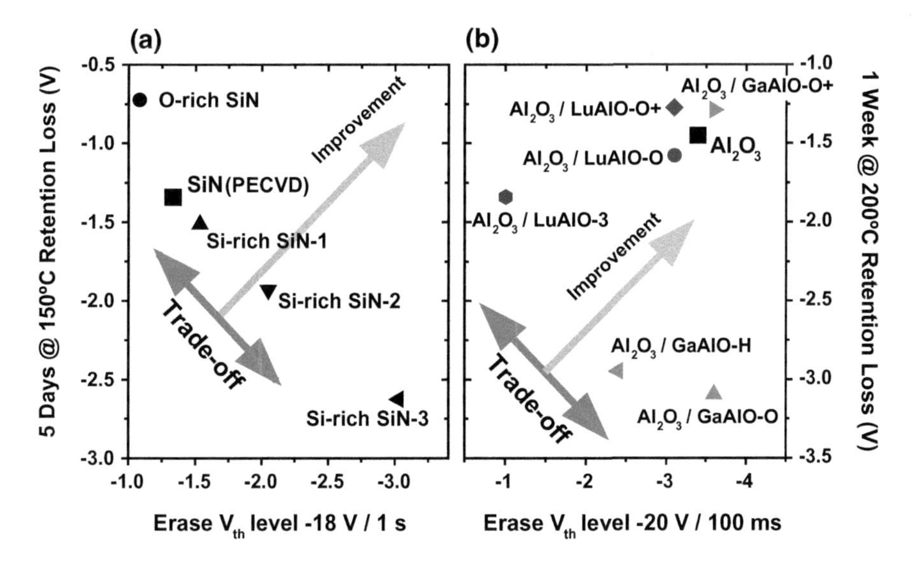

Fig. 4.35 TANOS trade-off between erasability and retention performance on large memory cells (μm range) with variation of the Si content in the SiN CTL (**a**) [50], and for different high-k layers on top of the Al_2O_3 blocking oxide (**b**) [51]

The endurance behavior of TANOS or similar CT cells is also generally worse than that of floating gate cells. This might be correlated to the inevitable tunnel currents through the hole CT stack as mentioned before.

Besides, the charge storage in a non-conducting layer can lead to inhomogeneously distributed charges which adversely affect the erase performance of CT cells [52, 53] and can also be responsible for the worse retention performance of small ground rule CT cells compared to large CT cells [42, 54].

All described reliability issues (erase performance, retention, and endurance) of CT memory cells are responsible for the fact that TANOS cells not been able to replace floating gate cells in planar 2D NAND Flash applications.

The only planar memory cell which has appeared on the market in a planar 2D memory array so far is the planar FG cell technology [55] as shown in Fig. 4.36. However, the TEM analysis of the cell structure does not show a dual layer floating gate. Instead the planar FG cell has a thin poly-Si FG layer with a quite thick inter gate dielectric (IGD) stack including some high-k dielectric layers and on top a high work function metal gate.

Since the IGD includes a SiO2 layer of similar thickness as the TOX plus additional layers, the gate coupling ratio must be significantly below 0.5, which makes this cell quite difficult to operate as a traditional FG cell.

Nevertheless, the published program and erase characteristics are very ideal with program and erase slopes ~1 and a large P/E window [55]. Such a characteristic would not be the case either for a traditional planar FG cell or a TANOS like charge trapping (CT) cell.

Most likely the working principle of the planar FG Micron cell is a combination of a FG and a CT cell. Besides the conducting poly-Si FG, the IGD stack is the one of a traditional CT cell. The advantage of the conducting FG introduction could be the fact that this layer provides a conduction band where the electrons can tunnel to

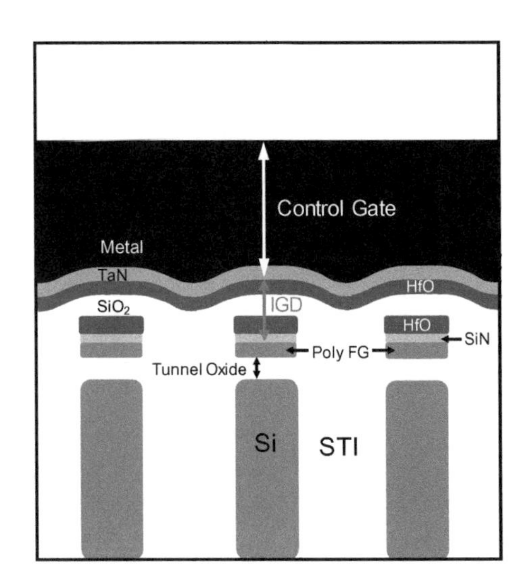

Fig. 4.36 Details of the of a 20 nm planar FG NAND cell technology with multi layer inter gate dielectric (IGD) [55]. This IGD includes a SiN layer directly on top of the Poly-Si FG layer which acts as an additional charge trapping layer together with the FG

under program conditions. This avoids the fly-through effect (which is visible in a reduced ISPP slope [56]), because the electrons don't need to be captured in discrete trap states and thermalize into the deep energy states of the traps. Under erase conditions, the FG layer provides a large number of free electrons that can tunnel towards the cell channel and therefore avoids the erase saturation.

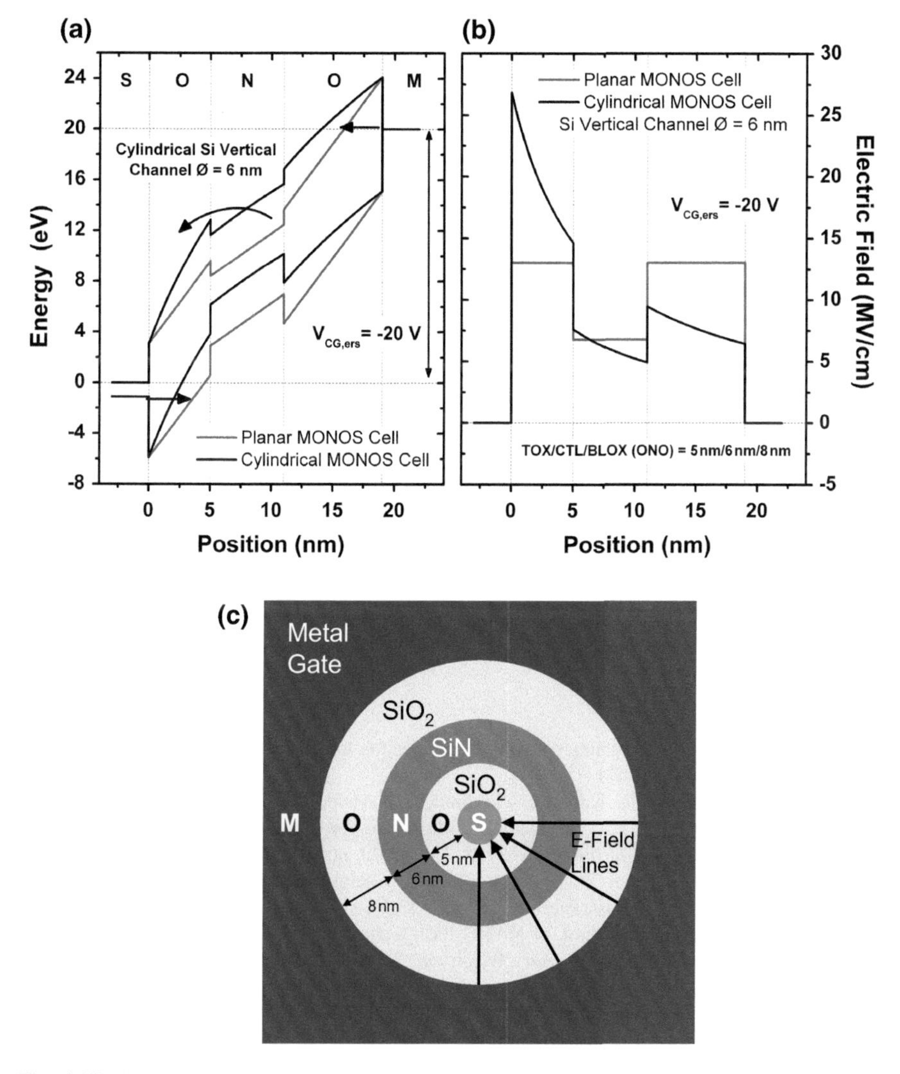

Fig. 4.37 Comparison of uncharged (no electrons stored in the SiN CTL) planar MONOS cells and cylindrical MONOS cells with an inner Si channel diameter of 6 nm. The band diagram (**a**) and the electric field conditions (**b**) show strongly increased fields in the tunnel oxide and significantly reduced fields in the BLOX of the cylindrical SONOS cell. The SONOS CT stack dimension (ONO) used in the simulations was TOX/CTL/BLOX = 5 nm/6 nm/8 nm (**c**)

When the program and erase operations are finished, the stored charges are most likely redistributed between the FG and the traps in the SiON charge trapping layer. As a result, it is not entirely correct to call this planar FG cell a floating gate cell. It is rather most likely a charge trapping cell with an additional conducting layer charge trapping layer (FG poly-Si). Therefore it could be a “hybrid FG-CT cell”.

However, common to both, the conventional CT and the “hybrid FG-CT cell” as presented in the working 2D FG cell [55], is the fact that they work better in a cylindrical cell geometry. The cylindrical shape of the memory cells in 3D cell approaches have one major advantage over fully planar memory cells, namely the electric field enhancement in the TOX and the field reduction in the BLOX or IGD [57]. The band diagram and the electric fields under erase conditions ($V_{CG,ers}$ = −20 V) for a planar MONOS cell vs. a cylindrical MONOS with a 6 nm inner-channel diameter are shown in Fig. 4.37a, b. The ONO stack dimensions used in the field calculations were t_{TOX} = 5 nm, t_{SiN} = 6 nm, and t_{BLOX} = 8 nm. It is clearly visible that the cylindrical cell geometry with an inner cell channel position strongly increases the TOX field in relation to the BLOX field. Therefore, the cylindrical geometry effectively acts as an increased gate coupling ratio of a floating gate cell. The TOX electric field enhancement can also be seen in the form of denser E-field lines in Fig. 4.37c.

It will be seen in the next chapter that the advantage of the cylindrical cell geometry is used in most of the 3D NAND Flash memory arrays.

Acknowledgement The author would like to acknowledge the whole Flash development team of the former Qimonda Company. Special thanks are addressed to Torsten Müller, Nigel Chan, and Stefano Parascandola for discussions, provision of a 3D FG cell field simulator script, RTN measurements, program saturation evaluations, and pitch fragmentation process images.

References

1. D. Kahng, S.M. Sze, A floating gate and its application to memory devices. Bell Syst. Techn. J. **46**(6), 1288–1295 (1967)
2. N. Chan, M.F. Beug, R. Knoefler, T. Mueller, T. Meldc, M. Ackermann, S. Riedel, M. Specht, C. Ludwig, A.T. Tilke, Metal control gate for sub-30 nm floating gate NAND memory, in *Proceeding of the 9th NVMTS*, Nov 2008, pp. 82–85
3. A. Kolodny, S.T.K. Nieh, B. Eitan, J. Shappir, Analysis and modelling of floating gate EEPROM cells. IEEE Trans. Electron Devices **33**(6), 835–844 (1986)
4. K. Kim, J. Choi, Future outlook of NAND flash technology for 40 nm node and beyond, in *Non-Volatile Semiconductor Memory Workshop, 2006. 21st IEEE NVSMW, 2006*, pp. 9–11
5. M. Wong, D.K.-Y. Liu, S.S.-W. Huang, Analysis of the subthreshold slope and the linear transconductance techniques for the extraction of the capacitance coupling coefficients of floating gate devices. IEEE Electron Device Lett. **13**(11), 566–568 (1992)
6. M.F. Beug, Q. Rafhay, M.J. van Duuren, R. Duane, Investigation of back-bias capacitance coupling coefficient measurement methodology for floating gate non-volatile memory cells. IEEE Trans. Electron Devices **57**(6), 1253–1260 (2010)
7. R.H. Fowler, L. Nordheim, Electron emission in intense electric films. Proc. R. Soc. Lond. **119**, 173–181 (1928)

8. K.-D. Suh, B.-H. Suh, Y.-H. Um, J.-K. Kim, Y.-J. Choi, Y.-N. Koh, S.-S. Lee, S.-C. Kwon, B.-S. Choi, J.-S. Yum, J.-H. Choi, J.-R. Kim, H.-K. Lim, A 3.3 V 32 Mb NAND flash memory with incremental step pulse programming scheme, in *IEEE International Solid-State Circuits Conference*, Feb 1995, pp. 128–129
9. C. Friederich, J. Hayek, A. Kux, T. Muller, N. Chan, G. Kobernik, M. Specht, D. Richter, D. Schmitt-Landsiedel, Novel model for cell—system interaction (MCSI) in NAND flash, in *IEEE International Electron Devices Meeting (IEDM)*, Dec 2008
10. K. Prall, K. Parat, 25 nm 64 Gb MLC NAND technology and scaling challenges, in *IEEE International Electron Devices Meeting (IEDM)*, Dec 2010, pp. 102–105
11. C.-H. Lee, S.-K. Sung, D. Jang, S. Lee, S. Choi, J. Kim, S. Park, M. Song, H.-C. Baek, E. Ahn, J. Shin, K. Shin, K. Min, S.-S. Cho, C.-J. Kang, J. Choi, K. Kim, J.-H. Choi, K.-D. Suh, T.-S. Jung, A highly manufacturable integration technology for 27 nm 2 and 3bit/cell NAND flash memory, in *IEEE International Electron Devices Meeting (IEDM)*, Dec 2010, pp. 98–101
12. D.J. DiMaria, E. Cartier, Mechanism for stress-induced leakage current in thin silicon dioxide films. J. Appl. Phys. **78**(6), 3883–3894 (1995)
13. M.F. Beug, N. Chan, T. Hoehr, L. Mueller-Meskamp, M. Specht, Investigation of program saturation in scaled interpoly dielectric floating gate memory devices. IEEE Trans. Electron Devices **56**(8), 1698–1704 (2009)
14. D. Wellekens, J. De Vos, J. Van Houdt, K. van der Zanden, Optimization of Al2O2 interpoly dielectric for embedded flash memory applications, in *Proceedings of Joint NVSMW/ICMTD*, May 2008, pp. 12–15
15. T.-S. Jung, Y.-J. Choi, K.-D. Suh, B.-H. Suh, J.-K. Kim, Y.-H. Lim, Y.-N. Koh, J.-W. Park, K.-J. Lee, J.-H. Park, K.-T. Park, J.-R. Kim, J.-H. Yi, H.-K. Lim, A 117-mm2 3.3-V only 128-Mb multilevel NAND flash memory for mass storage applications. IEEE J. Solid-State Circuits **31**(11), 1575–1583 (1996)
16. T. Cho, Y.-T. Lee, E.-C. Kim, J.-W. Lee, S. Choi, S. Lee, D.-H. Kim, W.-G. Han, Y.-H. Lim, J.-D. Lee, J.-D. Choi, K.-D. Suh, A dual-mode NAND flash memory: 1-Gb multilevel and high-performance 512-Mb single-level modes. IEEE J. Solid-State Circuits **36**(11), 1700–1706 (2001)
17. R. Micheloni, L. Crippa, A. Marelli, *Inside NAND Flash Memories*. Springer (2010)
18. R. Cernea, D.J. Lee, M. Mofidi, E.Y. Chang, Wy-Yi Chien, L. Goh, Y. Fong, J.H. Yuan, G Samachisa, D.C. Guterman, S. Mehrotra, K. Sato, H. Onishi, K. Ueda, F. Noro, K. Mijamoto, M. Morita, K. Umeda, K. Kubo, A 34 Mb 3.3 V serial flash EEPROM for solid-state disk applications, in *IEEE International Solid-State Circuits Conference (ISSCC)*, San Francisco, Feb 1995, pp. 126–127
19. A. Chimenton, P. Olivo, Fast identification of critical electrical disturbs in nonvolatile memories. IEEE Trans. Electron Devices **54**(9), 2438–2444 (2007)
20. J.H. Stathis, Reliability limits for the gate insulator in CMOS technology. IBM J. Res. Dev. **46** (2/3), 265–286 (2002)
21. P. Olivo, T.N. Nguyen, B. Ricco, High-field-induced degradation in ultrathin SiO2 films. IEEE Trans. Electron Devices **35**(12), 2259–2267 (1988)
22. S. Lai, Electrical properties of nitrided-oxide systems for use in gate dielectrics and EEPROM, in *Proceeding of the International Non-Volatile Memory Technology Conference,* 1998, pp. 6–7
23. J. Hwang, J. Seo, Y. Lee, S. Park, J. Leem, J. Kim, T. Hong, S. Jeong, K. Lee, H. Heo, H. Lee, P. Jang, K. Park, M. Lee, S. Baik, J. Kim, H. Kkang, M. Jang, J. Lee, G. Cho, J. Lee, B. Lee, H. Jang, S. Park, J. Kim, S. Lee, S. Aritome, S. Hong, Sungwook Park, A middle-1X nm NAND flash memory cell (M1X-NAND) with highly manufacturable integration technologies, in *IEEE International Electron Devices Meeting (IEDM)*, Dec 2011, pp. 199–202
24. U. Ganguly, Y. Yokota, T. Jing, S. Shiyu, M. Rogers, J. Miao, K. Thadani, H. Hamana, L. Garlen, B. Chandrasekaran, S. Thirupapuliyur, C. Olsen, V. Nguyen, S. Srinivasan, Scalability enhancement of FG NAND by FG shape modification, in *IEEE International Memory Workshop (IMW)*, May 2010

25. D. Wellekens, J. De Vos, J. Van Houdt, K. van der Zanden, Optimization of Al2O3 interpoly dielectric for embedded flash memory applications, in *Proceedings of the Joint NVSMW/ ICMTD*, May 2008, pp. 12–15
26. P. Blomme, M. Rosmeulen, A. Cacciato, M. Kostermans, C. Vrancken, S. Van Aerde, T. Schram, I. Debusschere, M. Jurczak, J. Houdt, Novel dual layer floating gate structure as enabler of fully planar flash memory, in *Symposium on VLSI Technology (VLSIT)*, June 2010, pp. 129–130
27. J.-D. Lee, S.-H. Hur, J.-D. Choi, Effects of floating-gate interference on NAND flash memory cell operation. IEEE Electron Device Lett. **23**(5), 264–266 (2002)
28. M.F. Beug, S. Parascandola, T. Hoehr, T. Muller, R. Reichelt, L. Muller-Meskamp, P. Geiser, T. Geppert, L. Bach, U. Bewersdorff-Sarlette, O. Kenny, S. Brandl, T. Marschner, S. Meyer, S. Riedel, M. Specht, D. Manger, R. Knofler, K. Knobloch, P. Kratzert, C. Ludwig, K.-H. Kusters, Pitch fragmentation induced odd/even effects in a 36 nm floating gate NAND technology, in *Proceedings of the NVMTS*, Nov 2008, pp. 77–81
29. N. Shibata, H. Maejima, K. Isobe, K. Iwasa, M. Nakagawa, M. Fujiu, T. Shimizu, M. Honma, S. Hoshi, T. Kawaai, K. Kanebako, S Yoshikawa, H. Tabata, A. Inoue, T. Takahashi, T. Shano, Y. Komatsu, K. Nagaba, M. Kosakai, N. Motohashi, K. Kanazawa, K. Imamiya, H. Nakai, A 70 nm 16 Gb 16-level-cell NAND flash memory, in *IEEE Symposium on VLSI Circuits*, 14–16 June 2007, pp. 190–191
30. R. Cernea, L. Pham, F. Moogat, S. Chan, B. Le, Y. Li, S. Tsao, T.-Y. Tseng, K. Nguyen, J. Li, J. Hu, J. Park, C. Hsu, F. Zhang, T. Kamei, H. Nasu, P. Kliza, K. Htoo, J. Lutze, Y. Dong, M. Higashitani, J. Yang, H.-S. Lin, V. Sakhamuri, A. Li, F. Pan, S. Yadala, S. Taigor, K. Pradhan, J. Lan, J. Chan, T. Abe, Y. Fukuda, H. Mukai, K. Kawakamr, C. Liang, T. Ip, S.-F. Chang, J. Lakshmipathi, S. Huynh, D. Pantelakis, M. Mofidi, K. Quader, A 34 MB/ s-program-throughput 16 Gb MLC NAND with all-bitline architecture in 56 nm, in *IEEE International Solid-State Circuits Conference (ISSCC)*, Feb 2008, pp. 420–624
31. Y.S. Kim, D.J. Lee, C.K. Lee, H.K. Choi, S.S. Kim, J.H. Song, D.H. Song, J.-H. Choi, K.-D. Suh, C. Chung, New scaling limitation of the floating gate cell in NAND flash memory, in *IEEE International Reliability Physics Symposium (IRPS)*, May 2010, pp. 599–603
32. H.H. Mueller, D. Wörle, M. Schulz, Evaluation of the coulomb energy for single-electron interface trapping in sub-μm metal-oxide-semiconductor field effect transistors. J. Appl. Phys. **75**(6), 2970–2979 (1994)
33. H. Miki, T. Osabe, N. Tega, A. Kotabe, H. Kurata, K. Tokami, Y. Ikeda, S. Kamohara, R. Yamada, Quantitative analysis of random telegraph signals as fluctuations of threshold voltages in scaled flash memory cells, in *IEEE International Reliability Physics Symposium (IRPS),* 2007, pp. 29–35
34. K. Seidel, R. Hoffmann, D.A. Löhr, T. Melde, M. Czernohorsky, J. Paul, M.F. Beug, V. Beyer, Comparison and analysis of trap mechanisms responsible for random telegraph noise and erratic programming on sub-50 nm floating gate flash memories, in *Non-Volatile Memory Technology Symposium (NVMTS)*, Oct 2009, pp. 67–71
35. M.F. Beug, R. Ferretti, K.R. Hofmann, Analysis and modeling of the transient local tunneling in gate oxides. IEEE Trans. Device Mater. Reliab. **4**(1), 73–79 (2004)
36. M.C. Chiu, B. Szu-M. Lin, M.F. Tsai, Y.S. Chang, M.H. Yeh, T.H. Ying, C. Ngai, J. Jin, S. Yuen, S. Huang, Y. Chen, L. Miao, K. Tai, A. Conley, I. Liu, Challenges of 29 nm half-pitch NAND flash STI patterning with 193 nm dry lithography and self-aligned double patterning, In *Proceedings of the SPIE 7140, 714021*, 2008, https://doi.org/10.1117/12.804685
37. P. Xu, Y. Chen, Y. Chen, L. Miao, S. Sun, S.-W. Kim, A. Berger, D. Mao, C. Bencher, R. Hung, C. Ngai, Sidewall spacer quadruple patterning for 15 nm half-pitch. Proc. SPIE **7973**, 79731Q (2011). https://doi.org/10.1117/12.881547
38. C. Bencher, Y. Chen, H. Dai, W. Montgomery, L. Huli, 22 nm half-pitch patterning by CVD spacer self alignment double patterning (SADP). Proc. SPIE **6924**, 69244E (2008). https://doi.org/10.1117/12.772953
39. C. Ludwig, S. Meyer, *Double patterning for memory ICs,* in *Recent Advances in Nanofabrication Techniques and Applications*, ed. by Bo Cui (InTech, 2011), pp. 417–432.

ISBN: 978–953-307-602-7, http://www.intechopen.com/articles/show/title/double-patterning-for-memory-ics

40. S. Aritome, S. Satoh, T. Maruyama, H. Watanabe, S. Shuto, G. J. Hemink, R. Shirota, S. Watanabe, F. Masuoka, A 0.67 μm2 self-aligned shallow trench isolation cell (SA-STI cell) for 3 V-only 256 Mbit NAND EEPROMs, in *IEEE International Electron Devices Meeting (IEDM)*, Dec 1994, pp. 61–64
41. M. Specht, H. Reisinger, F. Hofmann, T. Schulz, E. Landgraf, R.J. Luyken, W. Rösner, M. Grieb, L. Risch, Charge trapping memory structures with Al2O3 trapping dielectric for high-temperature applications. Solid-State Electron. **49**(5), 716–720 (2005)
42. M.F. Beug, T. Melde, M. Czernohorsky, R. Hoffmann, J. Paul, R. Knoefler, A.T. Tilke, Analysis of TANOS memory cells with sealing oxide containing blocking dielectric. IEEE Trans. Electron Devices **57**(7), 1590–1596 (2010)
43. R. van Schaijk, M. van Duuren, W.Y. Mei, K. van der Jeugd, A. Rothschild, M. Demand, Oxide–nitride–oxide layer optimisation for reliable embedded SONOS memories. Microelectron. Eng. **72**(1–4), 395–398 (2004)
44. T.Y. Chan, K.K. Young, C. Hu, A true single-transistor oxide-nitride-oxide EEPROM device. IEEE Electron Device Lett. **8**(3), 93–95 (1987)
45. B. Eitan, P. Pavan, I. Bloom, E. Aloni, A. Frommer, D. Finzi, NROM: A novel localized trapping, 2-bit nonvolatile memory cell. IEEE Electron Device Lett. **21**(11), 543–545 (2000)
46. H. Bachhofer, H. Reisinger, E. Bertagnolli, Transient conduction in multidielectric silicon–oxide–nitride–oxide semiconductor structures. J. Appl. Phys. **89**(5), 2791–2800 (2001)
47. A. Goda, M. Noguchi, Improvement of erase saturation for a highly reliable monos memory cell, in *IEEE Non-Volatile Semiconductor Memory Workshop (NVSMW)*, Feb 2003, pp. 65–68
48. C.H. Lee, K.I. Choi, M.K. Cho, Y.H. Song, K.C. Park, K. Kim, A novel SONOS structure of SiO2-SiN-Al2O3 with TaN metal gate for multi-giga bit flash memories, in *IEEE International Electron Devices Meeting (IEDM)*, Dec 2003, pp. 613–616
49. S.-C. Lai, H.-T. Lue, J.-Y. Hsieh, M.-J. Yang, Y.-K. Chiou, C.-W. Wu, T.-B. Wu, G.-L. Luo, C.-H. Chien, E.-K. Lai, K.-Y. Hsieh, R. Liu, C.-Y. Lu, Study of the erase mechanism of MANOS (metal/Al2O3/SiN/SiO2/Si) device. IEEE Electron Device Lett. **28**(7), 643–645 (2007)
50. G. Van den bosch, A. Furnemont, M.B. Zahid, R. Degraeve, R. Breuil, L. Cacciato, A. Rothschild, C. Olsen, U. Ganguly, J. Van Houdt, Nitride engineering for improved erase performance and retention of TANOS NAND flash memory, in *Non-Volatile Semiconductor Memory Workshop, 2008 and 2008 International Conference on Memory Technology and Design. NVSMW/ICMTD 2008. Joint*, 18–22 May 2008, pp. 128–129
51. L. Breuil, C. Adelmann, G. Van Den Bosch, A. Cacciato, M.B. Zahid, M. Toledano-Luque, A. Suhane, A. Arreghini, R. Degraeve, S. Van Elshocht, I. Debusschere, J. Kittl, M. Jurczak, J. Van Houdt, Optimization of the crystallization phase of rare-earth aluminates for blocking dielectric application in TANOS type flash memories, in *2010 Proceedings of the European Solid-State Device Research Conference (ESSDERC)*, 14–16 Sept 2010, pp. 440–443
52. M.F. Beug, T. Melde, M. Isler, L. Bach, M. Ackermann, S. Riedel, K. Knobloch, C. Ludwig, Anomalous erase behavior in charge trapping memory cells, in *Proceedings of the Joint Non-Volatile Semiconductor Memory Workshop/ International Conference on Memory Technology and Design (NVSMW/ICMTD)*, May 2008, pp. 121–123
53. Y.-J. Chen, L.H. Chong, S.-W. Lin, T.-H. Yeh, K.-F. Chen, J.-S. Huang, C.-H. Cheng, S.-H. Ku, N.-K. Zous, I-J. Huang, T.-T. Han, T.-H. Hsu, H.-T. Lue, M.-S. Chen, W.-P. Lu, K.-C. Chen, C.-Y. Lu, Source/Drain dopant concentration induced reliability issues in charge trapping NAND flash cells, in *IEEE International Reliability Physics Symposium (IRPS)*, May 2010, pp. 634–638
54. M.F. Beug, T. Melde, J. Paul, R. Knoefler, TaN and Al2O3 side wall gate-etch damage influence on program, erase, and retention of sub-50 nm TANOS NAND flash memory cells. IEEE Trans. Electron Devices **58**(6), 1728–1734 (2011)

55. N. Ramaswamy, T. Graettinger, G. Puzzilli, H. Liu, K. Prall, Engineering a planar NAND cell scalable to 20 nm and beyond, in *International Memory Workshop*, 26–29 May 2013, pp. 5–8
56. A. Furnemont, M. Rosmeulen, A. Cacciato, L. Breuil, K. De Meyer, H. Maes, J. Van Houdt, A consistent model for the SANOS programming operation, in *Proceedings 22nd IEEE Non-Volatile Semiconductor Memory Workshop*, Aug 2007, pp. 96–97
57. E. Nowak, A. Hubert, L. Perniola, T. Ernst, G. Ghibaudo, G. Reimbold, B. De Salvo, F. Boulanger, In-depth analysis of 3D Silicon nanowire SONOS memory characteristics by TCAD simulations, in *IEEE International Memory Workshop*, May 2010

Chapter 5
3D NAND Flash Memories

Rino Micheloni, Seiichi Aritome and Luca Crippa

Nowadays, Solid State Drives consume an enormous amount of NAND Flash memories [1] causing a restless pressure on increasing the number of stored bits per mm^2. Planar memory cells have been scaled for decades by improving process technology, circuit design, programming algorithms [2], and lithography.

Unfortunately, when approaching a minimum feature size of 1x-nm, more challenges pop up: doping concentration in the channel region becomes difficult to control [3], RTN [4] and electron injection statistics [5] widen threshold distributions, thus causing a significant hit to both endurance and retention. Furthermore, by reducing the distance between memory cells, the intra-wordline electric field becomes higher, pushing the bit error rate to an even higher level.

3D arrays can definitely be considered as a breakthrough for fueling a further increase of the bit density. Identifying the right way for going 3D is not so easy though.

Historically, Flash memory manufacturers have leveraged lithography to shrink the 2-dimensional (2D) memory cell [6].

This chapter is a partial reprint of R. Micheloni, S. Aritome, L. Crippa, "Array architectures for 3D NAND Flash Memories" in Proceedings of the IEEE, vol. 105, no. 9, pp. 1634–1649, Sept. 2017. © 2017 IEEE.

R. Micheloni (✉) · L. Crippa
Storage Solutions, Microsemi Corporation, Vimercate, MB, Italy
e-mail: rino.micheloni@ieee.org

L. Crippa
e-mail: luca.crippa@ieee.org

S. Aritome
IPCC, Industrial Property Cooperation Center, Tokyo, Japan
e-mail: aritomes@ieee.org

R. Micheloni et al. (eds.), *Inside Solid State Drives (SSDs)*,
Springer Series in Advanced Microelectronics 37,
https://doi.org/10.1007/978-981-13-0599-3_5

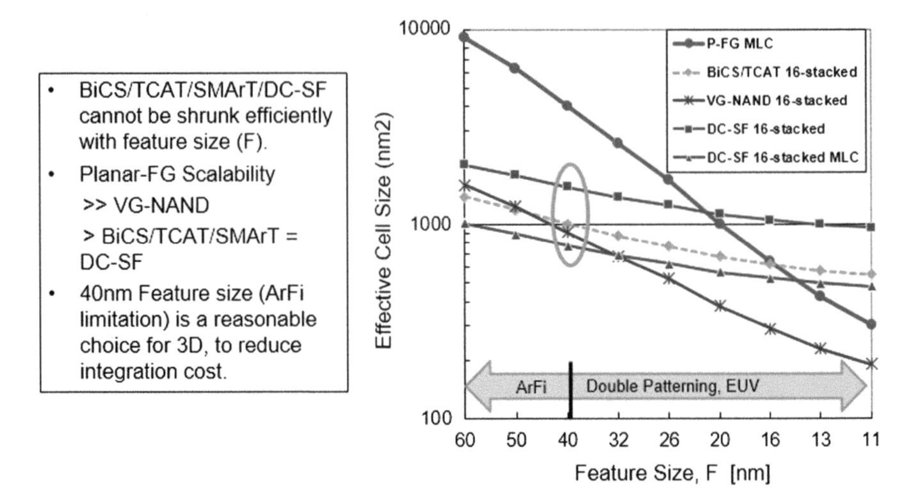

Fig. 5.1 3D NAND Flash scaling [7]

However, with 3D architectures, the "simple" reduction of the minimum feature size is running out of steam, as shown in Fig. 5.1 [7]: a higher number of stacked cells is the only hope for dramatically reducing the real estate of a stored bit.

3D arrays can leverage either *Floating Gate* (FG) or *Charge Trapping* (CT) technologies [8]. As a matter of fact, the vast majority of 3D architectures published to date are built with CT cells, mainly because of the simpler fabrication process. Nevertheless, Floating Gate is still around and there are commercial products who managed to integrate FG into a 3D array.

5.1 3D Charge Trap NAND Flash Memories

3D arrays can be efficiently built by vertically rotating the planar NAND Flash string of Fig. 5.2a, as displayed in Fig. 5.2b. The solution of choice is a conduction channel completely surrounded by the gate (Fig. 5.2c, d) [9]: indeed, the curvature effect helps increasing the electric field E_t across the tunnel oxide, and reduces the electric field E_b across the blocking oxide [10, 11], and this has a positive impact on oxide reliability and overall power consumption.

Vertical channel arrays have been historically driven by architectures known as BiCS, which stands for *Bit Cost Scalable* [12, 13] and P-BiCS, acronym for *Pipe-Shaped BiCS* [14–16], which are both leveraging CT cells. Let's get started with BiCS, which is sketched in Figs. 5.3 and 5.4 [17]. There is a stack of *Control Gates* (CGs), the lowest being the one of the *Source Line Selector* (SLS). The whole vertical stack is punched through and the resulting holes are filled with poly-silicon; each filled hole (a.k.a. pillar) forms a series of memory cells vertically connected in

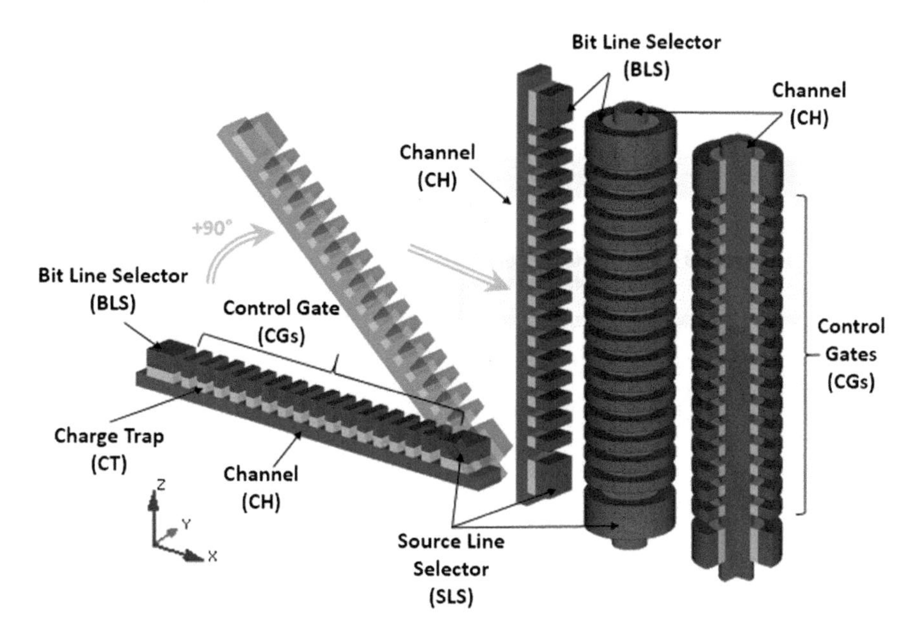

Fig. 5.2 The NAND Flash string goes vertical

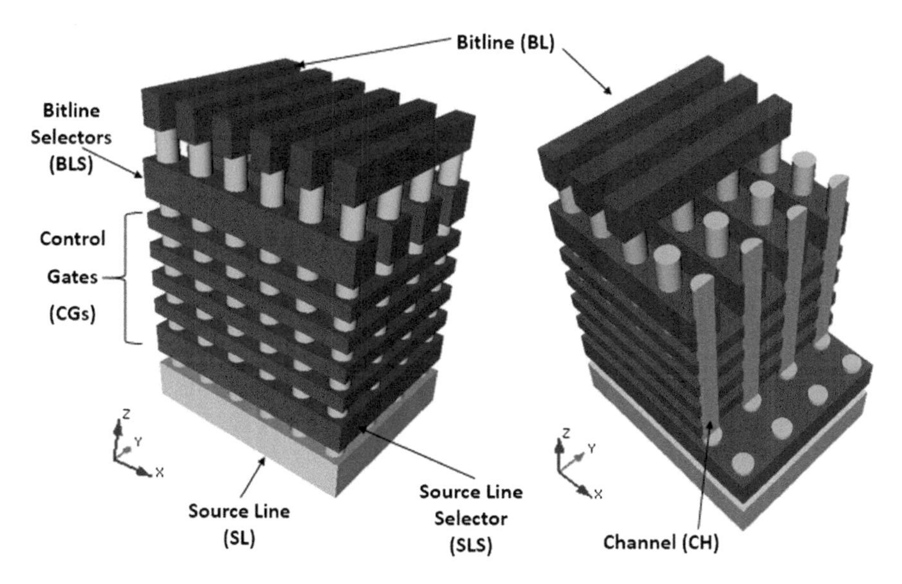

Fig. 5.3 BiCS architecture

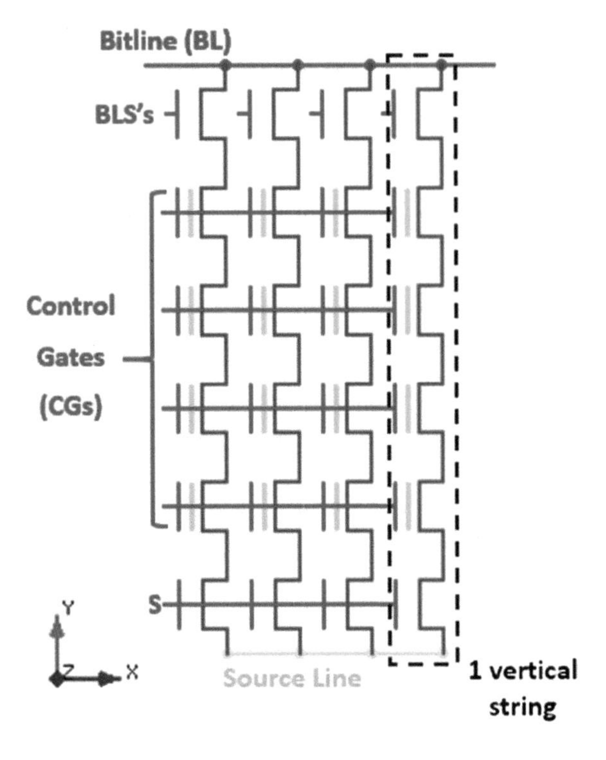

Fig. 5.4 Equivalent circuit of a BiCS array

a NAND fashion. *Bit Line Selectors* (BLS's) and *Bitlines* (BLs) are formed at the top of the structure [18].

The poly-silicon body of memory cells is not doped or lightly doped [10, 11]; indeed, considering the bad aspect ratio of the vertical polysilicon plug, p-n junctions cannot be easily realized by either diffusion or implantation in a trench structure. As usual, a select transistor (BLS) is used to connect each NAND string to a bitline; there is also another select transistor (SLS), which connects the other side of the string to the common source diffusion.

It is important to highlight that the number of critical and expensive lithography steps does not depend on the number of control gate plates because the whole 3D stack is drilled at one [19, 20].

As sketched in Fig. 5.5, vertical transistor have polysilicon body and this fact turned out to be one of the critical cornerstone of the 3D foundation. From a manufacturing perspective, the density of the traps at the grain boundary is very difficult to control, with such a vertical shape: the bad thing is that this poor control induces significant fluctuations of the characteristics of vertical transistors.

The recipe for fixing the trap density fluctuation problem is to manufacture a polysilicon body much thinner than the depletion width. In other words, by shrinking the polysilicon volume, the total number of traps goes down (Fig. 5.6). This particular structure is usually referred to as *Macaroni Body* [13]. A *filler layer*

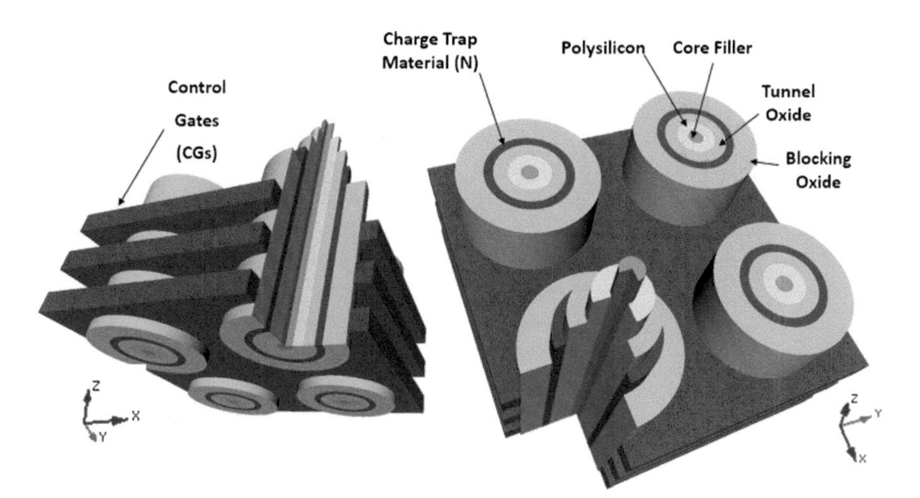

Fig. 5.5 BiCS memory cells

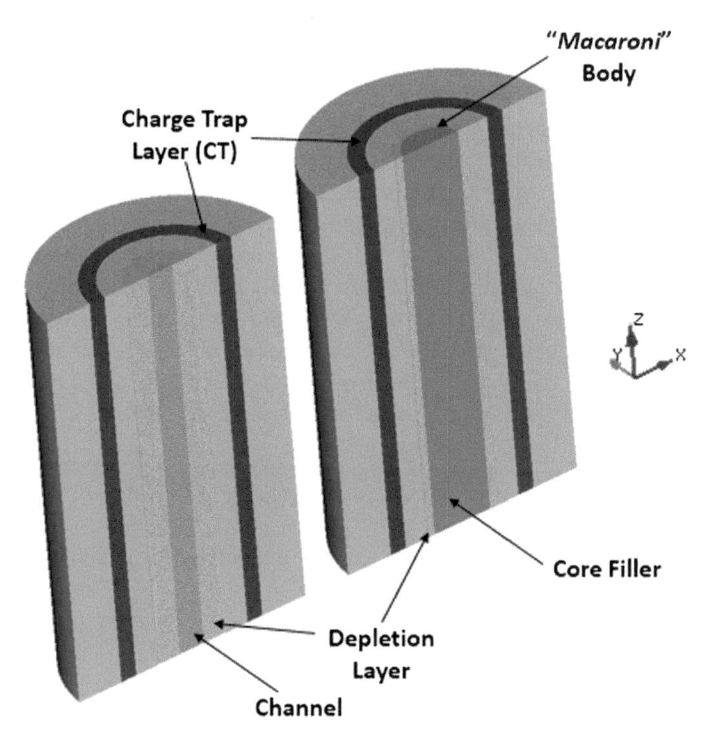

Fig. 5.6 A vertical transistor (right) modified with *Macaroni* body (left)

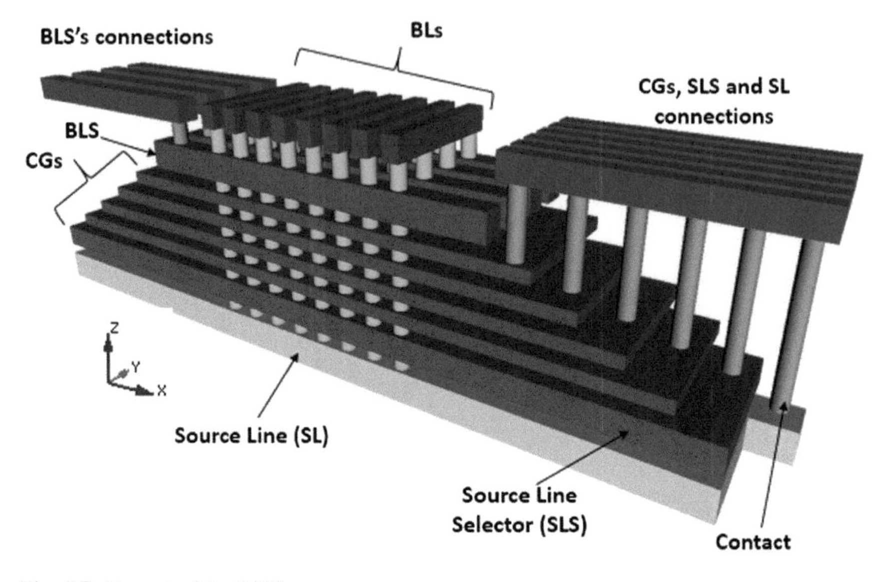

Fig. 5.7 Fan-out of the BiCS array

(i.e. a dielectric film) is used in the central part of the macaroni structure, essentially because it makes the manufacturing process easier.

The fabrication sequence of the BiCS array [21] starts from building the layers for control gates and selectors. Then, BLS stripes are defined. After forming pillars, bitlines are laid out by using a metal layer.

Control gate edges are extended to form a ladder to connect to the fan-out region, as sketched in Fig. 5.7 [12, 13, 21, 22]. Actually, there are 2 ladders: one of the 2 can't be used because it is masked by the metals biasing the bitline selectors.

Over time BiCS became P-BiCS, mainly to improve the Source Line resistance [23, 24]. In a nutshell, two vertical NAND strings are shorted together at the bottom of the 3D structure: in this way, they form a single NAND string and the 2 edges are connected to the bitline and to the Source Line, respectively (Fig. 5.8). Thanks to its U-shape, P-BiCS has few advantages over BiCS:

- retention is better because manufacturing creates less damages in the tunnel oxide;
- being at the top, the Source Line can be connected to a metal mesh, thus lowering its parasitic resistance;
- Source Line and bitline selectors are at the same height of the stack and, therefore, they can be equally optimized and controlled, thus obtaining a better string functionality.

Figure 5.9 shows a P-BiCS array [25].

One of the biggest drawbacks of P-BiCS is the fact that at the same height of the stack there are two different control gates which, of course, can't be biased together;

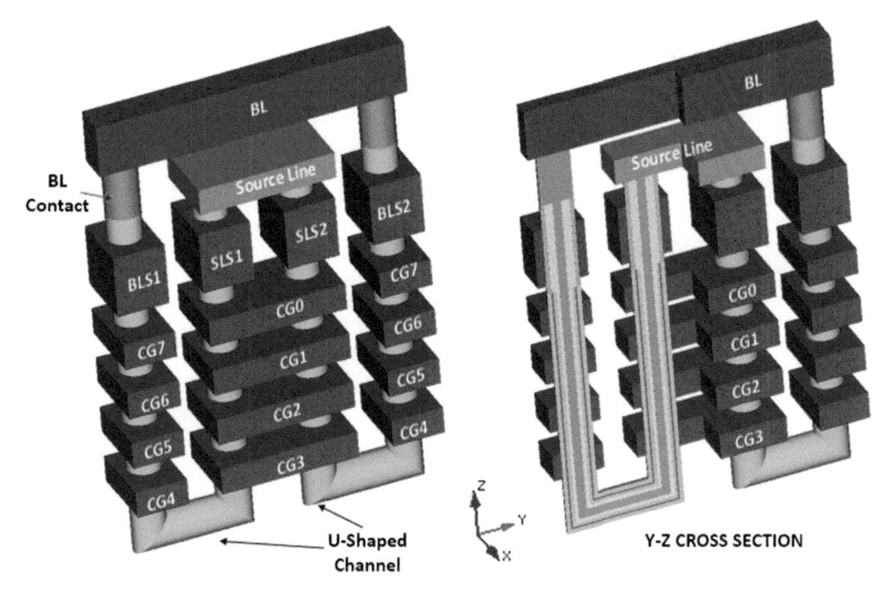

Fig. 5.8 P-BICS NAND strings

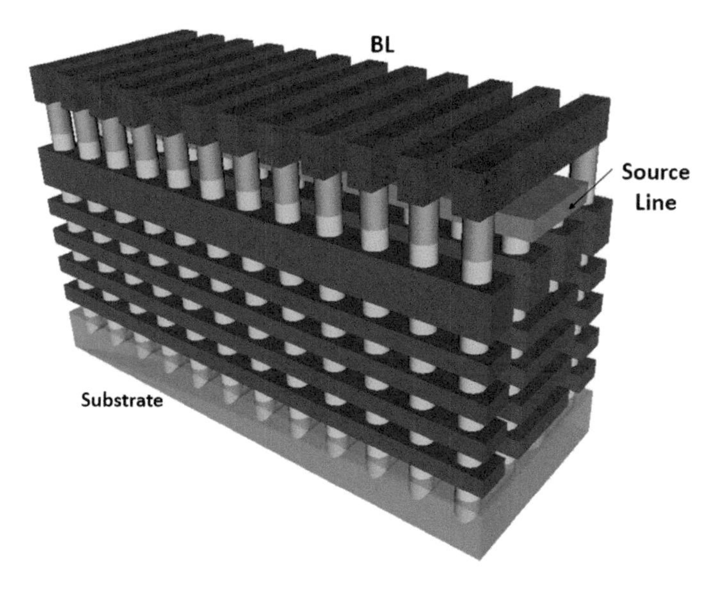

Fig. 5.9 P-BICS NAND Flash array

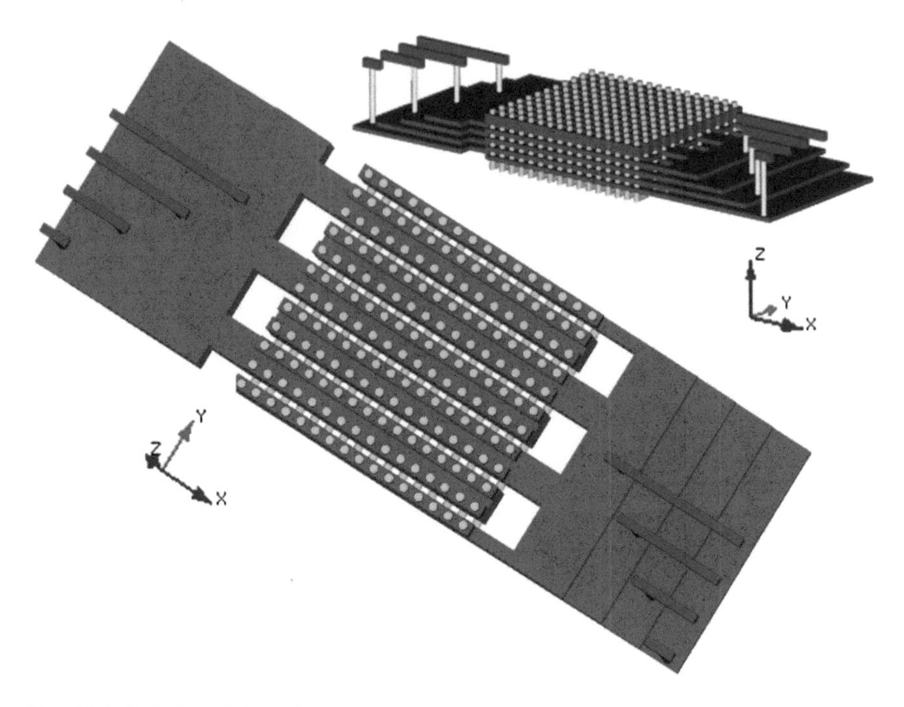

Fig. 5.10 Fork-shaped fan-out

therefore, the two layers can't be simply shorted together. As a result, compared to BiCS, a totally different and more complex fan-out is required [25], as displayed in Fig. 5.10: basically, a fork-shaped gate is adopted, such that each branch acts on two NAND pages.

A major advantage is the easier connection of the source line [14] through the "Top Level Source Line" of Fig. 5.11. This additional metal mesh guarantees a much better noise immunity for circuits.

Besides BiCS and P-BiCS, many other approaches were tried, including VRAT (*Vertical Recess Array Transistor*) [26], Z-VRAT (*Zigzag* VRAT) [26], and VSAT (*Vertical Stacked Array Transistor*) [27], and 3D-VG (*Vertical Gate*) NAND [28] which is a unique architecture where the channel runs along the horizontal direction.

TCAT (*Terabit Cell Array Transistor*) was disclosed in 2009 [29] and it was the foundation for V-NAND (Fig. 5.12), which is the first 3D memory device who reached the market. Except for SL+ regions which are n+ diffusions, the equivalent circuit of TCAT is the same of BiCS (Fig. 5.4). All SL+ lines are connected together to form the common Source Line (Fig. 5.13). There are 2 metal layers for decoding wordlines and NAND strings, respectively.

TCAT is based on *gate-replacement* [29], whereas BiCS is *gate-first*. Gate-replacement begins with the deposition of multiple oxide/nitride layers. After the stack formation, nitride is removed through an etching process. Afterwards,

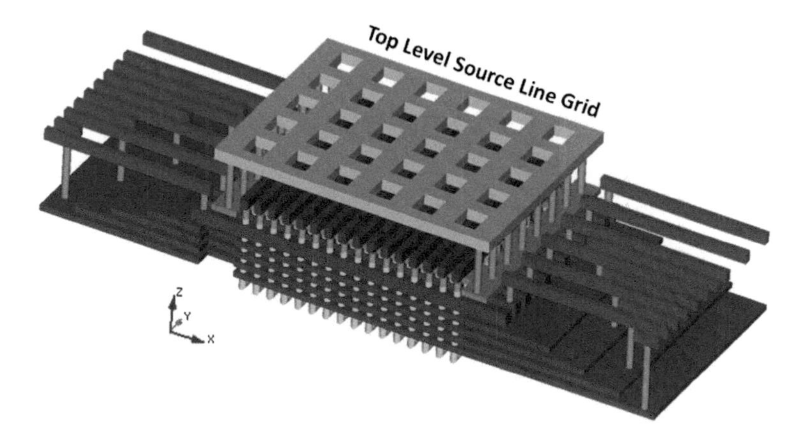

Fig. 5.11 P-BiCS: source line metal mesh

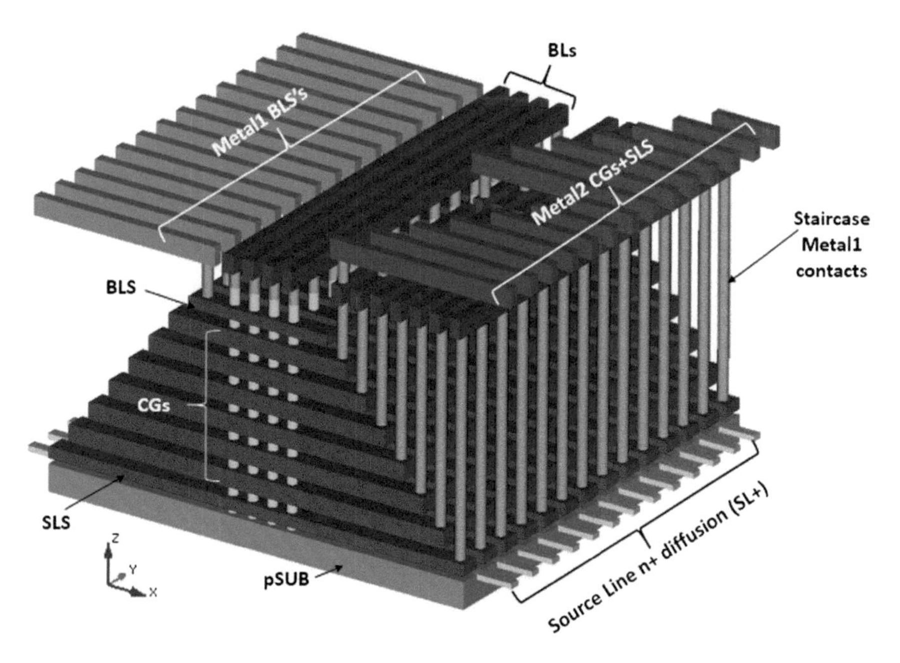

Fig. 5.12 TCAT NAND Flash array

tungsten metal gates are deposited and, finally, gates are separated by using another etching step. Metal gates translate into a lower wordline parasitic resistance, resulting in faster programming and reading operations.

The bulk erase operation is another significant difference compared to BiCS. Because NAND strings are close to n+ areas, during erasing, holes can come

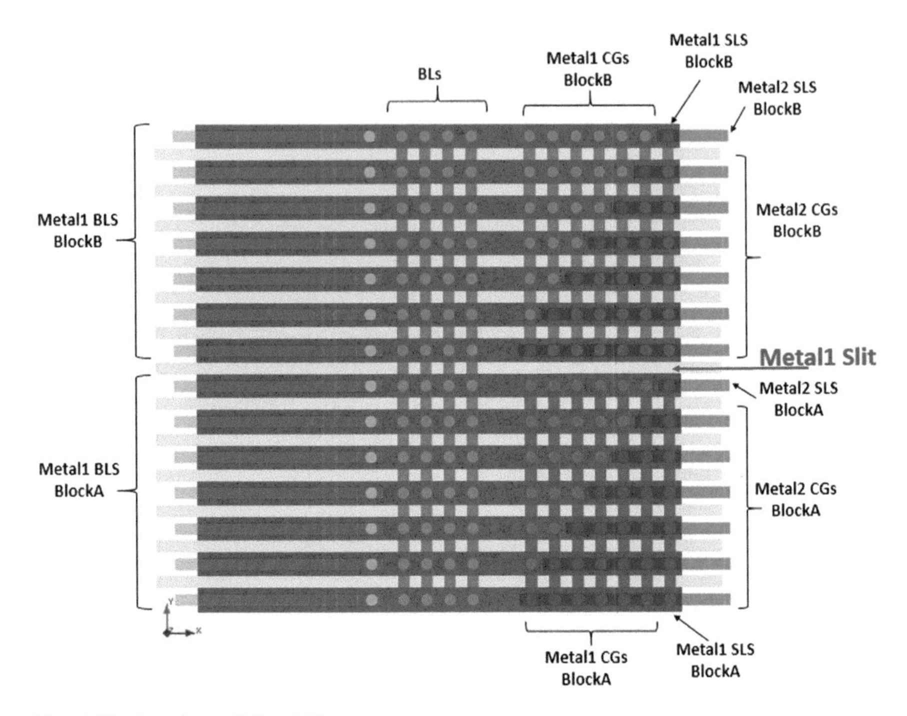

Fig. 5.13 Top view of Fig. 5.12

straight from the substrate, thus avoiding the GIDL (*Gate Induced Drain leakage*) on the source side, which is a well-known problem for BiCS.

BiCS and TCAT are compared in Fig. 5.14 [30]. Being TCAT based on a gate-last process, the charge trap layer is biconcave, and thanks to this particular shape it is much harder for charges to spread out. On the contrary, BiCS is

Fig. 5.14 BiCS versus TCAT

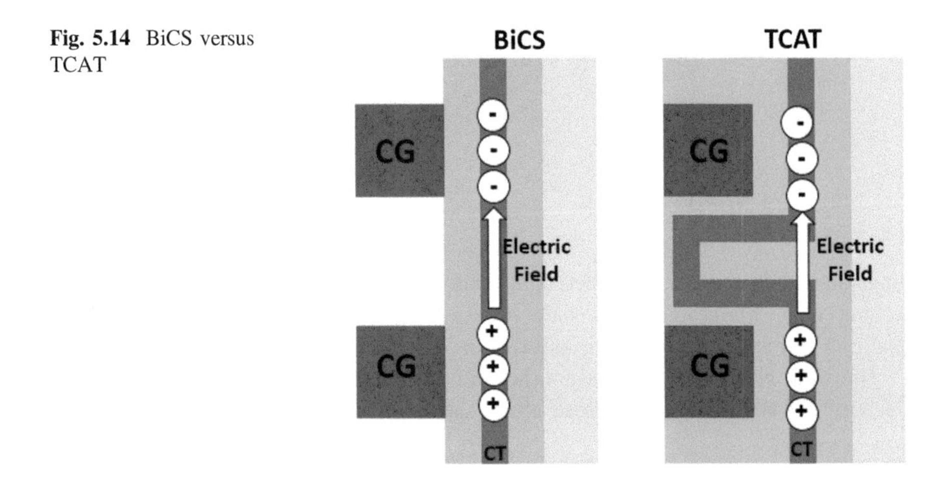

characterized by a charge trapping layer going through all gate plates, thus acting as a charge spreading path: of course, the main consequence of this layout is a degradation of data retention.

TCAT evolved into another architecture called V-NAND [31]. As depicted in Fig. 5.15, the first generation, V-NAND Gen1, had 24 wordline layers, plus additional dummy wordline layers (dummy CG) [32, 36, 37].

Why dummy layers? Mainly because of the floating body of the memory cells with vertical channel. In fact, during the programming operations, hot carriers are generated by the high lateral electric field located at the edge of the NAND string. Therefore, these hot carriers keep the voltage on the channel low during the programming operation of the first wordline (i.e. Program Disturb). Dummy wordlines before the first WL are an effective and simple solution to this problem [38, 39].

A 128 Gb TLC (3 bit/cell) device manufactured by using V-NAND Gen2 was published in 2015 [33, 40]. Gen2 had 32 memory layers instead of the previous 24 and introduced the concept of Single-Sequence Programming. Conventional (mainly 2D) TLC programming techniques go through the programming sequence multiple times. To be more specific, each wordline is programmed 3 times, such that V_{TH} distributions can be progressively tightened. Because of the smaller cell-to-cell interference (compared to FG), CT cells exhibit an intrinsic narrower native V_{TH} distribution. As a result, V-NAND Gen2 could write 3 pages of logic data in a single programming sequence. There are 2 benefits to this approach: reduced power consumption and faster programming.

V-NAND Gen3 appeared in 2016 [34], in the form of a 48 layer TLC device. With such a high number of gate layers, the very high aspect ratio of the pillar becomes a serious challenge for the etching technology. To mitigate this problem, the easiest solution is to shrink the thickness of gate layers. The downside of this approach is that the parasitic RC of the wordline gets higher, thus slowing access operations to the memory array. Moreover, channel's size fluctuations become critical. Indeed, pillars are holes drilled in the gate layer and they represent a barrier for charges flowing along the wordline: in essence, a distribution of the holes diameters generates a distribution of the parasitic resistances of gate layers. In addition, pillars, once manufactured, have the conic shape sketched in Fig. 5.16. The overall result is that the same voltage applied to different gate layers translàtes into a waveform per layer. An adaptive program pulse scheme can fix the problem. In a nutshell, the program pulse duration has to be tailored to the characteristics of the wordline layer. As the number of layers increases, the pillar becomes longer with a negative impact on the aspect ratio of the pillar. To compensate for that, V-NAND Gen4 [35], which is built on a stack of 64 layers, had to shrink both the layer thickness and the intra-layer distance (spacing). The downside is an increased wordline parasitic capacitance which adversely affects cell's reliability and timings. Improved circuits and programming algorithms can be used to tackle this problem [35].

As discussed, both BiCS [41] and V-NAND use CT cells, but Floating Gate still exists, as explained in the next section.

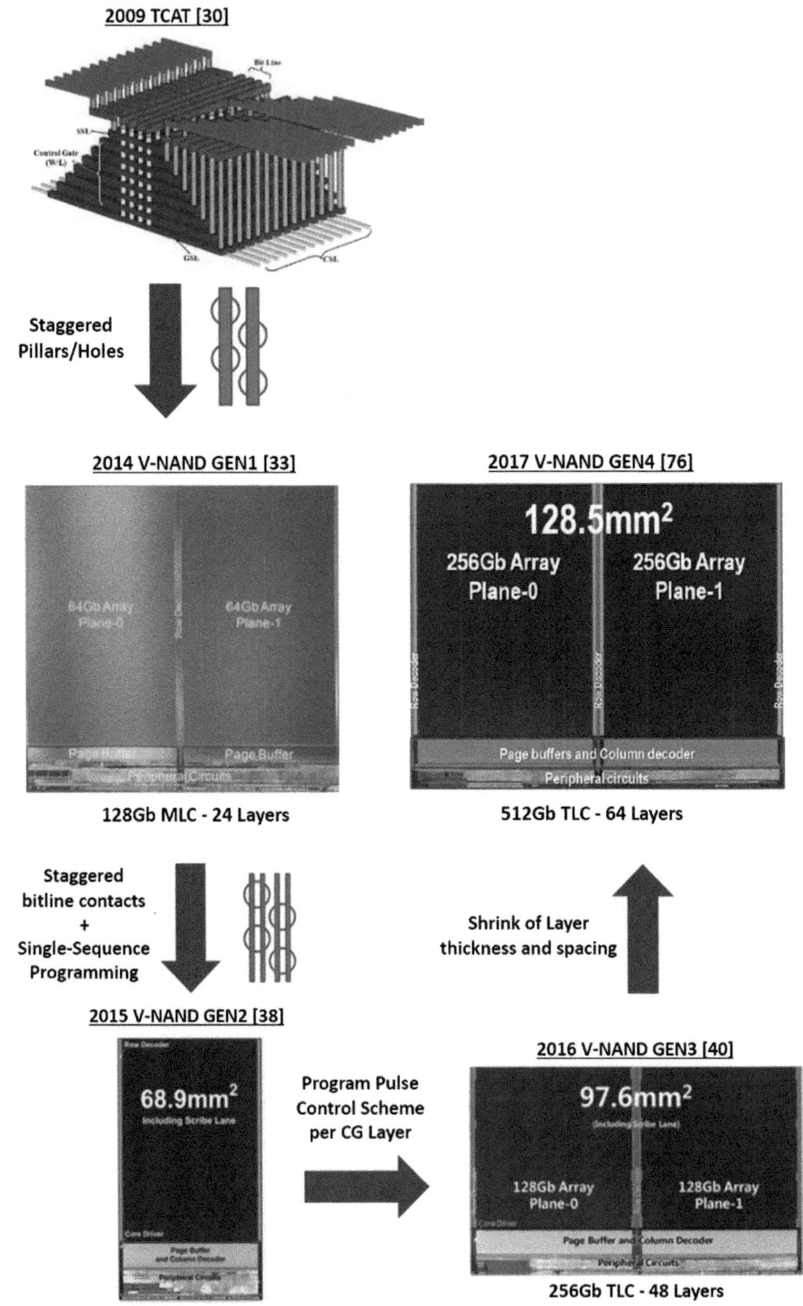

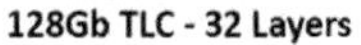

Fig. 5.15 Evolution from TCAT to V-NAND (reproduced with permission from [29, 32–35])

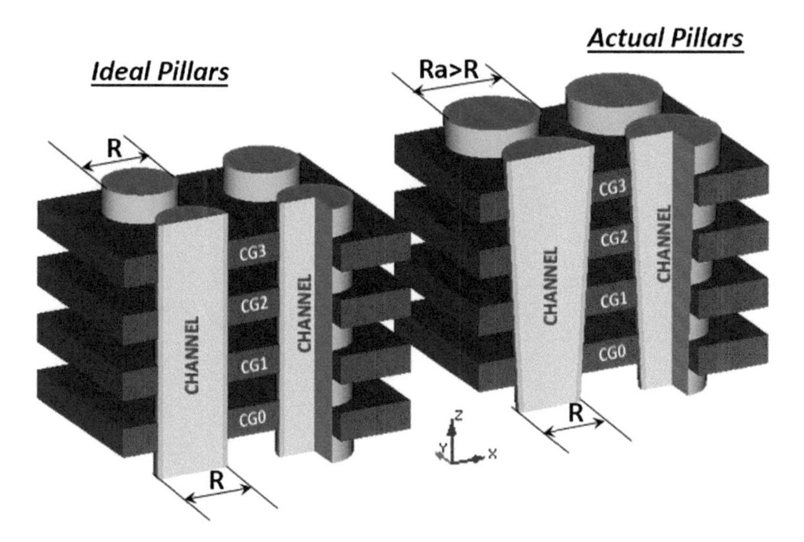

Fig. 5.16 Ideal versus actual shape of pillars

5.2 3D Floating Gate NAND Flash Memories

2D NAND Flash memories use FG cells which have been, improved and optimized for decades. Of course, there have been many attempts to reuse this know-how in 3D.

The first 3D attempt is known as *3D Conventional FG* (C-FG) or S-SGT (*Stacked-Surrounding Gate Transisto*r) [42–44], and it is sketched in Fig. 5.17.

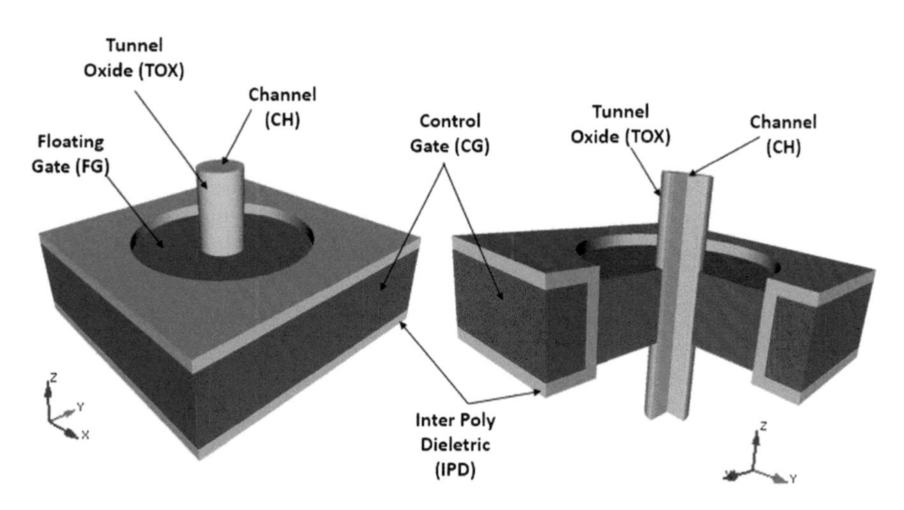

Fig. 5.17 3D C-FG cell

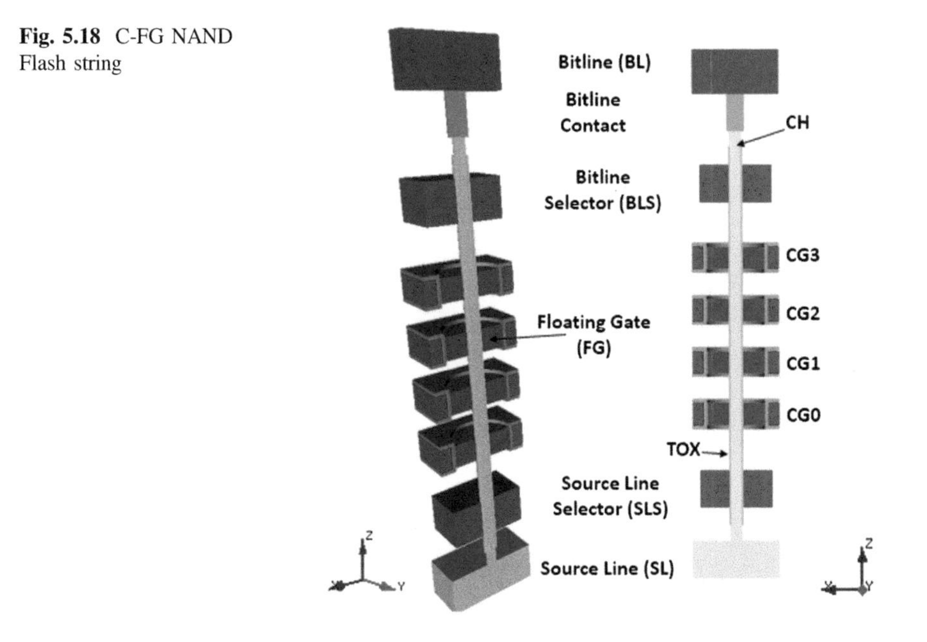

Fig. 5.18 C-FG NAND Flash string

A C-FG NAND string is shown in Fig. 5.18, including select transistors. Please note that both string selectors are manufactured as standard transistors, i.e. they haven't any floating gate. Figure 5.19 shows a C-FG array and Fig. 5.20 adds the fan-out region. While all wordlines at the same height of the stack are connected, BLS lines can't, because they need to be page selective per each CG layer. On the contrary, SLS transistors can be shorted together, thus saving both power and silicon area.

As already discussed in the previous Section, the Source Line is the *local ground* of memory cells. A big single Source Line plate laid out at the bottom of the stack, with a limited number of contacts, simply doesn't work: when tens of thousands of cells sink current, managing the voltage on the source side becomes a real challenge. Having more contacts to the Source plate is not an option. The *Source Line Metal Grid* sketched in Fig. 5.21 fixes this problem.

As already discussed, slits between NAND blocks are the most common way for reducing program/read disturbs and parasitic loads. Of course, there is no need to cut bitlines and Top Source Lines. This is fundamentally the same approach adopted in BiCS.

Because we are talking about FG cells, FG coupling between neighboring cells is the main hurdle for vertical scaling. With enhancement-mode operations, the high resistance of source/drain (S/D) regions should also be carefully considered. In fact, these regions need high-doping and this is not very easy to accomplish when the conduction channel is made of polysilicon. The solution to this problem is to electrically invert the S/D layer by using higher voltages during read. This simple solution is hardly manageable by C-FG cells because of the thin FG.

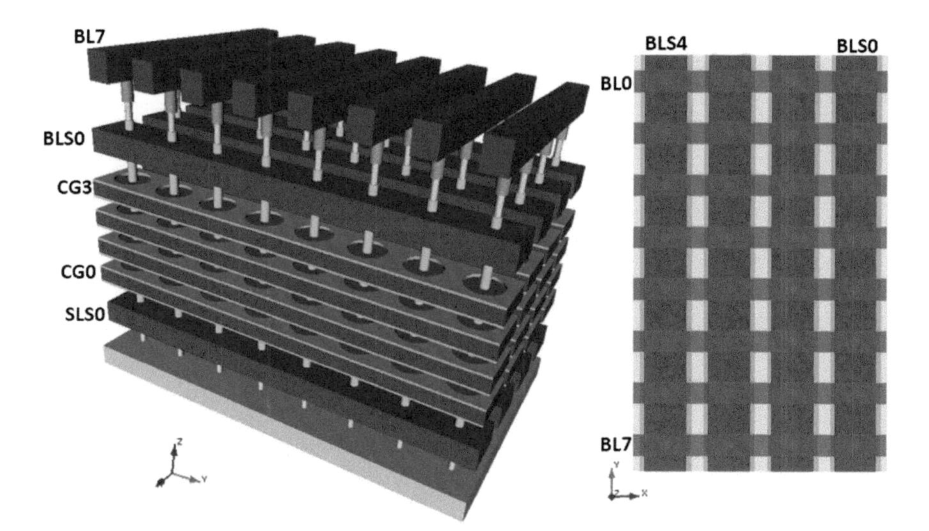

Fig. 5.19 C-FG NAND Flash array

Fig. 5.20 C-FG NAND Flash array with fan-out

The *Extended Sidewall Control Gate* (ESCG) structure, Fig. 5.22 [45], is another FG option and it was developed to contain the interference effect. Moreover, by applying a positive voltage to the ESCG structure, density of electrons on the surface of the pillar can be much higher than C-FG (even one order of magnitude): a highly inverted electrical source/drain can significantly lower the S/D resistance.

In addition, the ESCG shielding structure reduces the FG–FG coupling capacitance: the ESCG region is biased as CG, and the CG coupling capacitance (C_{CG}) is significantly increased because of the increased overlap area between CG and FG.

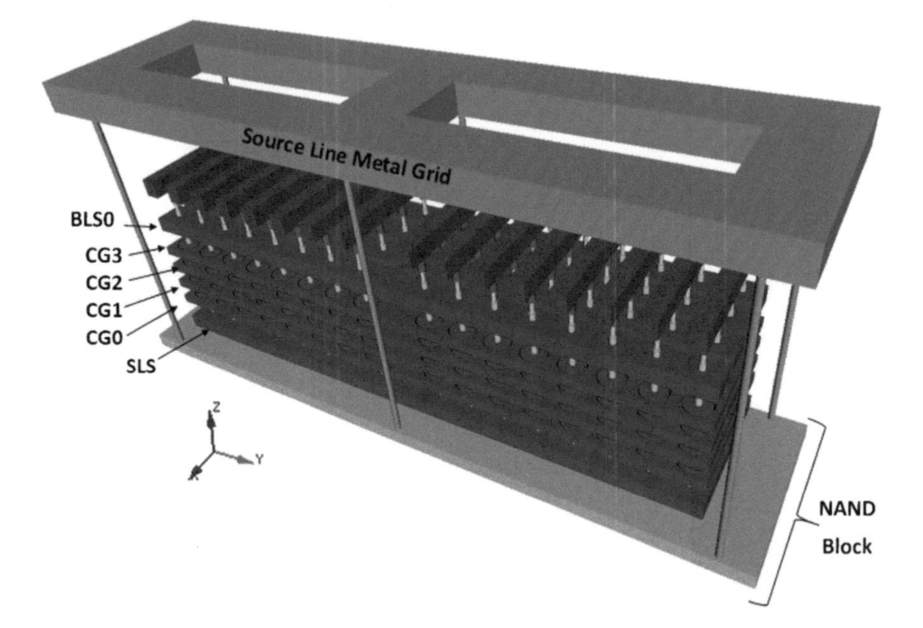

Fig. 5.21 C-FG array with source line metal grid

Fig. 5.22 ESCG NAND Flash cell

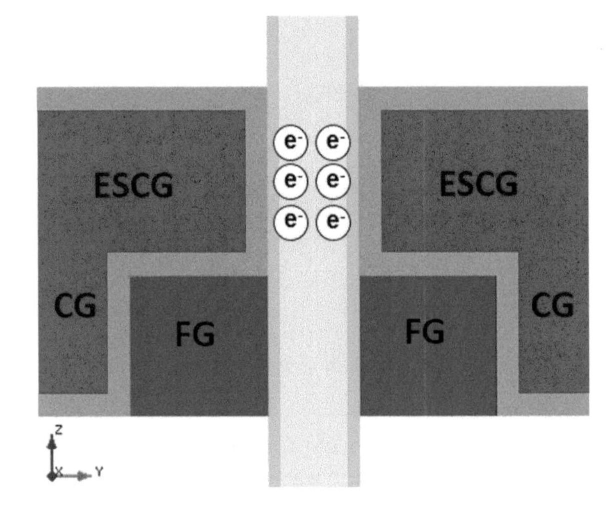

A higher CG coupling ratio is one of the key ingredients for achieving effective NAND Flash operations [46].

Another FG cell is DC-SF (*Dual Control-Gate with Surrounding Floating Gate,* Fig. 5.23) [47]. This time FG is controlled by two CGs. The impact on the FG/CG coupling ratio is remarkable, thanks to the enlargement of the FG/CG overlap area. Another positive aspect is the reduction of the voltages required for programming

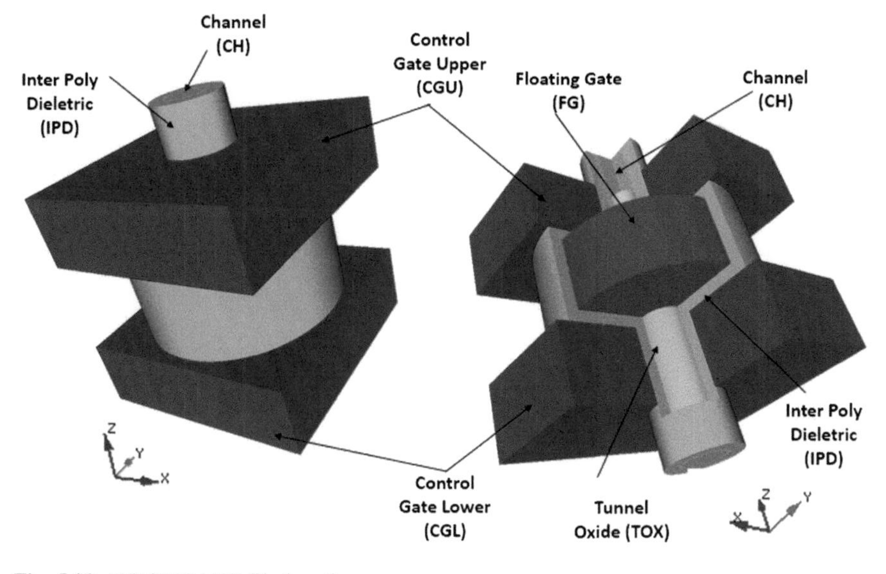

Fig. 5.23 DC-SF NAND Flash cell

and erasing. DC-SF eliminates the FG-FG interference because the CG between two adjacent FGs plays the role of an electrostatic shield [48].

FG is fully isolated by IPD (*Inter Poly Dielectric*) and capacitive coupled to upper and lower control gates, CGU and CGL, respectively. The tunnel oxide is located between the channel CH and FG, while IPD is on the sidewall of the CG. In this way, free charges cannot tunnel to the control gates.

BiCS and DC-SF NAND strings are sketched in Fig. 5.24. In BiCS the nitride layer, going across all gates, makes the cell prone to data retention issues. On the contrary, the surrounding FG is totally isolated: it is much easier for DC-SF to retain electrons [49, 50]. Of course, the downside of DC-SF is the fact there are two gate layers instead of one, coupled with much more complex biasing schemes [51, 52].

The *Separated Sidewall Control Gate* (S-SCG) Flash cell [53] displayed in Fig. 5.25 is another 3D FG option developed around the sidewall concept.

One of major drawbacks of this cell is the "direct" disturb to the neighboring passing cells, caused by the high SCG/FG coupling capacitance. We define it as "direct" because the sidewall CG is shared between adjacent cells: as a matter of fact, biasing SCG means biasing both FGs.

To minimize the decoding complexity, all SCGs belonging to one block adopt a common SCG scheme; besides their electrostatic shield functionality, sidewall gates can help all memory operations [54]. For instance, the common SCG is biased at 1 V during read operations, thus electrically inverting the channel (same as ESCG). Compared to ESCG, the electrical inversion happens simultaneously on source and drain, exactly because of the sidewall gates (Fig. 5.26). Same thing happens during

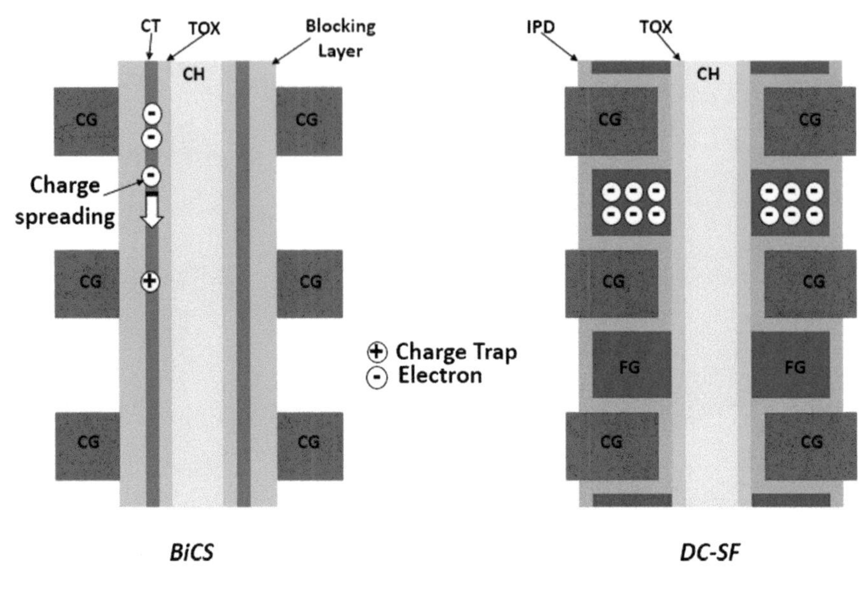

Fig. 5.24 BiCS versus DC-SF

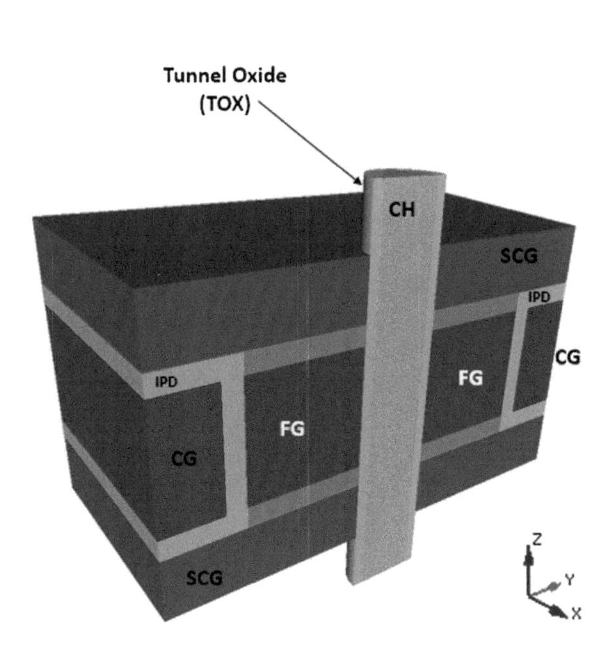

Fig. 5.25 S-SCG NAND Flash cell

programming: the common SCG is biased at a medium voltage to improve the channel boosting efficiency.

Besides the direct disturb, another problem of Sidewall Gates is the limitation of vertical scaling to around 30 nm; indeed, the thicknesses of SCG and IPD can't be scaled too much, otherwise they would breakdown when voltages are applied.

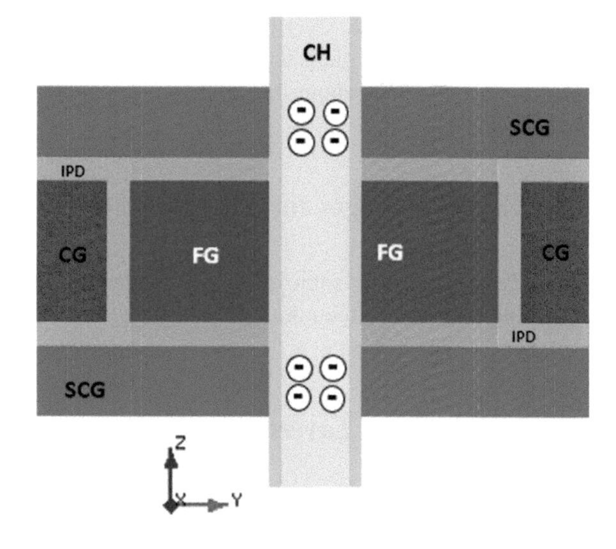

Fig. 5.26 Common SCG approach to enable Source/Drain inversion

Let's now take a look at examples of 3D FG NAND memory arrays of hundreds of Gb. As shown in Fig. 5.27, the first 3D FG device was published in 2015 [55], in the form of a 384 Gb TLC NAND based on C-FG. This memory device was built on stack of 32 (+dummy) memory layers.

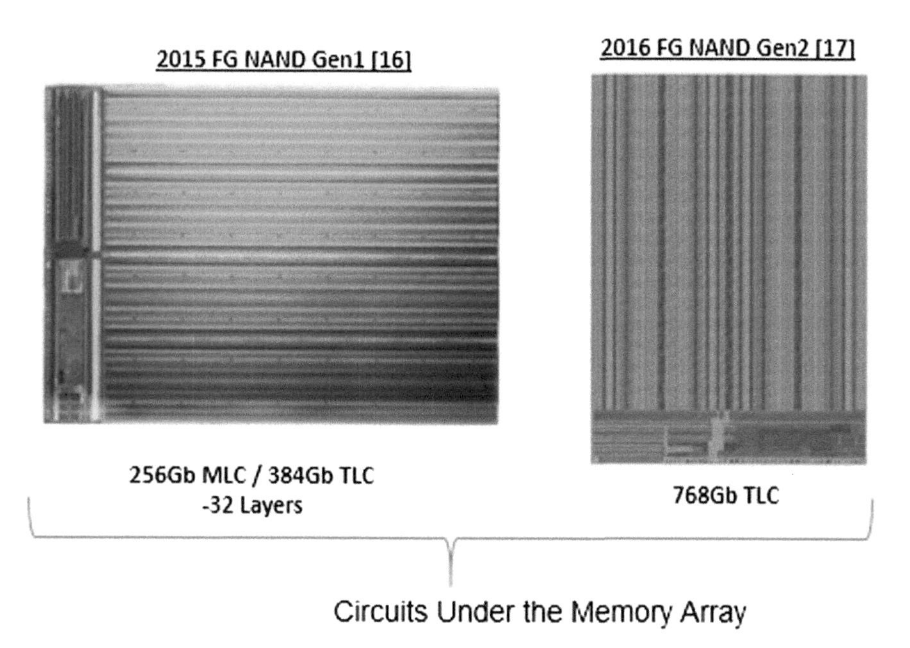

Fig. 5.27 3D FG NAND devices [14, 15]

A 768 Gb 3D FG NAND became public in the following year [56]. What is unique in this case is the fact that the area underneath the array was used for circuitry. More details about this approach are provided in Sect. 3.3.

5.3 Key Challenges for 3D Flash Development

In this Section we cover some of the key challenges that technologists and designers are facing to push 3D memories even further.

5.3.1 *Number of Layers*

To reduce the bit size, the number of stacked cells needs to go up, but this causes a bunch of problems hard to solve, as shown in Fig. 5.28 [6].

Pillar's *Aspect Ratio* (AR) is definitely the first challenge to overcome; in a stack of 32 cells AR can already be as high as 30. In this context, hole etching and gate patterning are extremely difficult, but of paramount importance.

A possible solution to this problem is to divide the stacking process in more steps to reduce the corresponding AR. For example, a NAND string made of 128 cells can be divided in 4 groups of 32 cells each, as shown in Fig. 5.28. The

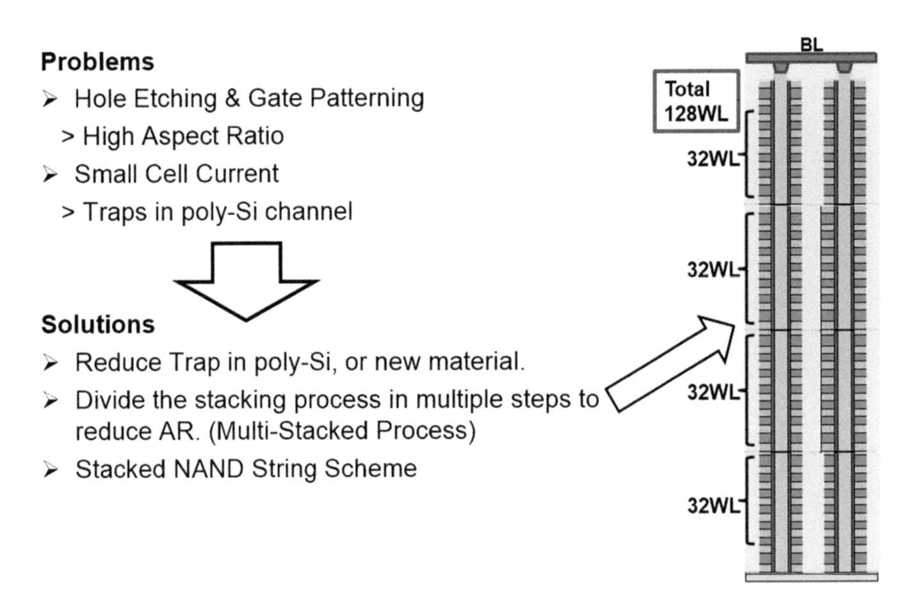

Fig. 5.28 Challenges for increasing the number of 3D layers [6]

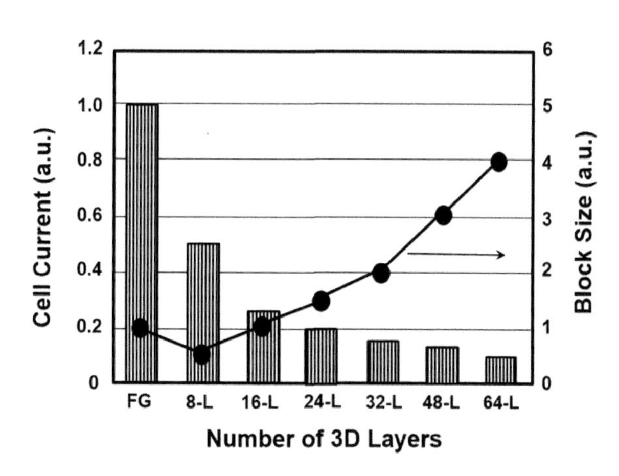

Fig. 5.29 Cell current and block size versus the number of 3D layers [57]

downside of this solution is the cost of the stacking process (in this example, 4 times higher than the cost of the plain solution).

Second problem is the small cell current [57]. With 2D sensing schemes, a 200 nA/cell saturation current is considered the right value because it gives a reasonable sensing margin. Unfortunately, as shown in Fig. 5.29, already with a stack of 24 layers, the cell current is just ~20% of FG cell. And it becomes lower and lower as the number of cells in the vertical stack increases. There are a couple of possible paths to solve this problem: sensing schemes with higher sensitivity, and the introduction of new materials enabling a higher cell mobility in the poly-Si channel (i.e. a higher current) [58–61].

All the above mentioned problems can be fixed if entire NAND strings could be stacked one on top of each other. In this case, either bitlines or source lines are fabricated between NAND strings. This special architecture can simultaneously reduce the aspect ratio and increase the sensing current at same time.

5.3.2 *Peripheral Circuits Under Memory Arrays*

In the first 3D generations [62, 63], peripheral circuits (charge pumps, logic, etc.) and core circuits (like Page Buffers and Row decoders) are located outside the memory matrix, like in a conventional 2D chip floorplan, as sketched in Fig. 5.30a. However, 3D memory cells are vertically stacked: in other words, memory transistors are not formed on the Si substrate; on the contrary, they are built around a deposited poly-Si (vertical pillar). Therefore, 3D architectures allow placing some circuits directly on the Si substrate under the memory array. Of course, this solution offers a significant reduction of the chip size.

Figure 5.30b shows a layout of a Flash memory with *Core Circuits Under the Array* (CCuA) [64] in addition, Fig. 5.30c displays the case where both *Core* and

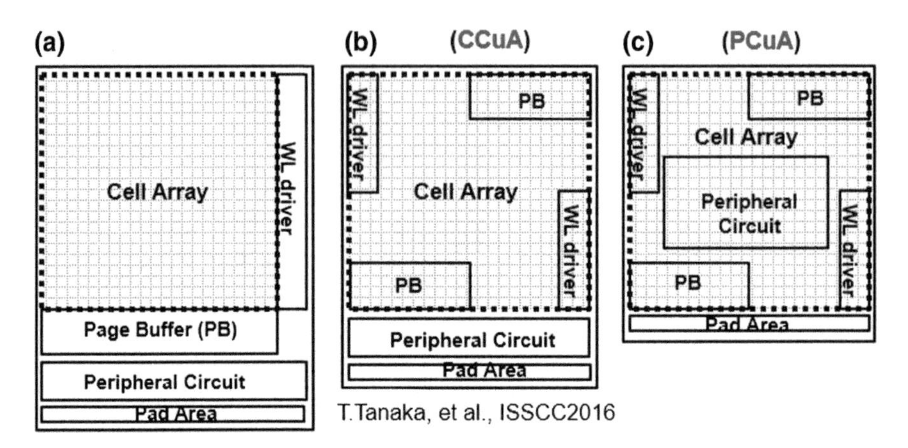

Fig. 5.30 3D NAND Flash memory layout: **a** conventional, **b** CCuA, and **c** PCuA [64]

Peripheral Circuits are manufactured on the Si substrate *under the Array* (PCuA) [65].

Efficiency of 2D and conventional 3D are between 60 and 81%. If CCuA is used, then the cell efficiency can be as high as 85%. In the extreme case, when both peripheral and core circuits sits under the memory matrix (PCuA), the cell efficiency can reach around 95%, because peripheral circuits usually occupy more than 10% of the whole chip.

This big area saving doesn't come for free. The most important challenge is manufacturing low resistance metal layers under the array: this is absolutely critical for a reliable circuit functionality. Usually, metal layers used in 2D NAND flash memories are made of Cu. However, when circuits are under the array, the high temperature processes (i.e. >800 °C) that 3D requires can seriously degrade the resistance of metal layers. Therefore, circuits under the array require 3D "low" temperature fabrication processes.

5.3.3 *Data Retention*

3D CT cells and 2D FG cells are completely different in terms of data retention properties. Generally speaking, 2D SONOS (*Silicon-Oxide-Nitride-Oxide-Silicon*, which is one variant of CT) cells exhibit larger V_{TH} shifts than 2D FG cells: this is caused by a fast charge detrapping through the tunnel oxide [66]. Figure 5.31 compares data retention of two different cells: (a) 3D SONOS cell and (b) 2D 2y-nm FG cell [57]. Both cells have been cycled 3,000 times. After 3k cycles 3D SONOS has a V_{TH} distribution width narrower than 2D FG; however, after baking at *High Temperature* (HT), the V_{TH} distribution becomes wider, and it has a bigger V_{TH} shift. For 3D SONOS cells, data retention is definitely one of the hottest topics.

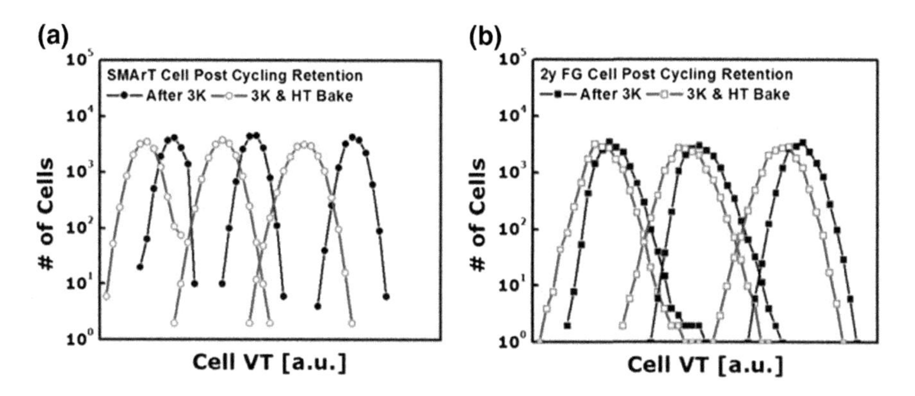

Fig. 5.31 Vth distribution of cycled cells after high temperature retention for **a** 3D SMArT cell and **b** 2D 2y-nm FG cell [57]

Another important retention issue for 3D SONOS is the fact that the relationship between charge loss and temperature is different from 2D FG, as shown in Fig. 5.32 [57], thus impacting the way accelerated tests should be performed. For 2D FG cells V_{TH} shift is linearly dependent upon the bake temperature, which says that the mechanism governing data loss remains constant. However, in 3D SONOS cell V_{TH} shift exhibits a non-linear relationship with respect to the bake temperature; in other words, the data loss mechanism changes from low to high temperature. The data loss mechanisms are dominated by band-to-band tunneling at low temperature and by thermal emission at high temperature [57]. As a consequence, simple temperature accelerated tests, which have been used for decades, should be used very carefully: retention below 90 °C has to be evaluated by extrapolating from data collected over at least 3 weeks at relatively low temperatures. It is worth highlighting that there multiple variations of CT cells; for example, BE-SONOS (*Bandgap Engineered*) can be used to optimize the bandgap structure of the SONOS cell [67].

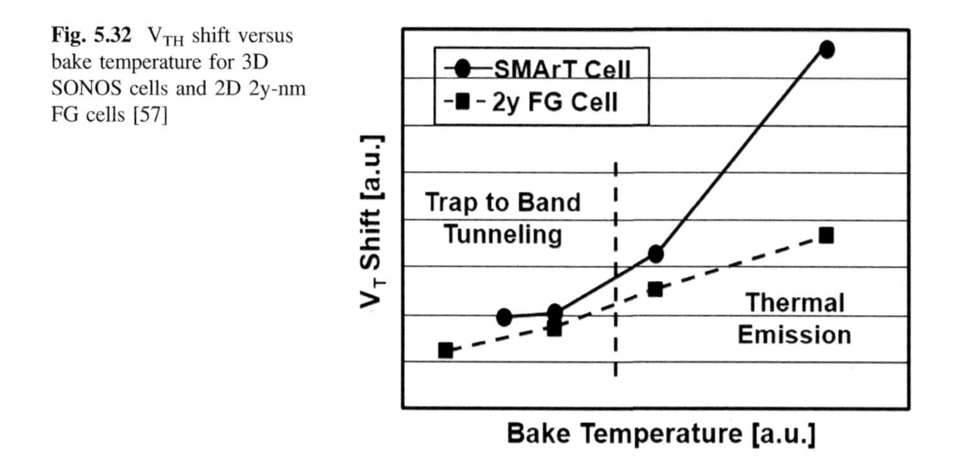

Fig. 5.32 V_{TH} shift versus bake temperature for 3D SONOS cells and 2D 2y-nm FG cells [57]

5.3.4 3D Program Disturb

Figure 5.33a shows one 3D NAND block [68, 69]. In each block, N strings are connected to the same bitline by means of N select transistors, namely DSL_1 to DSL_N. In a 2D NAND block, there is a 1:1 correspondence between strings and bitlines. As a matter of fact, 3D architectures introduce new program disturb modes, as sketched in Fig. 5.33b.

When DSL_1 is activated, strings (STRs) along DSL_1 are either being programmed or they suffer "X" disturb, depending on the BL bias. When we look at "X" disturb, bitlines are biased at Vcc and there is no difference with respect to 2D NAND. But in 3D, DSL_2 to DSL_N are turned off. We can distinguish two different situations, which we call "Y" and "XY" program disturbs. In the "Y" case bitlines are biased at ground and drain select transistors (DSL) are off; for "XY" we have bitlines at Vcc and DSL off.

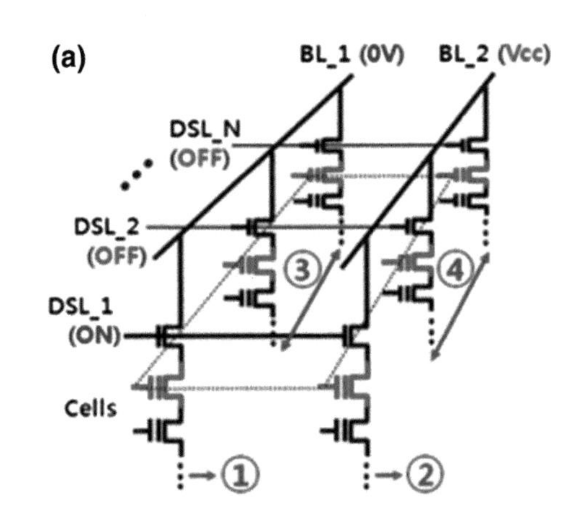

(b) Conventional modes

Mode	DSL	BL
①: STR in PGM	ON	0V
②: STR in X mode	ON	Vcc

Additional modes in 3D

Mode	DSL	BL
③: STRs in Y mode	OFF	0V
④: STRs in XY mode	OFF	Vcc

X-Disturbance Mode
Vdsl (Vg)
Vcc-Vg
Channel
DSL
Vbl (Vcc)
Boosting (Vch)
Vch-Vcc
Y-Disturbance Mode
Vdsl (0V)
0V
Channel
DSL
Vbl (0V)
Boosting (Vch)
Vch

Fig. 5.33 **a** Program disturb in a 3D NAND array. **b** 3D introduces two new program disturbs, Y and XY [69]

"XY" disturb mode is not severer than "X" mode. Being DSL off and BL at Vcc, the self-boosting voltage cannot cause a leakage current through DSL. On the contrary, in the "Y" mode BL is at ground, thus open the door to a possible leakage through DSL. In addition, DSL of 3D NAND shows a larger leakage current compared to 2D NAND [57, 69]. Moreover, in 2D the leakage current through DSL is prevented by the fact that V_{TH} of DSL becomes higher during programming thanks to a strong body effect. This is not the case with 3D NAND. Several approaches to suppress the above mentioned leakage current have been proposed over time [68]. These include: (1) DSL with high V_{TH}, (2) DSL negative bias, and (3) dummy wordlines between DSL and edge wordlines. Dummy wordlines can reduce the voltage drop going from the self-boosting voltage to the voltage applied to the DSL; on top of that, they are helpful for inhibiting the hot carrier generation that might take place on the edge wordline (in practice, they reduce the lateral electric field). Indeed, dummy WLs have to be carefully designed (biasing, V_{TH}, number of wordlines) given all the above mentioned functions. A detailed analysis of 3D program disturb mechanism can be found in [69].

5.4 Future Trend for 3D NAND Flash

Figure 5.34 shows cell's size scaling trend, based on published die photographs. 2D became flat below 20 nm, while 3D cell showed a significant reduction going from 24 to 64 layers. This 3D scaling speed will continue by increasing the height of the memory stack, and exploiting technological innovations like Multi-stacked and Stacked NAND string [70].

3D NAND arrays based on CT vertical channel were selected for volume production because the fabrication process is simpler than other 3D architectures. Volume production of 3D NAND Flash started in late 2013 with a 24 layer MLC (2 bit/cell) V-NAND [62, 71]. Year after year, the number of stacked cells grew up, as shown in [7, 63, 72], thus reducing the cost per bit and fueling an even more pronounced diffusion of Solid State Drives.

In this chapter we have presented many architectural options for building a 3D NAND array, including some of the latest and greatest layout options, but the 3D evolution is just at the beginning. In fact, two fundamentally different technologies, Floating and Charge Trap, are fighting each other, trying to prove that they can win in the long run, i.e. when scaling will be pushed to the limit. Flash manufactures are already shooting for 100 vertical layers with multi-level capabilities, including 4 bit/cell. No doubt that we'll see a lot of innovations in the near future: engineers and scientists are called to give their best effort to make this vertical evolution happen.

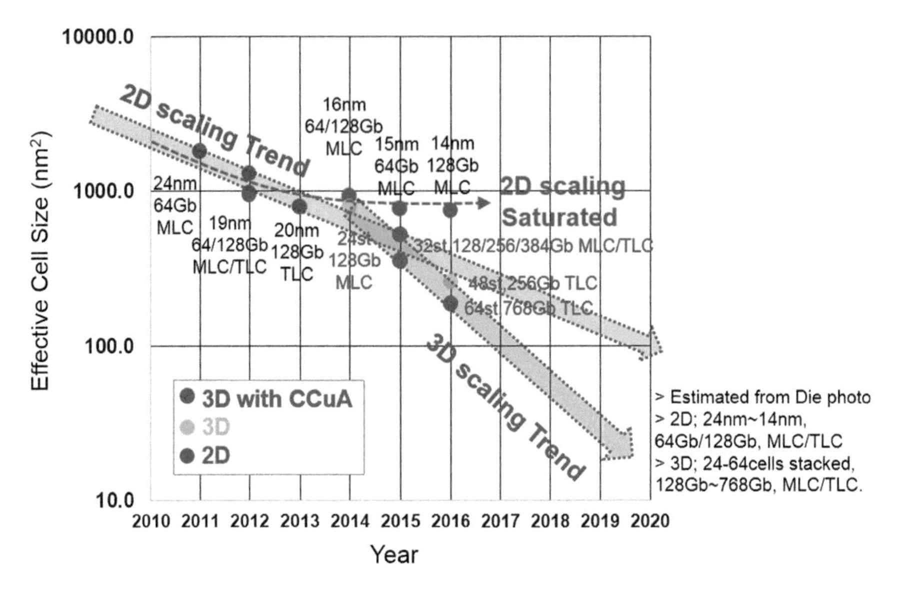

Fig. 5.34 Effective cell size trend

References

1. F. Masuoka, M. Momodomi, Y. Iwata, R. Shirota, New ultra high density EPROM and flash EEPROM with NAND structure cell, in *International Electron Devices Meeting*, vol. 33 (1987), pp. 552–555
2. R. Micheloni, L. Crippa, A. Marelli, *Inside NAND Flash Memories* (Chap. 6) (Springer, 2010)
3. T. Mizuno et al., Experimental study of threshold voltage fluctuation due to statistical variation of channel dopant number in MOSFET's. IEEE Trans. Electron Devices **41**(11), 2216–2221 (1994)
4. H. Kurata et al., The impact of random telegraph signals on the scaling of multilevel flash memories, in *Symposium on VLSI Technology* (2006)
5. C.M. Compagnoni et al., Ultimate accuracy for the NAND flash program algorithm due to the electron injection statistics. IEEE Trans. Electron Devices **55**(10), 2695–2702 (2008)
6. S. Aritome, *NAND Flash Memory Technologies*. IEEE Press Series on Microelectronics System, Wiley-IEEE Press, Published on Dec 2015
7. S. Aritome, 3D flash memories, in *International Memory Workshop 2011 (IMW 2011)*, short course
8. R. Micheloni, L. Crippa, A. Marelli, *Inside NAND Flash Memories* (Chap. 5) (Springer, 2010)
9. http://www.samsung.com/us/business/oem-solutions/pdfs/VNAND_technology_WP.pdf. White Paper, Sept 2014
10. R. Micheloni, L. Crippa, Multi-bit NAND flash memories for ultra high density storage devices (Chap 3), in *Advances in Non-volatile Memory and Storage Technology*, ed. by Y. Nishi (Woodhead Publishing, Sawston, 2014)
11. R. Micheloni et al., High-capacity NAND flash memories: XLC storage and single-die 3D (Chap 7), in *Memory Mass Storage*, ed. by G. Campardo et al. (Springer, 2011)
12. H. Tanaka et al., Bit cost scalable technology with punch and plug process for ultra high density flash memory, in *VLSI Symposium Technical Digest* (2007), pp. 14–15

13. Y. Fukuzumi et al., Optimal integration and characteristics of vertical array devices for ultra-high density, bit-cost scalable flash memory, in *IEDM Technical Digest* (2007), pp. 449–452
14. M. Ishiduki et al., Optimal device structure for pipe-shaped BiCS flash memory for ultra high density storage device with excellent performance and reliability, in *IEDM Technical Digest* (2009), pp. 625–628
15. T. Maeda et al., Multi-stacked 1G cell/layer pipe-shaped BiCS flash memory, in *Digest Symposium on VLSI Circuits*, June 2009, pp. 22–23
16. R. Katsumata et al., Pipe-shaped BiCS flash memory with 16 stacked layers and multi-level-cell operation for ultra high density storage devices, in *2009 Symposium on VLSI Technology* (2009), pp. 136–137
17. Y. Fukuzumi et al., Optimal integration and characteristics of vertical array devices for ultra-high density, bit-cost scalable flash memory, in *IEDM Technical Digest* (2007), pp. 449–452
18. H. Aochi, BiCS flash as a future 3-D non-volatile memory technology for ultra high density storage devices, in *Proceedings of International Memory Workshop* (2009), pp. 1–2
19. Y. Yanagihara et al., Control gate length, spacing and stacked layers number design for 3D-Stackable NAND flash memory 2, in *IEEE IMW* (2012), pp. 84–87
20. K. Takeuchi, Scaling challenges of NAND flash memory and hybrid memory system with storage class memory and NAND flash memory, in *IEEE Custom Integrated Circuits Conference (CICC)* (2013), pp. 1–6
21. A. Nitayama et al., Bit cost scalable (BiCS) flash technology for future ultra high density storage devices, in *2010 International Symposium on VLSI Technology Systems and Applications (VLSI TSA)*, Apr 2010, pp. 130–131
22. Y. Komori et al., Disturbless flash memory due to high boost efficiency on BiCS structure and optimal memory film stack for ultra high density storage device, in *IEDM Technical Digest* (2008), pp. 851–854
23. M. Ishiduki et al., Optimal device structure for pipe-shaped BiCS flash memory for ultra high density storage device with excellent performance and reliability, in *IEDM Technical Digest* (2009), pp. 625–628
24. T. Maeda et al., Multi-stacked 1G cell/layer pipe-shaped BiCS flash memory, in *Digest Symposium on VLSI Circuits*, June 2009, pp. 22–23
25. R. Katsumata et al., Pipe-shaped BiCS flash memory with 16 stacked layers and multi-level-cell operation for ultra high density storage devices, in *2009 Symposium on VLSI Technology* (2009), pp. 136–137
26. J. Kim et al., Novel 3-D structure for ultra high density flash memory with VRAT (vertical-recess-array-transistor) and PIPE (planarized integration on the same plane), in *2008 IEEE Symposium on VLSI Technology* (2008)
27. J. Kim et al., Novel vertical-stacked-array-transistor (VSAT) for ultra-high-density and cost-effective NAND flash memory devices and SSD (solid state drive), in *2009 IEEE Symposium on VLSI Technology* (2009)
28. H.T. Lue, T.H. Hsu et al., A highly scalable 8-layer 3D Vertical-Gate (VG) TFT NAND flash using junction-free buried channel BE-SONOS device, in *VLSI Symposia on Technology* (2010)
29. J. Jang et al., Vertical cell array using TCAT (terabit cell array transistor) technology for ultra high density NAND flash memory, in *2009 IEEE Symposium on VLSI Technology* (2009)
30. W. Cho et al., Highly reliable vertical NAND technology with biconcave shaped storage layer and leakage controllable offset structure, in *2010 Symposium on VLSI Technology (VLSIT)* (2010), pp. 173–174
31. J. Elliott, E.S. Jung, Ushering in the 3D memory era with V-NAND, in *Proceedings of Flash Memory Summit* (Santa Clara, CA, 2013), www.flashmemorysummit.com
32. K.-T. Park, Three-dimensional 128 Gb MLC vertical NAND flash memory with 24-WL stacked layers and 50 MB/s high-speed programming, in *IEEE ISSCC, Digest Technical Papers*, Feb 2014, pp. 334–335

33. J.-W. Im, 128 Gb 3b/cell V-NAND flash memory with 1 Gb/s I/O rate, in *IEEE International Solid-State Circuits Conference*, Feb 2015, pp. 130–131
34. D. Kang et al., 256 Gb 3b/Cell V-NAND flash memory with 48 stacked WL layers, in *IEEE International Solid-State Circuits Conference (ISSCC)*, Digest Technical Papers, Feb 2016, pp. 130–131
35. C. Kim et al., A 512 Gb 3b/cell 64-Stacked WL 3D V-NAND flash memory, in *2017 IEEE International Solid-State Circuits Conference Digest of Technical Papers (ISSCC)*, Feb 2017, pp. 202–203
36. K.-T. Park, Three-dimensional 128 Gb MLC vertical NAND flash memory with 24-WL stacked layers and 50 MB/s high-speed programming. IEEE J. Solid-State Circuit **50**(1) (2015)
37. K.T. Park, A world's first product of three-dimensional vertical NAND flash memory and beyond, in *NVMTS*, 27–29 Oct 2014
38. E. Choi et al., Device considerations for high density and highly reliable 3D NAND flash cell in near future, in *IEEE International Electron Devices Meeting* (2012), pp. 211–214
39. K. Shim et al., Inherent issues and challenges of program disturbance of 3D NAND flash cell, in *IEEE International Memory Workshop* (2012), pp. 95–98
40. J.-W. Im, 128 Gb 3b/cell V-NAND flash memory with 1 Gb/s I/O rate. J. Solid-State Circuit **51**(1) (2016)
41. R. Yamashita et al., A 512 Gb 3b/cell flash memory on 64-Word-Line-Layer BiCS technology, in *2017 IEEE International Solid-State Circuits Conference Digest of Technical Papers (ISSCC)*, Feb 2017, pp. 196–197
42. T. Endoh et al., Novel ultra high density flash memory with a stacked-surrounding gate transistor (S-SGT) structured cell, in *IEDM Technical Digest* (2001), pp. 33–36
43. T. Endoh et al., Novel ultra high density flash memory with a stacked-surrounding gate transistor (S-SGT) structured cell. IEEE Trans. Electron Devices **50**(4), 945–951 (2003)
44. T. Endoh et al., Floating channel type SGT flash memory, in *The 1999 Joint International Meeting*, Hawaii, vol. 99-2, Abstract No. 1323, 17–22 Oct 1999
45. M.S. Seo et al., The 3-dimensional vertical FG nand flash memory cell arrays with the novel electrical S/D technique using the extended sidewall control gate (ESCG), in *Proceedings of IEEE International Memory Workshop* (2010), pp. 1–4
46. M.S. Seo et al., 3-D vertical FG NAND flash memory with a novel electrical S/D technique using the extended sidewall control gate. IEEE Trans. Electron Devices **58**(9) (2011)
47. S. Whang et al., Novel 3-dimensional dual control gate with surrounding floating-gate (DC-SF) NAND flash cell for 1 Tb file storage application, in *Proceedings of International Electron Devices Meeting (IEDM)* (2010), pp. 668–671
48. Y. Noh et al., A new metal control gate last process (MCGL process) for high performance DC-SF (dual control gate with surrounding floating gate), in *3D NAND flash memory in Symposium on VLSI Technology* (2012), pp. 19–20
49. R. Micheloni, L. Crippa, Multi-bit NAND flash memories for ultra high density storage devices (Chap 3), in *Advances in Non-volatile Memory and Storage Technology*, ed. by Y. Nishi (Woodhead Publishing, 2014)
50. R. Micheloni et al., High-capacity NAND flash memories: XLC storage and single-die 3D (Chap 7), in *Memory Mass Storage*, ed. by G. Campardo et al. (Springer, 2011)
51. H. Yoo et al., New read scheme of variable Vpass-read for dual control gate with surrounding floating gate (DC-SF) NAND flash cell, in *Proceedings of 3rd IEEE International Memory Workshop* (2011), pp. 1–4
52. S. Aritome et al., Advanced DC-SF cell technology for 3-D NAND flash. IEEE Trans. Electron Devices **60**(4), 1327–1333 (2013)
53. M.S. Seo et al., A novel 3-D vertical FG nand flash memory cell arrays using the separated sidewall control gate (S-SCG) for highly reliable MLC operation, in *Proceedings of 3rd IEEE International Memory Workshop (IMW)* (2011), pp. 1–4
54. M.S. Seo et al., Novel concept of the three-dimensional vertical FG nand flash memory using the separated-sidewall control gate. IEEE Trans. Electron Devices **59**(8), 2078–2084 (2012)

55. K. Parat, C. Dennison, A floating gate based 3D NAND technology with CMOS under array, in *Conference on International Electron Devices Meeting (IEDM)* (San Francisco, USA, Dec 2015)
56. T. Tanaka et al., A 768 Gb 3 b/cell 3D-floating-gate NAND flash memory, in *2016 IEEE International Solid-State Circuits Conference (ISSCC), Digest of Technical Papers* (San Francisco, USA, 2016), pp. 142–143
57. Eun-Seok Choi; Sung-Kye Park, Device considerations for high density and highly reliable 3D NAND flash cell in near future, in *2012 IEEE International Electron Devices Meeting (IEDM)*, 10–13 Dec 2012, pp. 9.4.1–9.4.4
58. Subirats et al., Impact of discrete trapping in high pressure deuterium annealed and doped poly-Si channel 3D NAND macaroni, in *2017 IEEE International Reliability Physics Symposium (IRPS)*
59. L. Breuil, Improvement of poly-Si channel vertical charge trapping NAND devices characteristics by high pressure D2/H2 annealing, in *2016 IEEE 8th International Memory Workshop (IMW)*
60. E. Capogreco et al., MOVPE In1-xGaxAs high mobility channel for 3-D NAND Memory, in *2015 IEEE International Electron Devices Meeting (IEDM)*
61. J.G. Lisoni et al., Laser thermal anneal of polysilicon channel to boost 3D memory performance, in *2014 Symposium on VLSI Technology (VLSI-Technology), Digest of Technical Papers*
62. Ki-Tae Park et al., Three-dimensional 128 Gb MLC vertical nand flash memory with 24-WL stacked layers and 50 MB/s high-speed programming. IEEE J Solid-State Circuits **50**(1), 204–213 (2015)
63. J. Im et al., A 128 Gb 3b/cell V-NAND flash memory with 1 Gb/s I/O rate, in *2015 IEEE International Solid-State Circuits Conference, Digest of Technical Papers (ISSCC)*, Feb 2015, pp. 23–25
64. T. Tanaka et al., 7.7 A 768 Gb 3b/cell 3D-floating-gate NAND flash memory, in *2016 IEEE International Solid-State Circuits Conference (ISSCC)* (San Francisco, CA, 2016), pp. 142–144
65. S. Aritome, NAND flash memory revolution, in *2016 IEEE 8th International Memory Workshop (IMW)* (Paris, 2016), pp. 1–4
66. C.-P. Chen et al., Study of fast initial charge loss and its impact on the programmed states Vt distribution of charge-trapping NAND Flash, in *2010 IEEE International Electron Devices Meeting (IEDM)*, 6–8 Dec 2010, pp. 5.6.1, 5.6.4
67. H.-T. Lue, S.-Y. Wang, E.-K. Lai, K.-Y. Hsieh, R. Liu, C. Y. Lu, A BESONOS (Bandgap Engineered SONOS) NAND for post-floating gate era flash memory, in *Symposium on VLSI Technology* (2007)
68. K.-S. Shim et al., Inherent issues and challenges of program disturbance of 3D NAND flash cell, in *2012 4th IEEE International Memory Workshop (IMW)*, 20–23 May 2012, pp. 1–4
69. H.S. Yoo et al., Modeling and optimization of the chip level program disturbance of 3D NAND Flash memory, in *2013 5th IEEE International Memory Workshop (IMW)*, 26–29 May 2013, pp. 147–150
70. R. Micheloni (ed.), *3D Flash Memories* (Springer, 2016)
71. K.-T. Park et al., 19.5 three-dimensional 128 Gb MLC vertical NAND Flash-memory with 24-WL stacked layers and 50 MB/s high-speed programming, in *2014 IEEE International Solid-State Circuits Conference Digest of Technical Papers (ISSCC)*, pp. 334–335,Feb 9-13, 2014
72. S. Aritome, Scaling challenges beyond 1Xnm DRAM and NAND Flash, in *Joint Rump Session in VLSI Symposium 2012*

Chapter 6
NAND Flash Design

Luca Crippa and Rino Micheloni

Abstract A Solid-State-Disk is made up by a Flash controller plus a bunch of NAND Flash devices. This chapter focuses on design aspects of NAND chips. The information stored in each memory cell is fully analog because it is related to the number of electrons stored in the floating gate. When we program, erase or read, electrons must be injected, extracted and counted, respectively. All these operations require a mix of analog and digital circuits that need to be properly and timely driven. Starting from a generic floorplan of a NAND memory, we guide the reader through the main building blocks. First of all, we describe the logic part of the chip, from the embedded microcontroller, who is in charge of running all the internal algorithms, to the fast DDR interface. Counting the number of electrons in the floating gate is definitely one of the most challenging task, considering that has to be performed with few transistors: sensing techniques are described in Sect. 6.5. Programming and erasing floating gate cells require voltages higher than the chip power supply. Therefore, charge pumps are used to generate all the needed voltages within the chip. In multilevel storage, cell's gate biasing voltages need to be very accurate and voltage regulators become a must. All these circuits are described in the High Voltage Management section. Last but not least, the row decoder is introduced. This circuit has the task of properly biasing each single wordline in the NAND array, transferring the regulated high voltages to the gate of the memory cell.

6.1 NAND Flash Memories

A NAND chip contains a lot of different circuits, both digital and analog. Figure 6.1 sketches a floorplan of a Flash device. The basic architecture of the NAND array has already been presented in Chap. 2. With reference to Fig. 6.1, the memory array

L. Crippa (✉) · R. Micheloni
Storage Solutions, Microsemi Corporation, Vimercate, MB, Italy
e-mail: luca.crippa@ieee.org

R. Micheloni
e-mail: rino.micheloni@ieee.org

R. Micheloni et al. (eds.), *Inside Solid State Drives (SSDs)*,
Springer Series in Advanced Microelectronics 37,
https://doi.org/10.1007/978-981-13-0599-3_6

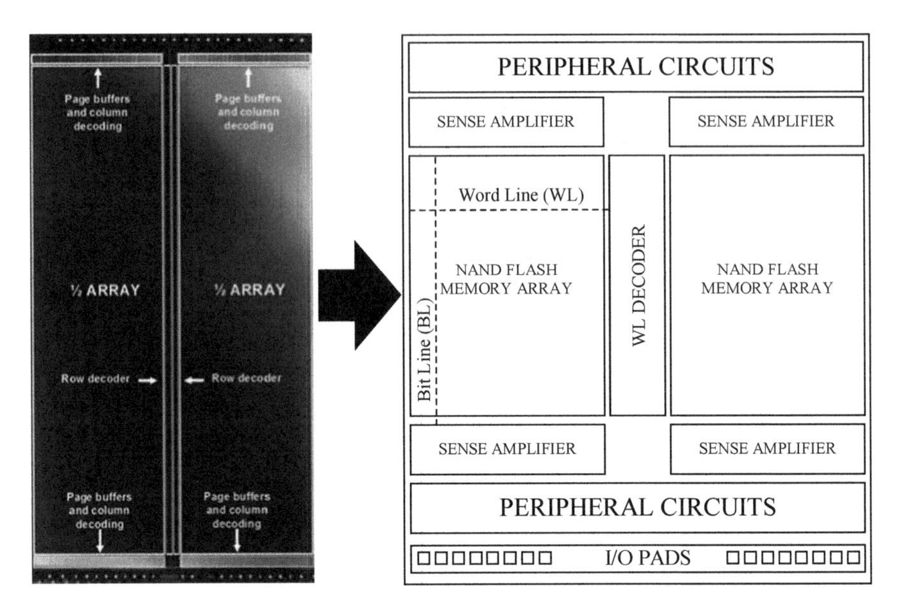

Fig. 6.1 A typical NAND Flash floorplan [1]

has been split in two independent planes. On the horizontal direction a wordline (WL) is highlighted, while a bitline (BL) is shown in the vertical direction. The Row Decoder is the block in charge of addressing and biasing each single wordline and it is located between the planes. BLs are connected to a sensing circuit (Sense Amp). The purpose of sense amplifiers is to read the analog information stored in the memory cell. In the periphery, we find charge pumps, voltage regulators, reference circuits, digital circuits, and redundancy structures. This chapter gives an overview of all the above mentioned circuits.

6.2 Logic Device View

Let's start our analysis from the peripheral circuits. First of all, we have the "Logic", a set of digital gates which enables the communication to the external host and manages data inside the device. In other words, it is the real brain of the memory.

We can identify some basic logic blocks, as shown in Fig. 6.2.

1. Control Interface (CI) [2–4]. It is the command interface between the NAND Flash and the external user;
2. Microcontroller. It stores and executes all the internal algorithms, such as read, program, erase and testmode operations.

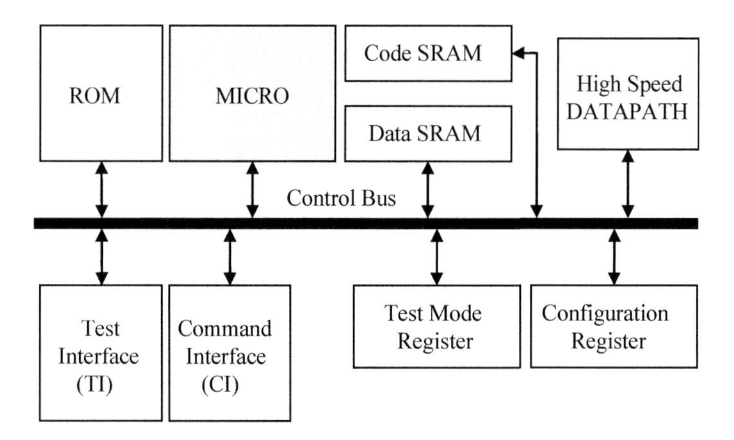

Fig. 6.2 Logic view of a NAND device

3. Error Correction Code (ECC) [5] could be embedded in the memory device. ECC improves the reliability of the read operation.
4. Memory testing is a fundamental functionality. For this reason, there is a Test Interface (TI) block, i.e. the interface to the user when device is in test mode.
5. Datapath. Basically, it is the fast link between I/Os and read circuits.
6. There are also a lot of registers, mainly for storing the configurations of the analog circuitry.
7. Redundancy: it can be managed by the microcontroller or it can be implemented as a finite state machine (FSM). This logic is used to increase the wafer yield.

6.2.1 Command Interface

In order to talk with the external user, Flash memory has to understand commands, take data and output data.

The logic block implementing this functionality is basically a finite state machine and is represented by the Command Interface (CI) when the device is in user mode and by the Test Interface (TI) when the device is in test mode.

CI understands legal or illegal command sequences, defined in the device specifications and interacts with other logic blocks as datapath and microcontroller. Control signals have been already described in Chap. 2. CI is composed by a huge finite state machine clocked by WE# and driven by all I/O signals such as ALE or CLE. Figure 6.3 represents CI and its interaction blocks.

1. I/Os are all control signals: R/B#, CLE, ALE, WP#, WE#, RE#, CE#, DQ[7:0].
2. Reset Interface exchanges reset information with logic global reset.

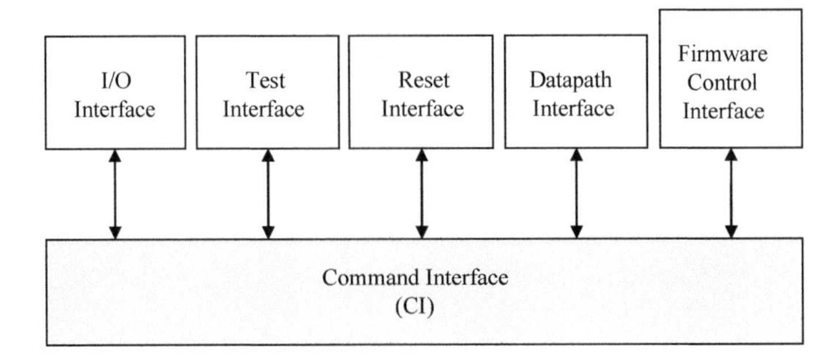

Fig. 6.3 Command interface and its interaction blocks

3. Datapath interface controls input and output datapaths.
4. Test interface toggles between user mode and test mode.
5. Firmware Control Interface enables microcontroller to execute internal algorithms.

CI is made up by multiple finite state machines, one for each basic function. The Command Interface Controller disables a specific FSM if that specific command is not allowed. During power up, CI Controller disables every commands, so that all the FSMs are disabled too. There is also a FSM that recognizes if a specific command is a read, a program or an erase and enables the correct sub-FSM. Every time the Controller receives an illegal sequence, the device goes into an IDLE state.

When the internal microcontroller executes a specific algorithm, the device is busy. In this situation, the only commands that the CI can accept are a reset and a testmode entry command.

6.2.2 Test Interface

Test Interface (TI) is used when we want to test some particular features, usually not accessible during normal operations (usermode). Test Interface is enabled by a specific command sequence, called testmode entry. Generally speaking, a NAND device can have these modes:

- Usermode that represents the standard functionality, where commands described in the device specification are available;
- Usertestmode that represents the standard functionality plus some particular commands;
- Testmode that is the test operational mode.

Figure 6.4 represents how it is possible to change the operational modes with proper command sequences recognized by the CI Controller.

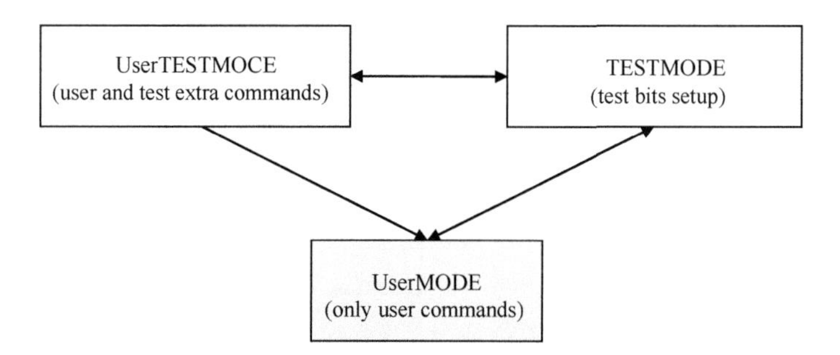

Fig. 6.4 Flow diagram used to change operational modes among usermode, usertestmode and testmode

Once TI is enabled, it substitutes CI: TI recognizes the command set and drives input and output data/address on the logic bus. Test Interface is allowed to access the different registers and different memory circuits without the aid of the microcontroller.

TI is built as a finite state machine in a similar way to the Command Interface.

Let's now explain what testmode registers are. All the circuits added for test purposes can't influence the standard user mode functionality and can't worsen performances. The adopted solution is sketched in Fig. 6.5. A TM register is associated with a UM register: when the signal TESTMODE is high, the output takes the value contained in the register TM, influencing the behavior of the circuitry downstream. When the signal TESTMODE goes low, the standard usermode functionality is enabled.

6.2.3 Datapath

Till few years ago, NAND memories had an asynchronous interface and it was very difficult to run frequencies higher that 40 MHz for data download/upload [6]. NAND chips have linear dimensions easily higher than 10 mm so that data have to flow through a long path with an unavoidable impact on the transmission time through the chip. One of the most adopted solutions to overcame this problem is the use of a pipeline on the datapath [7].

In the following we will describe datapath structure for a NAND memory with double side architecture and with control pads on the opposite side with respect to data pads.

With reference to Fig. 6.6 the data input sequence is here described.

1. During the low-phase of WE#, input buffers on I/O PADS block and latches on DP_UP and DP_DW blocks are enabled. In this way, input data flow to the latches placed in DP_UP and DP_DW blocks.

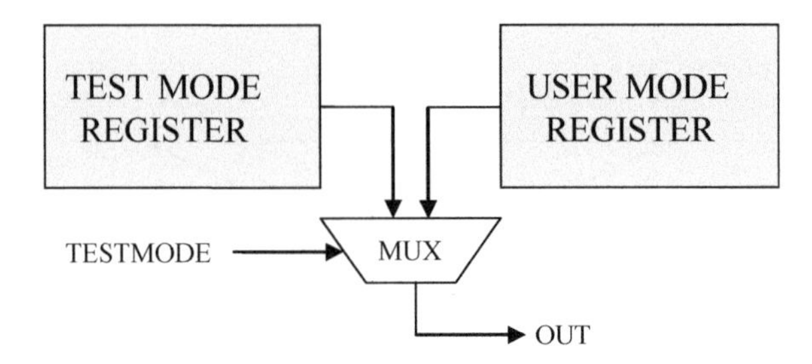

Fig. 6.5 Testmode registers

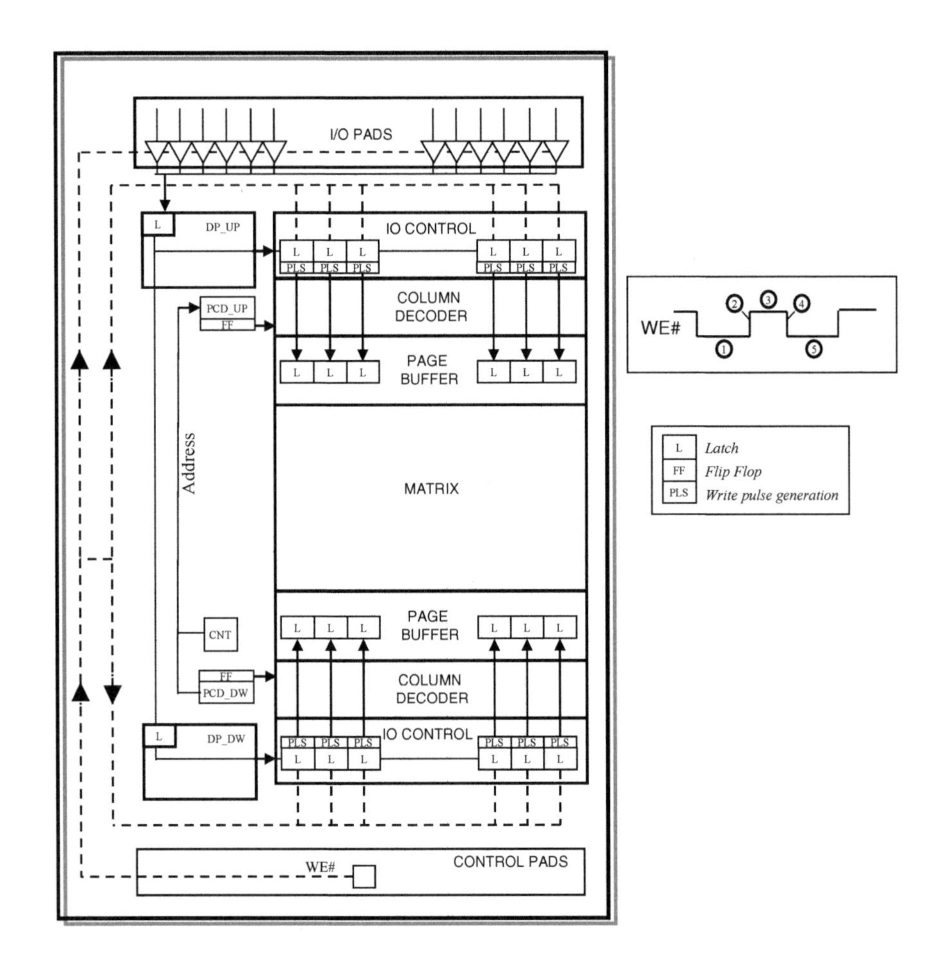

Fig. 6.6 Input datapath

2. On the rising edge of WE#, I/O PADS input buffers are disabled. Data are latched in DP_UP latches till the next falling edge of WE#. The counter addresses the appropriate page buffers for the following write operation.
3. On the high-phase of WE#, IO CONTROL latches are open and the COLUMN DECODER is addressing the right page buffers.
4. On the falling edge of WE#, data are latched in the IO CONTROL latches.
5. On the next low-phase of WE#, while I/O PADS input buffers and DP_UP latches receive new data from the user (as in phase 1), IO CONTROL generates write pulses for loading the latched data into the page buffer latches.

A similar approach is adopted for data output.

Performance driven applications like Solid-State-Disks (SSDs) are now forcing the NAND towards the adoption of a DDR interface, as described in Sect. 6.3.

6.2.4 Microcontroller

As already said, the microcontroller inside the memory is the "brain" of the device. Microcontroller implements the needed algorithms for a Flash memory. In order to be able to perform the necessary operations, these conditions must hold true:

- each sequence of operations that must be executed for a specific algorithm (read, program, erase etc.) has to be non-volatile;
- the microcontroller needs to perform arithmetical, logical and output operations.

Usually, microcode (FW) is stored in a ROM memory (Fig. 6.7). There could also be a Code RAM memory containing the specific firmware for testing and debugging.

The microcontroller contains a number of different blocks. First of all there is the Program Counter. It stores the address of the memory location containing the instruction that must be executed. It is also able to handle the address increment, the absolute or relative jumps and the calls to subroutines with different stack levels. The levels of stack indicate how a subroutine is far away from the main program.

Another important block is composed by the Internal Registers: they are necessary for the execution of an operation or a sequence of operations. A register can be either loaded with a constant value or with a value read from the ROM, and it can also be the result of an operation.

The microcontroller computational center is the Arithmetic Logic Unit or ALU. The ALU executes an operation associated with a specific opcode and implemented in the microcontroller. The operations can be with one or two operands. The operands can be internal registers, flags or constants read from the ROM. The result of the operation is stored in the internal registers, with the exception of test and compare operation.

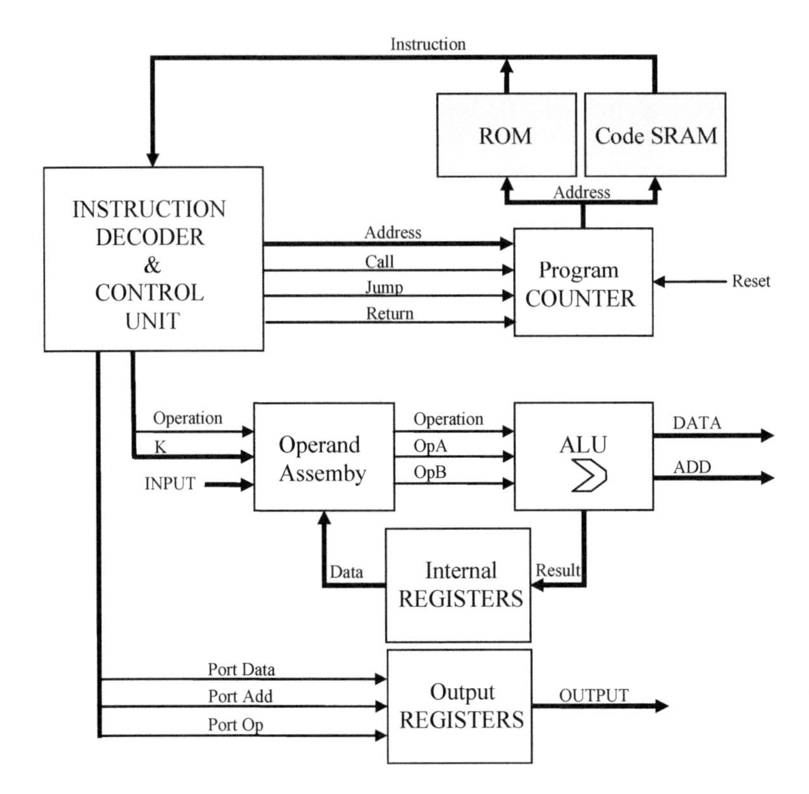

Fig. 6.7 Microcontroller structure with ROM and RAM memories

Finally, the last block of the microcontroller is constituted by the Output Registers. Each register is made up by a number of latches. The most advantageous structure for the output registers is based on the dual ports concept.

With this structure, the registers are handled by two independent ports called port A and port B. For instance, port A operates over all the outputs, while port B operates only over some output registers.

The dual ports structure allows the use of two different bank registers at the same time, so that it is possible to move more control signals at each clock cycle.

Apart from the internal structure, the characterizing feature of a microcontroller is what it is able to do, that is its Instruction Set. Before designing a microcontroller, we need to understand the must-have operations. In fact, general purpose microcontrollers are not useful in the NAND memory environment, because they are generally bigger and slower, in order to guarantee a full flexibility not needed in the device. In other words, it is useless to implement operations not used, but it is better to optimize the used ones.

6.3 NAND DDR Interface

Flash based systems are made up by several NAND memory devices and one controller. The controller has the primary function to communicate with NANDs and conveys data from/towards the external interface. Especially, SSDs call for a higher Read/Write throughputs; in other words, SSDs need to manage more NAND dies in parallel. Basically, there are a couple of options.

The first one is to increase the number of dies per channel as shown in Fig. 6.8a. This solution encounters limitations from channel parasitic loading. It has the advantage of lower pin count and lower hardware cost, especially for the controller, but it might not satisfy the requirements of Write throughput.

The second option is to increase the number of channels (Fig. 6.8b). This solution shifts all the problems inside the memory controller which has to manage the parallel data flow coming from all the memory channels. The drawback is that

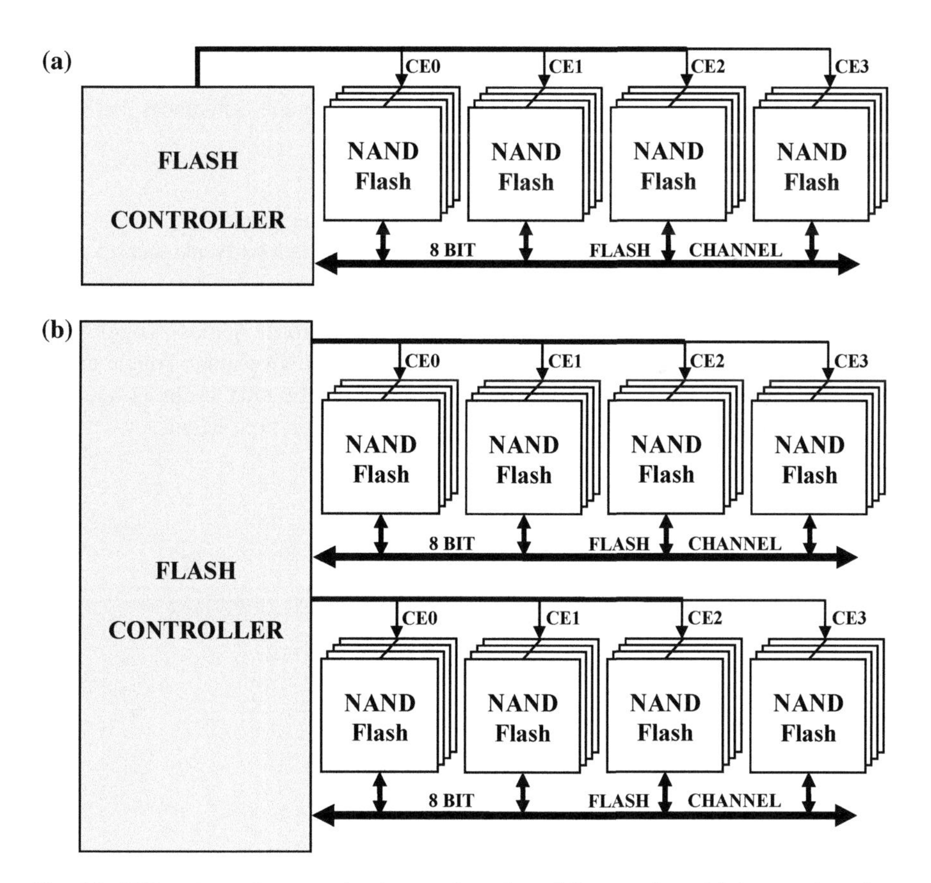

Fig. 6.8 SSD system enhancement: **a** increased number of dies per channel **b** increased number of channels

the controller has to manage the ECC for each channel and have the need of dedicated SRAM. On the positive side, this solution is scalable and flexible and allows to reach very high Read/Write throughput. Nowadays, multiple channel architectures are quite common in SSD design.

In every case, power and signal integrity must be addressed with careful interface design considerations. In this section, we mainly deal with the I/O bottleneck problem which must first be solved by a proper interface roadmap.

6.3.1 DDR Interface

High speed NAND introduced a Double Data Rate (DDR) interface in year 2008. As a matter of fact, NAND memories are now following the same path that DRAMs experienced from year 2000.

The challenge in the coming years will be the standardization of the interface among vendors. Two solutions are available in the market, as draft in Fig. 6.9. On one side, ONFI organization [8] introduced an interface with a clock and data strobe, ready for a DRAM-like evolutionary path. Pinout differences between legacy and ONFI 2.0 interfaces are:

- WE# becomes a fast CLK;
- RE# handles data direction by becoming W/R# (Write/Read#);
- I/O[7:0] renamed to DQ[7:0] (name change only, functionally identical);
- DQS, a new bi-directional signal, is enabled.

On the other side, Samsung decided for a different approach named "Toggle" [9] where only data strobe has been added to the legacy NAND pinout; Toggle mode adds DQS data strobe signals; RE# is used to trigger the read cycle as done in asynchronous interface; DQS is used to strobe the data on both edges.

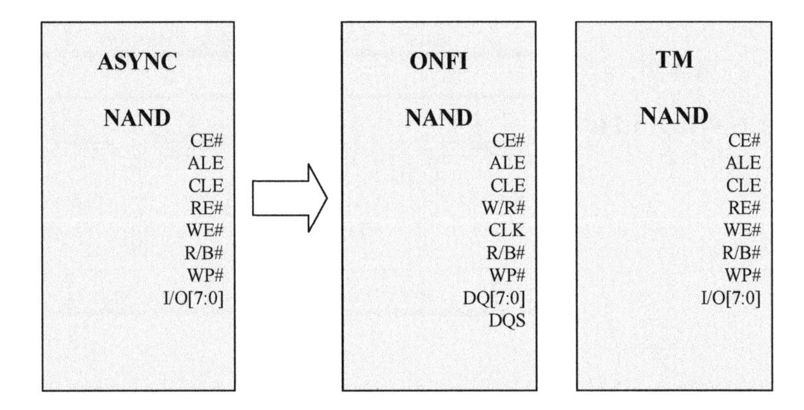

Fig. 6.9 Legacy NAND versus ONFI 2 and Toggle-mode synchronous NAND interface

As usual, JEDEC is now working on combining the above interfaces in a single standard.

ONFI has already released the third generation of specifications where they target 400 MB/s throughput, and Toggle is targeting the same speed. The interface roadmap stays with LVTTL bus driving style as long as possible in order to ease integration, but some design tricks have to be introduced in order to sustain higher bandwidths. This will include the proper scaling of interface voltage, the use of a specific termination type, On-Die Terminations, differential strobes and, going beyond, synchronization circuit. Finally, it will include DLL/PLL and the change to a SSTL class of terminated bus.

The DDR protocol diagrams are sketched in Fig. 6.10. A Synchronous clock must be provided to the memory chips (not needed in Toggle-mode interface). Bidirectional Data bus DQ is driven at every clock edge. Therefore, data throughput is doubled compared to a Single data rate system, assuming the same clock frequency.

Data strobe signal DQS behaves like all other DQs and it is used as data capture signal on the receiver side. Systems scalability benefits from this approach since DQS load always matches that of DQ lines, ensuring same timings: this is very important in SSD design because the parasitic load of a Flash channel changes when more dies are used.

6.3.2 *Power*

Let's consider an SSD where multiple Flash channels are used. Due to the channel parasitic capacitance, each time a single NAND die is written or read, the entire capacitance of data lines needs to be driven.

I/O power consumption in a DDR system can be written as [10]:

$$P = 9 \cdot \eta \cdot f \cdot C \cdot V^2 \tag{6.1}$$

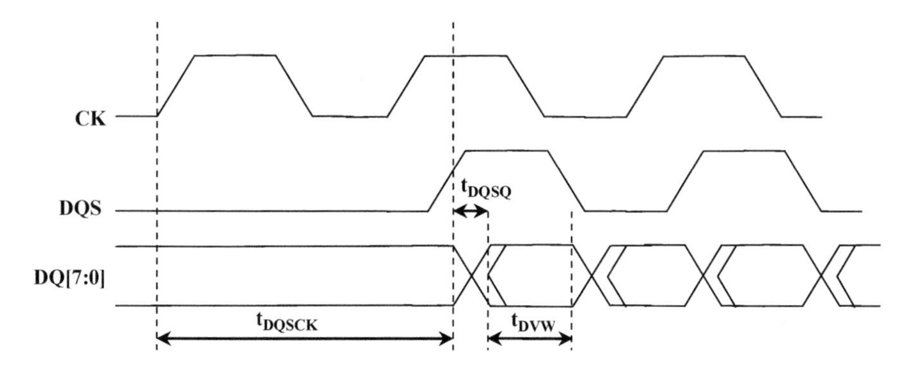

Fig. 6.10 DDR timing diagram

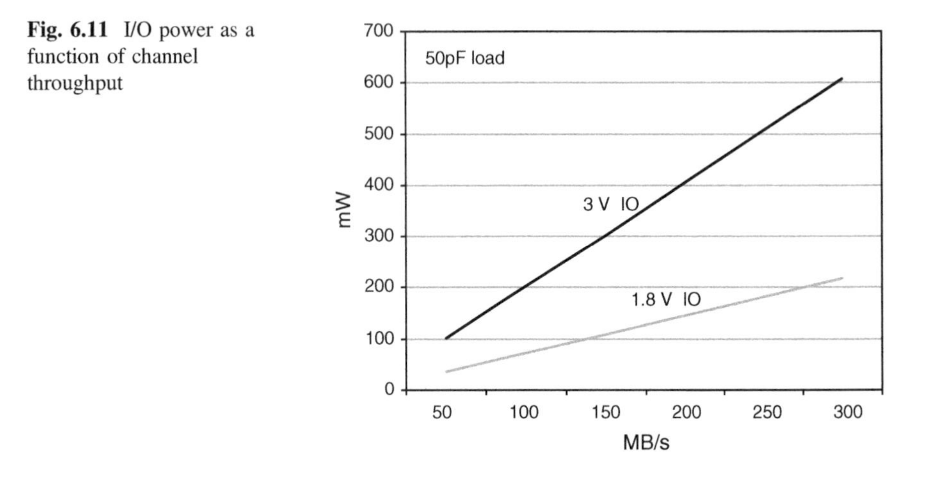

Fig. 6.11 I/O power as a function of channel throughput

where η is the bit activity ratio, f is the DDR frequency, C is the capacitance of a single line and V is the supply voltage of the interface. Figure 6.11 shows the impact of I/O power supply. Therefore, scaling the I/O interface voltage becomes a must, especially looking at higher clock frequencies.

6.3.3 Capacity

SSD storage capacity can be increased in two ways:

- by increasing the number of Flash channels;
- by increasing the number of NAND dies connected to a single channel.

As already mentioned, the first solution has been widely adopted, even if it increases the hardware complexity of the SSD controller.

The adoption of the second solution is mainly limited by the resulting I/O parasitic capacitance of the Flash channel. To partially overcome this limitation, it is possible to use advanced System in Package technologies such as *Through Silicon Vias* (TSV) [11]. TSV creates interesting opportunities for stacking, thanks to its low parasitic capacitance.

Figure 6.12 depicts a system in which memory chips are stacked and connected using a Local Interconnect Bus. The Interface Chip provides data translation from local interconnect bus to the external bus (i.e. Flash channel) by means of a standard off chip driver (OCD). It is worth mentioning that the local bus can be driven by standard CMOS buffers instead of OCD ESD-compliant structures. Furthermore, by using simplified ESD structures, the bus parasitic capacitance can become even lower.

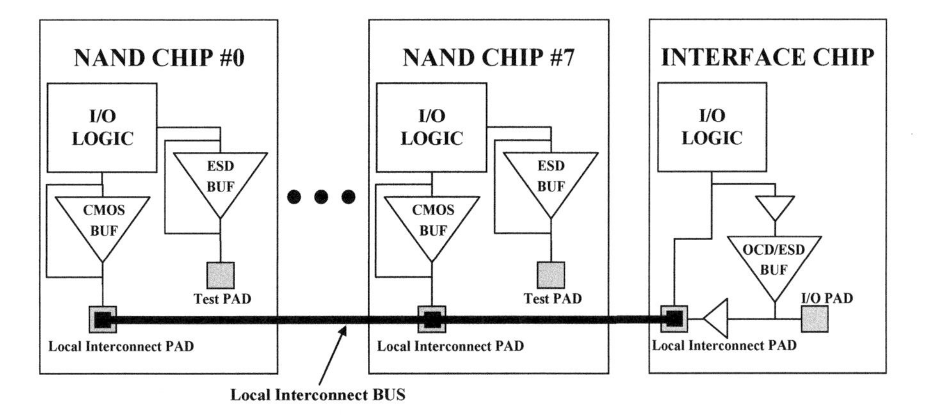

Fig. 6.12 Local interconnect bus architecture

6.4 I/O Design

This section starts with an overview of I/O design problems in legacy asynchronous NAND products available in the market. Design of high-speed I/O is then reviewed.

6.4.1 Basic CMOS Output Buffer Design

Usually, NAND output buffers need to drive large capacitive loads, in the range of 50–100 pF. In this situation the output capacitance transition is very long compared to the buffer switching time. The buffer conductance is usually made very large to reduce the charge/discharge time and match the specifications.

The memory data bus can be 8/16 bits: the current sunk by the parasitic capacitor of a single output buffer has to be multiplied by the number of switching data bits. Moreover, the inductance of the bonding wire (5–10 nH in TSOP packages) might generate bounces on internal power supply lines that could affect the functionalities of analog circuits [12]. This effect is called *Simultaneous Switching Noise* (SSN) and will be treated in more details later.

A basic output buffer with push-pull architecture is shown in Fig. 6.13.

In order to reduce the current peak, switching time of push-pull drivers have to be carefully controlled. As a consequence, if gates of PMOS and NMOS are driven at a lower speed, crowbar current becomes an issue. Crowbar occurs when both PMOS and NMOS are ON at the same time. To avoid this situation, the buffer structure of Fig. 6.14 can be adopted [3, 13]. In this configuration the pull-up is switched-off before the pull down is turned on (and vice versa).

NAND and NOR gates can be tuned to obtain a fast switching-off and a proper switching-on time. In the figure it is also shown the output enable signal OE that is used to turn the output stage in high impedance: in this way, data bus can be driven by somebody else.

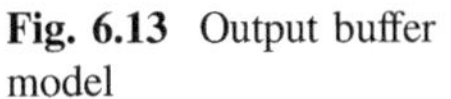

Fig. 6.13 Output buffer model

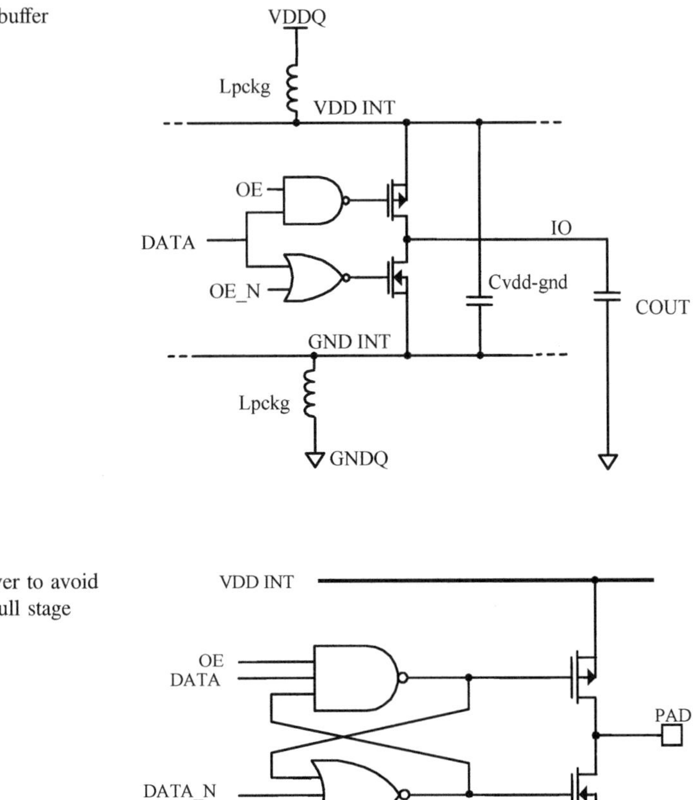

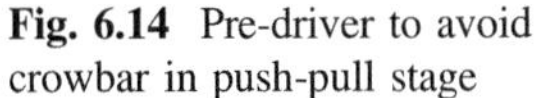

Fig. 6.14 Pre-driver to avoid crowbar in push-pull stage

Another important design constraint is the slew rate of the output driver. In asynchronous devices, the slew rate is generally controlled by acting on the pre-driver, so that the pull-up and pull-down transistors are gradually switched on/off [13, 14].

Generally, this is optimized in the slow corner and the result is a big variation with Process/Voltage/Temperature (PVT). The pre-driver *RC* output constant must be much smaller than the data window, otherwise there is a risk to have a data dependent jitter. If a wide data bus is used, it could be beneficial to consider skewing the output enable by a proper small delay and consequently spreading in time the current requests.

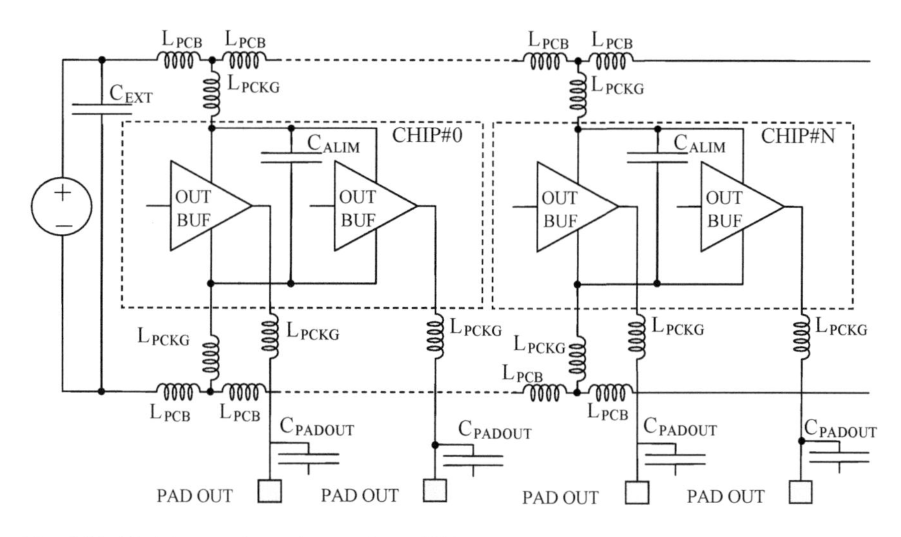

Fig. 6.15 Model example used to evaluate SSN

6.4.2 Simultaneous Switching Noise (SSN)

One of the main responsible for data window margins degradation is the simultaneous switching noise [13, 15–18]. SSN is an inductive noise caused by several outputs switching at the same time. One single buffer could have a good transient behavior, but, when all the data buffers are switching at the same time, the data AC behavior could be corrupted. The problem is serious in output buffer memory design because of two effects:

- jitter and signal bounces are increased and data window margin is reduced;
- the generated noise could affect other circuits, especially analog circuits and memory sense amplifiers, reducing operating margin or creating systematic non-working windows.

With a large capacitive load, a large current is requested to charge the load and the power network must supply that current. The current flows in inductances, typically in the bonding wires or leads of the package, and the resulting noise is injected into power and ground supplies. This noise is transferred to the output and the output AC characteristics are affected.

The simultaneous switching noise is determined, in principle, by the following equation:

$$V_{SSN} = N \cdot L \cdot \frac{\partial I}{\partial t} \tag{6.2}$$

where N is the number of switching outputs, L the equivalent inductance in which current must flow, and I the current per driver.

Since this mechanism is dependent on the number of output switching N, this makes the noise dependent also on the data sequence.

To deal correctly with SSN it is necessary to understand the complete signal current paths in the memory. In Fig. 6.15 a complete path is shown. Local metal resistances are omitted but they should be evaluated as possible sources of interference. It is straightforward to understand that the problem is really connected with the package. When TSOP packages are used, very long bonding wires can be present leading to high inductance values. Moving to higher data rates requires to leave such packages for more controllable Ball Grid Arrays.

6.4.3 High Speed NAND I/O Design

Output buffer in high speed signal transmission is often named *Off-Chip Driver* (OCD). In addition to the task of being the interface circuit between inside and outside, OCD in high speed memories has to accomplish several additional tasks.

- Translate data flow between single data rate (SDR) and DDR domains.
- Voltage domain change. The core of the memory could operate at a different voltage level than the I/O interface and the data signals have the need to be shifted from the core level to the interface voltage.
- Provide the AC/DC requirements such as V_{OL}/V_{OH}, slew rate or impedance matching.
- Provide the On-Die-Termination (ODT).
- ESD protection.

Various types of OCD are used in memory design depending on the interface type and speed. In order to introduce all the basic concepts, we focus here on the single ended CMOS buffer, which is widely used in DDR designs.

6.4.4 Double Data Rate OCD

A DDR OCD is a synchronous output buffer. In synchronous systems, OCD includes a register stage used to synchronize the output with the internal data bus. In DDR design a block named *serializer* is included in the buffer design as shown in Fig. 6.16. Serializer block performs the Single Data Rate (SDR) to DDR conversion: it receives 2n data at a given rate R (SDR) and multiplexes these data onto an internal line at a higher rate 2R (DDR).

We should highlight that the OCD is operating at a frequency higher than the one used by other blocks in the memory chip. Therefore, since we have to deal with

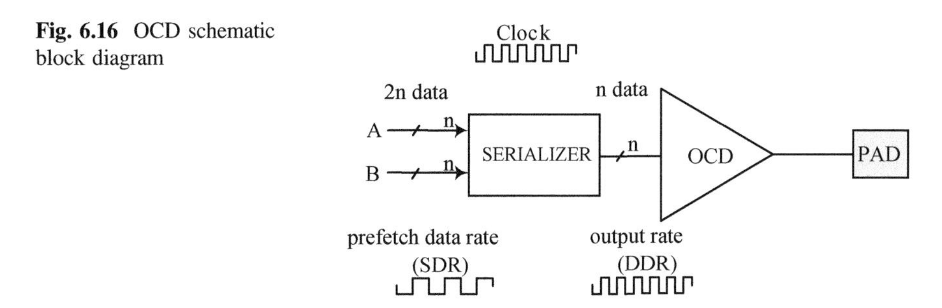

Fig. 6.16 OCD schematic block diagram

smaller delays inside the OCD, it is necessary to take more countermeasures in designing the block to avoid jitter eating almost all margins.

6.4.4.1 OCD Linearity: Push-Pull and Open-Drain Configurations

It is of primary importance to offer a linear behavior of the output characteristics because of the system signal integrity. In other words, OCD linearity is key for impedance matching with the external line.

6.4.4.2 Slew Rate Control and Bandwidth

Drivers should be designed in order to avoid driving frequencies greater than the signaling rate. Simple and sophisticated methods can be used, such as passive delays after the pre-driver or current control technique for the pre-driver stage. A time-split method is widely used. The basic principle is to split output pull-up and pull-down devices into branches and activate them serially with proper sequential delays. This time-distributed driver can be implemented in a simple analog form suitable for relative low operating frequency or digital form [13, 19, 20]. Figure 6.17 shows a basic implementation of the analog form where the pull-up/down branches are driven by a resistive line which contributes to define the *RC*

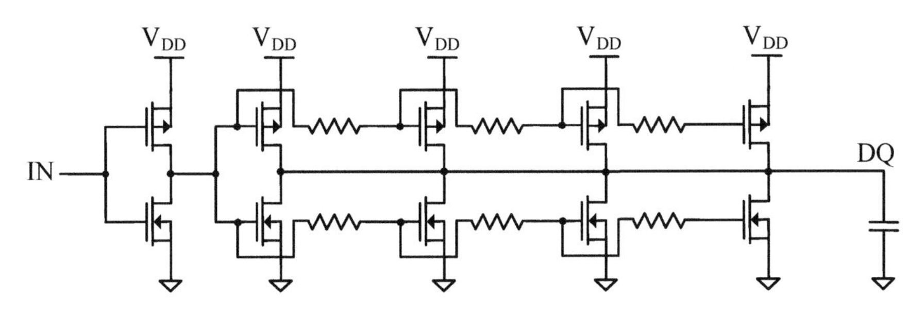

Fig. 6.17 Slew rate control by output driver time-distributed activation

delay element for each branch. Each branch can be "weighted" to obtain the best slew rate conditions.

6.4.4.3 Voltage Domain Change: Level Shifting

I/O voltage usually differs from the power supply of the NAND core. For example, the memory could internally operate at 1.5 V by means of a DC-DC down-converter, whereas the data interface needs a 3 V or 1.8 V driving. Voltage domain change occurs also when the memory has different power pins for core supply voltage and I/Os. This situation allows the use of independent supply generators to separate the noise coming from data bus and from the core region. In a simpler system design it is still possible to connect the pins to the same supply on the PCB. The OCD structure implements the level shifting function which consists in shifting the levels of the digital signals from the core voltage GND/VDD to the interface voltage GNDQ/VDDQ. Figure 6.18 shows a modified structure where NMOS transistors M5 and M6 are added in order to speed up the transition of nodes from low to high.

The level shifting circuit or, more generally, the point where the data change voltage domain, is critical in jitter generation. The two domains provide two different references for the signal detection; therefore, any disturbs on the power supply lines lead to the introduction of additional distortion.

6.4.4.4 Jitter Sources and Duty Cycle Distortion

Off-Chip Driver complexity implies that data is travelling along many gates before reaching the output stage. The design of the chain of inversions is fundamental in the control of duty cycle distortion. Duty cycle distortion occurs when:

- positive and negative slopes are different;
- number of inversion is odd;
- ground or power shifts.

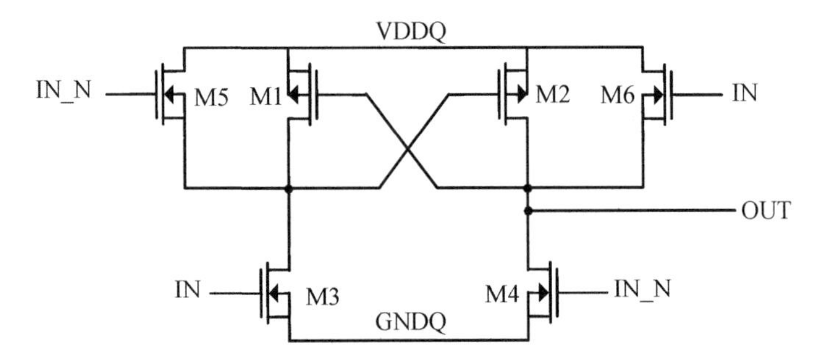

Fig. 6.18 Level shifter modified

To reduce the jitter in a chain of inverters it is necessary to keep the same slope in the chain, i.e. using the same ratio between the driver strength and the load, instead of trying to minimize the number of inverters in the chain. Another source of jitter is hidden in level shifters and voltage domain change. Level shifter sketched in Fig. 6.18 introduces asymmetric positive/negative slopes detected by a receiver gate with different time delay.

In conclusion, high-speed NANDs require a very sophisticated I/O design because of its impact on SSD's power, performances and signal integrity.

6.5 Read Operation: The Sense Amplifier

Let's now move in the core region. The reading operation is designed to address specific memory cells within the array and measure their information content. As in other types of Flash memories, the stored information is associated with the cell's threshold voltage V_{TH}: in Fig. 6.19 the threshold voltage distributions of cells containing one logic bit are shown. If the cell has a V_{TH} belonging to the erased distribution, it contains a logic "1", otherwise it contains a logic "0". Cells containing n bit of information have $2n$ different levels of V_{TH}.

Flash cells act like usual MOS transistors. Given a fixed gate voltage, the cell current is a function of its threshold voltage. Therefore, through a current measure, it is possible to understand which V_{TH} distribution the memory cell belongs to.

The fact that a memory cell belongs to a string made up by other cells has some drawbacks. First of all, the unselected memory cells must be biased in a way that their threshold voltages do not affect the current of the addressed cell. In other words, the unselected cells must behave as pass-transistors. As a result, their gate

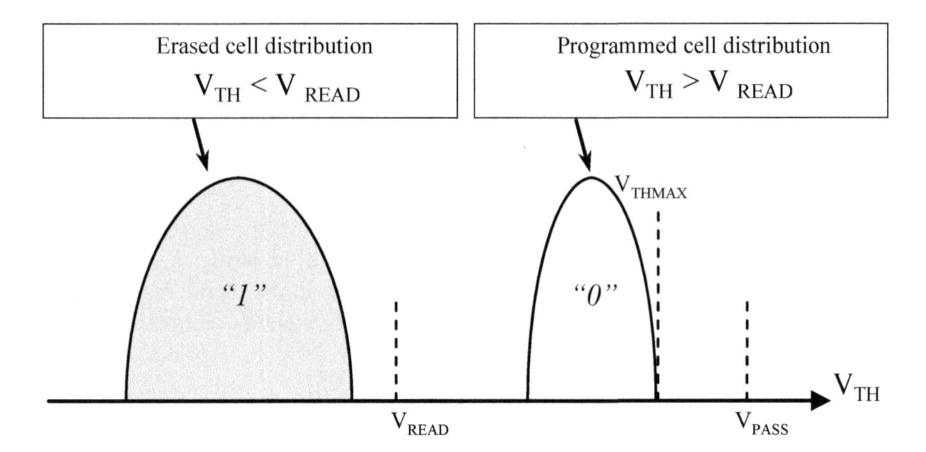

Fig. 6.19 Threshold voltage distributions of erased ("1") and programmed ("0") cells

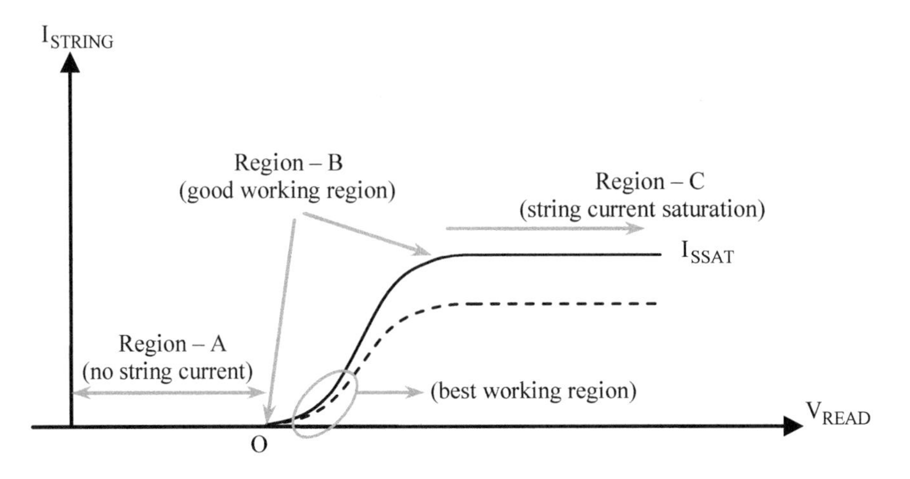

Fig. 6.20 Cell current characteristics versus gate voltage

must be driven to a voltage (commonly known as V_{PASS}) higher than the maximum possible V_{TH}. In Fig. 6.19 V_{PASS} has to be higher than V_{THMAX}.

However, the presence of $2^n - 1$ transistors in series has a limiting effect (saturation) on the current's maximum value; this maximum current is, therefore, much lower than the one available in NOR-type Flash memories.

Figure 6.20 shows the I–V (current-voltage) characteristic of a NAND cell (string): V_{READ} is applied to the selected gate while V_{PASS} bias the unselected gates. V_{PASS} is a fixed voltage. Three main string working-regions can be highlighted.

1. Region A: the addressed cell is not in a conductive state.
2. Region B: V_{READ} makes the addressed cell more and more conductive.
3. Region C: the cell is completely ON, but the series resistance of the pass transistors (unselected cells) limits the current to I_{SSAT}.

The string current in region C can be estimated as:

$$I_{SSAT} = \frac{V_{BL}}{(n-1)R_{ON}} \tag{6.3}$$

where R_{ON} is the series resistance of a single memory cell, V_{BL} is the voltage applied to the bitline and n is the number of the cells in the string. $R_{ON,}$ at a first approximation, is the resistance of a transistor working in the ohmic region.

For a MOS transistor in ohmic region the following equation holds true:

$$I_D = k \cdot \left[(V_{GS} - V_{TH}) \cdot V_{DS} - \frac{V_{DS}^2}{2} \right] \tag{6.4}$$

For small V_{DS} values, as in our case, (6.4) may be simplified as:

$$I_D = k[(V_{GS} - V_{TH}) \cdot V_{DS}] \tag{6.5}$$

Therefore, R_{ON} is equivalent to

$$R_{ON} = \frac{V_{DS}}{I_D} = \frac{1}{k(V_{GS} - V_{TH})} \tag{6.6}$$

Equation (6.6) shows that R_{ON} is a function of V_{TH}. In other words, I_{SSAT} depends on the V_{TH} values of the *n* cells in series. When all the cells are programmed to V_{THMAX}, R_{ON} takes its maximum value (dashed line in Fig. 6.20). R_{ON} influences the I–V characteristic also in region B but in a more negligible way. In order to reduce the dependency from $R_{ON,}$ the cell has to be read in region B as near as possible to point O.

The order of magnitude of the saturation current, in the state-of-the-art NAND technologies, is a few hundreds of nA, that means a reading current of some tens of nA. It is very hard to sense such small currents with the standard techniques used in NOR-type Flash memories, where the reading current is, at least, in the order of some μA. Moreover, in NAND devices, tens of thousands of strings are read in parallel. Therefore, tens of thousands of reading circuits are needed. Due to the multiplicity, a single reading circuit has to guarantee a full functionality with a very low area impact. As a matter of fact, the first memory NAND prototypes used traditional sensing methods, since the said currents were in the order of tens of μA [21].

The reading method of the Flash NAND memories consists in integrating the cell current on a capacitor in a fixed time (Fig. 6.21). The voltage ΔV_C across a capacitor *C*, charged by a constant current *I* for a time period ΔT, is described by the following equation:

$$\Delta V_C = \frac{I}{C} \Delta T \tag{6.7}$$

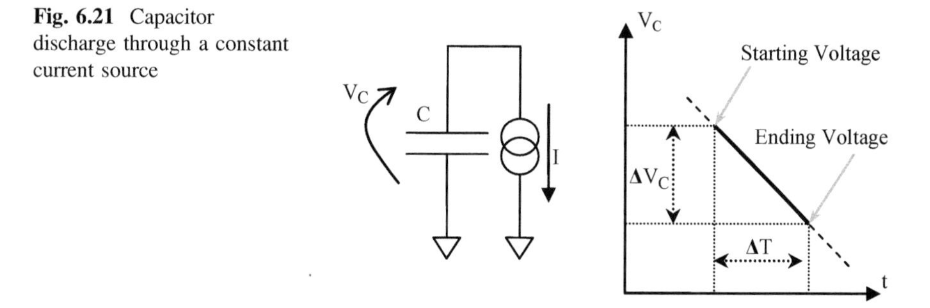

Fig. 6.21 Capacitor discharge through a constant current source

Since the cell current is related to its V_{TH}, the final voltage on the capacitor (ΔV) is a function of V_{TH} too.

There are different reading techniques, starting from the one using the bitline parasitic capacitor, ending with the most recent sensing technique which integrates the current on a little dedicated capacitor. The above mentioned techniques can be used both in SLC and MLC NAND memories. In the MLC case, multiple basic reading operations are performed at different gate voltages.

Historically, the first reading technique used the parasitic capacitor of the bitline as the element of the cell current integration [22–24].

In Fig. 6.22 the basic scheme is shown. V_{PRE} is a constant voltage. At the beginning, C_{BL} is charged up to V_{PRE} and then it is left floating (T_0). At T_1 the string is enabled to sink current (I_{CELL}) from the bitline capacitor. The cell gate is biased at V_{READ}. If the cell is erased, the sunk current is higher than (or equal to) I_{ERAMIN}. A programmed cell sinks a current lower than I_{ERAMIN} (it can also be equal to zero). C_{BL} is connected to a sensing element (comparator) with a trigger voltage V_{THC} equal to V_{SEN}. Since I_{ERAMIN}, C_{BL}, V_{PRE} and V_{SEN} are known, it follows that the shortest time (T_{EVAL}) to discharge the bitline capacitor is equal to:

$$T_{EVAL} = C_{BL}\frac{V_{PRE} - V_{SEN}}{I_{ERAMIN}} \tag{6.8}$$

If the cell belongs to the written distribution, the bitline capacitor will not discharge below V_{SEN} during T_{EVAL}. As a result, the output node (OUT) of the voltage comparator remains at 0. Otherwise, if the cell is erased, V_{BL} drops below V_{SEN} and the OUT signal is set to 1.

The basic sense amplifier structure is sketched in Fig. 6.23. During the precharge phase T_{PRE}, M_{SEL} and M_{PCH} are biased to V_{PRE} and $V_{DD} + V_{THN}$ respectively. V_{THN} is the threshold voltage of a NMOS transistor and V_{DD} is the device's power supply voltage.

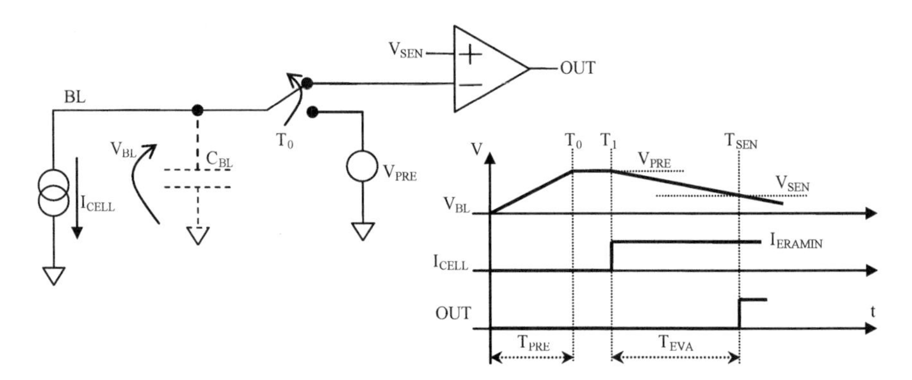

Fig. 6.22 Basic sensing scheme exploiting bitline capacitance and the related timing diagram

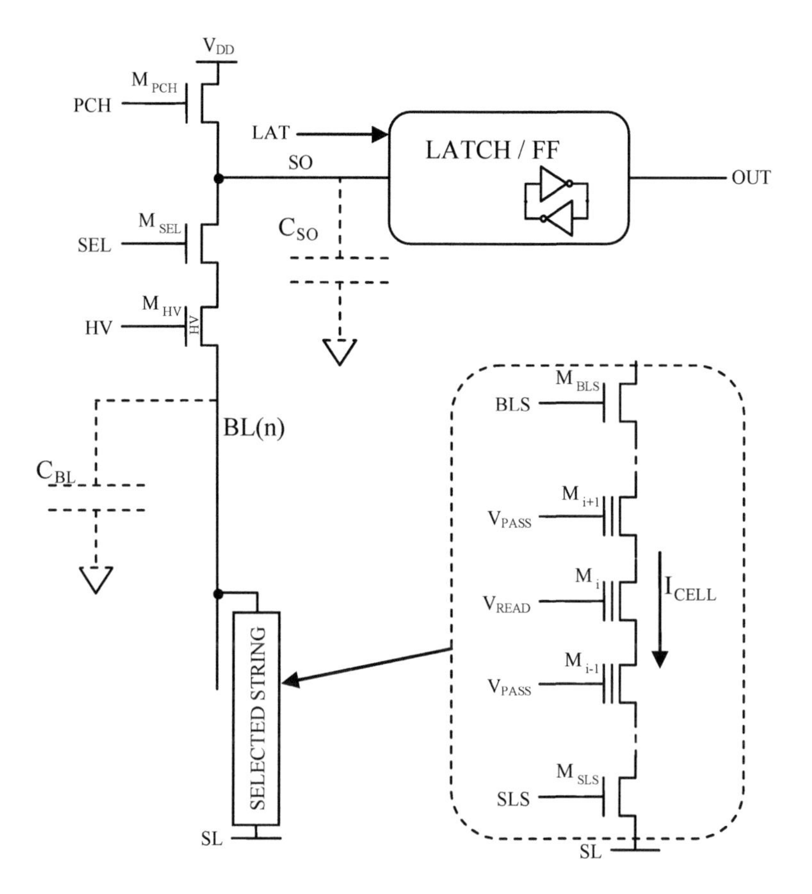

Fig. 6.23 Basic elements of the sense amplifier

As a consequence, C_{BL} is charged to the following value:

$$V_{BL} = V_{PRE} - V_{THN} \tag{6.9}$$

During this phase, the SO node charges up to V_{DD}. Since V_{GS} and V_{DS} can be higher than 20–22 V, M_{HV} has to be a high voltage (HV) transistor. In fact, during the erase phase, the bitlines are at about 20 V and M_{HV} acts as a protection element for the sense amplifier's low voltage components. Instead, during the reading phase, M_{HV} is biased at a voltage that makes it behave as pass-transistor. Moreover, during the precharge phase, the appropriate V_{READ} and V_{PASS} are applied to the string. M_{BLS} is biased to a voltage (generally V_{DD}) that makes it work as pass transistor. Instead, M_{SLS} is turned off in order to avoid cross-current consumption through the string.

Typically, V_{BL} is around 1 V. From (6.9), V_{PRE} values approximately 1.4–1.9 V, depending on the V_{THN} (NMOS threshold voltage). The bitline precharge phase usually lasts 5–10 μs, and depends on many factors, above all the value of the distributed bitline parasitic *RC*.

Sometimes this precharge phase is intentionally slowed down to avoid high current peaks from V_{DD}. In order to achieve this, the M_{PCH} gate could be biased with a voltage ramp from GND to $V_{DD} + V_{THN}$.

At the end of the precharge phase, PCH and SEL are switched to 0. As a consequence, the bitline and the SO node parasitic capacitor are left floating to a voltage of $V_{PRE} - V_{THN}$ and V_{DD} respectively. M_{SL} is then biased in order to behave as pass transistor. In this way the string is enabled to sink (or not) current from the bitline capacitor.

At this point, the evaluation phase starts. If the cell has a V_{TH} higher than V_{READ}, no current flows and the bitline capacitor maintains its precharged value.

Otherwise, if the cell has a V_{TH} lower than V_{READ}, the current flows and the bitline discharges.

6.5.1 Interleaving Architecture

Given the (6.8), it is clear that the bitline capacitance has a direct influence on the evaluation time. C_{BL} must fulfill the following requirements:

- it must be a known parameter;
- it must be immune to external noise.

Figure 6.24 is a bitline cross-section showing the different contributions to C_{BL}:

- C_{AD} is the parasitic capacitor between the bitline and the lower plane (usually it is the wordline plane);
- C_{AU} is the parasitic capacitor between the bitline and the upper plane (usually it is the source-line plane);

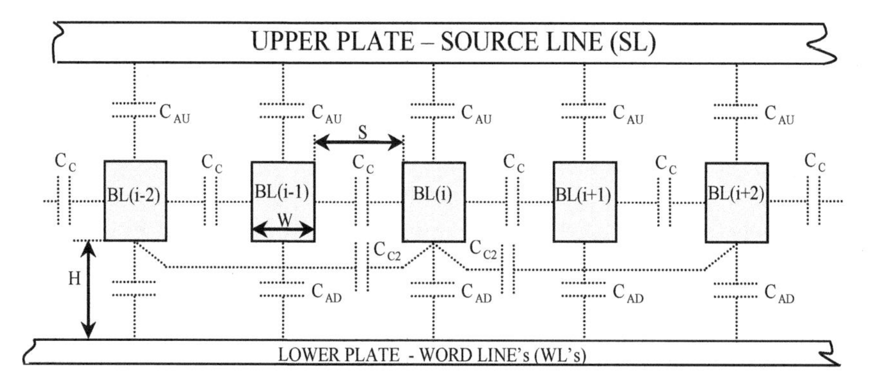

Fig. 6.24 Bitline parasitic capacitors

- C_C is the parasitic capacitor between two adjacent bitlines;
- C_{C2} is the parasitic capacitor between a bitline and its second nearest bitline.

Therefore, C_{BL} can be written as:

$$C_{BL} = C_{AU} + C_{AD} + 2C_C + 2C_{C2} \tag{6.10}$$

The above mentioned contributions depend on the bitline geometrical values (width W, height H and spacing S in Fig. 6.24), on the distance between upper and lower ground levels and on the oxide thickness. These parameters are not uniform among different wafers, dice and even within the same die. However, a correct reading must be ensured.

In all the explained theory, another important assumption is that the bitline capacitor has one of its terminals fixed to ground. Actually, looking at Fig. 6.24, C_{BL} ground terminal is physically distributed over four nodes:

1. the upper plate, usually the source-line;
2. the lower plate, usually the wordline or the source-line;
3. the left bitline;
4. the right bitline.

During the evaluation time the first two nodes are forced at a fixed voltage. Instead, the adjacent bitlines could be discharged by the strings connected to them.

With the continuous bitline shrinking (W and S in Fig. 6.24), the coupling capacitances play an important role. In sub-40 nm NAND technologies they contribute 80–90% of the total bitline capacitance. To overcome this issue, the interleaving architecture is introduced. While the even (or odd) bitlines are read, the odd (or even) bitlines are forced to a fixed voltage (generally ground), acting as electrical shield [22–24]. As shown in Fig. 6.25, M_{SLe} and M_{SLo} (bitline selectors) are placed between the bitlines and the page buffer PB(i). If the even bitlines BLe are read, M_{SELe} acts as a pass-transistor. Transistor M_{SELo} is turned off. The DISo signal turns on the M_{DISo} transistor, forcing the odd bitline B_{Lo} to the fixed BIAS voltage. M_{DISe} is turned off.

In order to minimize the power consumption, BIAS and the source line (SL) should be biased at the same voltage. In fact, these two nodes are shorted if a cell with $V_{TH} > V_{READ}$ belongs to the unselected bitlines. SL and BIAS are usually grounded during the reading operation.

With this architecture, the noise injection effect through the C_C coupling capacitors is eliminated. However, the coupling through C_{C2} (Fig. 6.24) is still in place. This contribution is not negligible: in the state-of-the-art technologies, C_{C2} contributes 5–10% of the total bitline capacitance. This problem is solved by the architecture described in the next section.

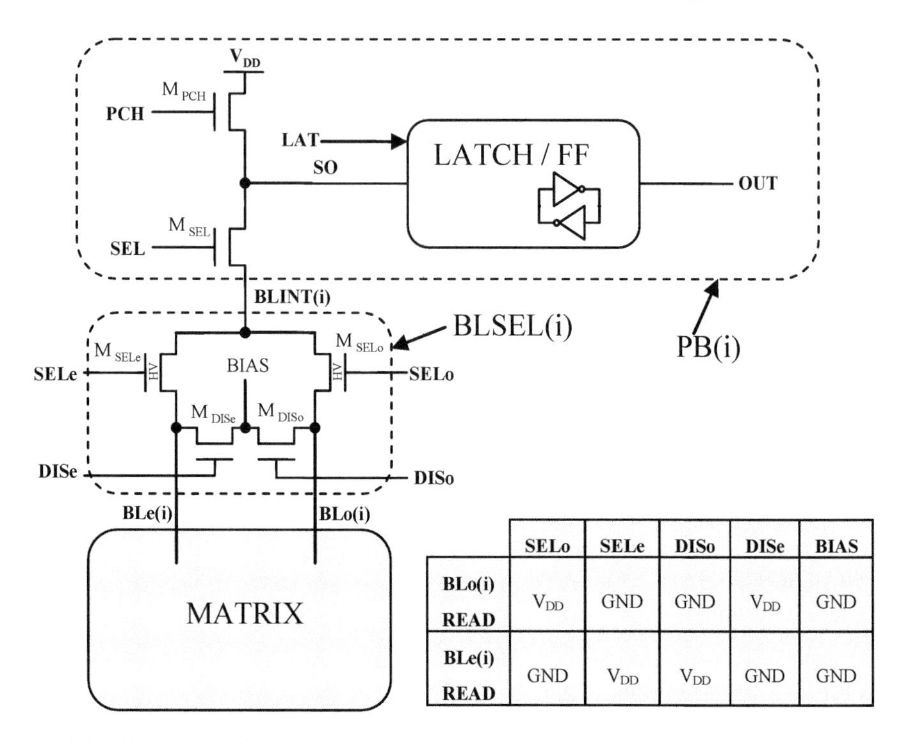

	SELo	SELe	DISo	DISe	BIAS
BLo(i) READ	V_{DD}	GND	GND	V_{DD}	GND
BLe(i) READ	GND	V_{DD}	V_{DD}	GND	GND

Fig. 6.25 Interleaving bitline architecture

6.5.2 All BitLine (ABL) Architecture

The sensing technique is basically the same used in the interleaving architecture. An intentionally placed capacitor is used instead of the C_{BL} bitline parasitic capacitor [25].

Figure 6.26 shows the main elements of the ABL sense amplifier. The latch is replaced by a voltage comparator with a V_{THSA} trigger voltage. The other elements are those ones already described in the interleaved architecture, but here used in a different way. The capacitor C_{SO} is involved in the integration of the cell current: it can be done using either MOS gates or poly-poly capacitors.

Figure 6.27 shows the timings used in a single read operation. The precharge phase is similar to that one described for the interleaving architecture, where M_{PCH} and M_{SEL} gates are biased to $V_{DD} + V_{THN}$ and V_{PRE} respectively. M_{HV} HVNMOS has the behavior already described and, during the single read operation phase, works as pass transistor. The signals which drive the string gates (V_{READ}, V_{PASS} and BLS) are activated as usually. Instead SLS signal is immediately activated in order to stabilize the bitlines during the precharge phase. In fact, if the SLS had been activated during the evaluation phase, there would have been a voltage drop on those bitlines with an associated sinking current string.

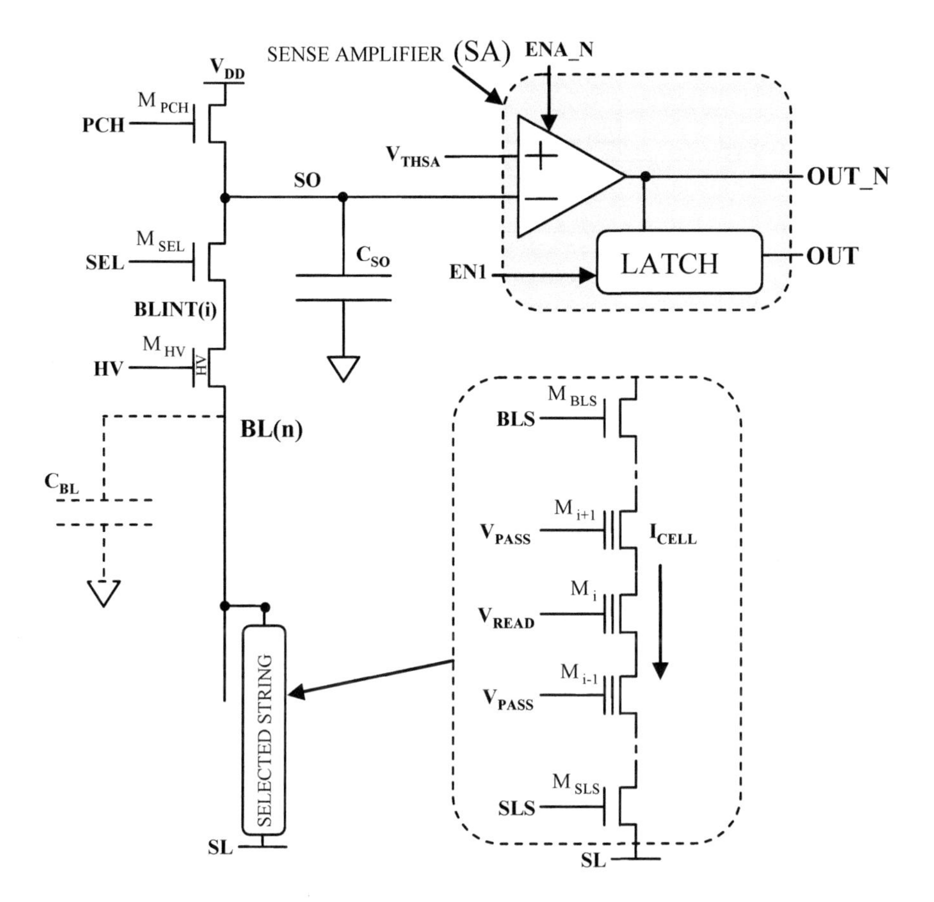

Fig. 6.26 ABL sense amplifier

The precharge final condition

$$V_{BL} = V_{PRE} - V_{THN} \tag{6.11}$$

is, therefore, valid only for the bitlines which have an associated string in a non conductive state.

Equation (6.11) should be replaced by:

$$V_{BL} = V_{PRE} - V_{THN} - \Delta \tag{6.12}$$

where Δ is the voltage drop on the bitlines resistance (typical values are in the order of hundreds of kΩ up to one MΩ).

At the end of the precharge phase (T_1), the bitlines are biased to a constant voltage and V_{SO} is equal to V_{DD}. At this point, M_{PCH} is switched off and the evaluation phase starts. Actually, M_{PCH} is biased to a V_{SAFE} voltage value in order

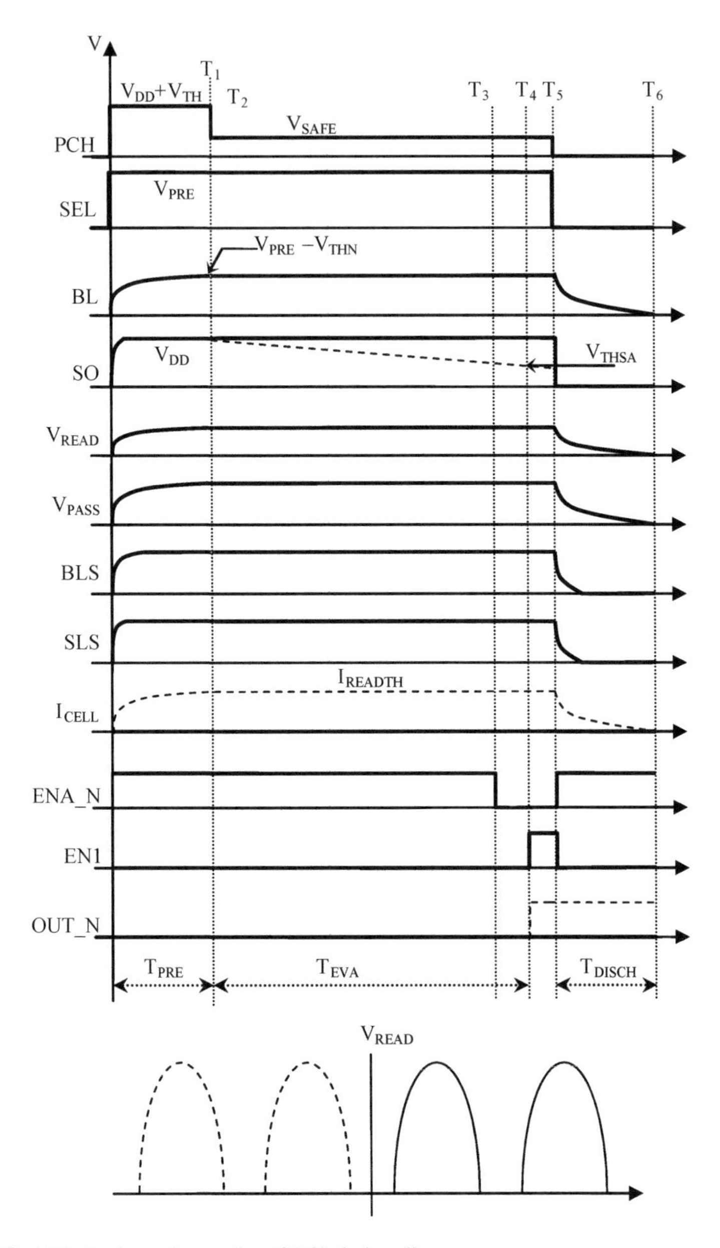

Fig. 6.27 ABL single read operation (SRO) timing diagram

to make M_{PCH} behave as a clamp transistor of the SO voltage. The following relation must be valid:

$$V_{SAFE} - V_{THN} \geq V_{PRE} - V_{THN} \Rightarrow V_{SAFE} > V_{PRE} \quad (6.13)$$

This clamp value must not influence the current integration on the SO capacitor, i.e. the clamping function can't take place above the V_{THSA} trigger voltage:

$$V_{SAFE} - V_{THN} \leq V_{THSA} \quad (6.14)$$

Therefore, from (6.13) and (6.14), the following conditions must hold true:

$$V_{PRE} - V_{THN} \leq V_{SAFE} - V_{THN} \leq V_{THSA} \quad (6.15)$$

When M_{PCH} is switched off, the cell current (through M_{PRE}) discharges the C_{SO} capacitor. If, during the evaluation time, $V_{SO} < V_{THSA}$ (trigger voltage of Fig. 6.26 comparator), than OUT_N switches (dotted lines in Fig. 6.27). The "threshold current" I_{READTH} is defined as:

$$I_{READTH} = \frac{\Delta V \cdot C_{SO}}{T_{EVAL}} \quad (6.16)$$

where

$$\Delta V = V_{DD} - V_{THSA} \quad (6.17)$$

Observe that, because the bitline is biased to a fixed voltage, a constant current I_{READTH} flows.

It is possible to extrapolate the evaluation time:

$$T_{EVAL} = \frac{\Delta V \cdot C_{SO}}{I_{READTH}} \quad (6.18)$$

Given the same read currents, it follows that the ratio between (6.8) and (6.18) is determined by the ratio between C_{BL} and C_{SO}. C_{BL} is a parasitic element and has a value of 2–4 pF. Instead, C_{SO} is a design element and has typical values around 20–40 fF, i.e. two orders of magnitude lower than C_{BL}. The reduction of the evaluation time from 10 μs to hundreds of ns is another advantage of the All Bitline architecture.

In addition, ABL architecture gives further advantages such as energy saving, bitline-coupling reduction and Floating-Gate-coupling reduction during program and read, and program stress reduction [2].

6.5.3 Read Voltage with Thermal Tracking

In a 2 bit-per-cell multilevel Flash NAND memory, four different threshold voltage (V_{TH}) distributions exist, as shown in Fig. 6.28. All the cells are in the "11" state after electrical erase. During programming phase, the threshold voltage of the cells is incremented in small steps until the desired value is reached. At the end of each program step, a verify operation is performed, in order to evaluate whether V_{TH} has gone above one of the verify voltages, V_{FY1}, V_{FY2} or V_{FY3}. Of course, verify voltage depends on which bits have to be stored in a given cell. For instance, in order to reach "00" logic value, threshold voltage has to go above V_{FY2}. Once target distribution is reached, further program pulses are not applied to that cell.

In order to univocally determine the logic value stored in the selected cell, read operation uses three voltage values, V_{READ0}, V_{READ1}, and V_{READ2} as shown in Fig. 6.28. Each read voltage is centered between two adjacent distributions so that read margins are maximized. For instance, the distance between V_{READ1} and the rightmost side of "10" distribution should be equal to the distance between V_{READ1} and the leftmost side of "00" distribution. With multilevel memories, the typical value for such distances is 300 mV.

In order to achieve the required precision, voltages to be applied to the cells are generated by means of voltage regulators which exploit band-gap techniques to generate a precise reference voltage. In this way, the voltages generated on-chip are independent from temperature, at least to a first approximation. On the other hand, the V_{TH} distributions of the memory cells are highly sensitive to temperature variations: as temperature increases, V_{TH} decreases and vice versa (see Fig. 6.29).

As a result, read margins are reduced when temperature varies, because the tails of the distributions get nearer and nearer to read voltages. For instance, as shown in

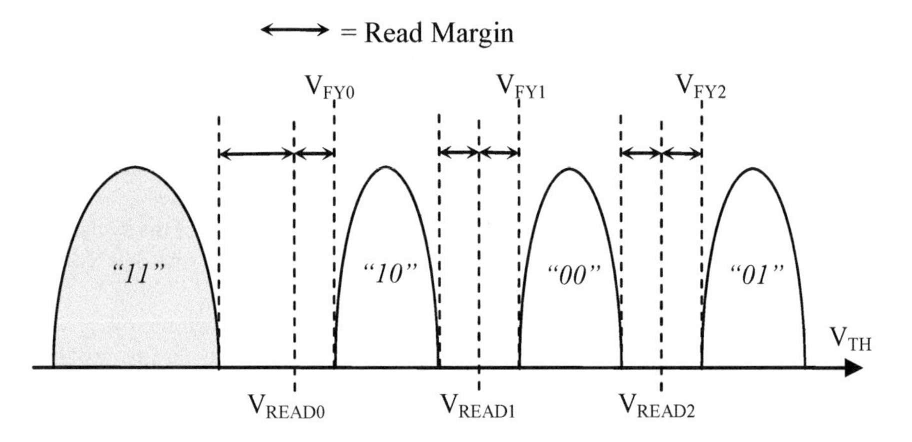

Fig. 6.28 Cell V_{TH} distributions in a 2 bit/cell NAND memory

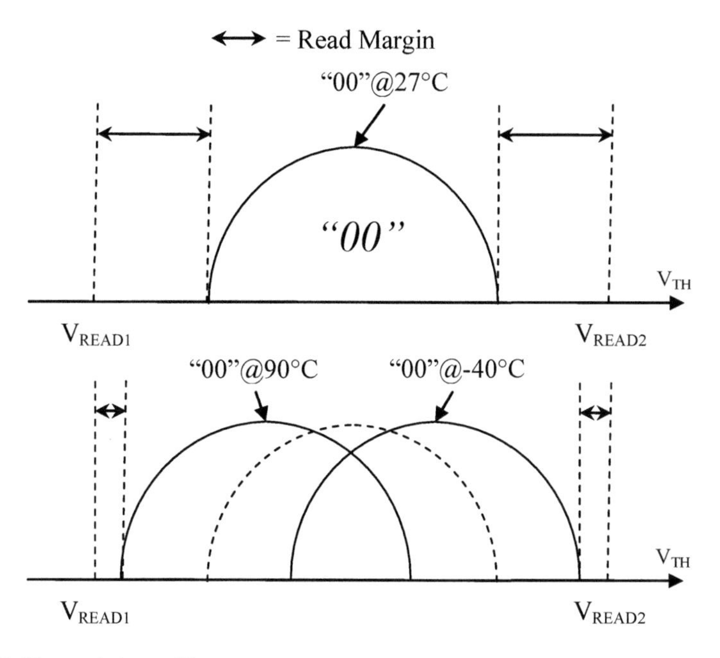

Fig. 6.29 V_{TH} variations with temperature

Fig. 6.29, "00" distribution gets nearer to V_{READ2} at low temperature, while it gets nearer to V_{READ1} at high temperature. The same is true for each distribution. Threshold voltage of the cell typically shifts of −1.5 mV/°C. As a consequence, overall variation is approximately 200 mV if a temperature range of −40 to 90 °C is considered.

Therefore, a specific type of read voltage regulator is needed [26–28]: that is, the thermal coefficient of its output voltage has to be as similar as possible to the coefficient of the cell's V_{TH}. In this way, read voltages rigidly shift with distributions, keeping the margins unaltered (Fig. 6.30). A similar constraint is true for verify voltages.

6.6 Program

As described in Chap. 5, V_{TH} is modified by means of the *Incremental Step Pulse Programming* (ISPP) algorithm (Fig. 6.31): a voltage step (whose amplitude and duration are predefined) is applied to the gate of the cell. Afterwards, a verify operation is performed, in order to check whether V_{THR} has exceeded a predefined voltage value (V_{VFY}). If the verify operation is successful, the cell has reached the desired state and it is excluded from the following program pulses. Otherwise

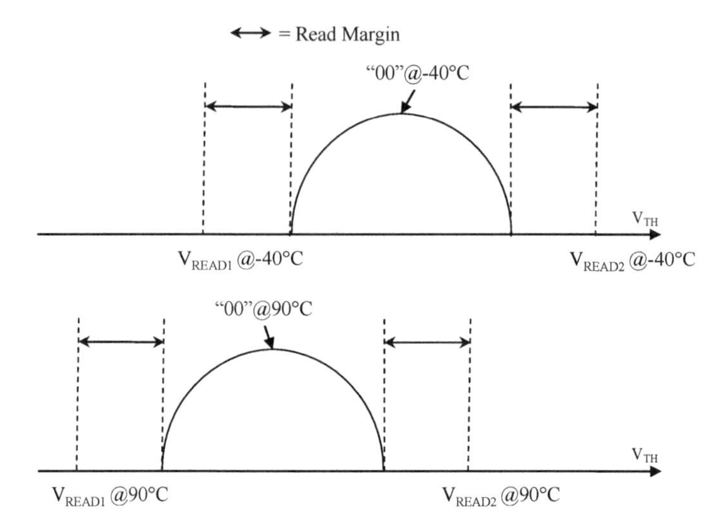

Fig. 6.30 V_{READ} tracking of V_{TH} variations with temperature

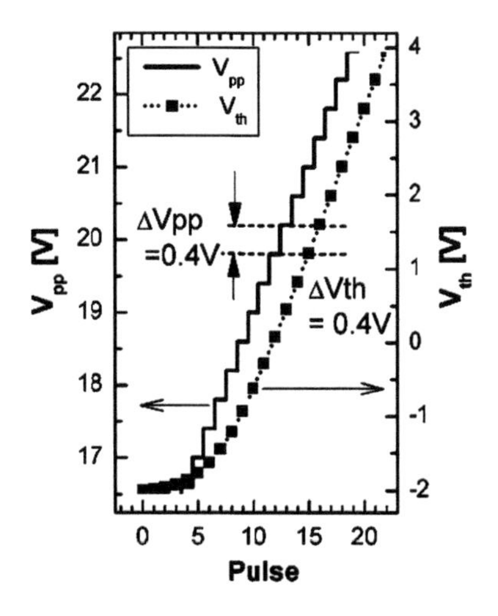

Fig. 6.31 Incremental step pulse programming (ISPP): constant V_{TH} shift

another cycle of ISPP is applied to the cell, where the program voltage is incremented by ΔVpp.

During the program operation, the cells share the high programming voltage on the selected wordline but the program operation has to be bit selective. Therefore, a high channel potential is needed to reduce the voltage drop across the tunneling dielectric and prevents the electrons tunneling from the channel to the floating gate as indicated by Fig. 6.32a. In the first NAND flash devices the channel was charged

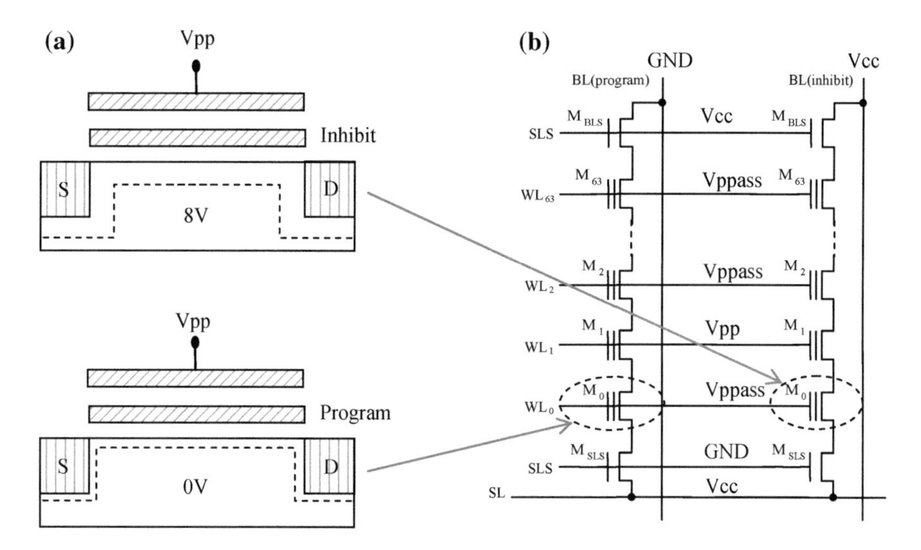

Fig. 6.32 Self boosted program inhibit scheme, **a** cell in program/inhibit state, **b** strings biasing in program/inhibit state

by applying 8 V to the bitlines of the program inhibited NAND strings. This method suffers from several disadvantages [29], especially power consumption and high stress on the oxide between adjacent bitlines.

The self boost program inhibit scheme is less power consuming. By charging the string select lines and the bitlines connected to inhibited cells to V_{cc}, the select transistors are diode connected (Fig. 6.32b). By raising the wordline potential (selected wordline to V_{pp} and unselected wordlines to V_{ppass}) the channel potential is boosted by the coupled series capacitance through the control gate, floating gate, channel and bulk.

In fact, when the voltage of the channel exceeds $V_{cc} - V_{TH,SSL}$, then SSL transistors are reverse biased and the channel of the NAND string becomes a floating node.

Two important typologies of disturbs are related to the program operation: the *Pass disturb* and the *Program disturb* as described in Chap. 5.

6.7 Erase

The erase operation resets the information of all the cells belonging to one block simultaneously.

Tables 6.1 and 6.2 summarize the erase voltages. During the erase pulse, all the wordlines belonging to the selected block are kept at ground, the matrix ip-well

Table 6.1 Electrical erase pulse voltages for the selected block

	T_0	T_1	T_2	T_3	T_4
BLeven	Float	Float	Float	Float	Float
BLodd	Float	Float	Float	Float	Float
DSL	Float	Float	Float	Float	Float
WLs	0 V	0 V	0 V	0 V	0 V
SSL	Float	Float	Float	Float	Float
SL	Float	Float	Float	Float	Float
ip-well	0 V	V_{ERASE}	V_{ERASE}	0 V	0 V

Table 6.2 Electrical erase pulse voltages for unselected blocks

	T_0	T_1	T_2	T_3	T_4
BLeven	Float	Float	Float	Float	Float
BLodd	Float	Float	Float	Float	Float
DSL	Float	Float	Float	Float	Float
WLs	Float	Float	Float	Float	Float
SSL	Float	Float	Float	Float	Float
SL	Float	Float	Float	Float	Float
ip-well	0 V	V_{ERASE}	V_{ERASE}	0 V	0 V

must rise (through a staircase) to 23 V and all the other nodes are floating. This phase lasts almost a millisecond and it is the phase when the actual electrical erase takes place.

Since the matrix ip-well (as well as the surrounding n-well) is common to all the blocks, it reaches high voltages also for the unselected blocks. In order to prevent an unintentional erase on those blocks, wordlines are left floating; in this way, their voltage can rise thanks to the capacitive coupling between the wordline layer and the underneath matrix layer. Of course, the voltage difference between wordlines and ip-well should be low enough to avoid Fowler-Nordheim tunneling.

After each erase pulse an erase verify (EV) follows. During this phase all the wordlines are kept at ground. The purpose is verifying if there are some cells that have a V_{TH} higher than 0 V, so that another erase pulse can be applied. If EV isn't successful for some columns of the block, there are some columns too programmed. If the maximum number of erase pulses is reached (typically 4), than the erase exits with a fail. Otherwise, the voltage applied to the matrix ip-well is incremented by ΔV_E and another erase pulse follows.

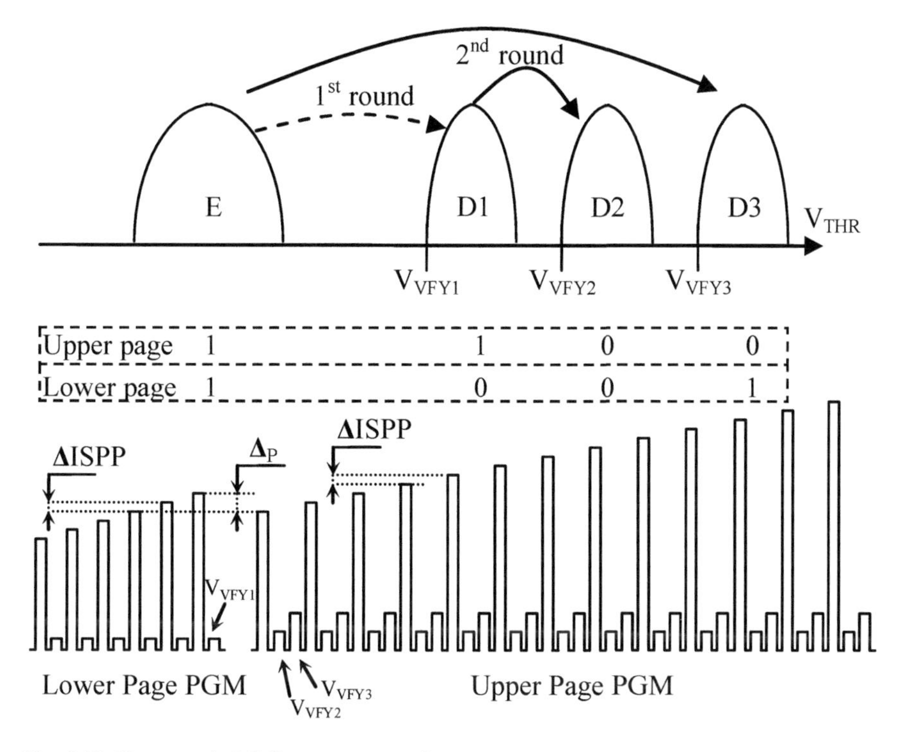

Fig. 6.33 Two rounds MLC program operation

6.8 MLC and XLC Storage

The obvious advantage of a 2 bit/cell implementation (MLC) with respect to a 1 bit/cell device (SLC) is that the area occupation of the matrix is half as much; on the other hand, the area of the periphery circuits, both analog and digital, increases. This is mainly due to the fact that the multilevel approach requires higher voltages for program (and therefore bigger charge pumps), higher precision and better performance in the generation of both the analog signals and the timings, and an increase in the complexity of the algorithms.

Figure 6.33 shows an example of how 2 bits are associated to the four read threshold distributions stored in the cell, and how the set of programmed distributions is built starting from the erased state "E". In this case the multilevel is achieved in two distinct rounds, one for each bit to be stored [2, 30, 31].

In the first round, the so-called lower-page (associated to the *Least Significant Bit*—LSB) is programmed. If the bit is "1", the read threshold of the cell V_{TH} does not change and, therefore, the cell remains in the erased state, E. If the bit is "0", V_{TH} is increased until it reaches the D1 state.

In the second round, the upper-page (associated to the *Most Significant Bit*—MSB) is programmed. If the bit is “1”, V_{TH} does not change and, therefore, the cell remains either in the erased state, E, or in the D1 state, depending on the value of the lower-page.

When MSB is “0”, V_{TH} is programmed as follows:

- if, during the first round, the cell remained in E state, then V_{TH} is incremented to D3;
- if, during the first round, the cell was programmed to D1, then, in the second round, V_{TH} reaches D2.

As usual, the program operation makes use of ISPP, and the verify voltages are V_{VFY2} and V_{VFY3}. Lower-page programming only needs the information related to LSB, while for the upper-page it is necessary to know both the starting distribution (LSB) and the MSB.

Because of technological variations, V_{TH} is not perfectly related to the amplitude of the program pulse (during ISPP): there are “fast” cells which reach the desired distribution with few ISPP pulses, while other “slow” cells require more pulses.

The amplitude of the first program pulse ($V_{PGMLSB0}$) of the lower-page should not allow the threshold V_{THR} of the “fastest” cell to exceed V_{VFY1}. If it happens, an undesired widening of distribution D2 occurs or, in the worst case scenario, V_{THR} might reach D2 distribution at once.

Typical $V_{PGMLSB0}$ is around 16 V. In case of program of “slow” cells from E to D1, the last programming step needs values as high as 19 V. Assuming ΔISPP equal to 250 mV, it takes 12 steps to move from 16 to 19 V.

Similarly, the starting pulse of the upper-page $V_{PGMMSB0}$ should have an amplitude such that the “fastest” cell does not go beyond V_{VFY2}.

$$V_{PGMMSB0} = V_{PGMLSB0} + (V_{VFY2} - V_{VFY1}) \tag{6.19}$$

The value of $V_{VFY2} - V_{VFY1}$ is typically around 1 V and, therefore, the initial voltage is about 17 V.

As shown in Fig. 6.33, the upper-page ISPP does not start from the last voltage used for the lower-page programming, but it begins at $V_{PGMLSB0} - \Delta_P$. For example, instead of starting at 19 V, it could start at 17 V, eight steps below.

Driven by cost, Flash manufacturers are now developing 3 bit/cell (8 V_{TH} distributions) and 4 bit/cell (16 V_{TH} distributions) [32–34]. Three and four bits per cell are usually referred to as XLC (8LC and 16LC, respectively). Unfortunately, due to reliability reasons, the V_{TH} window remains the MLC one; in fact, the highest verification level must be low enough to prevent bit failures caused by program disturb and read disturb. The more states a memory cell is made to store, the more finely divided is its V_{TH} window.

Of course, the main drawback is a slow program time. As the distribution width needs to be tighter, ISSP program step is smaller and the number of verify operations increases, as depicted in Fig. 6.34.

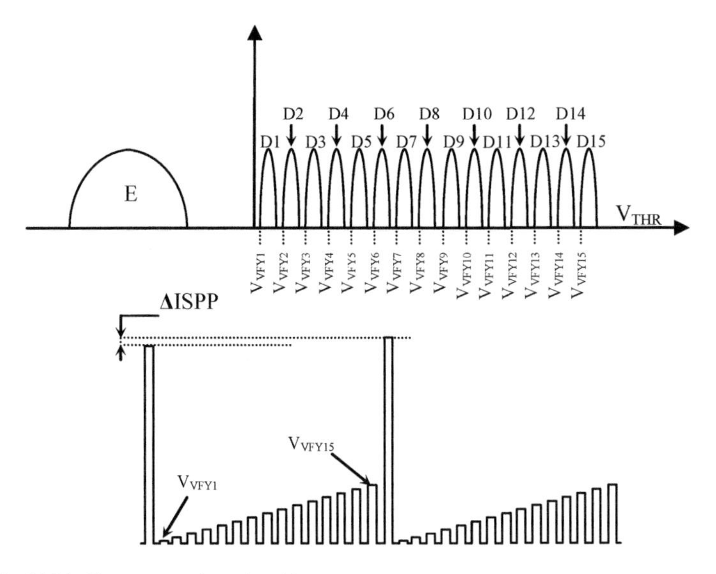

Fig. 6.34 4 bit/cell programming algorithm

6.9 High Voltage Management

Modifying or reading the number of electrons stored into the floating gate requires a big set of voltages. The High Voltage (HV) system has to provide all these voltages with the desired precision, timing and granularity. On top of that, many voltages have a value greater than the NAND power supply VDD, asking for an on-chip charge pump. This section deals with the HV basic building blocks.

6.9.1 Charge Pumps

In the NAND environment, one of the most used type of charge pumps is the Voltage Doubler [3]. The basic stage is shown in Fig. 6.35. It is a feedback system that can duplicate the input voltage and, essentially, it is made up by two n-channel transistors (MN1, MN2), two p-channel transistors (MP1, MP2) and two capacitors (C1, C2) of the same size.

In order to understand the principle of operation of this circuit, it can be assumed that, at the beginning, nodes A and B, as well as CK (pump clock) and its complement (CK#), are at GND. In this way, both transistors MN1 and MN2 are off. Voltage on the node IN (V_{IN}) is set to VDD (i.e. the chip power supply).

As soon as CK toggles from GND to VDD, V_A becomes VDD, activating transistor MN2. Since CK# remains at GND, the charge starts flowing from power

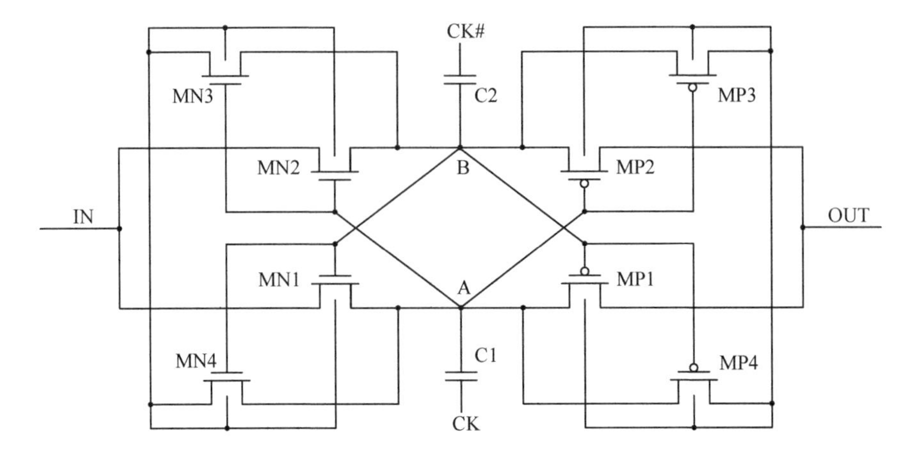

Fig. 6.35 Basic stage of a voltage doubler

supply to capacitor C2 until V_B reaches a value equal to VDD – $V_{TH,MN2}$. When CK goes to GND, transistor MN2 turns off.

At the same time, CK# gets to VDD and, therefore, V_B becomes (VDD – $V_{TH,MN2}$ + VDD), turning on transistor MN1. As a result, C1 is charged up to VDD. Of course, when CK# goes to GND again, V_B is, in principle, equal to VDD – $V_{TH,MN2}$. Since the signal CK is used as a clock, each capacitance is continuously charged and discharged between VDD and 2VDD. In other words, during each period of the clock either V_A or V_B is at 2VDD.

At this point, in order to build a real charge pump, voltages on nodes A and B have to be transferred to the next pump stage. Now MP1 and MP2 come into the game. When CK is at VDD, V_A is 2VDD and V_B is VDD. Transistor MN1 is, therefore, turned off while MP1 is active, transferring the voltage of node A to node OUT. In the meanwhile MP2 is off, MN2 is on and the capacitor C2 is charged up. When CK goes back to GND and CK# becomes VDD, then the circuit behaves in the opposite way: MN1 and MP2 are active (the former charges capacitor C1, the latter transfers the voltage of node B to the output) while MN2 and MP1 are turned off. It is worth to note that no active direct paths between IN and OUT are allowed: these paths would result in a loss of charge and, therefore, in a reduced output voltage.

As usual, when designing a charge pump, one issue to cope with is the biasing of the transistor body terminals. The easiest solution is to connect the body of the n-channel transistor to the power supply and the body of the p-channel transistor to the output node.

The drawback of this solution is that the output voltage is considerably reduced by the body-effect of the transistors itself. In Fig. 6.35a “dynamic biasing” has been chosen: bodies are continuously switched between V_A and V_B. As a result, the body of the NMOS transistors is always kept at the lowest voltage (through MN3 and

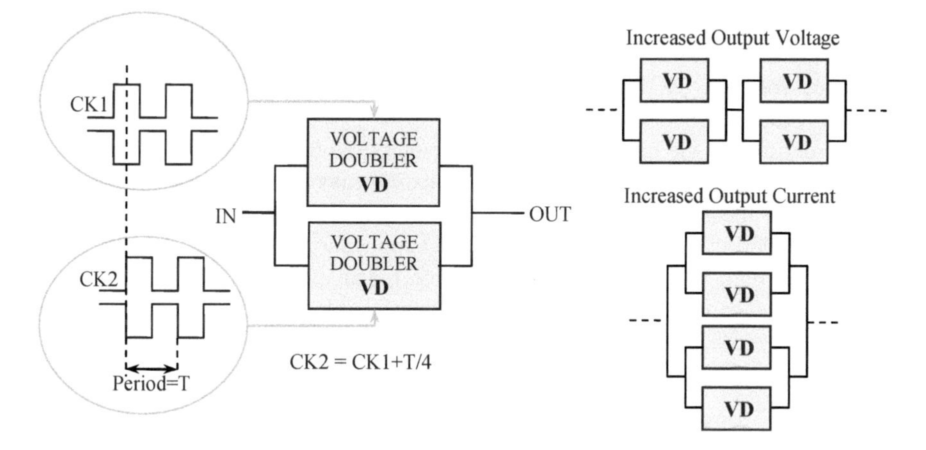

Fig. 6.36 Charge pump as a cascade of basic voltage doubler stages

MN4) while the body of the PMOS transistors is always at the highest voltage (through MP3 and MP4).

The basic stage of Fig. 6.35 can be used to build up more complex structures as depicted in Fig. 6.36. Usually, two stages are used in parallel in order to decrease the ripple of the output voltage.

In fact, due to the internal switching activity of the capacitors, the output of the pump can be more or less noisy. When talking about ripple, we generally refer to the height of the "peaks" that can be found in the output node waveform.

In order to properly control the output voltage, voltage doubler stages are inserted in a feedback loop as described in Fig. 6.37. A block called "Hireg" is used

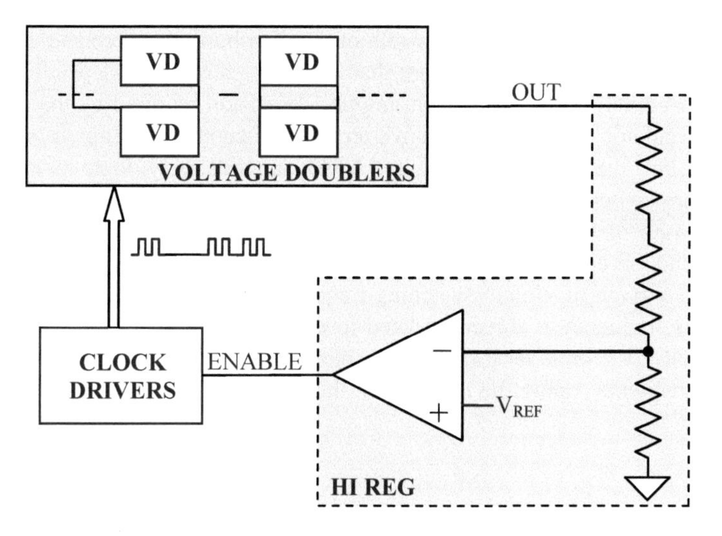

Fig. 6.37 Charge pump architecture

to limit the output voltage. Thanks to a resistive divider (it could also be made by CMOS diodes), the output voltage is compared with V_{REF} (usually a band-gap reference voltage). CK drivers are then enabled/disabled depending on the comparison result.

In order to find the best configuration, the output voltage of the charge pump is measured varying the CK period. A faster clock means higher output voltage, but faster clocks means bigger area of the CK drivers. The right trade-off has to be found considering that, in most of the NAND applications, silicon cost is the main driver. Optimum CK period is usually in the range of 60–80 ns considering an output resistance of around 10 kΩ. The voltage doubler pump can easily achieve voltages above 25 V starting from the chip VDD of 2.5 V. Power efficiency ηP can be as high as 20–30% if the current load remains in the range of few hundreds microAmpere.

$$\eta_P = \frac{V_{OUT} \cdot I_{OUT}}{V_{IN} \cdot I_{IN}} \tag{6.20}$$

6.9.2 *Internal Supply Voltage Regulator*

In many NAND devices, external supply voltage VDD is not directly applied to all the circuits [35, 36]. Some of them are powered by an internal supply (V_{INT}) filtered by a proper voltage regulator and this solution brings several advantages. For instance, in case of devices supplied at 3.6 V, a V_{INT} equal to 2 V allows the use of transistors whose oxide thickness is reduced, which are smaller and better performing. In the case of page buffers, by using V_{INT} it is possible to mitigate the dependency of the triggering threshold from VDD (i.e. several tens of milliVolt), which turns into a reduction of the width of the distributions. Of course, inside the NAND memory, there could be more than one V_{INT} regulators, depending on the design constraints (noise, power consumption, precision required by the circuits).

V_{INT} regulator is a DC-DC converter. Its conceptual scheme is shown in Fig. 6.38. For the sake of simplicity, VDD supplies only logic ports. When inverters are switching, voltage drop of V_{INT} is a function of the filtering capacitance C_{FILTER}, of the parasitic capacitance (gates, routing, junctions), and of the cross-conduction current.

Beyond a given maximum switching frequency of the logic, V_{INT} dramatically drops. This frequency is directly related to the cutoff frequency of the regulator. Since the DC-DC converter is designed using the same technology of the inverters, its cutoff frequency cannot be higher than the one of the plain inverter.

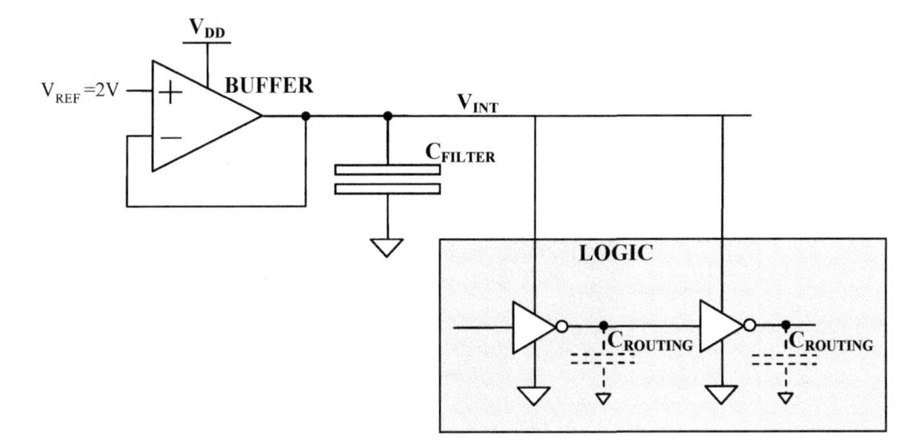

Fig. 6.38 Conceptual scheme of a DC-DC converter

6.9.3 Double-Supply Voltage Regulator

Both program and erase operations require voltages higher than VDD. For instance, the programming staircase voltage starts at 14–15 V and arrives at 25 V and beyond. High voltages are generated by a charge pump and filtered by a proper voltage regulator: in this way it is possible to reduce the ripple and obtain the desired output voltage value.

In 1 bit/cell Flash memories, voltage regulator is omitted and the output voltage of the pump is directly used, regulated by means of an on-off type of control. Typical ripple values are in the order of 1–2 V. In case of multilevel memories, the target voltage precision cannot be achieved without a voltage regulator.

NAND technology does not usually provide High Voltage (HV) PMOS transistor; therefore; it is not possible to implement traditional voltage regulators like the one shown in Fig. 6.39. In fact, the use of a low-voltage transistor for M_{POUT} would mean that the voltage drop across its terminals must be guaranteed not to exceed 4–5 V. This must be true both in static and in transient conditions. On top of that, all the required values for the staircase program pulse must be generated out of the pump output voltage (~30 V), beginning at 15 V: that is, M_{POUT} must be a HV transistor.

In order to solve the issue it is possible to design a voltage regulator [37] whose first differential stage is supplied by VDD, while the second one is supplied by a charge pump so that the HV value can be provided at the output (Fig. 6.40).

By supplying the first stage with VDD, PMOS LV transistors can be used to realize the current mirror ($M_{P1} - M_{P2}$). The second stage is instead designed using an NMOS HV (M_{NOUT}) together with a resistive pull-up (*R pull-up*).

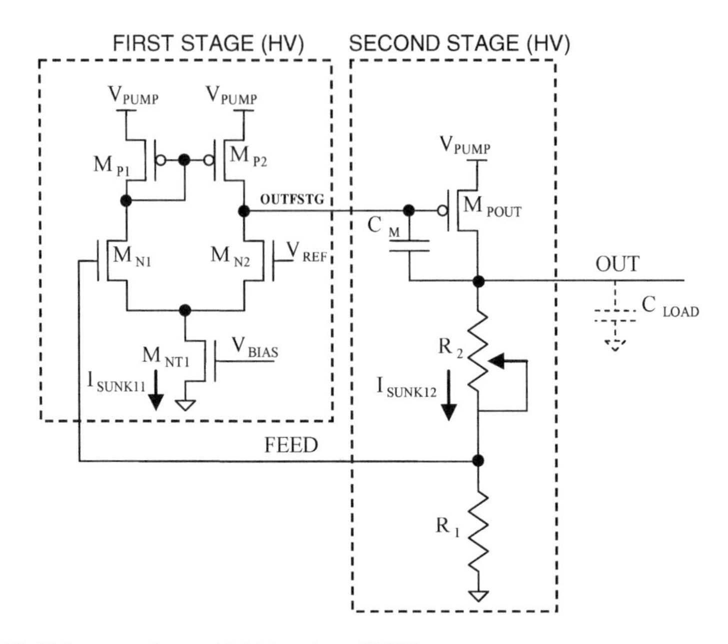

Fig. 6.39 Voltage regulator with high voltage PMOS

6.10 Wordline Decoder

One of the most critical circuits of the *High Voltage* (HV) system is the one used to bias the *WordLine* (WL). Actually, when it comes to NAND memories, a single wordline is not enough: all the wordlines belonging to the same NAND string must be properly biased at the same time. As a result, the Row Decoder, also called Wordline Decoder or Wordline Driver [4], has to provide a set of voltages: these values are defined by the algorithms described in Sects. 6.5 and 6.6.

When NAND technology provides only NMOS-type HV transistors, a possible implementation of the wordline driver is shown in Fig. 6.41. The wordline driver comprises:

- a *Pass-Transistor* (PT) for each wordline. These transistors are used to transfer voltages from the *Global WordLines* (GWLs), i.e. electrical signals, to the physical wordlines (WLs);
- a circuit to bias the gates of the above mentioned pass-transistors.

The biasing circuit of the gate of PTs consists of only one high voltage NMOS (M1). At first, all the gates are biased at a high voltage V_{PRECH} through M1. Then, M1 is switched off and, thanks to the gate-drain parasitic capacitance, the rising transient of GWL performs a boost of V_{BLC}, switching PTs on, as shown in Fig. 6.42.

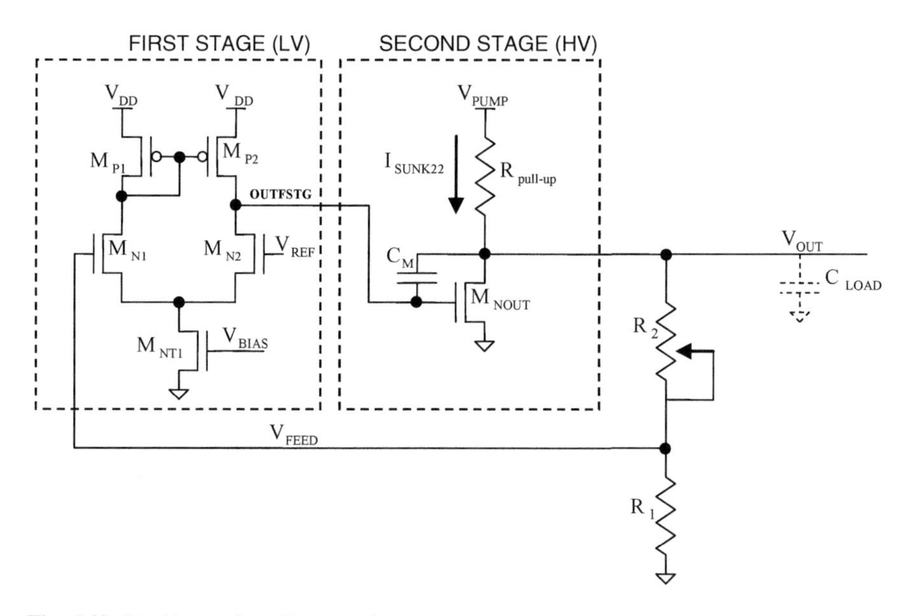

Fig. 6.40 Double-supply voltage regulator

However, there are several critical aspects to consider. First of all, the designer has to deal with a precharge phase of the PT gates: this phase must occur before biasing the global wordlines, otherwise the boost effect would be lost.

The precharge voltage V_{PRECH} has to match V_{MAX}, which is the maximum voltage required during each algorithm. V_{MAX} is not an issue during the read operation, when the voltages are relatively low, but it ends up being close to the breakdown voltage during the program operation. The duration of the precharge phase must be calibrated to allow V_{PRECH} reaching V_{MAX}: this time increases the overall operation time, especially during programming.

With reference to the circuit of Fig. 6.41, precharge is driven by the ENABLE signal. To fully exploit the precharge benefit, ENABLE has to be biased with a voltage greater than V_{PRECH}, in order to recover the threshold voltage $V_{TH,M1}$ of transistor M1.

Particular attention deserves the boost operation. Once the boost has occurred, V_{BLC} has to guarantee that, even varying temperature and technological parameters, each GWL and its corresponding WL are biased with the same voltage. Unfortunately, process and temperature variations mean that the V_{TH} of the pass transistors can vary as much as 100%. Therefore, the risk is to overcome the breakdown voltage of the oxide in some PVT (Process Voltage Temperature) corners allowed by the electrical specification of the NAND Flash memory.

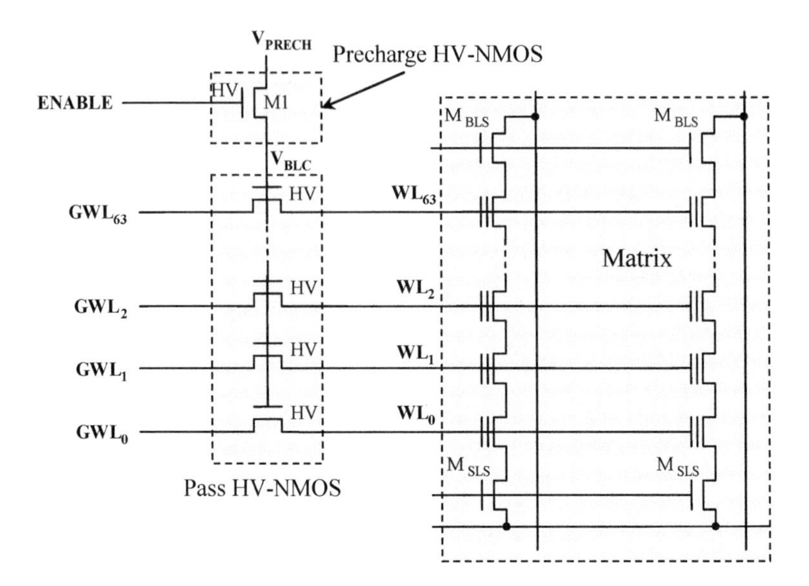

Fig. 6.41 All-NMOS wordline driver

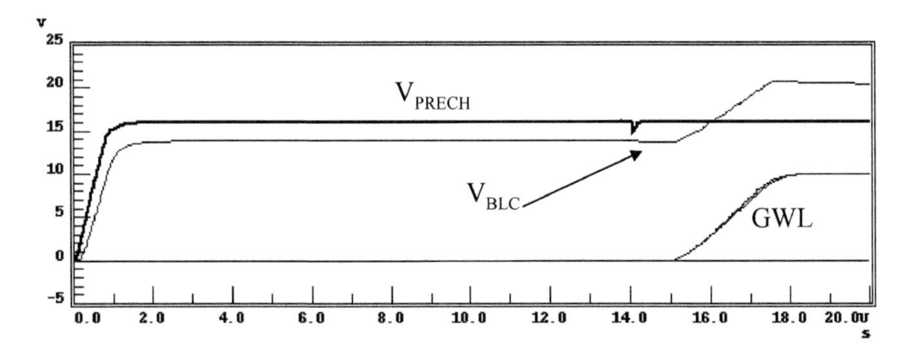

Fig. 6.42 Simulation of the circuit sketched in Fig. 6.39

Designers have developed a lot of different solutions for the row decoder, including a hierarchical approach [2, 3, 38]: due to the huge numbers of wordlines contained in a NAND array, the challenge is always to trade off performances with silicon area.

At this point the reader should be reasonably convinced that a NAND Flash memory is not a "pure" digital device: it is a real mix of digital and analog circuits, working at high and low voltages, and designed on a silicon technology developed for floating gate transistors…have fun!

References

1. R. Micheloni et al., A 4 Gb 2b/cell NAND Flash memory with embedded 5b BCH ECC for 36 MB/s system read throughput, in *IEEE International Solid-State Circuits Conference 2006, Digest of Technical Papers, ISSCC 2006*, Feb 2006, pp. 497–506
2. R. Micheloni, L. Crippa, A. Marelli, *Inside NAND Flash Memories* (Springer, New York, 2010)
3. G. Campardo, R. Micheloni, D. Novosel, *VLSI-Design of Non-volatile Memories* (Springer, New York, 2005)
4. P. Cappelletti, C. Golla, P. Olivo, E. Zanoni (eds.), *Flash Memories*, Chap. 5 (Kluwer, Boston, 1999)
5. R. Micheloni, A. Marelli, R. Ravasio, *Error Correction Codes for Non-volatile Memories* (Springer, Dordrecht, 2008)
6. G. Campardo et al., An overview of Flash architectural developments. Proc. IEEE **91**(4, April), 523–536 (2003)
7. M. Annaratone, *Digital CMOS Circuit Design* (Kluwer Academic Publishers, Boston, 1986)
8. www.onfi.org
9. https://www.denali.com/en/events/webcasts/2008/togglenand/
10. A. Chandrakasan, R. Brodersen (eds.), *Low Power CMOS Design* (Kluwer Academic Publishers, Boston, 1995)
11. H. Hikeda, A 3D packaging with 4 Gb chip-stacked DRAM and 3Gbps high-speed logic, in *3D-SIC 2007, International 3D-System Integration Conference 2007, Tokyo, Japan* (2007)
12. T. Wada, M.E. Kenji Mami, Simple noise model and low-noise data-output buffer for ultrahigh-speed memories. IEEE J. Solid-State Circuits **25**(6, December), 1586–1588 (1990)
13. S. Dabral, T. Maloney, *Basic ESD and I/O Design* (Wiley, New York, 1998)
14. E. Chioffi, F. Maloberti, High-speed, low-switching noise CMOS memory data output buffer. IEEE J. Solid-State Circuits **29**(11, November), 1359–1365 (1994)
15. S.H. HallGarrett, W. HallJames, A. McCall, *High-Speed Digital System Design—A Handbook of Interconnect Theory and Design Practices* (Wiley, New York, 2000)
16. P. Heydari, M. Pedram, Ground bounce in digital VLS circuits. IEEE Trans. VLSI Syst. **11**(2, April), 180–193 (2003)
17. R. Senthinathan, J. Prince, Simultaneous switching ground noise calculation for packaged CMOS devices. IEEE J. Solid-State Circuits **26**(November), 1724–1728 (1991)
18. R. Senthinathan, J.L. Prince, *Simultaneous Switching Noise of CMOS Devices and Systems* (Kluwer Academic Publisher, Boston, 1994)
19. S.J. Jou et al., Low switching noise and load-adaptive output buffer design techniques. IEEE JSSC **36**, 1239–1249 (2001)
20. B. Deutschmann, T. Ostermann, CMOS output driver with reduced ground bounce and electromagnetic emission, in *Solid-State Circuits Conference, ESSCIRC'03* (New York, 2003)
21. Y. Itoh et al., An experimental 4 Mb CMOS EEPROM with a NAND structured cell, in *36th IEEE International Solid-State Circuits Conference 1989, Digest of Technical Papers, ISSCC 1989, San Francisco*, Feb 1989, pp. 134–135
22. T. Tanaka et al., A quick intelligent page-programming architecture and a shielded bitline sensing method for 3 V-only NAND Flash memory. IEEE J. Solid-Stare Circuits **29**(11, November), 1366–1373 (1994)
23. T.-S. Jung et al., A 3.3 V 128 Mb multi-level NAND Flash memory for mass storage applications, in *43rd IEEE International Solid-State Circuits Conference 1996, Digest of Technical Papers, ISSCC 1996, San Francisco*, Feb 1996, pp. 32–33, 412
24. K. Imamiya et al., A 130 mm^2 256 Mb NAND Flash with shallow trench isolation technology, in *IEEE International Solid-State Circuits Conference 1999, Digest of Technical Papers, ISSCC 1999*, Feb 1999, pp. 112–113, 412

25. R.A. Cernea et al., A 34 MB/s MLC write throughput 16 Gb NAND with all bit line architecture on 56 nm technology. IEEE J. Solid-Stare Circuits **44**(1, January), 186–194 (2009)
26. L. Crippa, G. Ragone, M. Sangalli, R. Micheloni, Circuit and method for retrieving data stored in semiconductor memory cells, U.S. Patent No. 7474577, Assignee: STMicroelectronics/Hynix Semiconductor
27. T. Tanzawa, T. Tanaka, K. Takeuchi, Nonvolatile semiconductor memory with temperature compensation for read-verify referencing scheme, U.S. Patent No. 5864504, Assignee: Kabushiki Kaisha Toshiba (Kawasaki, JP)
28. T.-H. Cho, Y.-T. Lee, Multi-level Flash memory with temperature compensation, U.S. Patent No. 6870766, Assignee: Samsung Electronics Co., Ltd. (Suwon-si, KR)
29. K.-D. Suh et al., A 3.3 V 32 Mb NAND Flash memory with incremental step pulse programming scheme. IEEE J. Solid-State Circuits **30**(11, November), 1149–1156 (1995)
30. S. Lee et al., A 3.3 V 4 Gb four-level NAND Flash memory with 90 nm CMOS technology, in *IEEE International Solid-State Circuits Conference, ISSCC, Digest of Technical Papers, San Francisco*, vol. 1, Feb 2004, pp. 52–53, 513
31. D.-S. Byeon et al., An 8 Gb multi-level NAND Flash memory with 63 nm STI CMOS process technology, in *Solid-State Circuits Conference, ISSCC, Digest of Technical Papers, San Francisco*, vol. 1, Feb 2005, pp. 46–47
32. Y. Li et al., A 16 Gb 3b/cell NAND Flash memory in 56 nm with 8 MB/s write rate, in *IEEE International Solid-State Circuits Conference 2008, Digest of Technical Papers, ISSCC 2008, San Francisco*, Feb 2008, pp. 506–507, 632
33. N. Shibata et al., A 70 nm 16 Gb 16-Level-Cell NAND Flash memory. IEEE J. Solid-Stare Circuits **43**(4, April), 929–937 (2008)
34. C. Trinh et al. A 5.6 MB/s 64 Gb 4b/Cell NAND Flash memory in 43 nm, CMOS, in *IEEE International Solid-State Circuits Conference 2009, Digest of Technical Papers, ISSCC 2009, San Francisco*, Feb 2009, pp. 246–247
35. K. Takeuchi et al., A 56-nm CMOS 99-mm^2 8-Gb multi-level NAND Flash memory with 10-MB/s program throughput. IEEE J. Solid-Stare Circuits **42**(1, January), 219–232 (2007)
36. G.A. Rincon-Mora, *Analog IC Design with Low-Dropout Regulators*. Electronic Engineering (McGraw-Hill, New York, 2009)
37. L. Crippa, M. Sangalli, G. Ragone, R. Micheloni, Multistage regulator for charge-pump boosted voltage applications, not requiring integration of dedicated high voltage high side transistors, U.S. Patent App. 20070164811, Assignee: STMicroelectronics/Hynix Semiconductor
38. K. Kanda et al., A 120 mm^2 16 Gb 4-MLC NAND with 43 nm CMOS technology, in *2008 IEEE International Solid-State Circuits Conference (ISSCC), Digest of Technical Papers, San Francisco*, Feb 2008, pp. 430–431

Chapter 7
Memory Driven Design Methodologies for Optimal SSD Performance

L. Zuolo, C. Zambelli, Rino Micheloni and P. Olivo

7.1 Introduction

Solid State Drives (SSDs) are one of the electronic systems with the higher development rate in the last decade: they are widely used in hyper scale systems such as cloud computing and big data servers where performance is a constraint, as well as in consumer electronics by replacing traditional hard disk drives (HDDs) [1].

SSDs' design, in the last 5 years, faced an extraordinary evolution caused by the continuous development of NAND Flash memories representing their storage medium [2]. With this respect, as shown in Fig. 7.1, NAND Flash memories have completely transformed the way information is processed and stored. Starting as film and tape replacement for cameras and voice recorders, NAND Flash memories rapidly surpassed traditional magnetic storage supports and now they represent an obliged choice for high-performance storage solutions. The availability of NAND

This chapter is a partial reprint of L. Zuolo, C. Zambelli, R. Micheloni and P. Olivo, "Solid State Drives: Memory Driven Design Methodologies for Optimal Performance," in Proceedings of the IEEE, vol. 105, no. 9, pp. 1589–1608, Sept. 2017.

L. Zuolo · R. Micheloni
Microsemi Corporation, Vimercate, MB, Italy
e-mail: lorenzo.zuolo@microsemi.com

R. Micheloni
e-mail: rino.micheloni@microsemi.com

C. Zambelli (✉) · P. Olivo
Engineering Department, Università di Ferrara, Ferrara, Italy
e-mail: cristian.zambelli@unife.it

P. Olivo
e-mail: piero.olivo@unife.it

R. Micheloni et al. (eds.), *Inside Solid State Drives (SSDs)*,
Springer Series in Advanced Microelectronics 37,
https://doi.org/10.1007/978-981-13-0599-3_7

Flash-based SSDs also materialized as an astonishing proliferation of global-scaled corporations whose commercial strength is tightly coupled to the availability of SSDs engineered for big data centers and cloud computing. The previous developing strategy of SSDs, in fact, was based on a full compatibility with HDDs and therefore the SSDs' performance optimization was focused on that of the Flash Translation Layer (FTL), the firmware managing the basic memory operations [3–5]. FTL is responsible for a plug-and-play connection between the host system where the application is running and the SSD. To this respect, it must be considered that in the last 4 decades user applications have been designed to work with traditional magnetic HDDs, which are conceptually different from SSDs. Therefore, rather than redesign the whole architecture of the application, it is more convenient to leverage a command translation layer.

The development of SSDs was made possible by the use of sufficiently reliable Single Level Cells (SLC) NAND Flash memories [6], storing a single bit per cell in the traditional 0/1 digital paradigm with a low read error probability, thus requiring the design of simple engines for Error Correction Codes (ECC) [7]. The SATA protocol [8] interfacing the memory system and the host was sufficient to guarantee the requested Quality of Service (QoS), that is the ability of keeping a sustained performance over time within a defined threshold [9, 10]. As a whole, the SSD architecture optimization and the development of dedicated CAD tools for the exploration of the SSD design space were FTL-oriented, in a top-down approach.

In the last few years, the need for SSDs with higher storage capacities and performance joined to the availability of high density NAND Flash memories able to store 2, 3 or even 4 bits in a single cell [11], moved the design paradigm from a Top-Down to a Bottom-Up approach where the performance and the reliability of

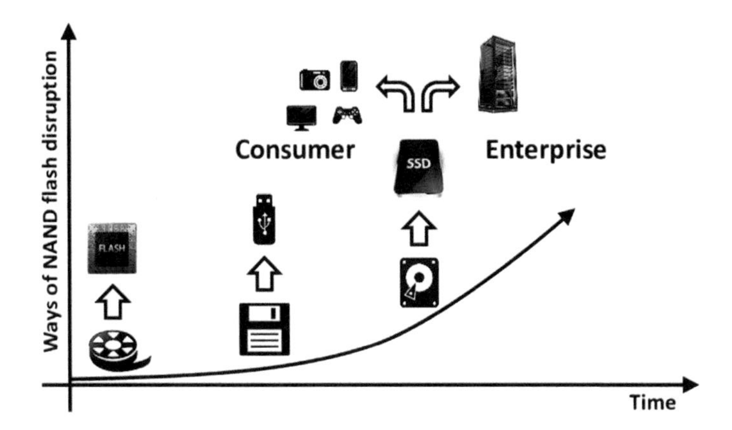

Fig. 7.1 Evolution of NAND Flash-based systems: from tape, film and floppy disk replacement to the explosive SSDs applications for cloud computing and big data centers. Reproduced with permission from L. Zuolo, C. Zambelli, R. Micheloni and P. Olivo, "Solid-State Drives: Memory Driven Design Methodologies for Optimal Performance," in Proceedings of the IEEE, vol. 105, no. 9, pp. 1589–1608, Sept. 2017. © 2017 IEEE

the storage medium dictate the design constraints. NAND Flash memories with scaled technologies, in fact, suffer from several physical mechanisms able to impact their reliability figures such as: (i) Endurance, that is the maximum number of Program/Erase (P/E) operations that the memory can withstand before leading to a failure; (ii) Data Retention, denoting the ability of a memory to keep a stored information over time with no biases applied; (iii) the immunity from Read Disturbs, representing the stress suffered by a memory cell when reading neighbor cells [12–14].

These reliability issues become more and more significant in Multi-Level Cells (MLC) [15], Triple-Level Cells (TLC) [16] and Quadruple-Level Cells (QLC) [17] storing 2, 3, and 4 bits per cell, respectively, where the undesired transfer of few electrons into/from the storage layer may alter significantly the memory information content. The basic parameter characterizing the NAND Flash memory reliability is the Raw Bit Error Rate (RBER), representing the fraction of erroneous bits retrieved during a read operation [14]. The knowledge of this parameter whose value increases with: technology scaling, the number of bits that a cell can store, the number of P/E operations, the time elapsed between two successive read operations, the number of repeated read operations on the same memory location, is now the driver for architectural and software design of present SSDs [18].

Multilevel NAND Flash memories require the availability of an ECC scheme able to correct the errors detected when reading the memory. The choice of the ECC code and the design of the correction engine represent the key points for present SSDs design since they must be carefully calibrated with respect to the figures of merit of the selected nonvolatile memories. A too simple ECC scheme may not be able to guarantee a suitable reliability, whereas a too complex one may reduce severely the read bandwidth because of the time required for error correction, with a consequent impact also on the system power consumption [19]. Based on the selected ECC code and of the designed ECC engine, an optimal error reduction algorithm for the memory read operation could be identified.

Once the ECC scheme has been designed, the Bottom-Up design flow rises to the memory controller, representing the interface towards the ECC engine and the memory storage system. The controller, to avoid that the design efforts devoted to optimize the ECC scheme vanish, must guarantee the bandwidth provided by the ECC block. With this respect, the SSD controller must be designed in order to manage a sufficient amount of commands to fully exploit the bandwidth of the underlying storage system. Similarly, also the interface towards the host must be able to guarantee the expected bandwidth. For this reason, SATA protocol is no longer able to deal with the performance made available by the other blocks in the SSD architecture so that SAS [20] and PCI-Express [21] are adopted for enterprise environments.

On the basis of this bottom-up SSDs design flow, from an accurate knowledge of the performance and limits of the selected NAND memories to the design of a suitable ECC engine and, successively to that of the controller and of the host interface, also CAD tools for SSD design must follow this Bottom-Up vision, while relaxing the efforts previously devoted to the FTL design [22].

7.2 The Impact of ECC on SSD Performance

As summarized in the previous section, because of endurance problems, poor data retention or read disturbs, the actual threshold voltage read in a cell may be different from the programmed one [14]. Therefore, when a page is read, some cells may return a wrong value, thus producing read errors. To overcome these problems, data-encoding guaranteeing a reconstruction of the correct read page data is mandatory in electronic systems using NAND Flash memories.

The correction capability of the code to be adopted is strictly related to the error probability. For a given technology node, since physical degrading mechanisms are the same independently of the different storage paradigms (SLC, ..., QLC), the error probability increases with the number of bits stored in a single cell.

In the first SLC memories, thanks to the large gap between the program and the erase voltage distributions, the error probability was very low, so that Bose-Chaudhuri-Hocquengham (BCH) codes able to correct few tens of bits in a 1 or 2 kB page were sufficient. With limited number of errors to be corrected, the correction time was not an issue and the read bandwidth and latency were marginally affected by the use of ECCs [23].

Figure 7.2a shows the typical blocks for ECC engines based on BCH codes: a high-speed encoder is connected to each one of the N_c SSD channels (that is a bus used to communicate with an array of N_d memory dies), whereas a reconfigurable parallel decoder (i.e. a multi-engine decoder) is shared among the channels [24]. The structure of the decoder is represented in Fig. 7.2b, where the Syndrome block determines whether an error is present, the Berlekamp-Massey block calculates the coefficients of the error locator polynomial, and the Chien machine locate the errors [25].

In multilevel architectures the number of errors to be corrected increases by an order of magnitude for any further bit stored in a single cell. Although ECC engines based on BCH codes are still used thanks to their simple hardware implementation, high numbers of bits to be corrected may affect significantly on the overall read time. Consequently, the correction time may become the bottleneck of the entire read procedure. In addition, because of the high number of errors, the probability of having uncorrectable pages (that are pages read with a number of wrong bits higher than the ECC correction capabilities) increases [26]. When a page is marked as uncorrectable, the read operation fails and the page content is irremediably lost. The adoption of parallel decoding architectures can reduce the bandwidth and latency degradation (at the expenses, however, of both area occupation and power consumption) but it cannot solve the problems caused by uncorrectable pages.

To deal with the presence of uncorrectable pages, two alternatives exist: (i) keep BCH codes and their ease of implementation while defining sophisticated read algorithms in order to reduce the number of errors [27]; (ii) develop ECC solutions based on different coding concepts, like Low Density Parity Check (LDPC) codes [28]. In the former case, the basic idea in the presence of uncorrectable pages consists in re-reading the page with different read reference voltages, in the attempt

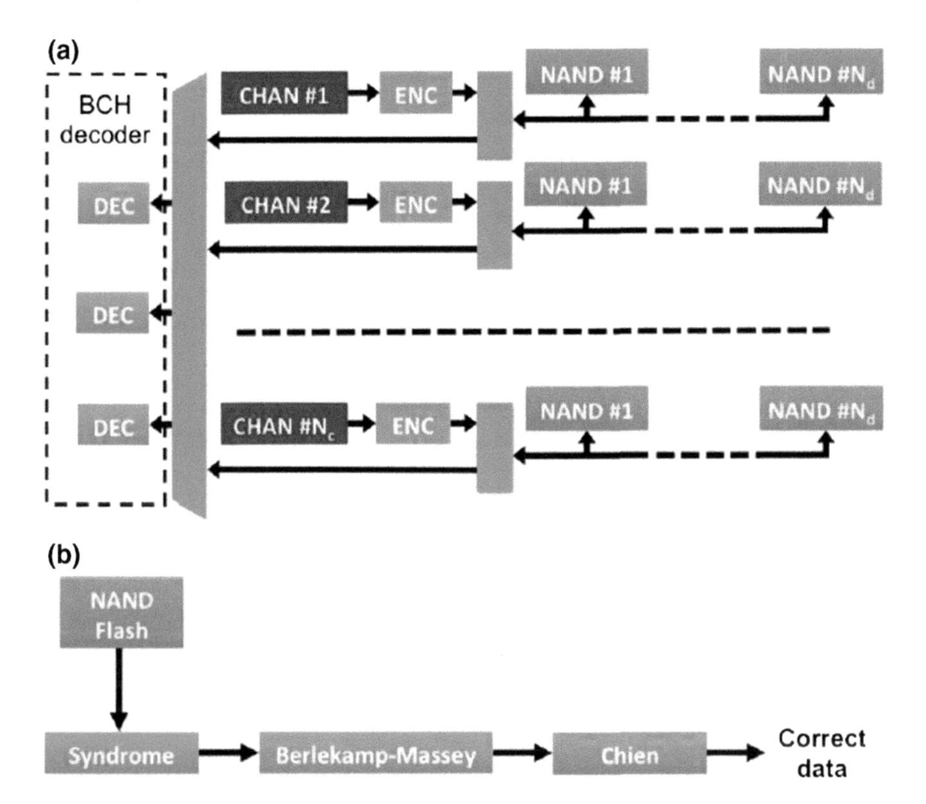

Fig. 7.2 **a** Schematic representation of an ECC architecture based on BCH codes. A high-speed encoder is connected to each SSD channel whereas a a reconfigurable parallel decoder is shared among the Nc channels. **b** Schematic representation of the BCH decoder. Reproduced with permission from L. Zuolo, C. Zambelli, R. Micheloni and P. Olivo, "Solid-State Drives: Memory Driven Design Methodologies for Optimal Performance", in Proceedings of the IEEE, vol. 105, no. 9, pp. 1589–1608, Sept 2017. © 2017 IEEE

of tracking the shift of the threshold voltage distributions. Such a solution led to the development of different read algorithms, generally defined as read retry [26]: the ECC engine automatically manages them and they call for (at least) a page re-reading with the unavoidable degradation of the read bandwidth. The latter solution adopts LDPC codes that, differently from BCH codes, present a much higher correction capability [28]. Figure 7.3 shows the typical blocks for ECC engines based on LDPC codes: the decoding engine is composed by two main blocks: the Hard Decoding (HD) and the Soft Decoding (SD).

From an operative point of view, LDPC decoding works as follows. Cells are read as '1' or '0' depending on their threshold voltage with respect to a fixed reference level. If during the ECC decoding phase the page is evaluated as uncorrectable, the LDPC decoding algorithm can be retried with the SD. To accomplish this second step, more information about the actual position of the NAND Flash threshold voltage distributions must be collected. The algorithm steps

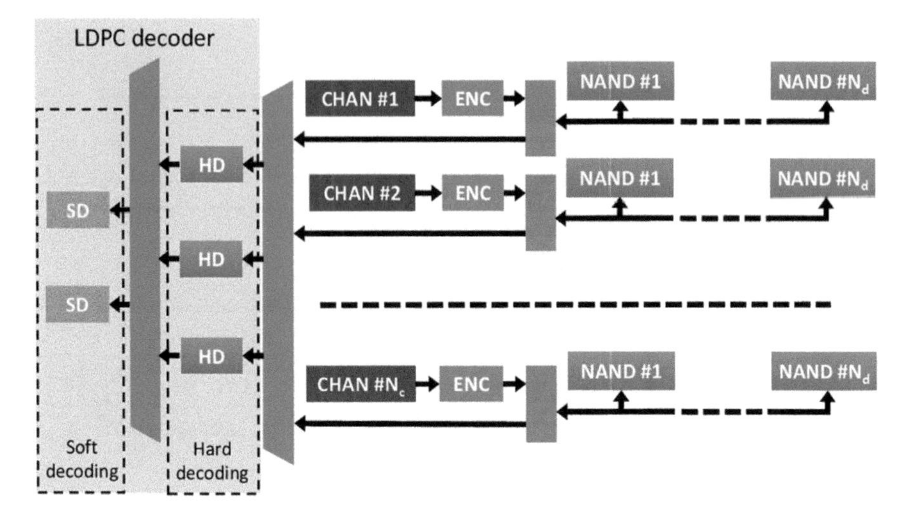

Fig. 7.3 Schematic representation of an ECC architecture based on LDPC codes. The decoding path is composed by two main blocks: the hard decoding, whose architecture is similar to that designed for BCH engines and the soft-level decoding. Reproduced with permission from L. Zuolo, C. Zambelli, R. Micheloni and P. Olivo, "Solid-State Drives: Memory Driven Design Methodologies for Optimal Performance", in Proceedings of the IEEE, vol. 105, no. 9, pp. 1589–1608, Sept 2017. © 2017 IEEE

sequentially the internal read references to lower and higher voltages thus reading the page twice. Data are transferred to the LDPC decoder and then they are bit-wise combined with those previously read with the first reference (i.e., called the HD reference). This step is possible because during the whole SD process the data read with the HD reference are stored in a dedicated buffer inside the SSD controller and used as a reference. The algorithm continues this process until the page is correctly read or the maximum number of soft-decoding operations is reached and the page is marked as uncorrectable [19].

LDPC is now the state-of-the-art in SSD products. However, to evaluate the optimal ECC engine design in terms of HD and SD implementation, the knowledge of the actual memory RBER is mandatory. With this respect, it is usual to leverage a worst-case design methodology where the correction strength figure of the HD is compared with the maximum percentage of uncorrectable pages measured at the end of the memory's lifetime. Figure 7.4 shows this process when a LDPC able to correct up to 100 bits in a 4320 Bytes codeword is considered for a TLC NAND Flash memory manufactured in a planar 1X technology node. Point A marks the maximum percentage of uncorrectable pages measured at the end of the memory's lifetime. As it can be seen, in this case switching from the HD to a one bit SD is sufficient to correct all the errors (point B). Other correction strategies like a two bits SD, become an over-design.

The above considerations are mandatory when it is required to design the optimum LDPC architecture (both in terms of correction strength and correction

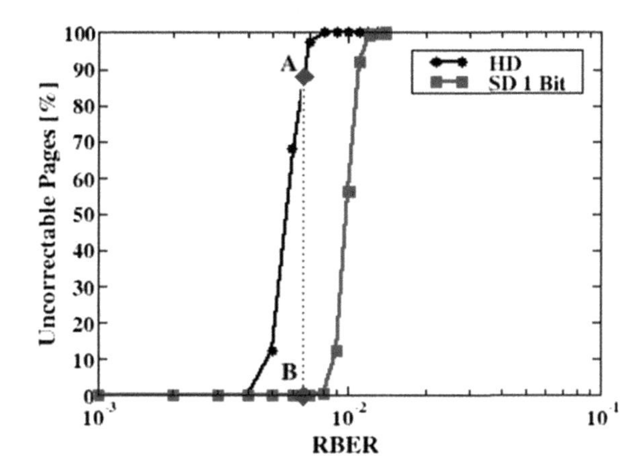

Fig. 7.4 Correction strength of both HD and SD when a LDPC able to correct up to 100 Bits in a 4320 Bytes codeword is considered for a 128-Gb TLC NAND Flash memory manufactured in a planar 1X technology node. Points A and B represent the maximum measured percentage of uncorrectable pages at the end of the memory lifetime, when HD and SD are used, respectively. Reproduced with permission from L. Zuolo, C. Zambelli, R. Micheloni and P. Olivo, "Solid-State Drives: Memory Driven Design Methodologies for Optimal Performance", in Proceedings of the IEEE, vol. 105, no. 9, pp. 1589–1608, Sept 2017. © 2017 IEEE

bandwidth) for the target SSD. In fact, since the SD directly affects the drive's bandwidth, once the correction strategy is defined (a one bit SD rather than a two bits SD) and the decoder's bandwidth is fixed, it is important to find the right balance between the number of HD and SD decoders. Figure 7.5 shows the read bandwidth obtained, for different HD implementations, in a 2 TB SSD featuring 16 channels each one connected to eight 128-Gbits TLC NAND Flash dies manufactured in a planar 1X technology node, as a function of the number of P/E cycles. Since each hard decoder in this example has a bandwidth of 1.2 GB/s and the SSD host interface is a PCI-Express GEN3x4 [21] with a maximum bandwidth of 4 GB/s, it is clear that a coarse design choice (that neglects the actual RBER evolution) requires 4 HD decoders. To this extent, any higher number would result in a cost ineffective overdesign.

However, since RBER increases with the number of P/E cycles, the percentage of uncorrectable pages detected by the HD increases as well. Consequently, SD is triggered and the read bandwidth rapidly decreases when the memory rated endurance is approached. To guarantee the expected performance and to extend the SSD working window, it is necessary to increase the number of HD decoders (see Fig. 7.5) as well as that of SD decoders. Figure 7.6a shows the calculated read bandwidth degradation with respect to the beginning of life) by implementing 8 HD decoders and different numbers of SD decoders. As it can be seen, to reduce the read bandwidth degradation at twice the rated endurance, 2 SD decoders can be used, while any larger number of decoders would result in an overdesign. Figure 7.6b shows the results obtained by using 16 HD decoders and different

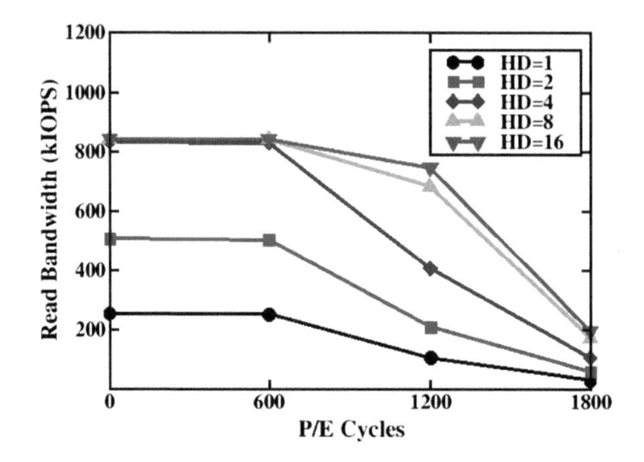

Fig. 7.5 Read bandwidth evolution as a function of the number of P/E cycles sustained by NAND Flash in a 2 TB SSD featuring a PCI-Express GEN3x4 host interface. The ECC engine is composed by a variable pool of HD decoders and a single SD decoder. The NAND Flash rated endurance is 900 P/E cycles. Reproduced with permission from L. Zuolo, C. Zambelli, R. Micheloni and P. Olivo, "Solid-State Drives: Memory Driven Design Methodologies for Optimal Performance", in Proceedings of the IEEE, vol. 105, no. 9, pp. 1589–1608, Sept 2017. © 2017 IEEE

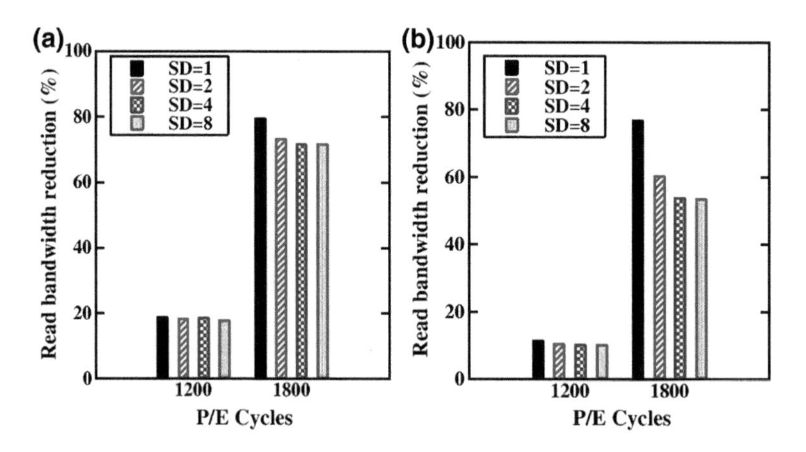

Fig. 7.6 Read bandwidth degradation with respect to the beginning of life at different endurance considering different SD levels. 8 and 16 HD decoders have been considered in (**a**) and (**b**), respectively. Reproduced with permission from L. Zuolo, C. Zambelli, R. Micheloni and P. Olivo, "Solid-State Drives: Memory Driven Design Methodologies for Optimal Performance", in Proceedings of the IEEE, vol. 105, no. 9, pp. 1589–1608, Sept 2017. © 2017 IEEE

numbers of SD decoders, showing a significant performance improvement thanks to a much higher hardware cost. From a designer point of view, an accurate trade-off evaluation between performance (i.e. read bandwidth reduction) and hardware cost must be based on the actual knowledge of the memory RBER evolution.

7.3 SSD Controller Design

The main block diagram of an SSD controller is shown in Fig. 7.7. Once the SSD's specifications have been fixed, and hence the maximum device bandwidth has been defined, the SSD controller design follows a simple rule of thumb to calculate N_c and N_d needed to meet the requirements. To calculate the actual controller bandwidth B_{cont}, it is sufficient to sum the bandwidth contributions B_{ch} of each channel:

$$B_{cont} = \sum_{i=1}^{N_c} B_{ch,i}$$

The maximum channel bandwidth $B_{ch,i}$ is obtained under the assumption that all the memory dies connected to channel i are addressed at the same time. By defining B_d as the bandwidth of each memory die, the theoretical controller bandwidth is given by:

$$B_{cont}^{th} = \sum_{i=1}^{N_c} B_{ch,i}^{max} = \sum_{i=1}^{N_c} N_{d,i} B_d$$

Previous equation represents, however, the theoretical condition under the hypothesis that all single dies can communicate simultaneously with the controller

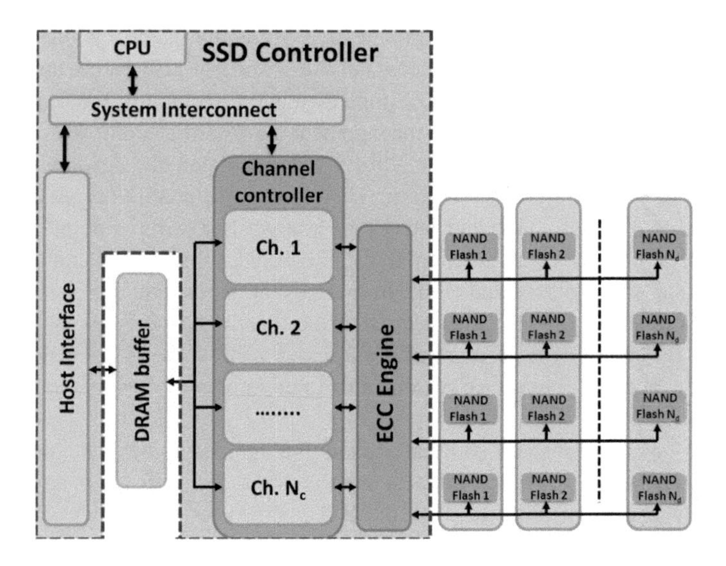

Fig. 7.7 Schematic representation of the SSD controller, considering N_c channels and N_d memory dies connected to each channel. Reproduced with permission from L. Zuolo, C. Zambelli, R. Micheloni and P. Olivo, "Solid-State Drives: Memory Driven Design Methodologies for Optimal Performance", in Proceedings of the IEEE, vol. 105, no. 9, pp. 1589–1608, Sept 2017. © 2017 IEEE

and, therefore, it represents the maximum achievable value. Unfortunately, for several reasons (e.g., access request to the same die, die's response time slowed down by a read retry operation, die busy for a program operation whose latency is much higher with respect to read latency, etc.), the probability that all dies can communicate simultaneously with the controller is generally <1. Taking into account that a number n of dies in a channel cannot serve new requests since they are processing other commands, the actual controller bandwidth is given by:

$$B_{cont} = \sum_{i=1}^{N_c} (N_{d,i} - n_i) B_d \leq B_{cont}^{th}$$

The above equation calculates the controller bandwidth in a fresh condition (i.e., at the beginning of the drive's lifetime). However, as previously described in the former sections of this chapter, the actual performance of the SSD is strongly affected by the reliability phenomena associated with the storage layer. Therefore, to take into account these effects, the equation can be modified as follows:

$$B_{cont}(PE, T, RD, WAF) = \sum_{i=1}^{N_c} [N_{d,i} - n_i(PE, T, RD, WAF)] B_d \leq B_{cont}^{th}$$

where *PE, T, RD* and *WAF* are the current Program/Erase cycle number of the drive, the working Temperature, the Read Disturb level of the memories, and the Write Amplification Factor, respectively. The *WAF* factor is defined as the ratio between the data written to the NAND Flash and the data written by the host. Generally, is a number greater than 1. It has been accurately described in [29] and it depends on several factors ascribed to the FTL implementation including Wear Leveling, Garbage Collection, and Bad Block management algorithms. Along with *WAF, P/E, T,* and *RD* introduce hard-to-model effects that complicate the description of the controller's bandwidth in a closed form. Therefore, to help SSD designers to calculate the actual performance and latency of a target SSD over time and use, the adoption of sophisticated simulation tools like SSDExplorer is mandatory [22]. Overall, what ultimately stands out from both previous equations is that, to approach as much as possible the ideal controller bandwidth, it is necessary to: (i) reduce the probability that a command addresses a busy die (i.e., a die already scheduled by another operation); (ii) maximize the number of dies that can process a new command.

This can be accomplished: (i) by increasing the number N_d of dies connected to each channel, which however impacts on the SSD cost; (ii) with an effective command management performed by the FTL; (iii) by using a DRAM as a data buffer.

7.3.1 Efficient Command Management

In nowadays SSDs, to efficiently manage the commands issued by the host, it is possible to leverage the Command Queue (CQ) concept [30]. This resource is usually implemented as a software routine shared between the host interface, which pushes host commands inside the CQ, and the SSD controller that manages the requested operations and pulls out the commands from the CQ.

Figure 7.8 shows the queuing hierarchy usually implemented in traditional SSD controllers [31]. Besides the external host CQ, it is common to have a dedicated small command queue for each NAND Flash memory die: the Target Command Queue (TCQ). Thanks to the TCQ, the host can continue to issue commands even when it tries to read or program a die that is in the busy state. In fact, when this condition is verified, the command is simply queued in the TCQ and the SSD controller can continue to fetch other commands from the host CQ. This technique allows maximizing B_{cont} since TCQs keep always-busy all the NAND Flash dies. It is thus clear that the main parameters controlling B_{cont} are the parallelism (i.e., N_c and N_d) and the queue depth (QD), that is the number of commands that the host interface can store.

The attempt of approaching the ideal performance in terms of bandwidth by increasing QD presents an unavoidable disadvantage: the increase of the service time (i.e. the time elapsed between the issue and the execution of a command) and,

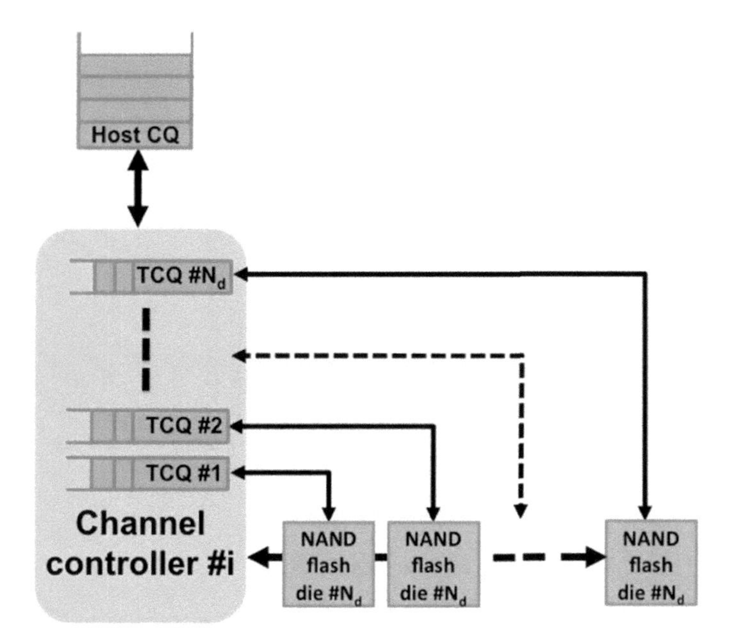

Fig. 7.8 Queueing hierarchy implemented inside the SSD controller for a generic channel. Reproduced with permission from L. Zuolo, C. Zambelli, R. Micheloni and P. Olivo, "Solid-State Drives: Memory Driven Design Methodologies for Optimal Performance", in Proceedings of the IEEE, vol. 105, no. 9, pp. 1589–1608, Sept 2017. © 2017 IEEE

consequently, of the SSD latency. Therefore, QD has a severe impact on QoS, that defines the maximum acceptable latency of the drive and it is calculated as the 99.99th percentile of the SSD latencies cumulative distribution. To this extent, QoS is used to quantify how the SSD behaves in the worst-case conditions [9]. By using this metric, it is possible to understand if the target SSD architecture is suitable for a specific application, such as real-time and safety-critical systems [32]. Figure 7.9 shows an example of how B_{cont} and QoS scale with the host QD. As expected, both B_{cont} and QoS increase with QD. This behavior, however, is in contrast with the requirements of high performance SSDs, which ask for achieving the target bandwidth with the lowest QoS. In fact, state-of-the-art user applications such as financial transactions or cloud platforms [33] are designed to work with storage devices, which have to serve an I/O operation within a specific period, which is usually upper-bounded, by the QoS requirement.

To deal with this requirement it is possible to use the Head-of-Line (HoL) blocking concept, whose effect is to limit the number of outstanding commands inside the SSD, thus partially solving the latency issue [34]. The HoL blocking is managed by the controller firmware implementing a FIFO stack whose dimensions can be dynamically defined. When the number of commands queued in a TCQ exceeds a predefined threshold, it is possible to trigger a blocking state inside the SSD controller that stops the submission of a new command from the host CQ. In such a way, depending on the HoL threshold value, it is possible to avoid long command queues inside the TCQs and, hence, the device QoS can be limited within a defined window.

The fine-grained QoS calibration made available by the HoL blocking, however, does not come free. If, besides B_{cont} and QoS, the average SSD latency is taken into

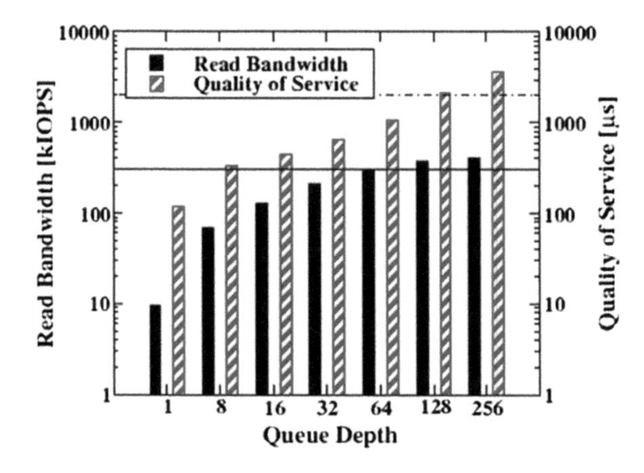

Fig. 7.9 B_{cont} and QoS as a function of the host Queue Depth. The full line and the dashed-dotted line represent the target B_{cont} and the target QoS, respectively. Simulations refer to an SSD featuring $N_c = 8$ and $N_d = 8$ TLC NAND Flash manufactured in a planar 1X technology node. Average read time is 86 µs and workload is 100% 4 kB random read. Reproduced with permission from L. Zuolo, C. Zambelli, R. Micheloni and P. Olivo, “Solid-State Drives: Memory Driven Design Methodologies for Optimal Performance”, in Proceedings of the IEEE, vol. 105, no. 9, pp. 1589–1608, Sept 2017. © 2017 IEEE

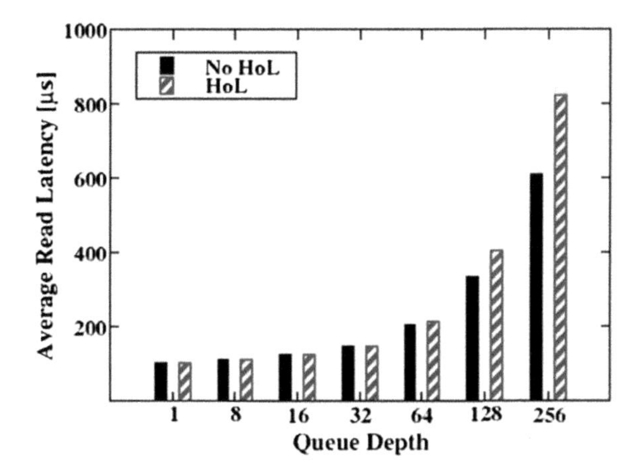

Fig. 7.10 Average SSD latency evaluated as a function of the host queue depth, for the same case of Fig. 7.9, with and without the HoL blocking. Reproduced with permission from L. Zuolo, C. Zambelli, R. Micheloni and P. Olivo, "Solid-State Drives: Memory Driven Design Methodologies for Optimal Performance", in Proceedings of the IEEE, vol. 105, no. 9, pp. 1589–1608, Sept 2017. © 2017 IEEE

account, it is clear that the HoL blocking effect has to be wisely used (see Fig. 7.10). When the HoL blocking is triggered it trades the QoS reduction with an increase of the average latency. Moreover, this behavior becomes more pronounced when high QDs are used (i.e., when a higher QoS reduction is required).

7.3.2 DRAM Data Caching

To increase the controller bandwidth and to approach as much as possible the theoretical bandwidth, it is possible to use a DRAM as data cache buffer [35]. As shown in Fig. 7.7, this block is located between the host interface and the channel controller. Standard data caching algorithms can be adopted, such as Least Recently Used (LRU) or Least Frequently Used (LFU) [36], to decrease the number of accesses to the Flash memories. Since data are addressed in a much faster memory, the access time can be reduced with respect to a standard NAND Flash read/program operation. In addition, since part of the data to be read/written are stored in the DRAM buffer, the number of accesses to the NAND Flash dies are reduced, thus limiting the number of busy dies.

These effects positively affect the SSD bandwidth and the average latency. Moreover, the reduction of the number of accesses to the NAND Flash dies increases their reliability. This point is strictly related to the smaller number of write operations, thus limiting endurance effects and, possibly, leading to a reduced read disturb issue.

Table 7.1 shows the cache-hit probability, the read bandwidth, the average latency, and the QoS calculated for the "no cache" case (i.e., a case where the

Table 7.1 NAND/DRAM size ratio and SSD performance for a configuration where LBA space is uniformly distributed

NAND/DRAM size ratio	No cache	256	50	15
Cache hit (%)	0	0.6	2.7	8.2
Read bandwidth (kIOPS)	301	312	318	337
Average Latency (μs)	206	204	200	189
QoS (ms)	1.07	1.19	1.13	1.03

DRAM data cache buffer is not present, assumed as reference) and for different ratios between the total NAND and the DRAM sizes. The number of cache hits (i.e. the percentage of memory accesses to the DRAM buffer with respect to the total number of data accesses) depends on the probability of addressing any single non-volatile memory page. All data have been collected considering a uniformly distributed Logical Block Address (LBA) space of the SSD and a LRU eviction policy is used as caching algorithm.

As it can be seen, the performance metrics of the simulated drive are not significantly influenced by the DRAM size. This is because the LBA space is uniformly distributed across all the SSD pages, therefore all data locations have the same probability to be addressed.

A uniformly distributed LBA space, however, represents the worst-case condition for the assessment of the benefits materialized by a caching algorithm. In general, real user workloads tend to follow different LBA distributions, which are more similar to a Gaussian or a Lognormal with a mode around a specific address. Consequently, if the I/O address profile of the target application is known, it is possible to optimize the DRAM cache size depending on the statistical parameters presented by the LBA profile itself.

Suppose to have Gaussian distributed workloads spanning across the whole LBA space of the drive. By considering a standard deviation around the average of the total SSD LBA address space, it is possible to design the proper DRAM size ratio in two different ways: (i) reducing the DRAM capacity while keeping the same cache hit probability and drive performance; (ii) increasing the DRAM capacity maximizing the number of cache hits and, therefore, boosting the drive performance.

Table 7.2 shows, for three different standard deviation values, the NAND/DRAM size ratio, the cache hit probability, the read bandwidth, the average latency, and the QoS of the target SSD architecture. As it can be seen, the performance metrics are almost similar with a significant reduction of the DRAM size for the tightest workload distribution.

Table 7.3 shows, for the (b) case, the NAND/DRAM size ratio, the cache hit probability, and the performance metrics of the target SSD architecture. With respect to the (b) case of Table 7.2, the NAND/DRAM size ratio has been reduced from 50 to 15. As it can be seen, it is possible to almost triplicate the cache-hit probability thus increasing the read bandwidth while reducing the average latency. It is worth to highlight that this performance improvement marginally influences the

Table 7.2 NAND/DRAM size ratio and SSD performance as a function of the LBA space distributions (assumed Gaussian)

Standard deviation	2%	10%	30%
NAND/DRAM size ratio	256	50	15
Cache hit (%)	15.3	15.3	15.3
Read bandwidth (kIOPS)	367	364	365
Average Latency (μs)	173	175	175
QoS (ms)	0.98	1.27	1.29

Table 7.3 NAND/DRAM size ratio and SSD performance for a configuration where the LBA space is that of case (**b**) in Table 7.2

NAND/DRAM size ratio	50	15
Cache hit (%)	15.3	42.1
Read bandwidth (kIOPS)	364	536
Average Latency (μs)	175	118
QoS (ms)	1.27	1.19

QoS, since it is related to the worst case (usually a read operation performed on a NAND Flash die). Summing up, the use of a DRAM cache offers advantages in terms of bandwidth, latency, and reliability. The design of an application specific SSD, in addition, can be optimized if the LBA space distribution is known, in order to reduce the DRAM size. Therefore, the drive design must be done concurrently with the application for which it represents the storage element.

7.4 Criteria for Optimal Host Interface Selection

The host interface represents the link between the SSD controller and the host where the application is running. Differently from the SSD controller that is fully customized, the physical structure of the communication interface follows consolidated standards. Now, the used interfaces are SATA [8] (mainly for consumer applications), SAS [20], and PCIe [21] (for enterprise environments).

The correct choice of the host interface represents a crucial aspect along the drive design phase since it allows guaranteeing that the SSD controller is used in optimal conditions. In a traditional design approach for general purpose SSDs, where both controller and host interface are chosen separately without any knowledge of the final application, the constraint of selecting a host interface able to guarantee a bandwidth $B_{hi} \geq B_{cont}$ (where B_{hi} is the maximum bandwidth of the host interface) at the lowest cost represents the standard approach, whereas a host interface whose $B_{hi} < B_{cont}$ would act as a bottleneck limiting the SSD performance. A detailed analysis of the impact of the host interface on the SSD's performance has been presented in [22].

If the application to be run on the host is known, a different approach can be adopted. It must be taken into account that the design of a fully customized SSD controller is much more expensive with respect to that of the host interface, which follows well-defined standards. By considering this economic aspect, it is convenient to design an SSD controller with top performance (rather than a family of controllers with different quality metrics) and to operate at the host interface level to satisfy the application requirements. As an example, if the controller has been designed to sustain a certain theoretical bandwidth and the application requires a lower bandwidth B_{app}, an interface satisfying the condition

$$B_{app} \leq B_{hi} \leq B_{cont}^{th}$$

can be selected, confirming that the ideal host interface must be chosen on the basis of the application and, therefore, on the drive use. In such a way, with a single SSD controller design, different application requirements can be satisfied by using different host interfaces. Such methodology allows reducing the controller bandwidth to match that of the application and lowering the design cost of the SSD controller. In addition, it allows also reducing the drive power consumption since, operating at a lower throughput, a lower number of NAND Flash dies are activated simultaneously.

An evolution of this design methodology, envisaging a single controller associated to different interfaces as a function of the application, considers a unique combination of SSD controller and host interface. In this case, each block is able to provide the maximum theoretical performance. The effective performance, however, can be tuned dynamically at software level by acting on the SSD's firmware and especially on the command queue depths, which can be modified during the normal execution. An example of this methodology can be found in [37, 38] where the SSD controller is able to automatically limit the performance of the drive depending on the allowed power consumption or on the thermal dissipation level. Such an approach that calls for the design of a single block embedding the SSD controller and the host interface, however implies a higher design cost for the development of a controller whose hardware resources can be programmed by the user.

7.5 Future Applications Opened by Hardware-Software Co-design for High-performance SSDs

In the last 40 years, all software applications and Operating Systems (OS), which make use of persistent storage architectures, have been designed to work with HDDs [1]. However, SSDs are physically and architecturally different from HDDs so that they need to execute the FTL algorithm to translate host commands [3–5]. The main role of FTL is to mimic the behavior of a traditional HDD and to enable the usage of SSDs in any electronic system without acting on the software stack. Besides this translation operation, SSD controllers have to run garbage collection, command-scheduling algorithms, data placement schemes, wear leveling, and errors correction. All these routines, even if on the one hand allow a "plug and play" connection of the SSD with traditional hardware and software, on the other hand they limit actual SSD performance. The main drawback of FTL is the Garbage Collection (GC) that is performed when valid pages belonging to a block to be erased are read and written in a different block. Such an operation, that is time and power consuming, reduces both drive bandwidth and NAND Flash reliability [29]. In the enterprise market and hyper scale data centers, performance and reliability losses induced by GC are not tolerable.

To deal with the above-mentioned challenges, software developers in data centers have shown, in the past few years, a growing interest for Software-Defined Flash (SDF) [39]. In this kind of environments, the driving forces in the design of computational nodes are reliability and high performance: therefore, even the I/O management has to be re-architected. SDF leverages a new SSD design approach called Host-Based FTL (HB-FTL) which allows the host system to: (i) optimize the host payload, i.e., the amount of data read/written with a single command and hence relieve the SSD from any host command translation or manipulation; (ii) remove the GC related to FTL execution; (iii) execute the FTL directly on top of its computational node (Open-Channel architecture [40]).

7.5.1 HB-FTL

HB-FTL considers the migration of all FTL routines from the SSD to a more powerful processor located outside the SSD. To this purpose, the processor must be able to issue commands to be interpreted directly by the NAND Flash dies, such as read, program and, especially, erase [41]. In this context, a new protocol called Light NVME (LNVME) [42] allows a native communication between NAND memories and the external processor. Thanks to this protocol, the FTL can be implemented and executed by the external processor such as the host where the application is running.

A first advantage provided by this approach concerns the optimization of the host payload. With this respect, since ECC coding/decoding operate on an entire memory page, read/write operations on a NAND Flash page must follow the constrains imposed by the ECC itself. As an example, consider a NAND Flash memory whose page size is 4 kB and a host reading/writing data on a 512 B basis.

Write operations are performed on the NAND memories only when the host has transferred eight 512 B data chunks. However, the host considers as accomplished a write operation when the SSD has acknowledged the data acquisition. If a power fail occurs between the data load and the effective storage in the nonvolatile layer, data are considered as lost. To avoid this occurrence, dedicated solutions such as supercapacitors [43] or the introduction of emerging non-volatile technologies, such as MRAM, replacing DRAM buffers can be adopted [44]. On the contrary, a NAND memory page is read every time the host requires even a single chunk. Therefore, even if only 512 B are requested by the host, the entire 4 kB page is read and decoded by the ECC. It is clear that, in this case, the SSD is operating at 1/8 of its theoretical read bandwidth.

To improve the SSD performance and to better exploit its internal resources, it is convenient to co-design the application payload with the ECC engine. The optimal solution is achieved by data chunks that are an integer multiple of the actual ECC codeword.

A more powerful approach takes into account that in HB-FTL-based SDF both the application and the FTL are processed in the same software environment [45]. Therefore, they can be co-designed in order to optimize the access pattern to the nonvolatile memory. As an example, the application can be designed to perform only sequential accesses to the storage medium, respecting the physical in-order-program of NAND Flash memories. By following this approach, the actual access to the NAND Flash dies is block-based rather than page-based which is typical of random write accesses. By moving the write granularity from pages to blocks, GC is no longer necessary. In addition, by serializing the write traffic to the NAND Flash memories, the write bandwidth is maximized.

7.5.2 *The Open-Channel Architecture*

The Open-Channel architecture [40] allows implementing the management of HB-FTL-based SDF.

Figure 7.11 sketches a template architecture that can be modeled by Open-Channel. Thanks to the PCI-Express interconnection and the LNVME protocol, a bunch of NAND Flash cards can establish a peer-to-peer communication with the host processor without requesting any specific management to the SSD controller [46]. In this architecture, “NAND Flash cards” are not standard SSDs because, besides a simple I/O processor, a channel controller for NAND addressing and an ECC engine, they do not embody any complex processor, DRAM or even FTL. Consequently, data read/write from/to these cards have to be considered as the raw output/input of NAND memories without any further manipulation.

Figure 7.12 shows the effectiveness of HB-FTL with respect to a standard FTL in increasing the SSD performance. To this purpose the HGST SN150 Ultrastar SSD [47], has been compared with a simulated drive featuring a HB-FTL approach and the same SSD configuration.

The comparison has been performed for different mixed workloads, from a 100% 4 kB random read, 0% random write to a 0% random read, 100% 4kB random write. All results show that in a standard FTL-based SSD performance decreases with the write percentage, whereas in a HB-FTL-based SSD performance is mostly independent from the write percentage. This result is due to the absence of the GC algorithm that strongly affects standard FTL-based SSDs.

Another architecture that can fully exploit the Open-Channel concept and the LNVME protocol relies on the usage of a dedicated accelerator in the form of a Multi-Purpose Processing Array (MPPA) [48, 49], as shown in Fig. 7.13. This

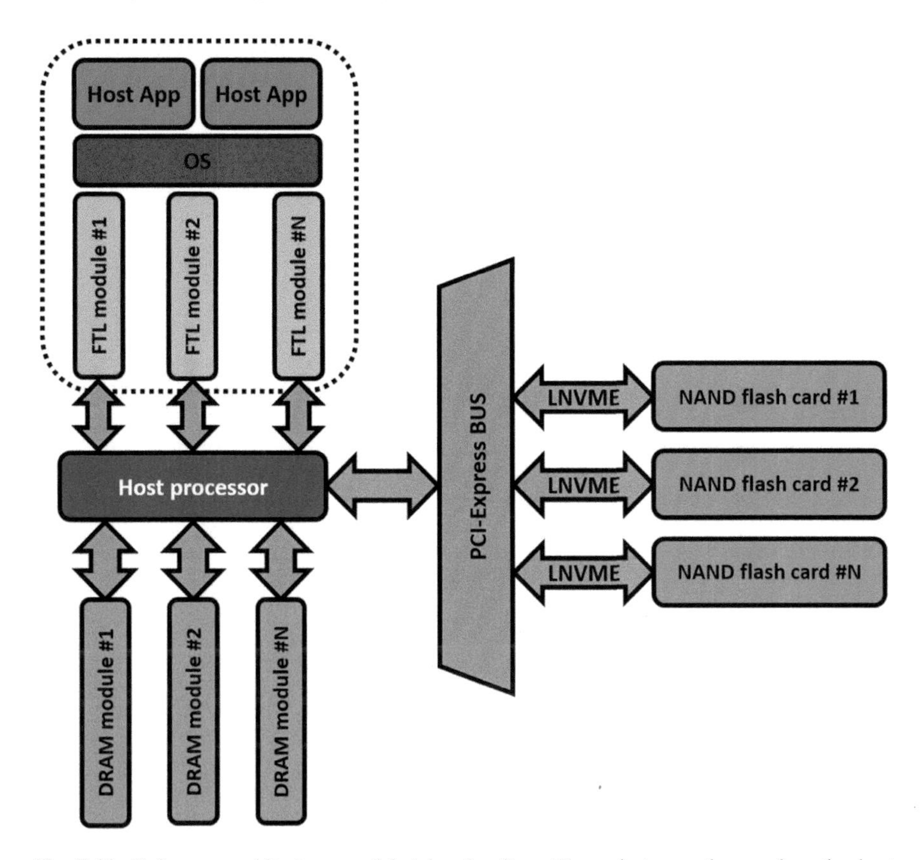

Fig. 7.11 Reference architecture modeled by the Open-Channel storage layer when the host processor is used for HB-FTL execution. More than one NAND Flash card are connected to the PCI-Express bus. The host processor executes different FTL modules. Reproduced with permission from L. Zuolo, C. Zambelli, R. Micheloni and P. Olivo, "Solid-State Drives: Memory Driven Design Methodologies for Optimal Performance", in Proceedings of the IEEE, vol. 105, no. 9, pp. 1589–1608, Sept 2017. © 2017 IEEE

solution allows the reduction of the host I/O command submission/completion timings.

These delays are strictly related to the host's processing capabilities, they represent the time spent by the host to execute the LNVME driver, and the OS file system for each submitted/completed I/O. It has been demonstrated that the performance of nowadays SSDs is heavily affected by the I/O submission/completions timings [50]. Moreover, in most recent architectures like the one based on the 3D Xpoint technology [51], these delays can even represent the actual bottleneck of the whole storage layer, whose IOPS are limited by the host system itself. Therefore, reducing these timings is the key for designing ultra-high performance storage systems.

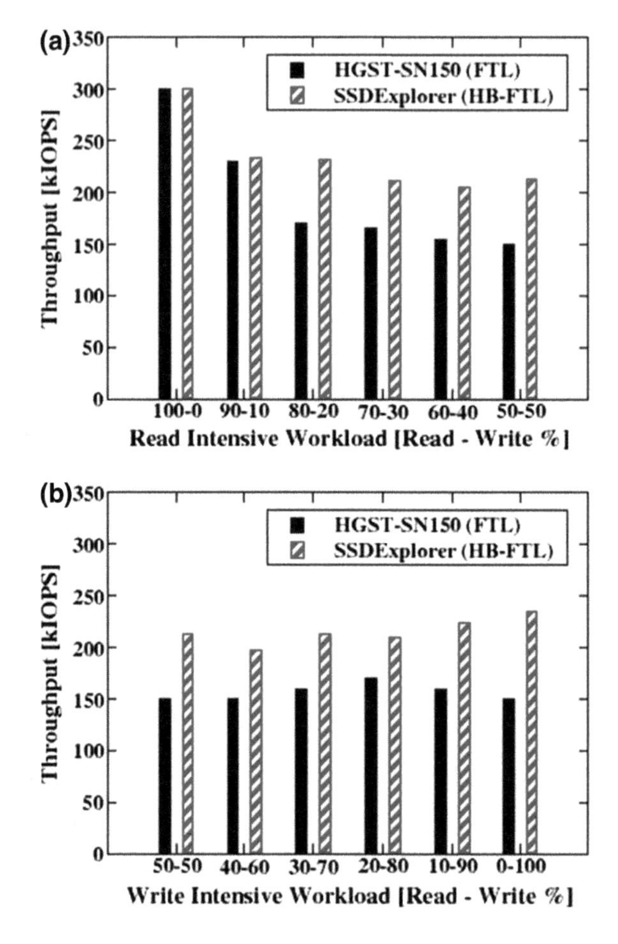

Fig. 7.12 Throughput (expressed in kIOPS) of HGST SN150 Ultrastar SSD architecture compared to that of a simulated HB-FTL-based drive with the same configuration: (top) read intensive and (bottom) write intensive workloads. A queue depth of 32 commands is used. Simulations have been performed with SSDExplorer [22]. Reproduced with permission from L. Zuolo, C. Zambelli, R. Micheloni and P. Olivo, "Solid-State Drives: Memory Driven Design Methodologies for Optimal Performance", in Proceedings of the IEEE, vol. 105, no. 9, pp. 1589–1608, Sept 2017. © 2017 IEEE

A possible solution to this problem is to switch the LNVME protocol from an interrupt-driven I/O completion mechanism to a polling-driven approach. In standard SSDs, when an I/O is completed, the Flash controller sends an interrupt to the host notifying that the transaction is ready to be transferred/processed. After that, the host can submit another command to the drive because the submission of an I/O is driven by a completion event. In theory, this approach requires that the host take action only when I/Os are submitted/completed, but in practice, it introduces long processing delays because of the OS interrupt service routines [50]. Polling the I/O completion events, on the contrary, can minimize the above-mentioned processing timings. It requires, however, that the host system monitors continuously the I/Os, thus wasting part of its processing capabilities. In light of all these considerations, moving the whole submission/completion process to a dedicated MPPA represents a good solution, which can offload the host system and, at the same time, exploit the full performance of the NAND Flash cards.

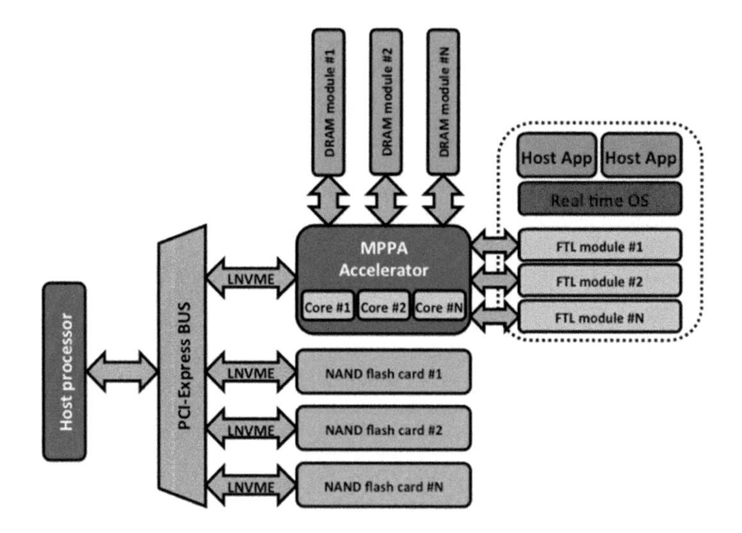

Fig. 7.13 Reference architecture modeled by the open-channel storage layer when a MPPA is used for HB-FTL execution. Besides the NAND Flash cards, the PCI-Express bus is connected to a MPPA accelerator executing different FTL modules. Reproduced with permission from L. Zuolo, C. Zambelli, R. Micheloni and P. Olivo, "Solid-State Drives: Memory Driven Design Methodologies for Optimal Performance", in Proceedings of the IEEE, vol. 105, no. 9, pp. 1589–1608, Sept 2017. © 2017 IEEE

These considerations push towards a new SSD design methodology: a complete virtualization of the storage backbone. In fact, both HB-FTL and Open-Channel allow to virtually separating the internal resources of the SSD (like channels and targets), providing a clear and straight path to OS data partitioning.

References

1. G. Wong, SSD Market Overview, in *Inside Solid State Drives (SSDs)*, ed. by R. Micheloni, A. Marelli, and K. Eshghi (Springer, 2012), pp. 1–17
2. Semiconductor Industry Association, International technology roadmap for semiconductors (2015), http://www.semiconductors.org/main/2015_international_technology_roadmap_for_semiconductors_itrs/
3. D. Liu, Y. Wang, Z. Qin, Z. Shao, Y. Guan, A space reuse strategy for flash translation layers in SLC NAND flash memory storage systems. IEEE Trans. VLSI Syst. **20**(6), 1094–1107 (2012)
4. T. Wang, D. Liu, Y. Wang, Z. Shao, FTL2: a hybrid flash translation layer with logging for write reduction in flash memory. ACM SIGPLAN Not. **48**(5), 91–100 (2013)
5. Y.H. Chang, P.C. Huang, P.H. Hsu, L.J. Lee, T.W. Kuo, D. Du, Reliability enhancement of flash-memory storage systems: an efficient version-based design. IEEE Trans. Comput. **62** (12), 2503–2515 (2013)
6. JEDEC Org., JESD 22-A 117 document, Oct 2011

7. R. Micheloni, A. Marelli, R. Ravasio, Basic coding theory, in *Error Correction Codes for Non-Volatile Memories*, ed. by R. Micheloni, A. Marelli, R. Ravasio (Springer, 2008), pp. 1–33
8. Serial ATA International Organization, SATA Revision 3.0 Specifications, www.sata-io.org
9. Intel Inc., Intel Solid-State Drive DC S3500 Series Quality of Service (2013), p. 9, http://www.intel.com/content/www/us/en/solid-state-drives/ssd-dc-s3500-spec.html
10. A. Grossi, L. Zuolo, F. Restuccia, C. Zambelli, P. Olivo, Quality-of-service implications of enhanced program algorithms for charge-trapping NAND in future solid-state drives, IEEE Trans. Dev. Mat. Reliab. **15**(3), 363–369 (2015)
11. S. Aritome, NAND flash memory technologies. Wiley-IEEE Press (2016)
12. J.D. Lee, J.H. Choi, D. Park, K. Kim, Degradation of tunnel oxide by FN current stress and its effects on data retention characteristics of 90 nm NAND flash memory cells, in *Proceedings International Reliability Physics Symposium*, Mar 2003, pp. 497–501
13. N. Mielke, H. Belgal, I. Kalastirsky, P. Kalavade, A. Kurtz, Q. Meng, N. Righos, J. Wu, Flash EEPROM threshold instabilities due to charge trapping during program/erase cycling, IEEE Trans. Dev. Mat. Reliab. **4**(3), 335–344 (2004)
14. N. Mielke, T. Marquart, N. Wu, J. Kessenich, H. Belgal, E. Schares, F. Trivedi, E. Goodness, L.R. Nevill, Bit error rate in NAND flash memories, in *Proceedings International Reliability Physics Symposium*, Apr 2008, pp. 9–19
15. K. Fukuda, Y. Watanabe, E. Makino, K. Kawakami, J. Sato, T. Takagiwa, N. Kanagawa, H. Shiga, N. Tokiwa, Y. Shindo, T. Ogawa, T. Edahiro, M. Iwai, O. Nagao, J. Musha, T. Minamoto, Y. Furuta, K. Yanagidaira, Y. Suzuki, D. Nakamura, Y. Hosomura, R. Tanaka, H. Komai, M. Muramoto, G. Shikata, A. Yuminaka, K. Sakurai, M. Sakai, H. Ding, M. Watanabe, Y. Kato, T. Miwa, A. Mak, M. Nakamichi, G. Hemink, D. Lee, M. Higashitani, B. Murphy, B. Lei, Y. Matsunaga, K. Naruke, T. Hara, A 151-mm^2 64-Gb 2 Bit/Cell NAND flash memory in 24-nm CMOS technology. IEEE J. Solid State Circuit **47**(1), 75–84 (2012)
16. K.T. Park, O. Kwon, S. Yoon, M.H. Choi, I.M. Kim, B.G. Kim, M.S. Kim, Y.H. Choi, S.H. Shin, Y. Song, J.Y. Park, J.E. Lee, C.G. Eun, H.C. Lee, H.J. Kim, J.H. Lee, J.Y. Kim, T.M. Kweon, H.J. Yoon, T. Kim, D.K. Shim, J. Sel, J.Y. Shin, P. Kwak, J.M. Han, K.S. Kim, S. Lee, Y.H. Lim, T.S. Jung, A 7 MB/s 64 Gb 3-Bit/Cell DDR NAND flash memory in 20 nm-node technology, in *IEEE International Solid-State Circuits Conference*, Feb 2011, pp. 212–213
17. C. Trinh, N. Shibata, T. Nakano, M. Ogawa, J. Sato, Y. Takeyama, K. Isobe, B. Le, F. Moogat, N. Mokhlesi, K. Kozakai, P. Hong, T. Kamei, K. Iwasa, J. Nakai, T. Shimizu, M. Honma, S. Sakai, T. Kawaai, S. Hoshi, J. Yuh, C. Hsu, T. Tseng, J. Li, J. Hu, M. Liu, S. Khalid, J. Chen, M. Watanabe, H. Lin, J. Yang, K. McKay, K. Nguyen, T. Pham, Y. Matsuda, K. Nakamura, K. Kanebako, S. Yoshikawa, W. Igarashi, A. Inoue, T. Takahashi, Y. Komatsu, C. Suzuki, K. Kanazawa, M. Higashitani, S. Lee, T. Murai, K. Nguyen, J. Lan, S. Huynh, M. Murin, M. Shlick, M. Lasser, R. Cernea, M. Mofidi, K. Schuegraf, K. Quader, A 5.6 MB/s 64 Gb 4b/Cell NAND flash memory in 43 nm CMOS, in *IEEE International Solid-State Circuits Conference*, Feb 2009, pp. 246–247
18. L. Zuolo, C. Zambelli, R. Micheloni, D. Bertozzi, P. Olivo, Analysis of reliability/performance trade-off in solid state drives, in *Proceedings International Reliability Physics Symposium*, June 2014, pp. 4B.3.1–4B.3.5
19. K. Zhao, W. Zhao, H. Sun, X. Zhang, N. Zheng, T. Zhang, LDPC-in-SSD: making advanced error correction codes work effectively in solid state drives, in *USENIX Conference on File and Storage Technologies* (2013), pp. 243–256
20. Seagate Technology LLC, Serial Attached SCSI (SAS) (2009), http://www.seagate.com/staticfiles/support/disc/manuals/Interface%20manuals/100293071c.pdf
21. PCI-SIG Ass., PCI Express Base 3.0 Specification (2013), http://www.pcisig.com/specifications/pciexpress/base3/
22. L. Zuolo, C. Zambelli, R. Micheloni, M. Indaco, S. Di Carlo, P. Prinetto, D. Bertozzi, P. Olivo, SSDExplorer: a virtual platform for performance/reliability-oriented fine-grained

design space exploration of solid state drives. IEEE Trans. Comput. Aided Design **34**(10), 1627–1638 (2015)
23. R. Micheloni, A. Marelli, R. Ravasio, Cyclic codes for non volatile storage, in *Error Correction Codes for Non-Volatile Memories*, ed. by R. Micheloni, A. Marelli, R. Ravasio, (Springer, 2008), pp. 167–198
24. Y. Lee, H. Yoo, I. Yoo, I.-C. Park, 6.4 Gb/s multi-threaded BCH encoder and decoder for multi-channel SSD controllers, in *IEEE International Solid-State Circuits Conference*, Feb 2012, pp. 426–428
25. R. Micheloni, A. Marelli, R. Ravasio, BCH hardware implementation in NAND flash memories, in *Error Correction Codes for Non-Volatile Memories*, ed. by R. Micheloni, A. Marelli, R. Ravasio (Springer, 2008), pp. 199–247
26. S.M. Jeff Yang, High-efficiency SSD for reliable data storage systems, in *Flash Memory Summit* (2012)
27. A. Cometti, L. Huang, A. Melik-Martirosian, Apparatus and method for determining a read level of a flash memory after an inactive period of time. US Patent 8,644,099, 4 Feb 2014
28. X. Wang, G. Dong, L. Pan, R. Zhou, Error correction codes and signal processing in flash memory, in *Flash Memories*, ed. by I. Stievano (2011), pp. 57–82
29. X. Hu, E. Eleftheriou, R. Haas, I. Iliadis, R. Pletka, Write amplification analysis in flash-based solid state drives, in *Proceedings ACM International Systems and Storage Conference*, May 2009, pp. 10:1–10:9
30. D. Rollins, Best practices for SSD performance measurement, in *Micron Technology, Inc., Technical Marketing Brief* (2011), https://www.micron.com/~/media/documents/products/technical-marketing-brief/briefssdperformancemeasure.pdf
31. K. Eshghi, R. Micheloni, SSD architecture and PCI express interface, in *Inside Solid State Drives (SSDs)*, ed. by R. Micheloni, A. Marelli, K. Eshghi (Springer, 2012), pp. 19–45
32. L.M. Grupp, J.D. Davis, S. Swanson, The bleak future of NAND flash memory, in *Proceedings Usenix International Conference on File and Storage Technologies* (2012), pp. 1–8
33. Avago Tech., Accelerating financial applications using solid state storage, (2011), http://docs.avagotech.com/docs/12353095
34. M. Karol, M. Hluchyj, S. Morgan, Input versus output queueing on a space-division packet switch. IEEE Trans. Commun. **35**(12), 1347–1356 (1987)
35. Intel, Intel X18-M X25-M SATA solid state drive. Enterprise Server/Storage Applications, http://cache-www.intel.com/cd/00/00/42/52/425265_425265.pdf
36. E.G. Coffman Jr., P.J. Denning, *Operating Systems Theory*. Prentice Hall Professional Technical Reference (1973)
37. S. Lee, T. Kim, K. Kim, J. Kim, Lifetime management of flash-based SSDs using recovery-aware dynamic throttling, in *Proceedings Usenix International Conference on File and Storage Technologies* (2012)
38. R.-S. Liu, C.-L. Yang, W. Wu, Optimizing NAND flash-based SSDs via retention relaxation, in *Proceedings Usenix International Conference on File and Storage Technologies* (2012)
39. J. Ouyang, S. Lin, S. Jiang, Z. Hou, Y. Wang, Y. Wang, SDF: software-defined flash for web-scale internet storage systems, in *Proceedings ACM International Conference on Architectural Support for Programming Languages and Operating Systems*, Mar 2014, pp. 471–484
40. Open-Channel Solid State Drives (2016), http://openchannelssd.readthedocs.org/en/latest/
41. A. Batwara, Leveraging host based flash translation layer for application acceleration, in *Flash Memory Summit*, Aug 2012
42. Open Channel Solid State Drives NVMe Specification (2016), http://bit.ly/2gfidpQ
43. Samsung Electronics Co., Power loss protection (PLP)—protect your data against sudden power loss (2014), http://www.samsung.com/semiconductor/minisite/ssd/downloads/document/SamsungSSD845DC05PowerlossprotectionPLP.pdf
44. C. Zambelli, G. Navarro, V. Sousa, I.L. Prejbeanu, L. Perniola, Phase change and magnetic memories for solid-state drive applications. Proc. IEEE **105**(9), 1790–1811 (2017)

45. J. Gonzalez, M. Bjrling, S. Lee, C. Dong, Y.R. Huang, Application-driven flash translation layers on open-channel SSDs, in *Non Volatile Memory Workshop*, Mar 2016, pp. 1–2
46. S. Bates, Accelerating data centers using NVMe and CUDA, in *Flash Memory Summit*, Aug 2014
47. HGST, Ultrastar SN150 Series NVMe PCIe x4 lane half-height half-length cardsolid-state drive product manual, https://www.hgst.com/sites/default/files/resources/USSN150_ProdManual.pdf
48. Kalray, The KalRay multi-purpose-processing-array (MPPA) (2016), http://www.kalrayinc.com/kalray/products/#processors
49. P. Couvert, High speed IO processor for NVMe over fabric (NVMeoF), in *Flash Memory Summit*, Aug 2016
50. J. Yang, D.B. Minturn, F. Hady, When polling is better than interrupt, in *USENIX Conference on File and Storage Technologies*, Feb 2012
51. F. Hady, Wicked fast storage and beyond, in *Non Volatile Memory Workshop*, Mar 2016

Chapter 8
SSD Reliability Assessment and Improvement

C. Zambelli and P. Olivo

8.1 Introduction

Solid State Drives (SSDs) are one of the electronic systems with the highest development rate in the last decade [1]. Their adoption as a hard disk drive (HDD) replacement in hyper scale environments like cloud computing and big data servers, as well as in consumer electronics, is relentless. SSDs' design faced an extraordinary evolution thanks to the continuous development of the storage medium integrated within, namely the NAND Flash memories [2]. SSDs performance and reliability figures of merit are intertwined with those of NAND Flash, although many other factors and components in the drive must be carefully analyzed to expose potential trade-offs. Such a consideration radically changed the design approach of SSDs, shifting from a design where the drive is seen as a mere replacement of a Hard Disk Drive (HDD) to a NAND Flash-centric approach [3]. The latter design paradigm allows achieving a high SSD reliability through a set of error mitigation techniques implemented at several levels (from NAND Flash physics and integrated circuit architecture to SSD firmware).

This chapter tackles the SSD reliability from different standpoints after the introduction, in Sect. 8.2, of the common terms used in its assessment. The Sect. 8.3 provides an overview of the physical mechanisms affecting the reliability of traditional planar NAND Flash technology as well as the 3-D integrated concepts. Proper reliability management solutions like the read retry and the soft decoding Error Correction Codes (ECCs) are introduced. Then, in Sect. 8.4, issues at die level like yield defects or extrinsic failures are exposed. Their mitigation is

C. Zambelli (✉) · P. Olivo
Engineering Department, Università di Ferrara, Ferrara, Italy
e-mail: cristian.zambelli@unife.it

P. Olivo
e-mail: piero.olivo@unife.it

R. Micheloni et al. (eds.), *Inside Solid State Drives (SSDs)*,
Springer Series in Advanced Microelectronics 37,
https://doi.org/10.1007/978-981-13-0599-3_8

addressed by techniques like the RAID (Redundant Array of Independent Disks) to improve the overall SSD reliability. Section 8.5 deals with non-NAND Flash failures like the DRAM, the SSD controller faults, or the sudden power down faults during drive operation. Finally, Sect. 8.6 will describe the SSD reliability qualification methods standardized in JEDEC JESD218 and JESD219 documents, where accelerated endurance and retention tests exploit the NAND Flash physics of failure to provide confident lifetime metrics to system designers.

8.2 Common Terms in SSD Reliability: HDD Heritage

Since SSDs have the basic functionality of state-of-the-art HDDs (i.e., read and write data sectors exchanged with a host system), their reliability terminology is the heritage of the metrics, requirements, and testing practices typical of the rotating magnetic drives [4]. When drive reliability is discussed, there are three typical concerns: (i) unrecoverable drive failures [5]; (ii) uncorrectable sector errors [6]; and (iii) silent errors [7]. A drive is considered failed up to an unrecoverable condition when either the drive ceases its functionality or its degradation level forces its replacement. An uncorrectable sector error is the condition experienced by the drive when the data required from a host system cannot be retrieved anymore and it is promptly warned through a signal. A silent error occurs if the drive returns a corrupted data to the host without providing an error signal.

The drive's reliability is specified in the datasheets with two metrics: the Mean Time Between Failures (MTBF) and the Uncorrectable Bit Error Rate (UBER). The former parameter is usually in the range of one or two million hours, practically meaning that a population of drives will average one failure every million-drive hours [4]. The MTBF is used for the calculation of the Annualized Failure Rate (AFR), that is the number of hours per year (i.e., 8760) divided by the MTBF. SSDs usually feature an AFR below 1% [8]. However, both MTBF and AFR formal definitions imply a constant failure rate, which is unrealistic due to the physics of failures (e.g., components wear-out, silicon aging, etc.). The UBER parameter on the other hand, specifies the number of uncorrectable bits divided by the number of read bits from the drive. SSDs' datasheets usually indicate this value to be in a range between 10^{-15} and 10^{-17}. The term "uncorrectable" is used since drives feature internal bit correction engines like ECC. The fraction of corrupted bits prior ECC is called the Raw Bit Error Rate (RBER), whose value is the foundation of the NAND-centric SSD design approach. The definition of UBER has many ambiguities that depend on the workload used to stress the drive (i.e., ratio between reads and writes), on the choice of the drive population, on the sectors count to be considered for error statistics, etc.

The JEDEC JESD218 [9] document defines a measurement method to separately evaluate the UBER and the AFR metrics, as we will see in the final section of this chapter. This distinction is mandatory, especially for RAID designers, since it allows discerning the magnitude of the data loss, from single files for UBER up to catastrophic drive failures for AFR.

Concerning silent errors, they are specified in units of errors per bit read [10]. Their occurrence cannot be covered by RAID systems since the data corruption is visible only by the host system. In enterprise scenario it is important to exploit design techniques that minimize their insurgence, whereas consumer scenario can sustain a more relaxed policy.

8.3 NAND Flash Reliability: Intrinsic Failures

The storage core of a SSD is the NAND Flash memory. Its concept is based on a metal oxide semiconductor device with a Floating Gate (FG) electrically isolated by means of a tunnel oxide and of an interpoly oxide as sketched in Fig. 8.1 [11]. The former oxide plays a basic role for the control of the device threshold voltage whose value represents, from a physical point of view, the stored information. Electrons transferred into the FG produce a threshold voltage (V_T) variation, thus varying the logic data stored within the memory. The charge quantity into FG modulates the current flowing through the device at fixed Control Gate (CG) bias [3]. The programming of a NAND Flash cell is performed by injecting electrons within the FG, whereas erasing is performed by removing that charge from the FG. In quiescent conditions, thanks to the two oxides, the charge stored does not leak away (theoretically), thus granting the non-volatile paradigm fulfillment.

The physical mechanism used for both injecting and extracting electrons to/from the FG is the Fowler-Nordheim (FN) tunneling [12]. High electric fields applied to the tunnel oxide allow for electron transfer across the thin insulator. The choice of using the tunneling mechanism for writing and erasing the information in NAND memories is due to the relatively high parallelism of the operation (i.e. thousands of cells belonging to the same group can be written or erased in parallel), although FN tunneling significantly impacts the reliability of the memory causing progressive degradation of the tunnel oxide. The cell programming operation requires an

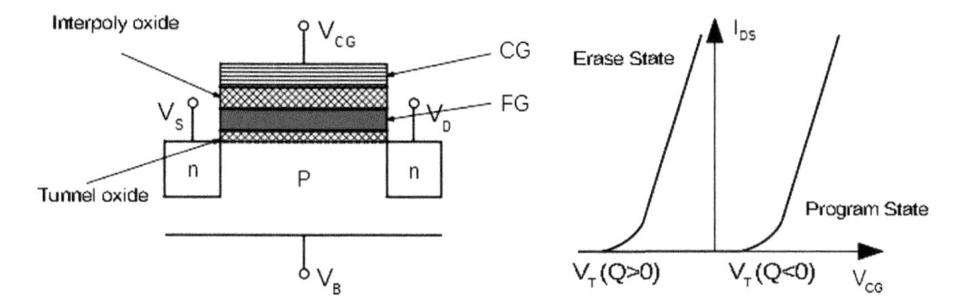

Fig. 8.1 NAND Flash cell structure and I-V characteristics dependent on the FG's charge in erase and programmed states

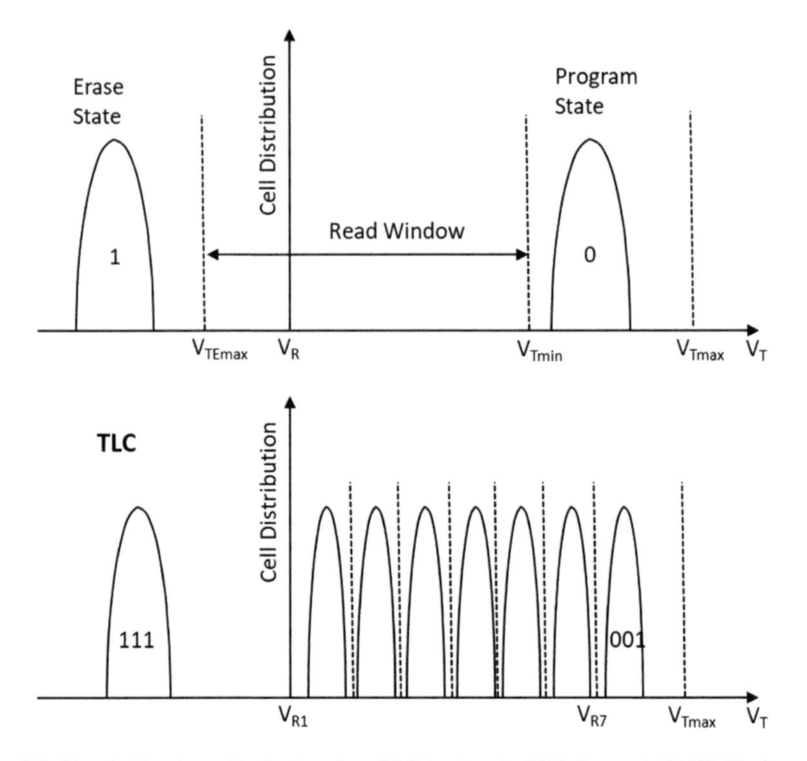

Fig. 8.2 Threshold voltage distributions in a SLC (top) and a TLC (bottom) NAND Flash array. V_{Tmin} and V_{Tmax} represent the minimum and the maximum target V_T for a programmed cell, respectively. V_{TEmax} represents the maximum V_T for an erased cell while V_R denotes the read voltage

accurate control of the electric field through the applied CG voltage (V_{CG}) in order to place the cell's V_T in a well-defined interval [V_{Tmin}, V_{Tmax}] (see Fig. 8.2, where the V_T distributions of a cell array are shown). A $V_T < V_{Tmin}$ would reduce the read margin guaranteeing a read operation immune from errors, whereas $V_T >$ V_{Tmax} could provoke read errors in other cells of the array due to the over-programming [3]. To this extent, the program operation is performed incrementally stepping the V_{CG} followed by a verify operation [13] that ends the program when the target V_T interval has been reached [14]. Read operation is performed by evaluating the current flowing through the cell when a fixed reference voltage V_R is applied to CG [16]. In a programmed cell (high V_T) the current is limited and the read circuitry produces a bit equal to 0, whereas in an erased cell (negative V_T) the high measured current is interpreted as a 1.

With the introduction of multilevel architectures (MLC, TLC, QLC) able to store 2, 3, and even 4 bits in a single cell, the programming and the reading operations become much more complex [3], since V_{Tmax} cannot be increased because of architectural and operating constrains [15]. The amplitude reduction of each interval calls for a very tight control of the charge injected within the FG.

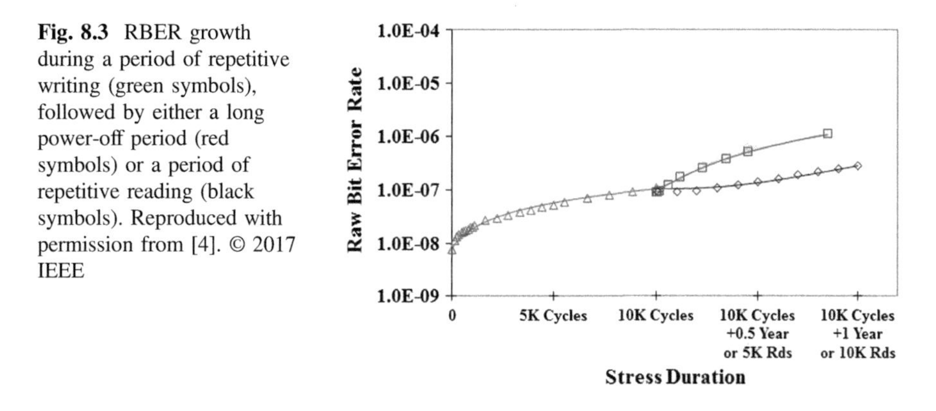

Fig. 8.3 RBER growth during a period of repetitive writing (green symbols), followed by either a long power-off period (red symbols) or a period of repetitive reading (black symbols). Reproduced with permission from [4]. © 2017 IEEE

8.3.1 *Raw Bit Errors*

Raw bit errors are most of the time a consequence of the memory's finite endurance and data retention capabilities. Errors occur when cells have incorrect threshold voltage values compared to their wanted placement. When raw bit errors occur immediately after reading a written NAND Flash block we are likely in the presence of write errors. If the errors appear after a time without biasing the cells or due to the repeated reads of the block we are in presence of data retention errors or read disturb errors, respectively.

Figure 8.3 shows that RBER is greater than zero since the very beginning of the NAND Flash blocks lifetime and progressively increases with the number of performed program/erase cycles. The resulting RBER either after retention or after read stress is a function of the cycling stress, pointing out that the overall RBER is the sum of the different error contributors [4]. RBER is a useful parameter for SSD reliability assessment, but must be evaluated with care since it is accurate only for a particular location at a particular moment in time. Indeed, the variability between different Flash pages within the same block or from a chip to another populating the SSD is so high that RBER has to be considered with its distribution rather than its punctual value. Error correction engines in the SSD are expected to cover all raw bit errors and must account the peculiarities of their insurgence to efficiently correct them.

8.3.2 *Reliability-Loss Mechanisms Affecting RBER*

Several physical mechanisms affect the intrinsic reliability of NAND Flash, although the degradation of the tunnel oxide electrical characteristics is still one of the fundamental process to keep in mind whenever the discussion on the memory lifetime is addressed. As said, because of the continuous charge transport through the cells insulator, traps can be created at the dielectric interfaces or within the

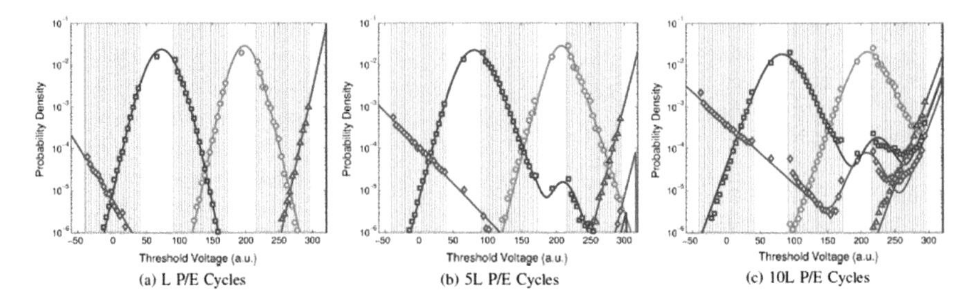

Fig. 8.4 V_T distribution of MLC (2bits/cell) NAND flash memory before (left) and after (right) endurance test. Reproduced with permission from [18]. © 2014 IEEE

oxide, which can alter the FN tunneling dynamics. Because of this, it becomes difficult to control the placement of the cells V_T: some of them can be slightly over-programmed, and their thresholds could end in an adjacent interval [17]. Because of this distribution broadening induced by the endurance stress (i.e., repeated program/erase cycles), read errors are produced and RBER increases in turn (see Fig. 8.4).

Oxide ageing and traps creation also reduce the data retention feature that is the ability of keeping unaltered the charge within the FG when the cell is in a quiescent state. Electrons may escape from the FG because of Trap-Assisted Tunneling (TAT) or Stress Induced Leakage Current (SILC) effects [19, 20], thus causing a modification of the threshold voltage distributions for the cells in the array (see Fig. 8.5). The risk that the threshold of a cell programmed in a given interval shifts to an adjacent interval increases significantly with the number of bits stored in a single cell. In MLC or TLC architectures, the number of electrons differentiating two adjacent intervals is few tens, therefore reducing data retention control [17]. Moreover, this phenomenon is strongly dependent on the temperature and on the cycling stress sustained by a memory block.

Other effects may worsen the RBER due to the inability of controlling the correct number of electrons to be transferred in the FG during a single programming pulse. Among them the most important are: (i) the Random Telegraph Noise (RTN) related to filling/empting of tunnel oxide traps affecting the V_T distributions stability [17]; (ii) the positive trapped charge in the tunnel oxide resulting in erratic

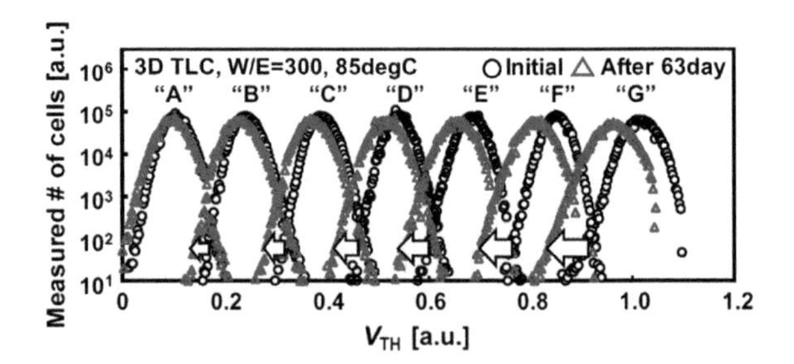

Fig. 8.5 V_T distribution of TLC (3bits/cell) NAND flash memory, before/after data-retention bake. Reproduced with permission from [21]. © 2017 IEEE

effects [22]; (iii) the electron injection statistics caused by the small number of carriers flowing to FG [17]. All of the mechanisms that may potentially affect any cell in the array, have a random and transient nature: they can occur during any programming pulse and they may produce threshold shifts larger than expected, with the risk of programming some cells with a threshold voltage larger than the desired one.

Besides endurance, retention, and placement-induced errors, there are some specific issues of the NAND Flash architecture that may lead to a RBER increase. The most common effects are the so called disturbs, that can be interpreted as the influence of an operation performed on a cell (Read or Write) on the charge content of a different cell. The read disturb may occur when reading many times the same cells without any erase operation of the entire block they belong to [23]. All the cells belonging to the same string of the cell to be read must be driven in an ON state, independently of their stored charge (see Fig. 8.6). The relatively high $V_{PASS} > V_{Tmax}$ applied to the CG of the unselected cells to turn on their conduction and the sequence of pulses applied during successive read operations may induce a charge gain due to SILC effects [20] or hot carrier effects [24]. These cells suffer a threshold voltage shift that may lead to read errors, when addressed. The probability of suffering from read disturb increases with the P/E number (i.e., towards the end of the memory useful lifetime) and it is higher in damaged cells. Read disturbs do not provoke permanent oxide damages: if erased and then reprogrammed, the correct charge content will be present within the FG.

A similar disturb occurs with the inhibit scheme during program operation to bias unselected wordlines for writing. In this case cells with enhanced tunneling characteristics because of TAT or fabrication variations can program unintentionally [4]; this effect is called program disturb. Other RBER-increase mechanisms do

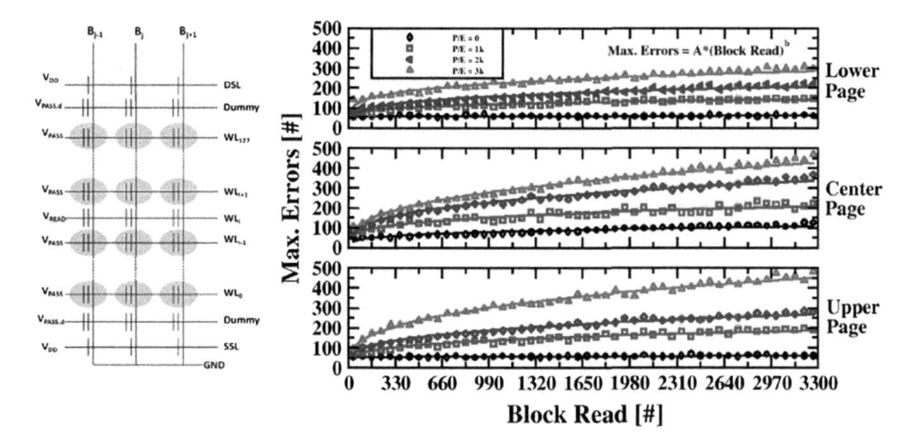

Fig. 8.6 Bias configuration for a read operation applied on a NAND Flash block (left). The cells in gray are those suffering the read disturb. Maximum read disturb errors number retrieved in all the wordlines of a specific NAND Flash page type at different P/E cycles (right). Reproduced with permission from [23]. © 2017 IEEE

not depend on traps and are induced by the NAND Flash technology scaling. Such issues are the cell-to-cell interference [25] and the Gate Induced Drain Leakage (GIDL) [26]. The FG coupling due to parasitic capacitances between cells mainly causes the former issue, thus it is greatly affected by cell scaling, and is well known to widen the V_T distributions. The latter effect is due to the usage of the self-boosting technique to inhibit unselected cells during programming [27]. 3-D NAND Flash technologies are expected to feature similar raw bit errors mechanisms, although different impact from the peculiar sources is expected due to the different materials used in their integration.

8.3.3 *Mitigating the Raw Bit Errors Through ECC*

The state-of-the-art in SSD data protection is to integrate an ECC engine in the SSD controller to handle the raw bit errors occurring throughout the entire lifetime of the drive. In NAND Flash, each page is split in different codewords, where a codeword is the sum of user data and some parity bytes to reconstruct the information in case of bit corruption [28]. The role of the SSD controller is to generate the parity when a write operation is issued to a specific NAND Flash page and later exploit that during a read operation. Given an ECC that can correct up to k failed bits in a codeword, we can calculate the codeword Failure Probability F_{CW} as:

$$F_{CW} = 1 - \sum_{i=0}^{k} \binom{n}{i} \cdot RBER^{i} \cdot (1 - RBER)^{n-i} \tag{8.1}$$

where n is the number of bits in the codeword [29]. This equation demonstrates that the relationship between the RBER and the reliability of the NAND Flash, measured in terms of a failure probability, is tight. Current NAND Flash technologies may require correction strengths up to 30 or more bits per codeword [4].

The ECC is sufficient to provide a target UBER within the specifications provided by the JEDEC standard [9], although care must be taken since the previous equation applies only for a punctual value of the RBER. As shown in Fig. 8.7, the RBER has significant variation from codeword-to-codeword due to many factors, so that the design of an ECC to mitigate raw bit errors must be performed not on an average basis, but rather on the extreme tails of the RBER distribution (i.e., highest codeword failure probabilities). The common ECC schemes implemented in SSD are the BCH and the LDPC [3], whose designs allow correcting a large number of failed bits providing at the same time a sufficient error detection capability to avoid silent errors or wrong corrections.

The latter feature is important for the application of the secondary correction mechanisms whenever the ECC is found to fail data correction from NAND Flash. SSD controllers can be instructed to retry the correction of a codeword by dynamically changing the read voltages and timings to reduce RBER up to a point

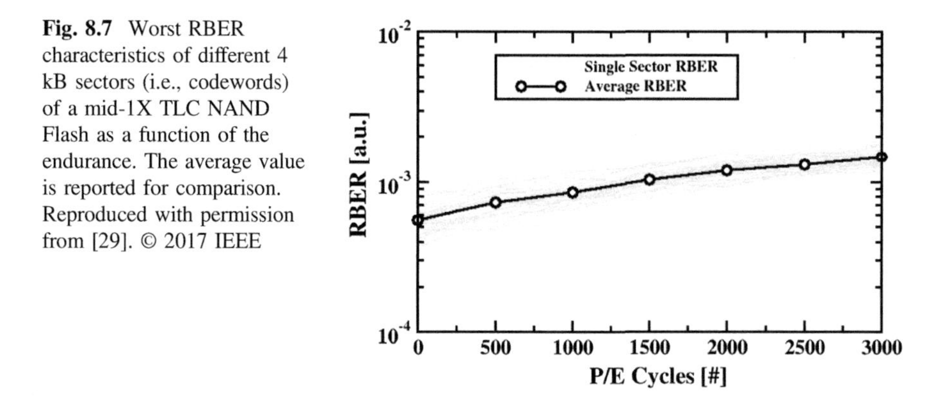

Fig. 8.7 Worst RBER characteristics of different 4 kB sectors (i.e., codewords) of a mid-1X TLC NAND Flash as a function of the endurance. The average value is reported for comparison. Reproduced with permission from [29]. © 2017 IEEE

where the ECC can actually correct the data. This strategy is called Read Retry or Moving Read [4]. The Read Retry algorithms used in NAND Flash have been designed to trade between reliability and performance features of the SSD. Indeed, the occurrence of an uncorrectable error event requires the intervention of the algorithm at the expense of a significant latency introduced by the ECC due to cascaded read retry operations, thus reducing the SSD bandwidth. This approach was devised for older NAND Flash technologies where the RBER was sufficiently low to guarantee a seldom intervention of the secondary error correction. As the technology continues to scale down, the memory cell storage distortion and noise sources become increasingly significant, leading to continuous degradation of raw bit errors features. As a result, the industry has been actively pursuing the transition of ECC from conventional BCH codes to more powerful soft-decision iterative coding solutions, in particular LDPC codes [30]. Nevertheless, since NAND flash memory read latency is proportional to the number of reads and the results must be transferred to the SSD controller through standard chip-to-chip links, a straightforward use of soft-decision ECC can result in significant read latency overhead. To this extent, Fig. 8.8 illustrates the intuitive progressive soft-decision sensing strategy exploited in common SSD controllers, which aims at using just-enough sensing precision for ECC decoding through a trial-and-error manner. This method can reduce the average read latency overhead [30].

Figure 8.9 quantitatively demonstrates the advantages of soft-decision ECC (in particular LDPC code) over existing BCH code. For the LDPC code, simulations were performed for both hard-decision sensing precision and different soft-decoding level precisions. As shown in the figure, although hard-decision decoding of LDPC code can slightly outperform the BCH code, soft-decision decoding can significantly improve the performance and advantage over BCH code [30].

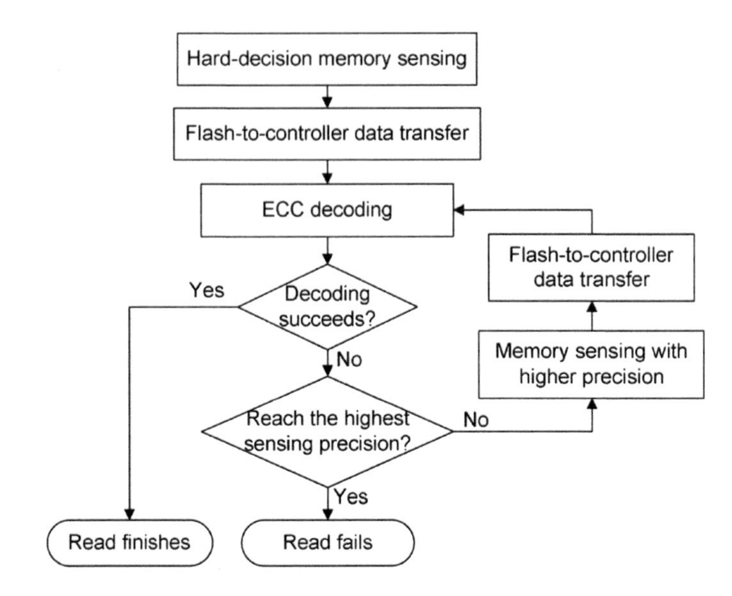

Fig. 8.8 Illustration of operational flow of progressive soft-decision sensing. Reproduced with permission from [30]. © 2013 IEEE

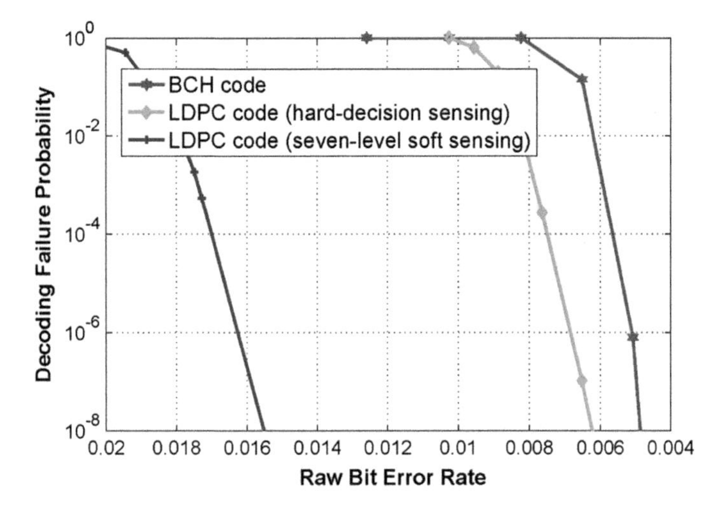

Fig. 8.9 Simulation results of BCH and LDPC codes to be exploited in NAND Flash raw bit errors mitigation. Reproduced with permission from [30]. © 2013 IEEE

8.3.4 Mitigating the Raw Bit Errors Through Firmware

MLC, TLC, and QLC NAND Flash architectures are the preferred solution in SSD when high storage density is required, but at the additional cost of increasing the overhead due to the appropriate RBER management policies adopted for mitigating the issues presented in the previous sections of this chapter.

Since most of the RBER sources depends on the endurance state of the memory, it is important to distribute the writing stress over the entire population of cells rather than on a single hot spot, thus avoiding that some blocks are updated continuously while the others keep unaltered their charge content. It is clear that blocks whose information is updated frequently are stressed with a large number of write-erase cycles. In order to keep the aging effects as uniform as possible, the number of both read and write cycles of each block must be monitored and stored in some firmware structures managed by the SSD controller. Those tables are part of an essential SSD's firmware component, namely the Flash Translation Layer (FTL). Wear leveling [31] is a process that reduces premature wear in NAND Flash devices by equalizing the endurance of a memory on its completely addressable user space. The most common implementation of wear leveling occurs in the FTL, which manages access to the memory device and determines how the NAND Flash blocks are used. Two types of data exist in NAND Flash devices: static and dynamic. A static data is information that is rarely, if ever, updated. It may be read frequently, but it seldom changes and it can theoretically reside in the same physical location for the life of the device. Dynamic data, on the other hand, is constantly changing and consequently requires frequent reprogramming. Dynamic wear leveling is a method of pooling the available blocks that are free of data and selecting the block with the lowest erase count for the next write. This method is most efficient for dynamic data because only the non-static portion of the NAND Flash array is wear-leveled. Static wear leveling utilizes all good blocks to evenly distribute wear, providing effective wear leveling and thereby extending the life of the device. This method tracks the cycle count of all good blocks and attempts to evenly distribute block wear throughout the entire device by selecting the available block with the least wear each time a program operation is executed. Each technique has its pros and cons, as described in [32].

Read disturb management is important as well for the RBER mitigation in firmware. This policy instructs the SSD controller to move highly read data to new blocks before the original blocks can be excessively disturbed [33]. Read disturb management is important, because a host computer can easily read a portion of NAND Flash codewords in the SSD many millions of times, which would be enough without refreshing of the data to result in RBER levels uncorrectable with any ECC scheme [4].

FTL also implement a policy called *scrubbing*. This methodology moves data to new blocks as for read disturb managements, but in this case it is done for resetting the retention time of memory cells. In this way, the SSD controller can create unlimited data retention lifetime while an SSD is powered on [4].

Finally, the FTL is also responsible for the randomization of the V_T values in the programming algorithm in order to reduce the RBER dependency on the interferences caused by specific programmed data patterns [34].

8.4 NAND Flash Reliability: Defects and Extrinsic Failures

Occasional failures can occur in NAND Flash even if not directly related to the intrinsic mechanisms of the cells. A NAND Flash is a complex Integrated Circuit (IC) system that includes, other than the array of cells, many heterogeneous circuits. A NAND Flash IC is composed by several macro blocks: the memory array, the data path circuitry that controls the input/output towards the external world, the decoders to select individual groups of cells in the array, and the high-voltage (HV) circuitry mandatory for all the read/write operations. The failure causes can be ascribed to yield defects (e.g., shorted wordlines or bitlines) or even to circuitry design flaws inside the peripheral sub-systems of the array. Most of these errors appear in fresh devices, although their behavior can be accelerated by temperature and electric field. As the Fig. 8.10 shows, the wafer process failures represent the major share of the failures. These are classified further into sub-classes based on the defect location or processing phase [35]. Although such process failures can occur in any IC in an SSD, defect issues in the NAND Flash ICs can be expected to

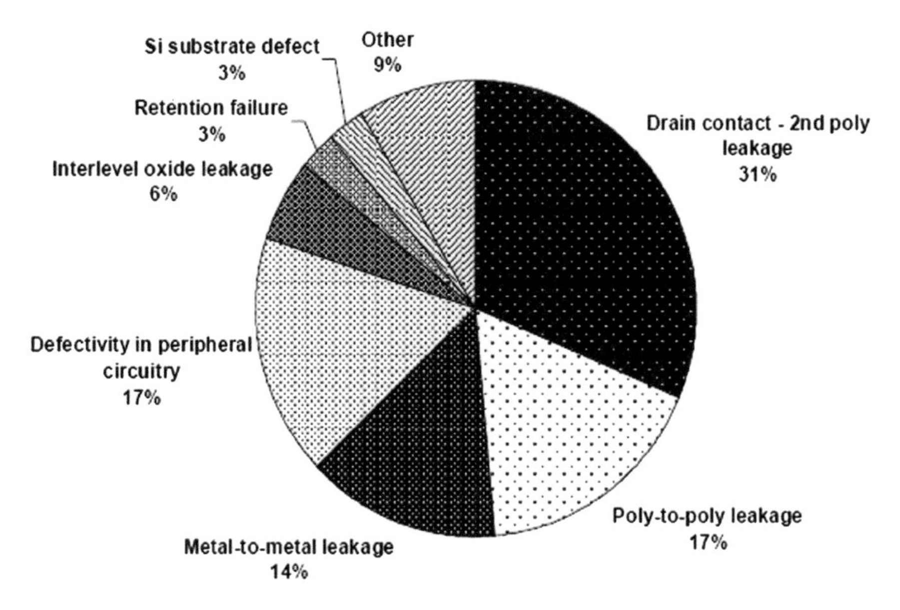

Fig. 8.10 Sub-classification of the wafer process related defects in Flash technology. Reproduced with permission from [35]. © 2006 IEEE

dominate because they occupy most of the silicon area of an SSD, they operate at very high voltages, and they contain many interconnects with the minimum-possible spacing [4]. Defects like two wordlines shorted together in a NAND Flash block due to insulation breakdown are an example of those issues. If any of the pages (MLC or TLC) on those wordlines are accessed, they will likely experience thousands of bit errors, overcoming the correction capabilities of any ECC even with secondary correction; if the shorting occurs early in a programming operation a program-status failure may result, signaling the failure to the SSD controller [4].

Some extrinsic failure mechanisms may not cause a NAND Flash IC failure, but can have an important impact on the raw bit errors. Among them, two phenomena related to voltage and temperature are considered: the power-supply induced errors and the temperature cross shifts. Design flaws in the high-voltage circuitry of a NAND Flash cause the former issue. In [36] it is found that the HV sub-system plays an important role on the reliability since its design affects sensitive analog circuits that control the behavior of the memory cells during read and write operations. Hence, raw bit errors are strongly dependent on the power-supply, and a different behavior of the memory reliability during its entire lifetime can be observed depending on the chosen power supply. Concerning the temperature cross shift, it is found that the current/voltage characteristics depicted in Fig. 8.1 heavily depends on the temperature, resulting in an additional instability of the threshold voltage distributions in NAND Flash. Although dedicated peripheral circuits in the memory IC are devoted to solve this issue, there is no complete mitigation of the errors generated by the phenomenon [37].

8.4.1 Mitigating Defects and Extrinsic Failures Through RAID

The mitigation of defects and NAND Flash IC extrinsic failures in SSDs is handled by a RAID technique performed within the drive [29]. Generally, in RAID architectures data are arranged following a specific pattern that mixes data and parity, where the latter represents the additional data required for the SSD to recover any of the pieces of stored information. A *stripe* is an ensemble of data and parity sectors representing the minimum unit for data reconstruction. The stripe length expresses how many user data elements are associated with parity elements. In case of RAID-5 approach within SSD, as shown in Fig. 8.11, the stripe length refers to a single parity element in a stripe (i.e., the notation N-to-1 is also used), whereas in RAID-6 approach it refers to double parity elements in a stripe (i.e., N-to-2). The choice of the stripe length and of the RAID level depends on a reliability/performance trade-off that an SSD wants to leverage on.

Let us assume the 3-to-1 RAID-5 configuration shown in Fig. 8.11 by considering the ($D0$; $D1$; $D2$; $P0$) stripe. If one of the data sectors Di with $i = 0 \ldots 2$ in the

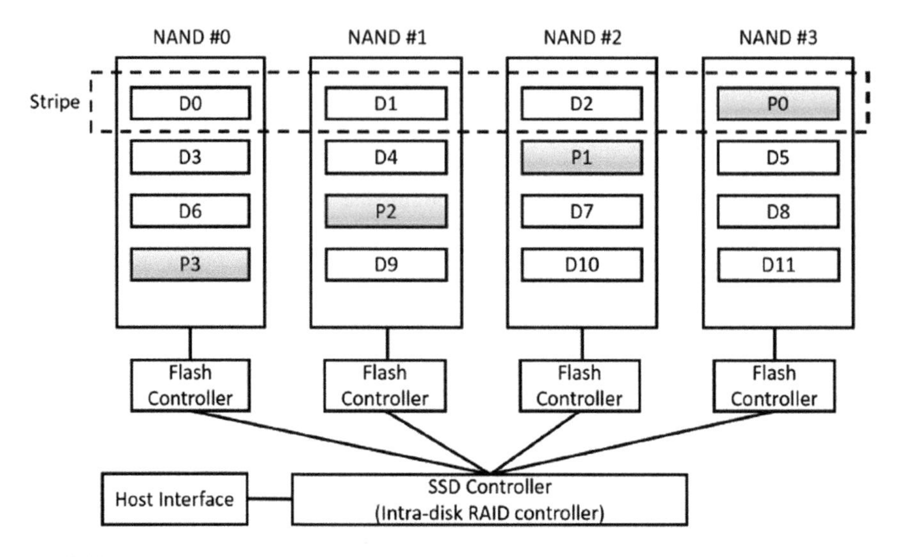

Fig. 8.11 Architecture example of a RAID-5 approach for SSD based on four NAND Flash devices. Reproduced with permission from [29]. © 2017 IEEE

stripe fails either due to unexpected NAND Flash IC failures or due to the impossibility to correct the data using the ECC, the RAID recovers the faulty sector via the parity *P0* by applying a XOR algorithm. The recovered data are then written on another sector of the SSD and the faulty one is marked as invalid and then retired by the SSD controller management firmware [29].

The correction strength of the RAID approach is different from SSD to SSD: in some cases, single codewords are protected, whereas in other cases the protection level goes up to the reconstruction of the entire failed NAND Flash IC.

8.5 SSD Reliability: Non-NAND Flash Failures

8.5.1 SSD Controller, DRAM Errors, and Firmware Failures

All SSDs must protect data inside the drive from the connection to the host system, through the circuits of the SSD, to the NAND Flash memory, and viceversa. While the NAND Flash memory employs its own ECC protection, there is the need to use additional state-of-the-art data protection methods, such as parity protection on internal buffers and checksum generation/checking. This additional level of protection will cover SSD failures in case of a faulty SSD controller, a noisy connection cable with the host system, a faulty DRAM, or a firmware bug.

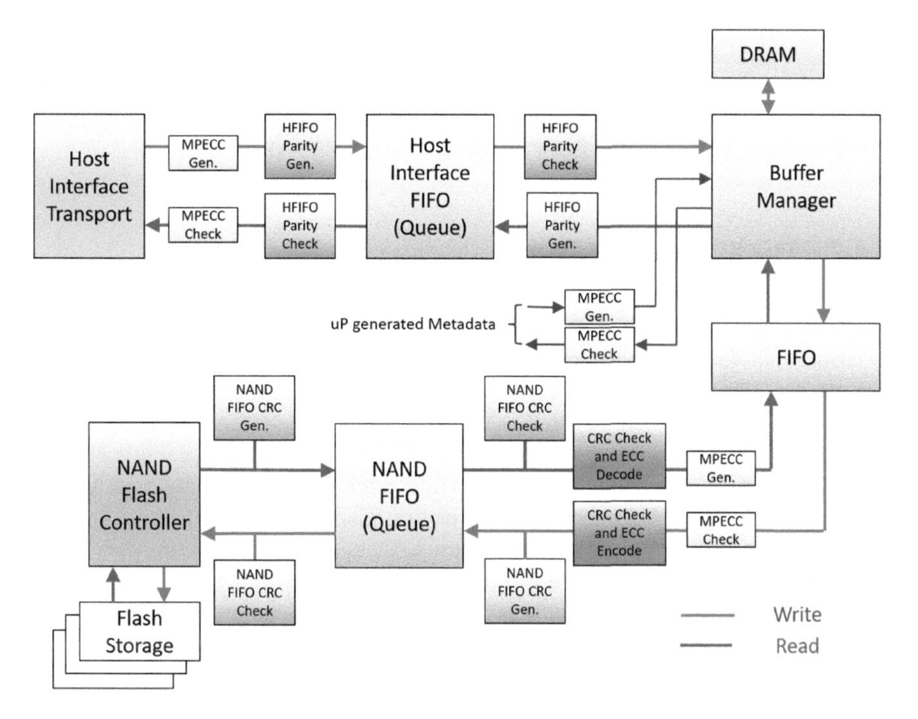

Fig. 8.12 End-to-End data protection in enterprise SSDs. The concept of the figure is based on the description provided in [38]

In client SSDs, as the datum passes from the host interface to the host FIFO where commands are queued, parity is generated. As the data exits the host FIFO, parity is checked. Next, a Cyclic Redundancy Check (CRC) and ECC are generated and stored with the data. Finally, a CRC is generated just before the data enters the NAND FIFO where specific NAND Flash commands are queued, and then it is checked when exiting. When data is read from the NAND, the process occurs in reverse order [38].

Enterprise drives build on the foundation of proven data path protection for client drives, but go one-step further, adding protection in the form of memory path error correction as shown in Fig. 8.12. An additional Memory Protection ECC (MPECC) is added. MPECC is designed to protect the host data by adding ECC coverage to the data as it enters the SSD. A multi-byte MPECC is generated on the host data in the physical layer of the host interface and it is independent of any ECC provided by the NAND devices themselves. This additional MPECC follows the host data through the SSD. As the MPECC and user data enter the host FIFO, parity is generated. As the data exits the host FIFO, that parity is checked. In the DRAM buffer manager, further MPECC protection is generated on the associated metadata for FTL structures. By adding MPECC protection to the metadata, both host data and metadata are protected. As the host data, its metadata, and the MPECC generated for both types of data exit the FIFO adjacent to the buffer manager, both are checked. Next, CRC and ECC are generated as with client drives. Finally, parity is generated before the data enters the NAND FIFO, and that parity is checked upon exit. On read commands, the process is carried out in reverse order [38].

8.5.2 The Power-Loss Issue

When programming a NAND Flash memory, the program operation must complete to ensure that data are stored reliably within the page. Data are at risk if power is lost when Flash memory cells are in the process of being programmed [39]. SSD have three causes of potential data loss or corruption when system power fails: (i) a loss of data; (ii) a loss of mapping information; (iii) a corruption of a single NAND Flash page within a wordline (in MLC architectures).

Most enterprise class SSDs rely on a power failure circuitry that monitors the supply voltage and generates an "early warning" signal to the SSD controller if the voltage drops below a predefined threshold. A secondary voltage hold-up-circuit is implemented to ensure the drive has power for a sufficient time to harden data whenever that warning is received. In addition, writes are not accepted by the drive until the secondary voltage source has been sufficiently charged to protect against loss of data upon power failures. The secondary voltage source can be a high capacity supercapacitor or a bank of discrete capacitors.

A supercapacitor is an electrolytic capacitive charge storage device. It is capable of storing a large amount of energy in a relatively small three-dimensional space. A generic supercapacitor-based voltage hold-up circuit is consistent with the block diagram shown in Fig. 8.13. Designing a supercapacitor-based power failure protection circuit is easy to do, and many SSDs employ this approach for this

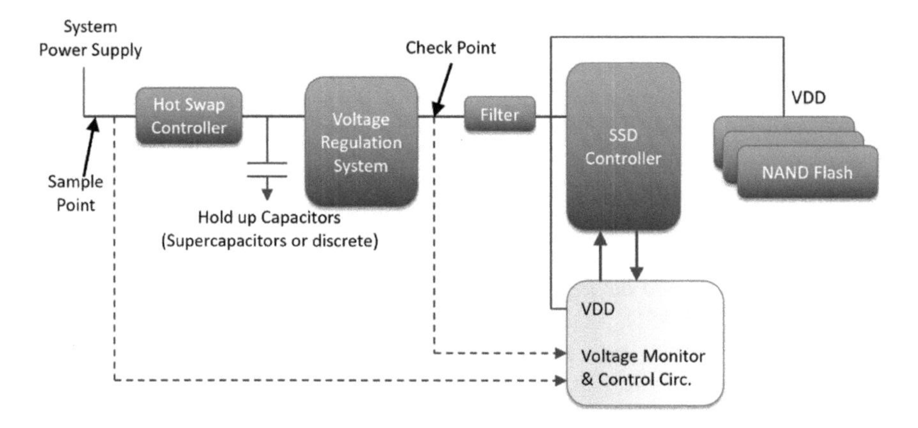

Fig. 8.13 Block diagram of a power failure circuit in a standard SSD. The concept of this figure is based on the description provided in [39]

reason. Unfortunately, there are a number of concerns related to long term supercapacitor reliability that makes the use of this component questionable for Enterprise-class SSDs. Supercapacitors are typically Aluminum Electrolytic Capacitors, featuring a high capacitance-to-size ratio and, therefore, they are an attractive choice for applications requiring large bulk capacitance like an SSD. However, like all electrolytic capacitors, supercapacitors suffer from a well-known set of deficiencies with regard to long-term reliability. In particular, supercapacitors "wear out", resulting in reduced capacitance over time. They use a wet electrolyte and the packaging is subject to ongoing losses via leakage and diffusion. The performance of the supercapacitor degrades slowly with electrolyte loss, until the onset of total failure occurs with little or no warning. In addition, loss rate increases with higher operating voltage, and in higher operating and non-operating temperature environments. For every 10 °C of ambient operating temperature rise, the life expectancy of a supercapacitor can be cut approximately in half.

Nowadays SSDs utilize either Niobium Oxide or Polymer Tantalum capacitors. These discrete capacitors do not employ a "wet" electrolyte and are not susceptible to the leakage related issues that plague supercapacitor technology. Niobium and Polymer Tantalum capacitors are rated to 85 °C, providing a higher temperature operating range with respect to supercapacitors (70 °C). Because of these factors, a discrete component based hold up circuit is more able to meet the demands of enterprise and industrial computing environments. Another advantage of discrete capacitors over supercapacitors is that they are highly predictable and reliable. However, lacking the compactness of supercapacitors, the capacitance-to-size ratio of a discrete solution is less space efficient and its implementation require a more careful design.

8.6 Assessing SSD Reliability Through Testing

When testing SSDs (and not single NAND Flash ICs) it is important to detect or estimate functional failures, errors in reading data, without considering the physical causes that produced such errors or failures. If the amount of errors or functional failures exceeds the acceptable limits, a successive failure analysis will try to investigate on the possible physical causes. Therefore, it is important to remind the basic difference between testing NAND Flash devices and verifying SSDs reliability: the former operation requires adopting all the possible test procedures to excite physical or architectural weaknesses, the latter consider the SSD as a black box where data are to be written, read and retained at their endurance and retention limits.

It must be observed, however, that the use of different technologies for NAND Flash memories produces different expectations in terms of both endurance and retention. To deal with different applications, NAND technologies, and producers, standard committees define the conditions of use and the corresponding endurance verification requirements. The following sections will refer to the JEDEC standard

JESD218A (Solid-State Drive Requirements and Endurance Test Method) [9], that defines parameters for standardized endurance rating so that the end user may consider the endurance rating as a factor in determining if an SSD is suitable for his particular application.

Since there are different levels of requirements for an SSD based on specific applications different levels of testing should be applied to verify the SSD suitability for the particular application. It is necessary to group different applications characterized by similar requirements in a limited number of classes: to this purpose, the JESD218A standard considers just two application classes: client and enterprise. These classes, of course, are not all-inclusive and it is clear that variations such as the operating systems and application architectures make a significant impact to the workload of an SSD, that represents the detailed sequence of host writes and reads (including data content and timing) applied during endurance testing. The actual workloads are defined in the JEDEC standard JESD219 [40] for the two considered classes and they are not reported in this text.

8.6.1 SSD Endurance and Retention Rating

A SSD manufacturer shall establish an endurance rating for an SSD that represents the maximum number of terabytes that may be written (TBW) by a host to the SSD, such that the following conditions are satisfied:

1. the SSD maintains its capacity;
2. the SSD maintains the required UBER for its application class;
3. the SSD meets the required Functional Failure Requirement (FFR) for its application class, that is the allowed cumulative number of failed drives that, over the TBW rating, fail to function properly in a way that is more severe than having a data error;
4. the SSD retains data with power off for the required time for its application class.

The requirements for standard classes of SSDs are based on a scenario in which the SSD are actively used for some periods of time during which the SSDs are written to their endurance ratings, followed by a power-down time period in which data must be retained. The requirements for the two SSD classes are reported in Table 8.1.

Table 8.1 SSD class and requirements

Application class	Client	Enterprise
Active use (power on)	8 h/day @ 40 °C	24 h/day @ 55°C
Retention use (power off)	1 year @ 30 °C	3 months @ 40°C
FFR	$\leq 3\%$	$\leq 3\%$
UBER requirement	$\leq 10^{-15}$	$\leq 10^{-16}$

SSD case temperatures are reported in Table 8.1 and they are intended to represent the relevant temperatures over the respective time periods, for the purpose of endurance and retention estimation, not the maximum and minimum specifications to be found on the SSD datasheets. For the client class, the retention temperature (30 °C) is also the temperature for the 16 h/day in which the SSD is powered down.

8.6.2 Endurance and Retention Stress Methods

There are two approaches for endurance verification: a direct method and an extrapolation method based on a HDD testing methodology. Both consist of endurance verification followed by retention verification. If the full TBW rating can be reached in a 1000-h stress, the direct method is to be followed. If this is not possible, then an extrapolation method is acceptable. If an SSD product from a qualification family has been qualified using the JESD218A standard, the subsequent products need only data from a 1000-h direct method evaluation, even if this results in those drives not being fully stressed to their endurance rating limits.

8.6.3 Direct Method

The endurance stress is to be performed both at high and low temperature; then, a retention test shall be performed. Since the retention time requirements are long (see Table 8.2), extrapolation or acceleration is required to validate the retention requirements.

8.6.3.1 Sample Size

For the first product to be qualified in a qualification family, the sample shall consist of SSDs from at least three nonconsecutive production lots and from all the fabrication plants responsible for the manufacture of the NAND memories used in the SSD. For subsequent products from a qualification family, a single production lot is sufficient. The number of SSD in the sample shall be sufficient to establish that both the FFR and UBER requirements are met at 60% confidence.

Table 8.2 Endurance stress temperatures by drive class

Application class	Client	Enterprise
Low temperature	≤25 °C	≤25 °C
High temperature	40 °C $\leq T \leq T_{max}$	60 °C $\leq T \leq T_{max}$

The sample size and acceptance criteria are defined by the following equations, which mathematically embody the 60% confidence requirement:

$$\mathrm{UCL(FF)} \leq \mathrm{FFR} \cdot \mathrm{SS} \tag{8.2}$$

$$\mathrm{UCL(DE)} \leq \min(\mathrm{TBW},\ \mathrm{TBR}) \cdot 8 \cdot 10^{12} \cdot \mathrm{UBER} \cdot \mathrm{SS} \tag{8.3}$$

where FF and DE are the acceptable numbers of Functional Failures and of Data Errors, respectively; TBR represents the number of TBytes Read; SS is the sample size in number of drives; FFR and UBER are expressed as fractions; UCL(x) is an upper confidence limit function that depends on the maximum number of accepted errors x.

For instance, for an accept-on-zero plan (no failures/error are accepted), UCL (0) = 0.92, while if 1 failure/error is accepted, UCL(1) = 2.03 and for 2 failures/errors accepted, UCL(2) = 3.11.

As an example, consider an accept-on-zero plan, FFR = 0.03 (corresponding to 3%); UBER = 10^{-16}, TWB = 100, all data read back and verified (therefore TBR = 100). Two sample sizes SS can be calculated from (8.2) and (8.3), respectively:

$$SS \geq UCL(0)/FFR = 0.92/0.03 = 30.1 \tag{8.4}$$

$$\begin{aligned} \mathrm{SS} &\geq \mathrm{UCL}(0)/\left[\min(\mathrm{TBW},\ \mathrm{TBR}) \cdot 8 \cdot 10^{12} \cdot \mathrm{UBER}\right] \\ &= 0.92/(100 \cdot 8 \cdot 10^{12} \cdot 10^{-16}) = 11.5 \end{aligned} \tag{8.5}$$

The required sample size is the larger of the two results and, therefore, at least 31 SSD must be tested. If the minimum sample size of 31 were chosen, than the verification test would pass if there were no functional failures in 31 drives. However, with SS = 31, from (8.3),

$$\mathrm{UCL(DE)} \leq 100 \cdot 8 \cdot 10^{12} \cdot 10^{-16} \cdot 31 = 2.48 \tag{8.6}$$

Since UCL(1) = 2.03 < 2.48 < 3.11 = UCL(2), up to one data error would be acceptable. Therefore, the verification would pass if there were no functional failures and no more than one data error.

It is important to notice that UBER is defined in terms of bits read, but for the purpose of endurance verification (8.3) counts the minimum of bits read and bits written. The rationale is twofold.

First, many data errors are transient with respect to rewriting of an SSD, but repeatable with respect to repeated reading. This means that a sector with corrupted data may pass without error if rewritten, however reading non-failing sectors multiple times is unlikely to detect additional errors. This means that if reads are less frequent than writes, then many errors will be missed. All data errors will be detected only if all written data are read before those sectors are rewritten. If the TBR is less than the TBW, then the UBER should be increased because of the likelihood that transient data errors went undetected. Using the TBR in place of the TWB accomplishes that goal.

Second, the JEDEC JESD218A standard is aligned to a reference read/write ratio of unity. If the TBR is equal to the TBW, then the UBER may be considered to be an error rate per bit read or per bit written: both are equivalent. If the TBR in the endurance stress is greater than the TBW the UBER must be TBW based.

It is important to remind that the previous criterion deals with *endurance* functional failures and *endurance* data errors. Failures that are not related to the act of writing data to its endurance limit, or by the subsequent retention stress, are to be excluded from the endurance verification, even it they must be considered in the drive qualification process. In some cases it is not easy to clearly identify *endurance* and *non-endurance* function failures. Failures that are not in the circuit path of the written data are clearly identified as *non-endurance* failures, while some failures that are in the circuit path of the written data *may* be considered as *non-endurance* failures if the cause of the failures were unrelated to the quantity of data written.

8.6.3.2 Endurance Stress

To verify the endurance capabilities, the drives are stressed to their full endurance specification (in TBW). The stress time depends on the drives performances and on those of the test equipment. If performance variations between test systems or the SSDs themselves cause some SSD to receive more writes than other in a given stress time, then the endurance specification must be reached by the average amount of data written. All data errors throughout the stress must be recorded, even if those errors are transient in nature. Testing the drive only at the end of the stress cannot be accepted.

Two approaches are acceptable for incorporating both high and low temperatures into the endurance stressing: the ramped-temperature approach and the split-flow approach.

In the ramped-temperature approach the temperature during the stress shall be switched periodically between the low and the high temperatures reported in Table 8.3, so that half of the test is at low and at high temperature, respectively. The ramp timing shall be such that no more than 25% of the stress is performed at

Table 8.3 Retention stress temperatures and times

Application class	Client	Enterprise
Stress duration and temperature	96 h @ T $\geq$ 66 °C or 500 h @ T $\geq$ 52 °C	96 h @ T $\geq$ 66 °C or 500 h @ T $\geq$ 52 °C

intermediate temperatures during the transition between the two limit temperatures. As for the temperature switching frequency, no more than 10% of the endurance stress can be performed within any single half-cycle.

In the split approach, the sample is divided in two groups. The former undergoes endurance testing at a fixed low temperature, the latter at a fixed high temperature. The two temperature ranges are the same as for the ramped approach (see Table 8.2).

The T_{max} values are chosen so that the endurance stress time would be equivalent to one year at the active-use temperature and hours/day shown in Table 8.2 assuming an activation energy of 1.1 eV. In fact, although an SSD would be expected to reach its TBW rating over a lifetime of several years, for the specific purpose of calculating T_{max}, the full TBW is assumed to occur within a single year. This is a conservative assumption, since a shorter time allows less relaxation between writes.

In addition, the endurance stress T_{max} values may also account for a realistic amount of delay for relaxation which would occur if the stress temperature were too high. These delays, consisting of the drive being powered down or being powered up but not being written to, combined with the effect of the elevated temperature endurance stressing, must stay within the one-year equivalent time.

The temperature T_{max} as well the additional delay time and temperatures may be extracted by solving

$$t_D e^{-\frac{E_a}{KT_D}} + t_s\left[FH_S e^{-\frac{E_a}{KT_{SH}}} + (1 - FH_S)e^{-\frac{E_a}{KT_{SL}}}\right] = t_U\left[FH_U e^{-\frac{E_a}{KT_{UH}}} + (1 - FH_U)e^{-\frac{E_a}{KT_{UL}}}\right] \tag{8.7}$$

where t_D, t_S, t_U are the delay time, the stress time and the use time, respectively; T_D is the temperature applied during the delay; T_{SH} and T_{SL} are the high and the low temperatures during the endurance stress in °K, respectively; T_{UH} and T_{UL} are the high and the low temperatures during the use conditions in °K, respectively; FH_S and FH_U are the fraction of time spent at high temperature during endurance stressing and use condition, respectively; K is the Boltzmann's constant equal to $8.6171 \cdot 10^{-5}$ eV/°K while E_a is the activation energy equal to 1.1 eV.

For example, consider the client application class from Table 8.1, $T_{UH} = 40$ °C = 313.15 °K; $T_{UL} = 30$ °C = 303.15 °K; $FH_U = 1/3$ (8 h/day); a 1000 h stress time using the ramped approach ($t_s = 1000$ h, $T_{SL} = 25$ °C = 298.15 °K and $FH_S = 1/2$) and no additional delays ($t_d = 0$). From (8.7) it possible to derive the endurance stress high temperature, considering that one year of normal use corresponds to $t_U = 8766$ h:

$$
\begin{aligned}
&t_s\left[\frac{1}{2}e^{-\frac{E_a}{KT_{SH}}}+\frac{1}{2}e^{-\frac{E_a}{KT_{SL}}}\right]=t_U\left[\frac{1}{3}e^{-\frac{E_a}{KT_{UH}}}+\frac{2}{3}e^{-\frac{E_a}{KT_{UL}}}\right]\\
&e^{-\frac{E_a}{KT_{SH}}}=\frac{2t_U}{t_S}\left[\frac{1}{3}e^{-\frac{E_a}{KT_{UH}}}+\frac{2}{3}e^{-\frac{E_a}{KT_{UL}}}\right]-e^{-\frac{E_a}{KT_{SL}}}\\
&T_{SH}=-\frac{E_a}{K}\frac{1}{\ln\left[\frac{2t_U}{t_S}\left(\frac{1}{3}e^{-\frac{E_a}{KT_{UH}}}+\frac{2}{3}e^{-\frac{E_a}{KT_{UL}}}\right)-e^{-\frac{E_a}{KT_{SL}}}\right]}\\
&T_{SH}=-\frac{1.1}{8.6171\cdot 10^{-5}}\frac{1}{\ln\left[\frac{2\cdot 8766}{1000}\left(\frac{1}{3}1.978\cdot 10^{-18}+\frac{2}{3}5.156\cdot 10^{-19}\right)-2.544\cdot 10^{-19}\right]}\\
&T_{SH}=330.76\,^{\circ}K=57.61\,^{\circ}C
\end{aligned}
\tag{8.8}
$$

Hence, the maximum temperature T_{max} for a 1000-h stress, for the client application class, ramped approach, no delays is 58 °C.

If it is chosen to perform the test at 50 °C instead of 58 °C, it is possible to add an additional delay, whose duration and temperature can also be derived from (8.7) by imposing T_{SH} = 50 °C = 323.15 °K.

$$
t_De^{-\frac{E_a}{KT_D}}=t_U\left[FH_Ue^{-\frac{E_a}{KT_{UH}}}+(1-FH_U)e^{-\frac{E_a}{KT_{UL}}}\right]-t_s\left[FH_Se^{-\frac{E_a}{KT_{SH}}}+(1-FH_S)e^{-\frac{E_a}{KT_{SL}}}\right]
\tag{8.9}
$$

For example, if a 100-h delay is added to the 1000-h endurance stress, a T_D = 67 °C can be directly calculated as in (8.10):

$$
e^{-\frac{E_a}{KT_D}}=\frac{8766}{100}\left[\frac{1}{3}e^{-\frac{E_a}{313.15k}}+\frac{2}{3}e^{-\frac{E_a}{303.15k}}\right]-\frac{1000}{100}\left[\frac{1}{2}e^{-\frac{E_a}{323.15k}}+\frac{1}{2}e^{-\frac{E_a}{298.15k}}\right]
\tag{8.10}
$$

Therefore, the endurance test would consist of 1000 h of active endurance stress with the temperature ramped between 25 and 50 °C, with an additional 100 h spent in a non-writing mode at a temperature not greater than 67 °C.

8.6.3.3 Retention Stress

After the endurance stress, SSDs are to be powered down and baked at elevated temperatures in order to establish the data retention capability. For the ramped-temperature approach, all drives in the sample are to be baked while, for the split approach, only the drives stressed at high temperatures are to baked. The SSDs are to be fully written with data prior to the bake and fully read after the test with internal error correction bypassed. The number of data errors resulting from the retention stress is to be added to that resulting from the endurance stress.

The temperatures required for the retention verification are reported in Table 8.3.

Two equivalent options are given for the bake temperature and durations and they are chosen to correspond to the required data retention times for the common temperature-accelerated mechanism responsible for data degradation in non-volatile memories, assuming an activation energy of 1.1 eV.

Not all mechanisms responsible for data loss, however, are accelerated by temperature and therefore a second evaluation is required at room temperature. This requirement holds only for the first product in a qualification family to be qualified; subsequent products are exempt. In the ramped-temperature approach the low temperature retention qualification is performed before the high temperature stress: in the split-flow approach, only the drives stressed at low temperatures undergo the low temperature retention test. Since time acceleration via higher temperatures is impossible, the room-temperature retention evaluation requires mathematical extrapolations based on drive-level or component-level bit-error-rate data.

When basing the extrapolation on drive-level bit-error-rate data, low-temperature retention tests require at least 500 h at a temperature between 10 and 30 °C. The bit error rate can be measured at several times (for instance, 48, 168 and 500 h) and then the trend can be extrapolated. The fraction of error bits with respect to the total bit number, called the Raw Bit Error Rate (RBER), depends on the program/erase cycle count and the retention time. For BER $\ll 1$,

$$\mathrm{RBER} = \mathrm{RBER}_0 + \mathrm{B}_0 \cdot \mathrm{t}^{\mathrm{m}} \tag{8.11}$$

where RBER_0 is the Bit Error Rate at the beginning of the retention period, B_0 is an arbitrary scale factor dependent on materials and processes, t is the retention time and m is a retention power low coefficient (typically 1 or 2).

To verify the useful retention lifetime, the RBER can be measured as a function of time and the parameters RBER_0, B_0 and m fit (8.11). The resulting fitted equation may then be used to estimate the RBER at the desired retention time of Table 8.1 and such a value must be below the ECC capability of the SSD controller, also considering a safety margin between the calculated ECC capability and the RBER.

Consider for example an SSD with a calculated ECC capability of 4×10^{-5} and a safety margin equal to 2. Also, consider that the RBER data of Table 8.4 have been obtained:

The extrapolated RBER at t = 8776 h (=1 year) must be below the ECC capability with safety margin whose value is $2 \cdot 10^{-5}$. As it can be seen in Fig. 8.14, the extrapolated RBER reaches the ECC capability with safety margin after 10,000 h and, therefore, the retention requirement of Table 8.2 is met.

Table 8.4 Calculated RBER (example)

Retention time (h)	RBER
0	0
48	$5.63 \cdot 10^{-8}$
168	$2.23 \cdot 10^{-7}$
500	$7.41 \cdot 10^{-7}$
1000	$1.59 \cdot 10^{-6}$

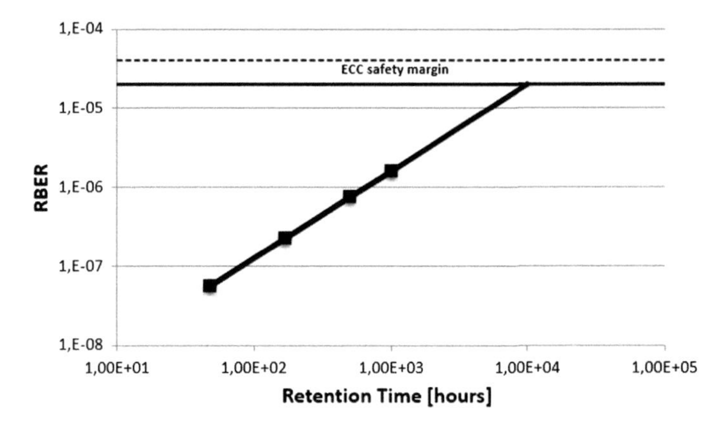

Fig. 8.14 Example of the verification of the retention requirements via extrapolation of the drive-level bit error rate. The concept of the figure is based on the description provided in [9]

The mathematical extrapolation can also be performed using raw bit error rate data from nonvolatile memory components, if available: at the end of the endurance stress, the room-retention evaluation can be derived by using the retention data calculated for the nonvolatile memory components inside the SSD for the specific number of program/erase cycles experienced during the extrapolation test.

8.6.4 Extrapolation Method

If the direct method would require more than 1000 h of endurance stress, an extrapolation method can be used. Some of the proposed methods require special access to SSD internal operation or to nonvolatile memory components information which make these methods possible only for the SSD manufacturer.

Independently of the extrapolation method used for endurance and retention verification, some general requirements are to be ensured:

- the SSD must meet the requirements of Table 8.1 for FFR and UBER, for the temperatures and times stated in the Table;
- the FFR and UBER requirements must be met for both low-temperature and high-temperature endurance stressing, with temperature ranges of at least 25–40 °C for client SSD and 25–60 °C for enterprise SSD;
- data retention is to be verified under the assumption that the endurance stressing in use takes place over no longer than 1 year at the endurance use temperature and hours per day of Table 8.1;
- data retention is to be verified both for a temperature-accelerated mechanism (assuming an activation energy of 1.1 eV) and a non-temperature-accelerated mechanism;
- all requirements are to be established at a 60% statistical confidence level.

8.6.4.1 Extrapolation of FFR and Bad-Block Trends

For the endurance evaluation, an SSD may be stressed to only some fraction of the TBW rating. During the endurance stress, functional failures may occur, as well as a certain number of blocks marked as “bad”. The increase in these two quantities may be plotted as a function of TBW in a lognormal or Weibull plot and extrapolated to the TBW rating to obtain estimates of the final levels of FFR and bad blocks.

This extrapolation method is not acceptable for verifying that the UBER requirements are met, because UBER may have a highly steep dependence on TBW that makes extrapolations from low TBW data quite unreliable.

8.6.4.2 FFR and UBER Estimation from Reduced-Capacity SSDs

The capacity of an SSD may be artificially reduced so that some nonvolatile memory components or blocks are not written, while the remaining ones are written more extensively than would be the case of the full-capacity SSD. In this context, an SSD will be considered to have reached its endurance rating limit if the stressed fraction of the nonvolatile memory components reaches the target program/erase cycles.

For this approach to be used, the manufacturer must ensure that the method of capacity reduction does not significantly distort the normal internal operation of the SSD. Simply reducing the logical span of written data is generally not sufficient, since the SSD controller and firmware make use of the full nonvolatile memory capacity, if not instructed.

A variation of this method is to extend the nonvolatile memory program/erase cycles beyond the target expected at the TBW rating, in order to generate functional failures and data errors. The resulting data can then be plotted and FFR and *UBER* can be extrapolated for the expected, lower, TBW rating.

References

1. http://www.storagesearch.com/chartingtheriseofssds.html. Accessed 2018
2. R. Micheloni, S. Aritome, L. Crippa, Array architectures for 3-D NAND flash memories. Proc. IEEE **105**(9), 1634–1649 (2017)
3. L. Zuolo, C. Zambelli, R. Micheloni, P. Olivo, Solid-state drives: memory driven design methodologies for optimal performance. Proc. IEEE **105**(9), 1589–1608 (2017)
4. N.R. Mielke, R.E. Frickey, I. Kalastirsky, M. Quan, D. Ustinov, V.J. Vasudevan, Reliability of solid-state drives based on NAND flash memory. Proc. IEEE **105**(9), 1725–1750 (2017)
5. W. Jiang, C. Hu, Y. Zhou, A. Kanevsky, Are disks the dominant contributor for storage failures?: a comprehensive study of storage subsystem failure characteristics. ACM Trans. Storage **4**(3), 7 (2008)
6. L. Bairavasundaram, G. Goodson, S. Pasupathy, J. Schindler, An analysis of latent sector errors in disk drives, in *Proceedings of the ACM SIGMETRICS International Conference on Measurement and Modeling of Computer Systems*, 2007, pp. 289–300

7. L. Bairavasundaram, A. Arpaci-Dusseau, G. Goodson, B. Schroeder, An analysis of data corruption in the storage stack. ACM Trans. Storage **4**(3), 7 (2008)
8. B. Schroeder, G. Gibson, Understanding disk failure rates: what does an MTTF of 1,000,000 hours mean to you? ACM Trans. Storage **3**(3), 8 (2007)
9. JEDEC, *JESD218B Solid-State Drive (SSD) Requirements and Endurance Test Method* (2016)
10. N. Mielke, Accelerated testing of radiation-induced soft errors in solid-state drives. IEEE Trans. Device Mater. Rel. **15**(4), 552–558 (2015)
11. F. Masuoka, M. Momodomi, Y. Iwata, R. Shirota, New ultra high density EPROM and flash EPROM cell with NAND structure, in *IEEE IEDM Technical Digest* pp. 552–555 (1987)
12. M. Lenzlinger, E.H. Snow, Fowler-Nordheim tunneling into thermally grown SiO_2. J. Appl. Phys. **40**, 273–283 (1969)
13. M. Momodomi, T. Tanaka, Y. Iwata, Y. Tanaka, H. Oodaira, Y. Itoh, R. Shirota, K. Ohuchi, F. Masuoka, A 4 Mb NAND EEPROM with Tight Programmed Vt Distribution. IEEE J. Solid State Circ. **26**(4), 492–496 (1991)
14. G.J. Hemink, T. Tanaka, T. Endoh, S. Aritome, R. Shirota, Fast and accurate programming method for multi-level NAND EEPROMs, in *VLSI Symposium on Technology and Circuits*, June 1995, pp. 129–130
15. A. Chimenton, P. Pellati, P. Olivo, Analysis of erratic bits in flash memories. IEEE Trans. Devices Mater. Reliab. **1**(4), 179–184 (2001)
16. M. Momodomi, Y. Itoh, R. Shirota, Y. Iwata, R. Nakayama, R. Kirisawa, T. Tanaka, S. Aritome, T. Endoh, K. Ohuchi, F. Masuoka, An Experimental 4-Mbit CMOS EEPROM with a NAND-structured cell. IEEE J. Solid State Circ. **24**(5), 1238–1243 (1989)
17. C. Monzio Compagnoni, A. Goda, A.S. Spinelli, P. Feeley, A.L. Lacaita, A. Visconti, Reviewing the evolution of the NAND flash technology. Proc. IEEE **105**(9), 1609–1633 (2017)
18. T. Parnell, N. Papandreou, T. Mittelholzer, H. Pozidis, Modelling of the threshold voltage distributions of sub-20 nm NAND flash memory, in *IEEE Global Communications Conference* (Austin, TX, 2014), pp. 2351–2356
19. K. Lee, M. Kang, S. Seo, D. Kang, D.H. Li, Y. Hwang, H. Shin, Separation of corner component in TAT mechanism in retention characteristics of Sub 20-nm NAND flash memory. IEEE Elect. Device Lett. **35**(1), 51–53 (2014)
20. G.J. Hemink, K. Shimizu, S. Aritome, R. Shirota, Trapped hole enhanced stress induced leakage currents in NAND EEPROM tunnel oxides, in *Proceedings of International Reliability Physics Symposium,* Apr 1996, pp. 117–121
21. K. Mizoguchi, T. Takahashi, S. Aritome, K. Takeuchi, Data-retention characteristics comparison of 2D and 3D TLC NAND flash memories, in 2017 *IEEE International Memory Workshop (IMW)* (Monterey, CA, 2017), pp. 1–4
22. A. Chimenton, C. Zambelli, P. Olivo, A statistical model of erratic behaviors in flash memory arrays. IEEE Trans. Electr. Devices **58**(11), 3707–3711 (2011)
23. C. Zambelli, P. Olivo, L. Crippa, A. Marelli, R. Micheloni, Uniform and concentrated read disturb effects in mid-1X TLC NAND flash memories for enterprise solid state drives, in *2017 IEEE International Reliability Physics Symposium (IRPS)*, (Monterey, CA, 2017), pp. PM-5.1–PM-5.4
24. H.H. Wang, P.S. Shieh, C.T. Huang, K. Tokami, R. Kuo, S.H. Chen, H.C. Wei, S. Pittikoun, S. Aritome, a new read-disturb failure mechanism caused by boosting hot-carrier injection effect in MLC NAND flash memory, in *IEEE International Memory Workshop*, May 2009, pp. 1–2
25. J. Lee, S. Hur, J. Choi, Effects of floating-gate interference on NAND flash memory cell operation. IEEE Elect. Device Lett. **23**(5), 264–266 (2002)
26. J. Lee, C. Lee, M. Lee, H. Kim, K. Park, W. Lee, A new programming disturbance phenomenon in NAND flash memory by source/drain hot-electrons generated by GIDL current, in *Non-volatile Semiconductor Memory Workshop*, Feb 2006, pp. 31–33

27. S. Satoh, H. Hagiwara, T. Tanzawa, K. Takeuchi, R. Shirota, A novel isolation-scaling technology for NAND EEPROMs with the minimized program disturbance, in *IEDM Technical Digest*, Dec 1997, pp. 291–294
28. N. Mielke et al., Bit error rate in NAND flash memories, in *Proceedings of IEEE International Reliability Physics Symposium Phoenix*, Apr 2008, (AZ, USA), pp. 9–19
29. C. Zambelli, A. Marelli, R. Micheloni, P. Olivo, Modeling the endurance reliability of intradisk RAID solutions for Mid-1X TLC NAND flash solid-state drives, in *IEEE Transactions on Device and Materials Reliability*, Dec 2017, vol. 17, no. 4, pp. 713–721
30. G. Dong, N. Xie, T. Zhang, Enabling NAND flash memory use Soft-decision error correction codes at minimal read latency overhead. IEEE Trans. Circ. Syst. I Regul. Paper **60**(9), 2412–2421 (2013)
31. R. Micheloni, A. Marelli, R. Ravasio, *Error Correction Codes for Non-Volatile Memories*, Springer (2008)
32. Micron Corporation, TN-29-42: Wear-Leveling Techniques in NAND Flash Devices, Application Note, 2008
33. H. Belgal, Apparatus, system, and method for improving read endurance for a nonvolatile memory. U.S. Patent 8954650B2, 10 Feb 2015
34. J. Cha, S. Kang, Data randomization scheme for endurance enhancement and interference mitigation of multilevel flash memory devices. ETRI J. **35**(1), 166–169 (2013)
35. P. Muroke, Flash memory field failure mechanisms, in *2006 IEEE International Reliability Physics Symposium Proceedings* (San Jose, CA, 2006), pp. 313–316
36. C. Zambelli, P. King, P. Olivo, L. Crippa, R. Micheloni, Power-supply impact on the reliability of mid-1X TLC NAND flash memories, in *2016 IEEE International Reliability Physics Symposium (IRPS)*, (Pasadena, CA, 2016), pp. 2B-3-1–2B-3-6
37. Y. Li, 3 Bit Per Cell NAND Flash Memory on 19 nm Technology, Flash Memory Summit, Aug 2012
38. Micron Corporation, Comparison of Client and Enterprise SSD Data Path Protection, Application Note (2011)
39. SMART Storage Systems, Power Failure Protection, Application Note (2012)
40. JEDEC, JESD219 Solid-State Drive (SSD) Endurance Workloads (2012)

Chapter 9
Reliability Issues in Flash-Memory-Based Solid-State Drives: Experimental Analysis, Mitigation, Recovery

Yu Cai, Saugata Ghose, Erich F. Haratsch, Yixin Luo and Onur Mutlu

Abstract NAND flash memory is ubiquitous in everyday life today because its capacity has continuously increased and cost has continuously decreased over decades. This positive growth is a result of two key trends: (1) effective process technology scaling; and (2) multi-level (e.g., MLC, TLC) cell data coding. Unfortunately, the reliability of raw data stored in flash memory has also continued to become more difficult to ensure, because these two trends lead to (1) fewer electrons in the flash memory cell floating gate to represent the data; and (2) larger cell-to-cell interference and disturbance effects. Without mitigation, worsening reliability can reduce the lifetime of NAND flash memory. As a result, flash memory controllers in solid-state drives (SSDs) have become much more sophisticated: they incorporate many effective techniques to ensure the correct interpretation of noisy data stored in flash memory cells. In this chapter, we review recent advances in SSD error characterization, mitigation, and data recovery techniques for reliability and lifetime improvement. We provide rigorous experimental data from state-of-the-art MLC and TLC NAND flash devices on various types of flash memory errors, to motivate the need for such techniques. Based on the understanding developed by the experimental characterization, we describe several mitigation and recovery techniques, including (1) cell-to-cell interference mitigation; (2) optimal multi-level cell sensing; (3) error correction using state-of-the-art algorithms and methods; and (4) data recovery when error correction fails. We quantify the reliability improvement provided by each of

O. Mutlu (✉)
ETH Zürich, Zürich, Switzerland
e-mail: omutlu@gmail.com

Y. Cai · S. Ghose (✉) · Y. Luo · O. Mutlu
Carnegie Mellon University, Pittsburgh, PA, USA
email: ghose@cmu.edu

E. F. Haratsch
Seagate Technology, Fremont, CA, USA

R. Micheloni et al. (eds.), *Inside Solid State Drives (SSDs)*,
Springer Series in Advanced Microelectronics 37,
https://doi.org/10.1007/978-981-13-0599-3_9

these techniques. Looking forward, we briefly discuss how flash memory and these techniques could evolve into the future.

Solid-state drives (SSDs) are widely used in computer systems today as a primary method of data storage. In comparison with magnetic hard drives, the previously dominant choice for storage, SSDs deliver significantly higher read and write performance, with orders of magnitude of improvement in random-access input/output (I/O) operations, and are resilient to physical shock, while requiring a smaller form factor and consuming less static power. SSD capacity (i.e., storage density) and cost-per-bit have been improving steadily in the past two decades, which has led to the widespread adoption of SSD-based data storage in most computing systems, from mobile consumer devices [91, 107] to enterprise data centers [67, 174, 199, 233, 257].

The first major driver for the improved SSD capacity and cost-per-bit has been *manufacturing process scaling*, which has increased the number of flash memory cells within a fixed area. Internally, commercial SSDs are made up of NAND flash memory chips, which provide nonvolatile memory storage (i.e., the data stored in NAND flash is correctly retained even when the power is disconnected) using *floating-gate (FG) transistors* [111, 172, 187] or *charge trap transistors* [65, 268]. In this paper, we mainly focus on floating-gate transistors, since they are the most common transistor used in today's flash memories. A floating-gate transistor constitutes a flash memory cell. It can encode one or more bits of digital data, which is represented by the level of charge stored inside the transistor's *floating gate*. The transistor traps charge within its floating gate, which dictates the *threshold voltage* level at which the transistor turns on. The threshold voltage level of the floating gate is used to determine the value of the digital data stored inside the transistor. When manufacturing process scales down to a smaller technology node, the size of each flash memory cell, and thus the size of the transistor, decreases, which in turn reduces the amount of charge that can be trapped within the floating gate. Thus, process scaling increases storage density by enabling more cells to be placed in a given area, but it also causes reliability issues, which are the focus of this paper.

The second major driver for improved SSD capacity has been the use of a single floating-gate transistor to represent *more than* one bit of digital data. Earlier NAND flash chips stored a single bit of data in each cell (i.e., a single floating-gate transistor), which was referred to as single-level cell (SLC) NAND flash. Each transistor can be set to a specific threshold voltage within a fixed range of voltages. SLC NAND flash divided this fixed range into two *voltage windows*, where one window represents the bit value 0 and the other window represents the bit value 1. Multi-level cell (MLC) NAND flash was commercialized in the last two decades, where the same voltage range is instead divided into *four* voltage windows that represent each possible 2-bit value (00, 01, 10, and 11). Each voltage window in MLC NAND flash is therefore much smaller than a voltage window in SLC NAND flash. This makes it more difficult to identify the value stored in a cell. More recently, triple-level cell (TLC) flash has been commercialized [7, 86], which further divides the range, providing *eight* voltage windows to represent a 3-bit value. Quadruple-level cell (QLC) flash, storing a 4-bit value per cell, is currently being developed [203]. Encoding more bits per cell increases the capacity of the SSD without increasing the chip size, yet it also decreases reliability by making it more difficult to correctly store and read the bits.

The two major drivers for the higher capacity, and thus the ubiquitous commercial success, of flash memory as a storage device, are also major drivers for its reduced reliability and are the causes of its scaling problems. As the amount of charge stored in each NAND flash cell decreases, the voltage for each possible bit value is distributed over a wider voltage range due to greater process variation, and the *margins* (i.e., the width of the gap between neighboring voltage windows) provided to ensure the raw reliability of NAND flash chips have been diminishing, leading to a greater probability of flash memory errors with newer generations of SSDs. NAND flash memory errors can be induced by a variety of sources [19], including flash cell wearout [19, 20, 162], errors introduced during programming [17, 23, 162, 212], interference from operations performed on adjacent cells [21, 23, 31, 75, 151, 182, 207, 209], and data retention issues due to charge leakage [19, 22, 29, 30, 182].

To compensate for this, SSDs employ sophisticated error-correcting codes (ECCs) within their controllers. An SSD controller uses the ECC information stored alongside a piece of data in the NAND flash chip to detect and correct a number of *raw bit errors* (i.e., the number of errors experienced before correction is applied) when the piece of data is read out. The number of bits that can be corrected for every piece of data is a fundamental tradeoff in an SSD. A more sophisticated ECC can tolerate a larger number of raw bit errors, but it also consumes greater area overhead and latency. Error characterization studies [19, 20, 75, 162, 182, 212] have found that, due to NAND flash wearout, the probability of raw bit errors increases as more *program/erase (P/E) cycles* (i.e., *write accesses*, or *writes*) are performed to the drive. The raw bit error rate eventually exceeds the maximum number of errors that can be corrected by ECC, at which point data loss occurs [22, 27, 174, 233]. The *lifetime* of a NAND-flash-memory-based SSD is determined by the number of P/E cycles that can be performed successfully while avoiding data loss for a minimum *retention guarantee* (i.e., the required minimum amount of time, after being written, that the data can still be read out without uncorrectable errors).

The decreasing raw reliability of NAND flash memory chips has drastically impacted the lifetime of commercial SSDs. For example, older SLC NAND-flash-based SSDs were able to withstand 150,000 P/E cycles (writes) to each flash cell, but contemporary 1x-nm (i.e., 15–19 nm) process-based SSDs consisting of MLC NAND flash can sustain only 3,000 P/E cycles [168, 212, 294]. With the raw reliability of a flash chip dropping so significantly, approaches to mitigating reliability issues in NAND-flash-based SSDs have been the focus of an important body of research. A number of solutions have been proposed to increase the lifetime of contemporary SSDs, ranging from changes to the low-level device behavior (e.g., [17, 20, 21, 287]) to making SSD controllers much more intelligent in dealing with individual flash memory chips (e.g., [22, 26, 28–31, 86, 161, 162]). In addition, various mechanisms have been developed to successfully recover data in the event of data loss that may occur during a read operation to the SSD (e.g., [21, 22, 26]).

In this chapter, we provide a comprehensive overview of the state of flash-memory-based SSD reliability, with a focus on (1) fundamental causes of flash memory errors, backed up by (2) quantitative error data collected from real state-of-the-art flash memory devices, and (3) sophisticated error mitigation and data recovery techniques developed to tolerate, correct, and recover from such errors. To this end, we first discuss the architecture of a state-of-the-art SSD, and describe mechanisms used in a commercial SSD to reduce the probability of data loss (Sect. 9.1). Next, we discuss the low-level behavior of the underlying NAND flash memory chip in an SSD, to illustrate fundamental reasons why errors can occur in flash memory (Sect. 9.2). We then discuss the root causes of these errors, quantifying the impact of each error source using experimental characterization data collected from real NAND flash memory chips (Sect. 9.3). For each of these error sources, we describe various state-of-the-art mechanisms that mitigate the induced errors (Sect. 9.4). We next examine several error recovery flows to successfully extract data from the SSD in the event of data loss during a read operation (Sect. 9.5). Then, we look to the future to foreshadow how the reliability of SSDs might be affected by emerging flash memory technologies (Sect. 9.6). Finally, we briefly examine how other memory technologies (such as DRAM, which is used prominently in a modern SSD, and emerging nonvolatile memory) suffer from similar reliability issues to SSDs (Sect. 9.7).

9.1 State-of-the-Art SSD Architecture

In order to understand the root causes of reliability issues within SSDs, we first provide an overview of the system architecture of a state-of-the-art SSD. The SSD consists of a group of NAND flash memories (or *chips*) and a *controller*, as shown in Fig. 9.1. A host computer communicates with the SSD through a high-speed host interface (e.g., AHCI, NVMe; see Sect. 9.1.3.1), which connects to the SSD controller. The controller is then connected to each of the NAND flash chips via memory *channels*.

9.1.1 Flash Memory Organization

Figure 9.2 shows an example of how NAND flash memory is organized within an SSD. The flash memory is spread across multiple flash chips, where each chip contains one or more flash *dies*, which are individual pieces of silicon wafer that are connected together to the pins of the chip. Contemporary SSDs typically have 4–16 chips per SSD, and can have as many as 16 dies per chip. Each chip is connected to one or more physical memory channels, and these memory channels are not shared across chips. A flash die operates independently of other flash dies, and contains between one and four *planes*. Each plane contains hundreds to thousands of flash *blocks*. Each block is a 2D array that contains hundreds of rows of flash cells (typically 256–1024 rows) where the rows store contiguous pieces of data. Much like banks in a multi-bank memory (e.g., DRAM banks [36, 130, 131, 143, 145, 147, 148, 188, 194, 195]), the planes can execute flash operations in parallel, but the planes within a

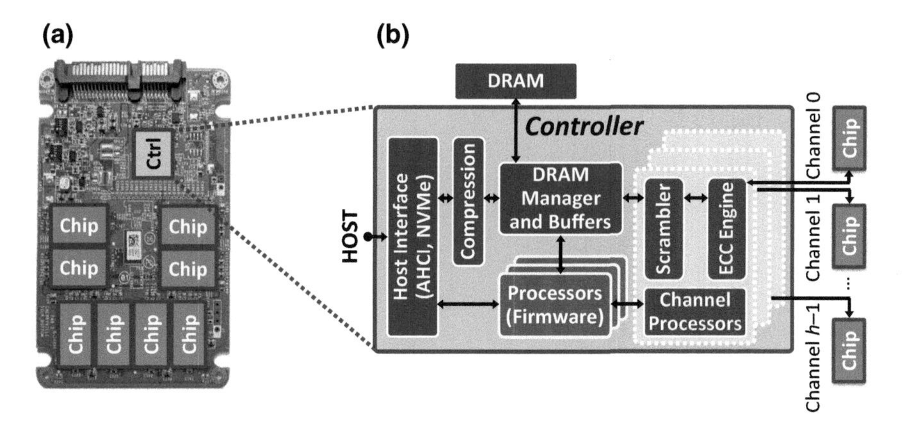

Fig. 9.1 **a** SSD system architecture, showing controller (Ctrl) and chips. **b** Detailed view of connections between controller components and chips. Adapted from [15]

die share a single set of data and control buses [1]. Hence, an operation can be started in a different plane in the same die in a pipelined manner, every cycle. Figure 9.2 shows how blocks are organized within chips across multiple channels. In the rest of this work, without loss of generality, we assume that a chip contains a single die.

Data in a block is written at the unit of a *page*, which is typically between 8 and 16 kB in size in NAND flash memory. All read and write operations are performed at the granularity of a page. Each block typically contains hundreds of pages. Blocks in each plane are numbered with an ID that is unique within the plane, but is shared across multiple planes. Within the block, each page is numbered in sequence. The controller firmware groups blocks with the same ID number across multiple chips and planes together into a *superblock*. Within each superblock, the pages with the same page number are considered a *superpage*. The controller *opens* one superblock (i.e., an empty superblock is selected for write operations) at a time, and typically writes data to the NAND flash memory one superpage at a time to improve sequential read/write performance and make error correction efficient, since some parity information is kept at superpage granularity (see Sect. 9.1.3.10). Having the ability to write to all of the pages in a superpage simultaneously, the SSD can fully exploit the internal parallelism offered by multiple planes/chips, which in turn maximizes write throughput.

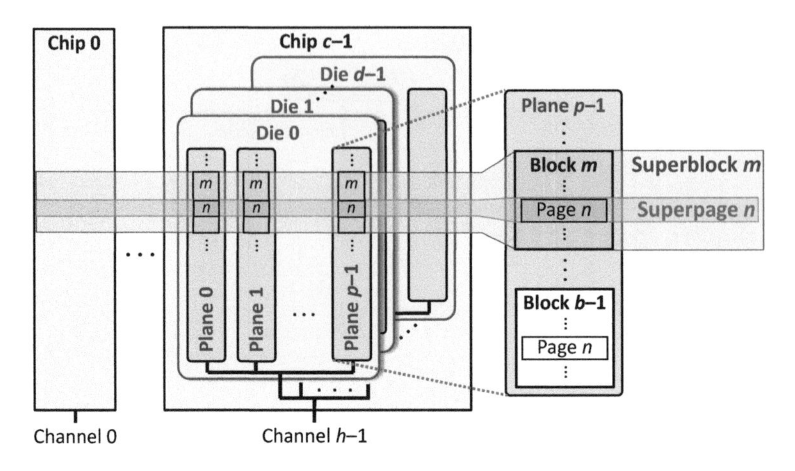

Fig. 9.2 Flash memory organization. Reproduced from [15]

9.1.2 Memory Channel

Each flash memory channel has its own data and control connection to the SSD controller, much like a main memory channel has to the DRAM controller [74, 87, 88, 100, 129, 130, 132, 135, 189, 191, 194, 195, 250–252]. The connection for each channel is typically an 8- or 16-bit wide bus between the controller and one of the flash memory chips [1]. Both data and flash commands can be sent over the bus.

Each channel also contains its own control signal pins to indicate the type of data or command that is on the bus. The *address latch enable* (ALE) pin signals that the controller is sending an address, while the *command latch enable* (CLE) pin signals that the controller is sending a flash command. Every rising edge of the *write enable* (WE) signal indicates that the flash memory should write the piece of data currently being sent on the bus by the SSD controller. Similarly, every rising edge of the *read enable* (RE) signal indicates that the flash memory should send the next piece of data from the flash memory to the SSD controller.

Each flash memory die connected to a memory channel has its own *chip enable* (CE) signal, which selects the die that the controller currently wants to communicate with. On a channel, the bus broadcasts address, data, and flash commands to all dies within the channel, but only the die whose CE signal is active reads the information from the bus and executes the corresponding operation.

9.1.3 SSD Controller

The SSD controller, shown in Fig. 9.1b, is responsible for (1) handling I/O requests received from the host, (2) ensuring data integrity and efficient storage, and (3) managing the underlying NAND flash memory. To perform these tasks, the controller runs firmware, which is often referred to as the *flash translation layer* (FTL). FTL tasks are executed on one or more embedded processors that exist inside the controller. The controller has access to DRAM, which can be used to store various controller metadata (e.g., how host memory addresses map to physical SSD addresses) and to cache relevant (e.g., frequently accessed) SSD pages [174, 229].

When the controller handles I/O requests, it performs a number of operations on both the requests and the data. For requests, the controller *schedules* them in a manner that ensures correctness and provides high/reasonable performance. For data, the controller *scrambles* the data to improve raw bit error rates, performs *ECC encoding/decoding*, and in some cases *compresses/decompresses* and/or *encrypts/decrypts* the data and employs *superpage-level data parity*. To manage the NAND flash memory, the controller runs *firmware* that maps host data to physical NAND flash pages, performs *garbage collection* on flash pages that have been invalidated, applies *wear leveling* to evenly distribute the impact of writes on NAND flash reliability across all pages, and manages bad NAND flash blocks. We briefly examine the various tasks of the SSD controller.

9.1.3.1 Scheduling Requests

The controller receives I/O requests over a *host controller interface* (shown as *Host Interface* in Fig. 9.1b), which consists of a system I/O bus and the protocol used to communicate along the bus. When an application running on the host system needs to access the SSD, it generates an I/O request, which is sent by the host over the host controller interface. The SSD controller receives the I/O request, and inserts the request into a queue. The controller uses a *scheduling policy* to determine the order in which the controller processes the requests that are in the queue. The controller then sends the request selected for scheduling to the FTL (part of the *Firmware* shown in Fig. 9.1b).

The host controller interface determines how requests are sent to the SSD and how the requests are queued for scheduling. Two of the most common host controller interfaces used by modern SSDs are the Advanced Host Controller Interface (AHCI) [99] and NVM Express (NVMe) [202]. AHCI builds upon the Serial Advanced Technology Attachment (SATA) system bus protocol [238], which was originally designed to connect the host system to magnetic hard disk drives. AHCI allows the host to use advanced features with SATA, such as *native command queuing* (NCQ). When an application executing on the host generates an I/O request, the application sends the request to the operating system (OS). The OS sends the request over the SATA bus to the SSD controller, and the controller adds the request to a single *command queue*. NCQ allows the controller to schedule the queued I/O requests in a different order than the order in which requests were received (i.e., requests are scheduled *out of order*). As a result, the controller can choose requests from the queue in a manner that maximizes the overall SSD performance (e.g., a younger request can be scheduler earlier than an older request that requires access to a plane that is occupied with serving another request). A major drawback of AHCI and SATA is the limited throughput they enable for SSDs [284], as the protocols were originally designed to match the much lower throughput of magnetic hard disk drives. For example, a modern magnetic hard drive has a sustained read throughput of 300 MB/s [237], whereas a modern SSD has a read throughput of 3500 MB/s [232]. However, AHCI and SATA are widely deployed in modern computing systems, and they currently remain a common choice for the SSD host controller interface.

To alleviate the throughput bottleneck of AHCI and SATA, many manufacturers have started adopting host controller interfaces that use the PCI Express (PCIe) system bus [217]. A popular standard interface for the PCIe bus is the NVM Express (NVMe) interface [202]. Unlike AHCI, which requires an application to send I/O requests through the OS, NVMe directly exposes multiple SSD I/O queues to the applications executing on the host. By directly exposing the queues to the applications, NVMe simplifies the software I/O stack, eliminating most OS involvement [284], which in turn reduces communication overheads. An SSD using the NVMe interface maintains a separate set of queues for *each* application (as opposed to the single queue used for all applications with AHCI) within the host interface. With more queues, the controller (1) has a larger number of requests to select from during scheduling, increasing its ability to utilize idle resources (i.e., *channels*, *dies*, *planes*; see Sect. 9.1.1); and (2) can more easily manage and control the amount of interference that an application experiences from other concurrently-executing applications. Currently, NVMe is used by modern SSDs that are designed mainly for high-performance systems (e.g., enterprise servers, data centers [283, 284]).

9.1.3.2 Flash Translation Layer

The main duty of the FTL (which is part of the *Firmware* shown in Fig. 9.1) is to manage the mapping of *logical addresses* (i.e., the address space utilized by the host) to *physical addresses* in the underlying flash memory (i.e., the address space for actual locations where the data is stored, visible only to the SSD controller) for each page of data [54, 80]. By providing this indirection between address spaces, the FTL can *remap* the logical address to a different physical address (i.e., move the data to a different physical address) *without* notifying the host. Whenever a page of data is written to by the host or moved for underlying SSD maintenance operations (e.g., garbage collection [40, 288]; see Sect. 9.1.3.3), the old data (i.e., the physical location where the overwritten data resides) is simply marked as invalid in the physical block's *metadata*, and the new data is written to a page in the flash block that is currently open for writes (see Sect. 9.2.4 for more detail on how writes are performed).

The FTL is also responsible for *wear leveling*, to ensure that all of the blocks within the SSD are evenly worn out [40, 288]. By evenly distributing the *wear* (i.e., the number of P/E cycles that take place) *across* different blocks, the SSD controller reduces the heterogeneity of the amount of wearout across these blocks, thereby extending the lifetime of the device. The wear-leveling algorithm is invoked when the current block that is being written to is full (i.e., no more pages in the block are available to write to), and it enables the controller to select a new block from the *free list* to direct the future writes to. The wear-leveling algorithm dictates which of the blocks from the free list is selected. One simple approach is to select the block in the free list with the lowest number of P/E cycles to minimize the variance of the wearout amount across blocks, though many algorithms have been developed for wear leveling [39, 71].

9.1.3.3 Garbage Collection

When the host issues a write request to a logical address stored in the SSD, the SSD controller performs the write *out of place* (i.e., the updated version of the page data is written to a different physical page in the NAND flash memory), because in-place updates cannot be performed (see Sect. 9.2.4). The old physical page is marked as *invalid* when the out-of-place write completes. *Fragmentation* refers to the waste of space within a block due to the presence of invalid pages. In a *fragmented* block, a fraction of the pages are invalid, but these pages are unable to store new data until the page is erased. Due to circuit-level limitations, the controller can perform erase operations only at the granularity of an *entire block* (see Sect. 9.2.4 for details). As a result, until a fragmented block is erased, the block wastes physical space within the SSD. Over time, if fragmented blocks are not erased, the SSD will run out of pages that it can write new data to. The problem becomes especially severe if the blocks are highly fragmented (i.e., a large fraction of the pages within a block are invalid).

To reduce the negative impact of fragmentation on usable SSD storage space, the FTL periodically performs a process called *garbage collection*. Garbage collection finds highly-fragmented flash blocks in the SSD and recovers the wasted space due to invalid pages. The basic garbage collection algorithm [40, 288] (1) identifies the highly-fragmented blocks (which we call the *selected blocks*), (2) migrates any valid pages in a selected block (i.e., each valid page is written to a new block, its virtual-to-physical address mapping is updated, and the page in the selected block is marked as invalid), (3) erases each selected block (see Sect. 9.2.4), and (4) adds a pointer to each selected block into the *free list* within the FTL. The garbage collection algorithm typically selects blocks with the highest number of invalid pages. When the controller needs a new block to write pages to, it selects one of the blocks currently in the free list.

We briefly discuss five optimizations that prior works propose to improve the performance and/or efficiency of garbage collection [1, 52, 80, 84, 90, 161, 222, 276, 288]. First, the garbage collection algorithm can be optimized to determine the most efficient frequency to invoke garbage collection [222, 288], as performing garbage collection too frequently can delay I/O requests from the host, while not performing garbage collection frequently enough can cause the controller to stall when there are no blocks available in the free list. Second, the algorithm can be optimized to select blocks in a way that reduces the number of page copy and erase operations required each time the garbage collection algorithm is invoked [84, 222].

Third, some works reduce the latency of garbage collection by using multiple channels to perform garbage collection on multiple blocks in parallel [1, 90]. Fourth, the FTL can minimize the latency of I/O requests from the host by pausing erase and copy operations that are being performed for garbage collection, in order to service the host requests immediately [52, 276]. Fifth, pages can be grouped together such that all of the pages within a block become invalid around the same time [80, 90, 161]. For example, the controller can group pages with (1) a similar degree of *write-hotness* (i.e., the frequency at which a page is updated; see Sect. 9.4.6) or (2) a similar *death time* (i.e., the time at which a page is overwritten). Garbage collection remains an active area of research.

9.1.3.4 Flash Reliability Management

The SSD controller performs many background optimizations that improve flash reliability. These flash reliability management techniques, as we will discuss in more detail in Sect. 9.4, can effectively improve flash lifetime at a very low cost, since the optimizations are usually performed during idle times, when the interference with the running workload is minimized. These management techniques sometimes require small metadata storage in memory (e.g., for storing the near-optimal read reference voltages [21, 22, 162]), or require a timer (e.g., for triggering refreshes in time [29, 30]).

9.1.3.5 Compression

Compression can reduce the size of the data written to minimize the number of flash cells worn out by the original data. Some controllers provide compression, as well as decompression, which reconstructs the original data from the compressed data stored in the flash memory [154, 300]. The controller may contain a *compression engine*, which, for example, performs the LZ77 or LZ78 algorithms. Compression is optional, as some types of data being stored by the host (e.g., JPEG images, videos, encrypted files, files that are already compressed) may not be compressible.

9.1.3.6 Data Scrambling and Encryption

The occurrence of errors in flash memory is highly dependent on the data values stored into the memory cells [19, 23, 31]. To reduce the dependence of the error rate on data values, an SSD controller first scrambles the data before writing it into the flash chips [32, 121]. The key idea of scrambling is to probabilistically ensure that the actual value written to the SSD contains an equal number of randomly distributed zeroes and ones, thereby minimizing any data-dependent behavior. Scrambling is performed using a reversible process, and the controller *descrambles* the data stored in the SSD during a read request. The controller employs a *linear feedback shift register* (LFSR) to perform scrambling and descrambling. An n-bit LFSR generates 2^{n-1} bits worth of pseudo-random numbers without repetition. For each page of data to be written, the LFSR can be seeded with the *logical* address of that page, so that the page can be correctly descrambled even if maintenance operations (e.g., garbage collection) migrate the page to another physical location, as the logical address is unchanged. (This also reduces the latency of maintenance operations, as they do not need to descramble and rescramble the data when a page is migrated.) The LFSR then generates a pseudo-random number based on the seed, which is then XORed with the data to produce the scrambled version of the data. As the XOR operation is reversible, the same process can be used to descramble the data.

In addition to the data scrambling employed to minimize data value dependence, several SSD controllers include data encryption hardware [55, 89, 271]. An SSD that contains data encryption hardware within its controller is known as a *self-encrypting drive* (SED). In the controller, data encryption hardware typically employs AES encryption [55, 59, 201, 271], which performs multiple rounds of substitutions and permutations to the unencrypted data in order to encrypt it. AES employs a separate key for each round [59, 201]. In an SED, the controller contains hardware that generates the AES keys for each round, and performs the substitutions and permutations to encrypt or decrypt the data using dedicated hardware [55, 89, 271].

9.1.3.7 Error-Correcting Codes

ECC is used to detect and correct the raw bit errors that occur within flash memory. A host writes a page of data, which the SSD controller splits into one or more chunks. For each chunk, the controller generates a *codeword*, consisting of the chunk and a correction code. The strength of protection offered by ECC is determined by the *coding rate*, which is the chunk size divided by the codeword size. A higher coding rate provides weaker protection, but consumes less storage, representing a key reliability tradeoff in SSDs.

The ECC algorithm employed (typically BCH [10, 92, 153, 243] or LDPC [72, 73, 167, 243, 298]; see Sect. 9.5), as well as the length of the codeword and the coding rate, determine the total *error correction capability*, i.e., the maximum number of raw bit errors that can be corrected by ECC. ECC engines in contemporary SSDs are able to correct data with a relatively high raw bit error rate (e.g., between 10^{-3} and 10^{-2} [103]) and return data to the host at an error rate that meets traditional data storage reliability requirements (e.g., a post-correction error rate of 10^{-15} in the JEDEC standard [105]). The *error correction failure rate* (P_{ECFR}) of an ECC implementation, with a codeword length of l where the codeword has an error correction capability of t bits, can be modeled as:

$$P_{ECFR} = \sum_{k=t+1}^{l} \binom{l}{k} (1 - \text{BER})^{(l-k)} \text{BER}^{k} \tag{9.1}$$

where BER is the bit error rate of the NAND flash memory. We assume in this equation that errors are independent and identically distributed.

In addition to the ECC information, a codeword contains cyclic redundancy checksum (CRC) parity information [229]. When data is being read from the NAND flash memory, there may be times when the ECC algorithm incorrectly indicates that it has successfully corrected all errors in the data, when uncorrected errors remain. To ensure that incorrect data is not returned to the user, the controller performs a CRC check in hardware to verify that the data is error free [219, 229].

9.1.3.8 Data Path Protection

In addition to protecting the data from raw bit errors within the NAND flash memory, newer SSDs incorporate error detection and correction mechanisms throughout the SSD controller, in order to further improve reliability and data integrity [229]. These mechanisms are collectively known as *data path protection*, and protect against errors that can be introduced by the various SRAM and DRAM structures that exist within the SSD.[1] Figure 9.3 illustrates the various structures within the controller that employ data path protection mechanisms. There are three data paths that require protection: (1) the path for data written by the host to the flash memory, shown as a red solid line in Fig. 9.3; (2) the path for data read from the flash memory by the host, shown as a green dotted line; and (3) the path for metadata transferred between the firmware (i.e., FTL) processors and the DRAM, shown as a blue dashed line.

In the write data path of the controller (the red solid line shown in Fig. 9.3), data received from the host interface (❶ in the figure) is first sent to a host FIFO buffer

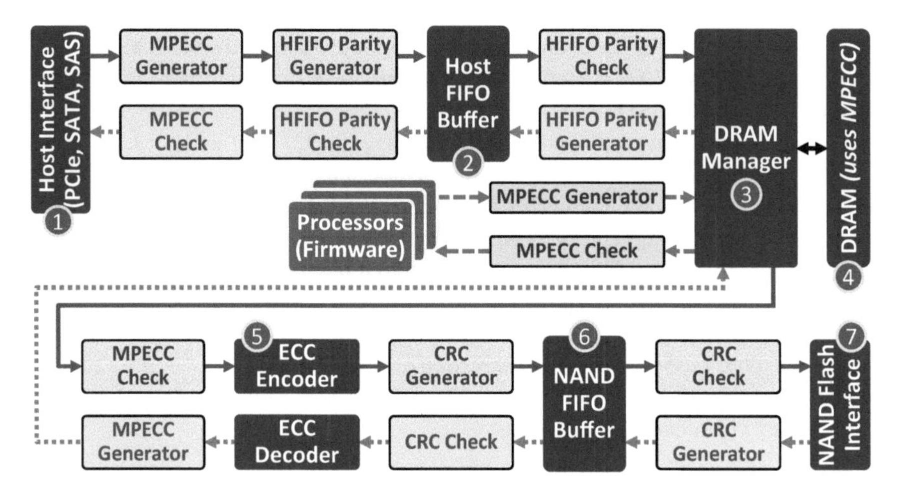

Fig. 9.3 Data path protection employed within the controller. Reproduced from [15]

[1] See Sect. 9.7 for a discussion on the possible types of errors that can be present in DRAM.

(❷). Before the data is written into the host FIFO buffer, the data is appended with *memory protection ECC* (MPECC) and *host FIFO buffer* (HFIFO) parity [229]. The MPECC parity is designed to protect against errors that are introduced when the data is stored within DRAM (which takes place later along the data path), while the HFIFO parity is designed to protect against SRAM errors that are introduced when the data resides within the host FIFO buffer. When the data reaches the head of the host FIFO buffer, the controller fetches the data from the buffer, uses the HFIFO parity to correct any errors, discards the HFIFO parity, and sends the data to the DRAM manager (❸). The DRAM manager buffers the data (which still contains the MPECC information) within DRAM (❹), and keeps track of the location of the buffered data inside the DRAM. When the controller is ready to write the data to the NAND flash memory, the DRAM manager reads the data from DRAM. Then, the controller uses the MPECC information to correct any errors, and discards the MPECC information. The controller then encodes the data into an ECC codeword (❺), generates CRC parity for the codeword, and then writes both the codeword and the CRC parity to a NAND flash FIFO buffer (❻) [229]. When the codeword reaches the head of this buffer, the controller uses CRC parity to detect any errors in the codeword, and then dispatches the data to the flash interface (❼), which writes the data to the NAND flash memory. The read data path of the controller (the green dotted line shown in Fig. 9.3) performs the same procedure as the write data path, but in reverse order [229].

Aside from buffering data along the write and read paths, the controller uses the DRAM to store essential metadata, such as the table that maps each host data address to a physical block address within the NAND flash memory [174, 229]. In the metadata path of the controller (the blue dashed line shown in Fig. 9.3), the metadata is often read from or written to DRAM by the firmware processors. In order to ensure correct operation of the SSD, the metadata must not contain any errors. As a result, the controller uses memory protection ECC (MPECC) for the metadata stored within DRAM [165, 229], just as it did to buffer data along the write and read data paths. Due to the lower rate of errors in DRAM compared to NAND flash memory (see Sect. 9.7), the employed memory protection ECC algorithms are not as strong as BCH or LDPC. We describe common ECC algorithms employed for DRAM error correction in Sect. 9.7.

9.1.3.9 Bad Block Management

Due to process variation or uneven wearout, a small number of flash blocks may have a much higher raw bit error rate (RBER) than an average flash block. Mitigating or tolerating the RBER on these flash blocks often requires a much higher cost than the benefit of using them. Thus, it is more efficient to identify and record these blocks as *bad blocks*, and avoid using them to store useful data. There are two types of bad blocks: *original bad blocks* (OBBs), which are defective due to manufacturing issues (e.g., process variation), and *growth bad blocks* (GBBs), which fail during runtime [259].

The flash vendor performs extensive testing, known as *bad block scanning*, to identify OBBs when a flash chip is manufactured [181]. Initially, all blocks are kept in the erased state, and contain the value 0xFF in each byte (see Sect. 9.2.1). Inside each OBB, the bad block scanning procedure writes a specific data value (e.g., 0×00) to a specific byte location within the block that indicates the block status. A good block (i.e., a block without defects) is not modified, and thus its block status byte remains at the value 0xFF. When the SSD is powered up for the first time, the SSD controller iterates through all blocks and checks the value stored in the block status byte of each block. Any block that does not contain the value 0xFF is marked as bad, and is recorded in a *bad block table* stored in the controller. A small number of blocks in each plane are set aside as *reserved blocks* (i.e., blocks that are not used during normal operation), and the bad block table automatically remaps any operation originally destined to an OBB to one of the reserved blocks. The bad block table remaps an OBB to a reserved block in the same plane, to ensure that the SSD maintains the same degree of parallelism when writing to a superpage, thus avoiding performance loss. Less than 2% of all blocks in the SSD are expected to be OBBs [204].

The SSD identifies growth bad blocks during runtime by monitoring the status of each block. Each superblock contains a bit vector indicating which of its blocks are GBBs. After each program or erase operation to a block, the SSD reads the *status reporting registers* to check the operation status. If the operation has failed, the controller marks the block as a GBB in the superblock bit vector. At this point, the controller uses superpage-level parity to recover the data that was stored in the GBB (see Sect. 9.1.3.10), and *all data in the superblock* is copied to a different superblock. The superblock containing the GBB is then erased. When the superblock is subsequently opened, blocks marked as GBBs are *not* used, but the remaining blocks can store new data.

9.1.3.10 Superpage-Level Parity

In addition to ECC to protect against bit-level errors, many SSDs employ RAID-like parity [63, 113, 180, 215]. The key idea is to store parity information within each superpage to protect data from ECC failures that occur within a single chip or plane. Figure 9.4 shows an example of how the ECC and parity information are organized within a superpage. For a superpage that spans across multiple chips, dies, and planes, the pages stored within one die or one plane (depending on the implementation) are used to store parity information for the remaining pages. Without loss of generality, we assume for the rest of this section that a superpage that spans c chips and d dies per chip stores parity information in the pages of a single die (which we call the *parity die*), and that it stores user data in the pages of the remaining $(c \times d) - 1$ dies. When all of the user data is written to the superpage, the SSD controller XORs the data together one plane at a time (e.g., in Fig. 9.4, all of the pages in Plane 0 are XORed with each other), which produces the parity data for that plane.

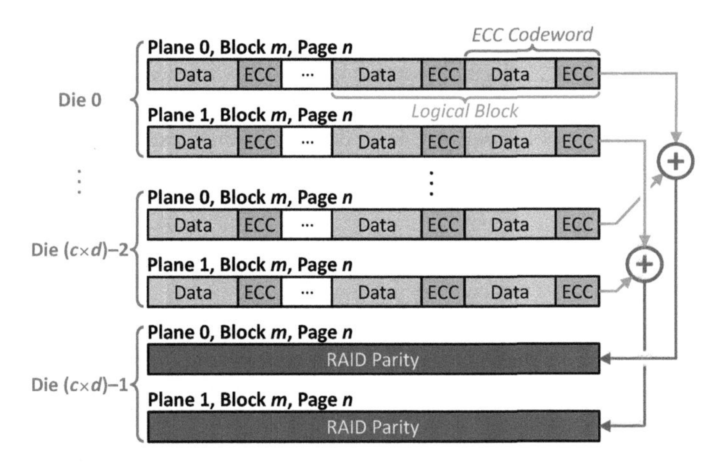

Fig. 9.4 Example layout of ECC codewords, logical blocks, and superpage-level parity for superpage n in superblock m. In this example, we assume that a logical block contains two codewords. Reproduced from [15]

This parity data is written to the corresponding plane in the parity die, e.g., Plane 0 page in Die $(c \times d) - 1$ in the figure.

The SSD controller invokes superpage-level parity when an ECC failure occurs during a host software (e.g., OS, file system) access to the SSD. The host software accesses data at the granularity of a *logical block* (LB), which is indexed by a *logical block address* (LBA). Typically, an LB is 4 kB in size, and consists of several ECC codewords (which are usually 512 BB to 2 kB in size) stored consecutively within a flash memory page, as shown in Fig. 9.4. During the LB access, a read failure can occur for one of two reasons. First, it is possible that the LB data is stored within a *hidden* GBB (i.e., a GBB that has not yet been detected and excluded by the bad block manager). The probability of storing data in a hidden GBB is quantified as P_{HGBB}. Note that because bad block management successfully identifies and excludes most GBBs, P_{HGBB} is much lower than the total fraction of GBBs within an SSD. Second, it is possible that at least one ECC codeword within the LB has *failed* (i.e., the codeword contains an error that cannot be corrected by ECC). The probability that a codeword fails is P_{ECFR} (see Sect. 9.1.3.7). For an LB that contains K ECC codewords, we can model P_{LBFail}, the overall probability that an LB access fails (i.e., the rate at which superpage-level parity needs to be invoked), as:

$$P_{LBFail} = P_{HGBB} + [1 - P_{HGBB}] \times [1 - (1 - P_{ECFR})^K] \tag{9.2}$$

In (9.2), P_{LBFail} consists of (1) the probability that an LB is inside a hidden GBB (left side of the addition); and (2) for an LB that is not in a hidden GBB, the probability of any codeword failing (right side of the addition).

When a read failure occurs for an LB in plane p, the SSD controller reconstructs the data using the other LBs in the same superpage. To do this, the controller reads the LBs stored in plane p in the other $(c \times d) - 1$ dies of the superpage, including the LBs in the parity die. The controller then XORs all of these LBs together, which retrieves the data that was originally stored in the LB whose access failed. In order to correctly recover the failed data, all of the LBs from the $(c \times d) - 1$ dies must be correctly read. The overall superpage-level parity failure probability P_{parity} (i.e., the probability that more than one LB contains a failure) for an SSD with c chips of flash memory, with d dies per chip, can be modeled as [215]:

$$P_{parity} = P_{LBFail} \times [1 - (1 - P_{LBFail})^{(c \times d) - 1}] \tag{9.3}$$

Thus, by designating one of the dies to contain parity information (in a fashion similar to RAID 4 [215]), the SSD can tolerate the *complete failure* of the superpage data in one die without experiencing data loss during an LB access.

9.1.4 Design Tradeoffs for Reliability

Several design decisions impact the SSD *lifetime* (i.e., the duration of time that the SSD can be used within a bounded probability of error without exceeding a given performance overhead). To capture the tradeoff between these decisions and lifetime, SSD manufacturers use the following model:

$$\text{Lifetime (Years)} = \frac{\text{PEC} \times (1 + \text{OP})}{365 \times \text{DWPD} \times \text{WA} \times R_{compress}} \tag{9.4}$$

In (9.4), the numerator is the total number of full drive writes the SSD can endure (i.e., for a drive with an X-byte capacity, the number of times X bytes of data can be written). The number of full drive writes is calculated as the product of PEC, the total P/E cycle *endurance* of each flash block (i.e., the number of P/E cycles the block can sustain before its raw error rate exceeds the ECC correction capability), and 1 + OP, where OP is the *overprovisioning factor* selected by the manufacturer. Manufacturers overprovision the flash drive by providing more physical block addresses, or PBAs, to the SSD controller than the *advertised capacity* of the drive, i.e., the number of logical block addresses (LBAs) available to the operating system. Overprovisioning improves performance and endurance, by providing additional free space in the SSD so that maintenance operations can take place without stalling host requests. OP is calculated as:

$$\text{OP} = \frac{\text{PBA count} - \text{LBA count}}{\text{LBA count}} \tag{9.5}$$

The denominator in (9.4) is the number of full drive writes per year, which is calculated as the product of days per year (i.e., 365), DWPD, and the ratio between the total size of the data written to flash media and the size of the data sent by the host (i.e., WA $\times R_{compress}$). DWPD is the number of full disk writes per day (i.e., the number of times per day the OS writes the advertised capacity's worth of data).

DWPD is typically less than 1 for read-intensive applications, and could be greater than 5 for write-intensive applications [29]. WA (*write amplification*) is the ratio between the amount of data written into NAND flash memory by the controller over the amount of data written by the host machine. Write amplification occurs because various procedures (e.g., garbage collection [40, 288]; and remapping-based refresh, Sect. 9.4.3) in the SSD perform additional writes in the background. For example, when garbage collection selects a block to erase, the pages that are remapped to a new block require background writes. $R_{compress}$, or the compression ratio, is the ratio between the size of the compressed data and the size of the uncompressed data, and is a function of the entropy of the stored data and the efficiency of the compression algorithms employed in the SSD controller. In (9.4), DWPD and $R_{compress}$ are largely determined by the workload and data compressibility, and cannot be changed to optimize flash lifetime. For controllers that do not implement compression, we set R compress to 1. However, the SSD controller can trade off other parameters between one another to optimize flash lifetime. We discuss the most salient tradeoffs next.

Tradeoff Between Write Amplification and Overprovisioning. As mentioned in Sect. 9.1.3.3, due to the granularity mismatch between flash erase and program operations, garbage collection occasionally remaps remaining valid pages from a selected block to a new flash block, in order to avoid block-internal fragmentation. This remapping causes additional flash memory writes, leading to *write amplification*. In an SSD with more overprovisioned capacity, the amount of write amplification decreases, as the blocks selected for garbage collection are older and tend to have fewer valid pages. For a greedy garbage collection algorithm and a random-access workload, the correlation between WA and OP can be calculated [62, 93], as shown in Fig. 9.5. In an ideal SSD, both WA and OP should be minimal, i.e., WA = 1 and OP = 0%, but in reality there is a tradeoff between these parameters: when one increases, the other decreases. As Fig. 9.5 shows, WA can be reduced by increasing OP, and with an infinite amount of OP, WA converges to 1. However, the reduction of WA is smaller when OP is large, resulting in diminishing returns.

In reality, the relationship between WA and OP is also a function of the storage space utilization of the SSD. When the storage space is *not* fully utilized, many more

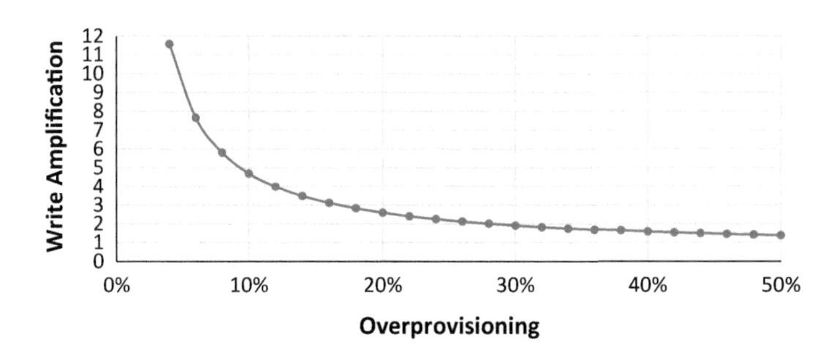

Fig. 9.5 Relationship between write amplification (WA) and the overprovisioning factor (OP). Reproduced from [15]

pages are available, reducing the need to invoke garbage collection, and thus WA can approach 1 without the need for a large amount of OP.

Tradeoff Between P/E Cycle Endurance and Overprovisioning. PEC and OP can be traded against each other by adjusting the amount of redundancy used for error correction, such as ECC and superpage-level parity (as discussed in Sect. 9.1.3.10). As the error correction capability increases, PEC increases because the SSD can tolerate the higher raw bit error rate that occurs at a higher P/E cycle count. However, this comes at a cost of reducing the amount of space available for OP, since a stronger error correction capability requires higher redundancy (i.e., more space). Table 9.1 shows the corresponding OP for four different error correction configurations for an example SSD with 2.0 TB of advertised capacity and 2.4 TB (20% extra) of physical space. In this table, the top two configurations use ECC-1 with a coding rate of 0.93, and the bottom two configurations use ECC-2 with a coding rate of 0.90, which has higher redundancy than ECC-1. Thus, the ECC-2 configurations have a lower OP than the top two. ECC-2, with its higher redundancy, can correct a greater number of raw bit errors, which in turn increases the P/E cycle endurance of the SSD. Similarly, the two configurations with superpage-level parity have a lower OP than configurations without superpage-level parity, as parity uses a portion of the overprovisioned space to store the parity bits.

When the ECC correction strength is increased, the amount of overprovisioning in the SSD decreases, which in turn increases the amount of write amplification that takes place. Manufacturers must find and use the correct tradeoff between ECC correction strength and the overprovisioning factor, based on which of the two is expected to provide greater reliability for the target applications of the SSD.

9.2 NAND Flash Memory Basics

A number of underlying properties of the NAND flash memory used within the SSD affect SSD management, performance, and reliability [9, 12, 182]. In this section,

Table 9.1 Tradeoff between strength of error correction configuration and amount of SSD space left for overprovisioning

Error correction configuration	Overprovisioning factor (%)
ECC-1 (0.93), no superpage-level parity	11.6
ECC-1 (0.93), with superpage-level parity	8.1
ECC-2 (0.90), no superpage-level parity	8.0
ECC-2 (0.90), with superpage-level parity	4.6

we present a primer on NAND flash memory and its operation, to prepare the reader for understanding our further discussion on error sources (Sect. 9.3) and mitigation mechanisms (Sect. 9.4). Recall from Sect. 9.1.1 that within each plane, flash cells are organized as multiple 2D arrays known as flash blocks, each of which contains multiple pages of data, where a page is the granularity at which the host reads and writes data. We first discuss how data is stored in NAND flash memory. We then introduce the three basic operations supported by NAND flash memory: read, program, and erase.

9.2.1 Storing Data in a Flash Cell

NAND flash memory stores data as the *threshold voltage* of each flash cell, which is made up of a *floating-gate transistor*. Figure 9.6 shows a cross section of a floating-gate transistor. On top of a flash cell is the *control gate* (CG) and below is the floating gate (FG). The floating gate is insulated on both sides, on top by an inter-poly oxide layer and at the bottom by a tunnel oxide layer. As a result, the electrons programmed on the floating gate do not discharge even when flash memory is powered off.

For *single-level cell* (SLC) NAND flash, each flash cell stores a 1-bit value, and can be programmed to one of two threshold voltage states, which we call the ER and P1 states. *Multi-level cell* (MLC) NAND flash stores a 2-bit value in each cell, with four possible states (ER, P1, P2, and P3), and *triple-level cell* (TLC) NAND flash stores a 3-bit value in each cell with eight possible states (ER, P1–P7). Each state represents a different value, and is assigned a *voltage window* within the range of all possible threshold voltages. Due to variation across *program* operations, the threshold voltage of flash cells programmed to the same state is initially distributed across this voltage window.

Figure 9.7 illustrates the threshold voltage distribution of MLC (top) and TLC (bottom) NAND flash memories. The x-axis shows the threshold voltage (V_{th}), which spans a certain voltage range. The y-axis shows the probability density of each voltage level across all flash memory cells. The threshold voltage distribution of each threshold voltage state can be represented as a probability density curve that spans over the state's voltage window.

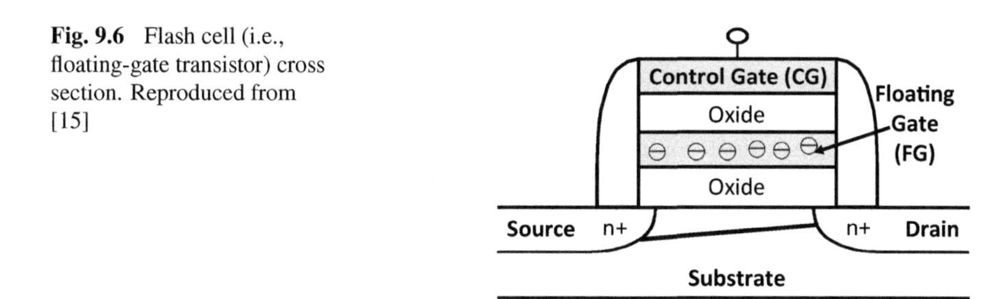

Fig. 9.6 Flash cell (i.e., floating-gate transistor) cross section. Reproduced from [15]

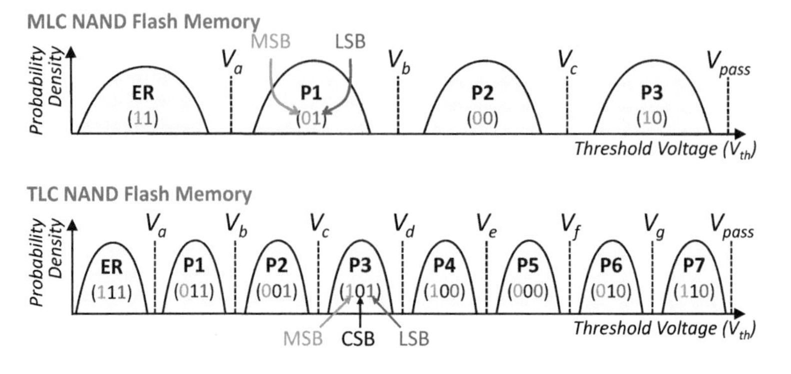

Fig. 9.7 Threshold voltage distribution of MLC (top) and TLC (bottom) NAND flash memory. Reproduced from [15]

We label the distribution curve for each state with the name of the state and a corresponding bit value. Note that some manufacturers may choose to use a different mapping of values to different states. The bit values of adjacent states are separated by a Hamming distance of 1. We break down the bit values for MLC into the most significant bit (MSB) and least significant bit (LSB), while TLC is broken down into the MSB, the center significant bit (CSB), and the LSB. The boundaries between neighboring threshold voltage windows, which are labeled as V_a, V_b, and V_c for the MLC distribution in Fig. 9.7, are referred to as *read reference voltages*. These voltages are used by the SSD controller to identify the voltage window (i.e., state) of each cell upon reading the cell.

9.2.2 Flash Block Design

Figure 9.8 shows the high-level internal organization of a NAND flash memory block. Each block contains multiple rows of cells (typically 128–512 rows). Each row of cells is connected together by a common *wordline* (WL, shown horizontally in Fig. 9.8), typically spanning 32–64 K cells. All of the cells along the wordline are logically combined to form a page in an SLC NAND flash memory. For an MLC NAND flash memory, the MSBs of all cells on the same wordline are combined to form an *MSB page*, and the LSBs of all cells on the wordline are combined to form an *LSB page*. Similarly, a TLC NAND flash memory logically combines the MSBs on each wordline to form an MSB page, the CSBs on each wordline to form a *CSB page*, and the LSBs on each wordline to form an LSB page. In MLC NAND flash memory, each flash block contains 256–1024 flash pages, each of which are typically 8–16 kB in size.

Within a block, all cells in the same column are connected in series to form a *bitline* (BL, shown vertically in Fig. 9.8) or *string*. All cells in a bitline share a common ground (GND) on one end, and a common *sense amplifier* (SA) on the other for

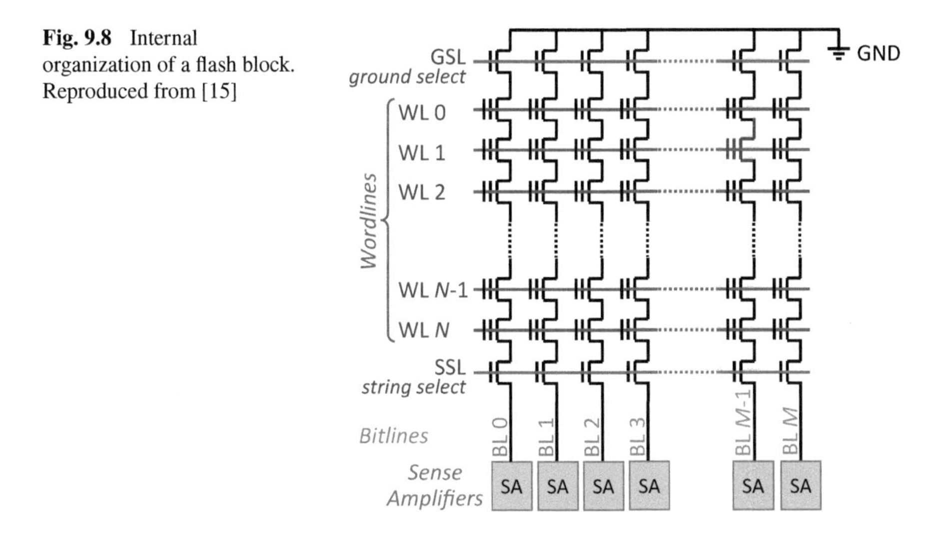

Fig. 9.8 Internal organization of a flash block. Reproduced from [15]

reading the threshold voltage of one of the cells when decoding data. Bitline operations are controlled by turning the *ground select line* (GSL) and *string select line* (SSL) transistor of each bitline on or off. The SSL transistor is used to enable operations on a bitline, and the GSL transistor is used to connect the bitline to ground during a read operation [184]. The use of a common bitline across multiple rows reduces the amount of circuit area required for read and write operations to a block, improving storage density.

9.2.3 Read Operation

Data can be read from NAND flash memory by applying read reference voltages onto the control gate of each cell, to sense the cell's threshold voltage. To read the value stored in a single-level cell, we need to distinguish only the state with a bit value of 1 from the state with a bit value of 0. This requires us to use only a single read reference voltage. Likewise, to read the LSB of a multi-level cell, we need to distinguish only the states where the LSB value is 1 (ER and P1) from the states where the LSB value is 0 (P2 and P3), which we can do with a single read reference voltage (V_b in the top half of Fig. 9.7). To read the MSB page, we need to distinguish the states with an MSB value of 1 (ER and P3) from those with an MSB value of 0 (P1 and P2). Therefore, we need to determine whether the threshold voltage of the cell falls between V_a and V_c, requiring us to apply each of these two read reference voltages (which can require up to two consecutive read operations) to determine the MSB.

Reading data from a triple-level cell is similar to the data read procedure for a multi-level cell. Reading the LSB for TLC again requires applying only a single read reference voltage (V_d in the bottom half of Fig. 9.7). Reading the CSB requires two read reference voltages to be applied, and reading the MSB requires four read reference voltages to be applied.

As Fig. 9.8 shows, cells from multiple wordlines (WL in the figure) are connected in series on a *shared* bitline (BL) to the sense amplifier, which drives the value that is being read from the block onto the memory channel for the plane. In order to read from a single cell on the bitline, *all of the other cells* (i.e., *unread* cells) on the same bitline must be switched on to allow the value that is being read to propagate through to the sense amplifier. The NAND flash memory achieves this by applying the *pass-through voltage* onto the wordlines of the unread cells, as shown in Fig. 9.9a. When the pass-through voltage (i.e., the maximum possible threshold voltage V_{pass}) is applied to a flash cell, the source and the drain of the cell transistor are connected, regardless of the voltage of the floating gate. Modern flash memories guarantee that all *unread* cells are *passed through* to minimize errors during the read operation [21].

9.2.4 Program and Erase Operations

The threshold voltage of a floating-gate transistor is controlled through the injection and ejection of electrons through the tunnel oxide of the transistor, which is enabled by the Fowler-Nordheim (FN) tunneling effect [9, 69, 216]. The tunneling current (J_{FN}) [12, 216] can be modeled as:

$$J_{FN} = \alpha_{FN} E_{ox}^2 e^{-\beta_{FN}/E_{ox}} \tag{9.6}$$

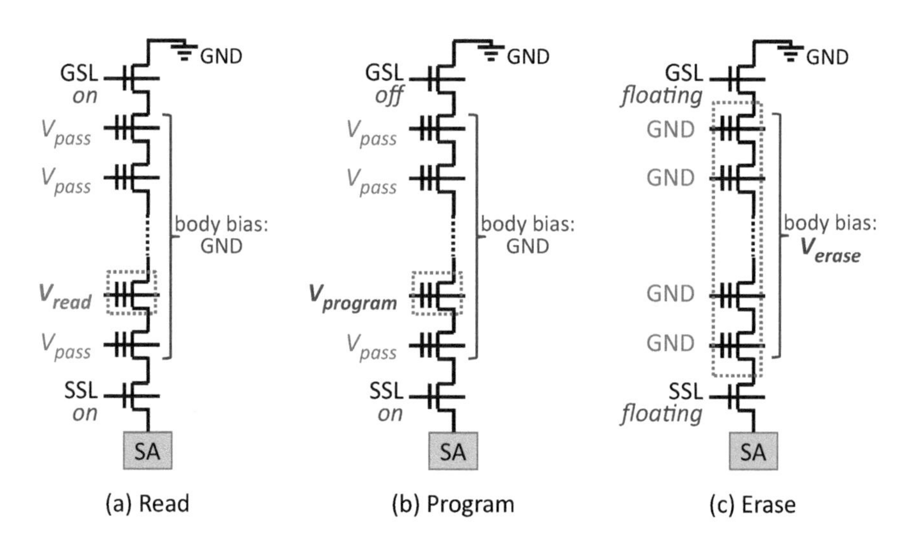

Fig. 9.9 Voltages applied to flash cell transistors on a bitline to perform **a** read, **b** program, and **c** erase operations. Reproduced from [15]

In (9.6), α_{FN} and β_{FN} are constants, and E_{ox} is the electric field strength in the tunnel oxide. As (9.6) shows, J_{FN} is exponentially correlated with E_{ox}.

During a program operation, electrons are injected into the floating gate of the flash cell from the substrate when applying a high positive voltage to the control gate (see Fig. 9.6 for a diagram of the flash cell). The pass-through voltage is applied to all of the other cells on the same bitline as the cell that is being programmed as shown in Fig. 9.9b. When data is programmed, charge is transferred into the floating gate through FN tunneling by repeatedly pulsing the programming voltage, in a procedure known as *incremental step-pulse programming* (ISPP) [9, 182, 253, 267]. During ISPP, a high programming voltage ($V_{program}$) is applied for a very short period, which we refer to as a *step-pulse*. ISPP then verifies the current voltage of the cell using the voltage V_{verify}. ISPP repeats the process of applying a step-pulse and verifying the voltage until the cell reaches the desired target voltage. In the modern all-bitline NAND flash memory, all flash cells in a single wordline are programmed concurrently. During programming, when a cell along the wordline reaches its target voltage but other cells have yet to reach their target voltage, ISPP *inhibits* programming pulses to the cell by turning off the SSL transistor of the cell's bitline.

In SLC NAND flash and older MLC NAND flash, *one-shot programming* is used, where all of the ISPP step-pulses required to program a cell are applied back to back until all cells in the wordline are fully programmed. One-shot programming does *not* interleave the program operations to a wordline with the program operations to another wordline. In newer MLC NAND flash, the lack of interleaving between program operations can introduce a significant amount of cell-to-cell program interference on the cells of immediately-adjacent wordlines (see Sect. 9.3.3).

To reduce the impact of program interference, the controller employs *two-step programming* for sub-40 nm MLC NAND flash [23, 209]: it first programs the LSBs into the erased cells of an unprogrammed wordline, and then programs the MSBs of the cells using a separate program operation [17, 20, 207, 209]. Between the programming of the LSBs and the MSBs, the controller programs the LSBs of the cells in the wordline immediately above [17, 20, 207, 209]. Figure 9.10 illustrates the two-step programming algorithm. In the first step, a flash cell is *partially programmed* based on its LSB value, either staying in the ER state if the LSB value is 1, or moving to a temporary state (TP) if the LSB value is 0. The TP state has a mean voltage that falls between states P1 and P2. In the second step, the LSB data is first read back into an internal buffer register within the flash chip to determine the cell's current threshold voltage state, and then further programming pulses are applied based on the MSB data to increase the cell's threshold voltage to fall within the voltage window of its final state. Programming in MLC NAND flash is discussed in detail in [17, 20].

TLC NAND flash takes a similar approach to the two-step programming of MLC, with a mechanism known as *foggy-fine programming* [156], which is illustrated in Fig. 9.11. The flash cell is first partially programmed based on its LSB value, using a *binary* programming step in which very large ISPP step-pulses are used to significantly increase the voltage level. Then, the flash cell is partially programmed again based on its CSB and MSB values to a new set of temporary states (these steps are

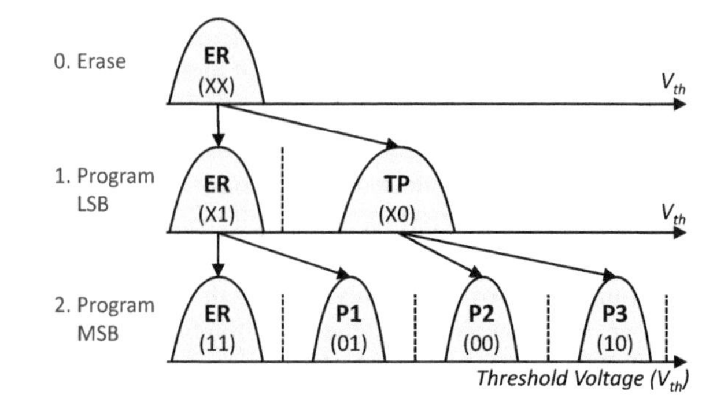

Fig. 9.10 Two-step programming algorithm for MLC flash. Reproduced from [15]

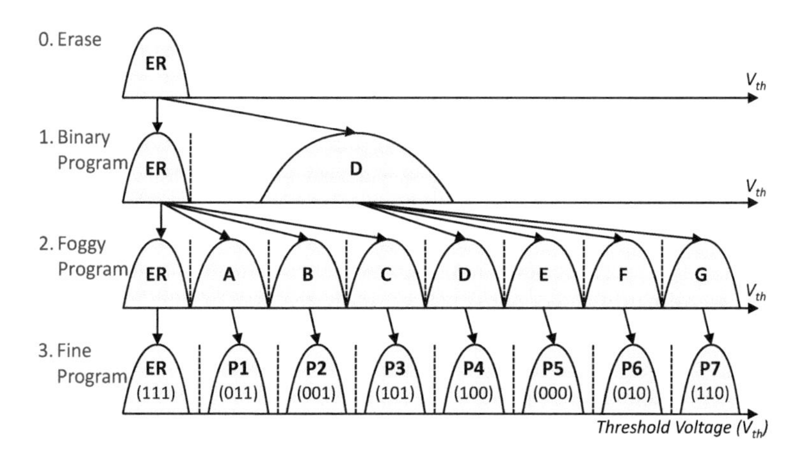

Fig. 9.11 Foggy-fine programming algorithm for TLC flash. Reproduced from [15]

referred to as *foggy* programming, which uses smaller ISPP step-pulses than binary programming). Due to the higher potential for errors during TLC programming as a result of the narrower voltage windows, all of the programmed bit values are buffered after the binary and foggy programming steps into SLC buffers that are reserved in each chip/plane. Finally, *fine* programming takes place, where these bit values are read from the SLC buffers, and the smallest ISPP step-pulses are applied to set each cell to its final threshold voltage state. The purpose of this last fine programming step is to fine tune the threshold voltage such that the threshold voltage distributions are tightened (bottom of Fig. 9.11).

Though programming sets a flash cell to a specific threshold voltage using programming pulses, the voltage of the cell can drift over time after programming. When no external voltage is applied to any of the electrodes (i.e., CG, source, and drain) of a flash cell, an electric field still exists between the FG and the substrate, generated by the charge present in the FG. This is called the *intrinsic electric field* [12], and it generates *stress-induced leakage current* (SILC) [9, 60, 200], a weak tunneling current that leaks charge away from the FG. As a result, the voltage that a cell is programmed to may not be the same as the voltage read for that cell at a subsequent time.

In NAND flash, a cell can be reprogrammed with new data *only after* the existing data in the cell is erased. This is because ISPP can only *increase* the voltage of the cell. The erase operation resets the threshold voltage state of *all cells in the flash block* to the ER state. During an erase operation, electrons are ejected from the FG of the flash cell into the substrate by inducing a high negative voltage on the cell transistor. The negative voltage is induced by setting the CG of the transistor to GND, and biasing the transistor body (i.e., the substrate) to a high voltage (V_{erase}), as shown in Fig. 9.9c. Because all cells in a flash block share a common transistor substrate (i.e., the bodies of all transistors in the block are connected together), a flash block must be erased in its entirety [184].

9.3 NAND Flash Error Characterization

Each block in NAND flash memory is used in a cyclic fashion, as is illustrated by the observed raw bit error rates seen over the lifetime of a flash memory block in Fig. 9.12. At the beginning of a *cycle*, known as a *program/erase (P/E) cycle*, an erased block is *opened* (i.e., selected for programming). Data is then programmed into the open block one page at a time. After all of the pages are programmed, the block is closed, and none of the pages can be reprogrammed until the whole block is erased. At any point before erasing, read operations can be performed on a *valid* programmed page (i.e., a page containing data that has not been modified by the host). A page is marked as invalid when the data stored at that page's logical address by the host is modified. As ISPP can only inject more charge into the floating gate but cannot remove charge from the gate, it is not possible to modify data to a new arbitrary value *in place* within existing NAND flash memories. Once the block is erased, the P/E cycling behavior repeats until the block is *worn out* (i.e., the block can no longer avoid data loss over the course of the minimum data retention period guaranteed by the manufacturer). Although the 5x-nm (i.e., 50–59 nm) generation of MLC NAND flash could endure ~10,000 P/E cycles per block before being worn out, modern 1x-nm (i.e., 15–19 nm) MLC and TLC NAND flash can endure only ~3,000 and ~1,000 P/E cycles per block, respectively [136, 168, 212, 294].

As shown in Fig. 9.12, several different types of errors can be introduced at any point during the P/E cycling process: *P/E cycling errors*, *program errors*, errors due to *cell-to-cell program interference*, *data retention errors*, and errors due to *read disturb*. As discussed in Sect. 9.2.1, the threshold voltage of flash cells programmed to the same state is distributed across a voltage window due to variation across program operations and across different flash cells. Several types of errors introduced during the P/E cycling process, such as data retention and read disturb, cause the threshold voltage distribution of each state to shift and widen. Due to the shift and widening, the tails of the distributions of each state can enter the margin that originally existed between each of the two neighboring states' distributions.

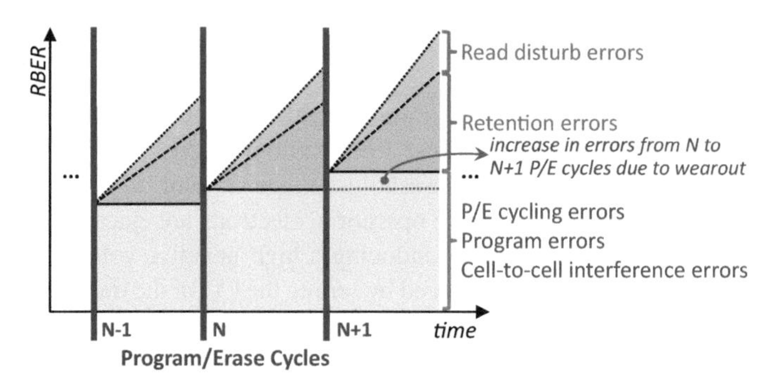

Fig. 9.12 Pictorial depiction of errors accumulating within a NAND flash block as P/E cycle count increases. Reproduced from [15]

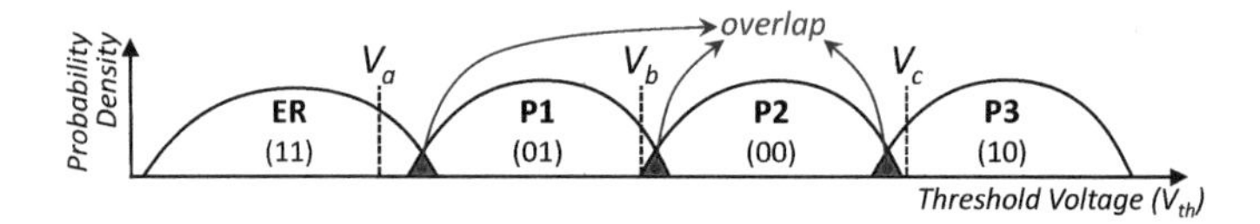

Fig. 9.13 Threshold voltage distribution shifts and widening can cause the distributions of two neighboring states to overlap with each other (compare to Fig. 9.7), leading to read errors. Reproduced from [15]

Thus, the threshold voltage distributions of different states can start overlapping, as shown in Fig. 9.13. When the distributions overlap with each other, the read reference voltages can no longer correctly identify the state of some flash cells in the overlapping region, leading to *raw bit errors* during a read operation.

In this section, we discuss the causes of each type of error in detail, and characterize the impact that each error type has on the amount of raw bit errors occurring within NAND flash memory. We use an FPGA-based testing platform [18] to characterize state-of-the-art TLC NAND flash chips. We use the read-retry operation present in NAND flash devices to accurately read the cell threshold voltage [20–23, 29, 31, 70, 162, 208] (for a detailed description of the read-retry operation, see Sect. 9.4.4). As absolute threshold voltage values are proprietary information to flash vendors, we present our results using normalized voltages, where the nominal maximum value of V_{th} is equal to 512 in our normalized scale, and where 0 represents GND. We also describe characterization results and observations for MLC NAND flash chips. These MLC NAND results are taken from our prior works [14, 17, 19–23, 29–31, 162], which provide more detailed error characterization results and analyses. To our knowledge, this paper provides the first experimental characterization and analysis of errors in real *TLC* NAND flash memory chips (Tables 9.2, 9.3 and 9.4).

We later discuss mitigation techniques for these flash memory errors in Sect. 9.4, and provide procedures to recover in the event of data loss in Sect. 9.5.

9.3.1 P/E Cycling Errors

A P/E cycling error occurs when either (1) an erase operation fails to reset a cell to the ER state; or (2) when a program operation fails to set the cell to the desired target state. P/E cycling errors occur because electrons become trapped in the tunnel oxide after stress from repeated P/E cycles. Errors due to such electron trapping (which we refer to as *P/E cycling noise*) continue to accumulate over the lifetime of a NAND flash block. This behavior is called *wearout*, and it refers to the phenomenon where, as more writes are performed to a block, there are a greater number of raw bit errors that must be corrected, exhausting more of the fixed error correction capability of the ECC (see Sect. 9.1.3.7).

Figure 9.14 shows the threshold voltage distribution of TLC NAND flash memory after 0 P/E cycles and after 3,000 P/E cycles, without any retention or read disturb errors present (which we ensure by reading the data *immediately* after programming). The mean and standard deviation of each state's distribution are provided in Table 9.5 in the Appendix (for other P/E cycle counts as well). We make two observations from the two distributions. First, as the P/E cycle count increases, each state's threshold voltage distribution systematically (1) shifts to the right and (2) becomes wider. Second, the amount of the shift is greater for lower-voltage states (e.g., the ER and P1 states) than it is for higher-voltage states (e.g., the P7 state).

The threshold voltage distribution shift occurs because as more P/E cycles take place, the quality of the tunnel oxide degrades, allowing electrons to tunnel through the oxide more easily [186]. As a result, if the same ISPP conditions (e.g., programming voltage, step-pulse size, program time) are applied throughout the lifetime of the NAND flash memory, more electrons are injected during programming as a flash memory block wears out, leading to higher threshold voltages, i.e., the right shift of the distribution. The distribution of each state widens due to the process variation present in (1) the wearout process, and (2) the cell's structural characteristics. As the distribution of each voltage state widens, more overlap occurs between neighboring distributions, making it less likely for a read reference voltage to determine the correct value of the cells in the overlapping regions, which leads to a greater number of raw bit errors.

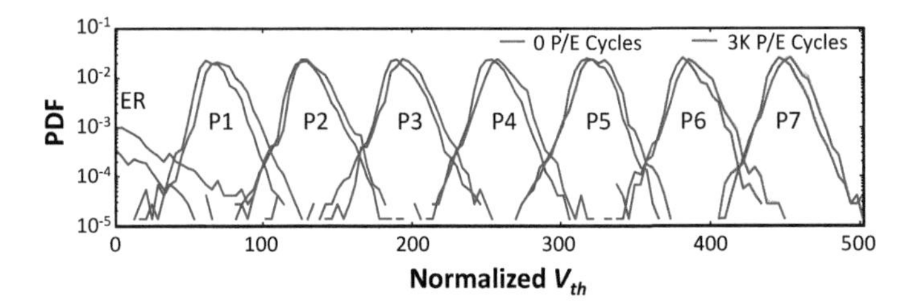

Fig. 9.14 Threshold voltage distribution of TLC NAND flash memory after 0 P/E cycles and 3,000 P/E cycles. Reproduced from [15]

The threshold voltage distribution trends we observe here for TLC NAND flash memory trends are similar to trends observed previously for MLC NAND flash memory [19, 20, 162, 212], although the MLC NAND flash characterizations reported in past studies span up to a larger P/E cycle count than the TLC experiments due to the greater endurance of MLC NAND flash memory. More findings on the nature of wearout and the impact of wearout on NAND flash memory errors and lifetime can be found in our prior work [14, 19, 20, 162].

9.3.2 Program Errors

Program errors occur when data read directly from the NAND flash array contains errors, and the erroneous values are used to program the new data. Program errors occur in two major cases: (1) partial programming during two-step or foggy-fine programming, and (2) *copyback* (i.e., when data is copied inside the NAND flash memory during a maintenance operation) [94]. During two-step programming for MLC NAND flash memory (see Fig. 9.10), in between the LSB and MSB programming steps of a cell, threshold voltage shifts can occur on the partially-programmed cell. These shifts occur because several other read and program operations to cells in *other* pages within the same block may take place, causing interference to the partially-programmed cell. Figure 9.15 illustrates how the threshold distribution of the ER state widens and shifts to the right after the LSB value is programmed (step 1 in the figure). The widening and shifting of the distribution causes some cells that were originally partially programmed to the ER state (with an LSB value of 1) to be misread as being in the TP state (with an LSB value of 0) during the *second* programming step (step 2 in the figure). As shown in Fig. 9.15, the misread LSB value leads to a program error when the final cell threshold voltage is programmed [17, 162, 212]. Some cells that should have been programmed to the P1 state (representing the value 01) are instead programmed to the P2 state (with the value 00), and some

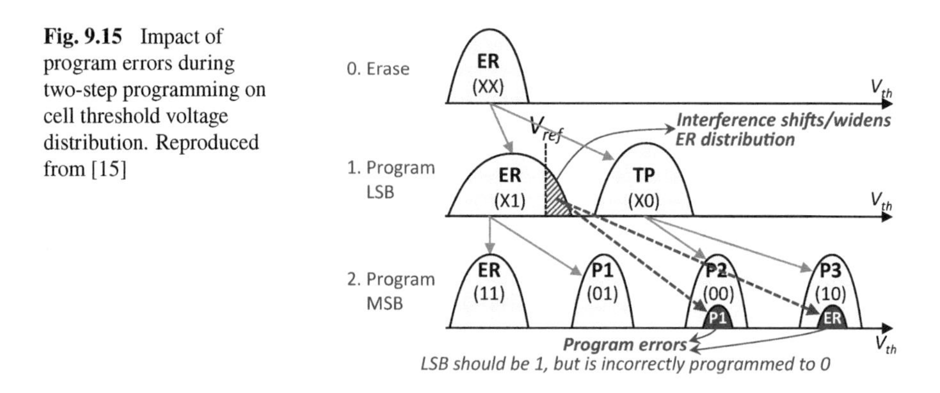

Fig. 9.15 Impact of program errors during two-step programming on cell threshold voltage distribution. Reproduced from [15]

cells that should have been programmed to the ER state (representing the value 11) are instead programmed to the P3 state (with the value 10).

The incorrect values that are read before the second programming step are *not* corrected by ECC, as they are read directly inside the NAND flash array, without involving the controller (where the ECC engine resides). Similarly, during foggy-fine programming for TLC NAND flash (see Fig. 9.11), the data may be read incorrectly from the SLC buffers used to store the contents of partially-programmed wordlines, leading to errors during the fine programming step. Program errors occur during *copyback* [94] when valid data is read out from a block during maintenance operations (e.g., a block about to be garbage collected) and reprogrammed into a new block, as copyback operations do *not* go through the SSD controller.

Program errors that occur during partial programming predominantly shift data from lower-voltage states to higher-voltage states. For example, in MLC NAND flash, program errors predominantly shift data that should be in the ER state (11) into the P3 state (10), or data that should be in the P1 state (01) into the P2 state (00) [17]. This occurs because MSB programming can only *increase* (and not reduce) the threshold voltage of the cell from its partially-programmed voltage (and thus cannot move a multi-level cell that should be in the P3 state into the ER state, or one that should be in the P2 state into the P1 state). TLC NAND flash is much less susceptible to program errors than MLC NAND flash, as the data read from the SLC buffers in TLC NAND flash has a much lower error rate than data read from a partially-programmed MLC NAND flash wordline [242].

From a rigorous experimental characterization of modern MLC NAND flash memory chips [17], we find that program errors occur primarily due to two types of errors affecting the partially-programmed data. First, cell-to-cell program interference (Sect. 9.3.3) on a partially-programmed wordline is no longer negligible in newer NAND flash memory compared to older NAND flash memory, due to manufacturing process scaling. As flash cells become smaller and are placed closer to each other, cells in partially-programmed wordlines become more susceptible to bit flips. Second, partially-programmed cells are more susceptible to read disturb errors than fully-programmed cells (Sect. 9.3.5), as the threshold voltages stored in these cells are no more than approximately half of V_{pass} [17], and cells with lower threshold voltages are more likely to experience read disturb errors.

More findings on the nature of program errors and the impact of program errors on NAND flash memory lifetime can be found in our prior work [17, 162].

9.3.3 Cell-to-Cell Program Interference Errors

Program interference refers to the phenomenon where the programming of a flash cell induces errors on adjacent flash cells within a flash block [23, 31, 58, 75, 151]. The interference occurs due to *parasitic capacitance coupling* between these cells. As a result, when the threshold voltage of an adjacent flash cell increases, the threshold voltage of the *victim cell* increases as well. The unintended threshold voltage shifts can eventually move a cell into a different state than the one it was originally programmed to, leading to a bit error.

We have shown, based on our experimental analysis of modern MLC NAND flash memory chips, that the threshold voltage change of the victim cell can be accurately modeled as a linear combination of the threshold voltage changes of the adjacent cells when they are programmed, using linear regression with least-square-error estimation [23, 31]. The cells that are physically located immediately next to the victim cell (called the *immediately-adjacent cells*) are the major contributors to the cell-to-cell interference of a victim cell [23]. Figure 9.16 shows the eight immediately-adjacent cells for a victim cell in 2D planar NAND flash memory.

The amount of interference that program operations to the immediately-adjacent cells can induce on the victim cell is expressed as:

$$\Delta V_{victim} = \sum_{X} K_X \Delta V_X \tag{9.7}$$

where ΔV_{victim} is the change in voltage of the victim cell due to cell-to-cell program interference, K_X is the *coupling coefficient* between cell X and the victim cell, and ΔV_X is the threshold voltage change of cell X during programming. Table 9.2 lists the coupling coefficients for both 2y-nm and 1x-nm NAND flash memory. We make two key observations from Table 9.2. First, we observe that the coupling coefficient is greatest for wordline neighbors (i.e., immediately-adjacent cells on the same bitline, but on a neighboring wordline) [23]. The coupling coefficient is directly related to the effective capacitance C between cell X and the victim cell, which can be calculated as:

$$C = \varepsilon S/d \tag{9.8}$$

where ε is the permittivity, S is the effective cell area of cell X that faces the victim cell, and d is the distance between the cells. Of the immediately-adjacent cells, the wordline neighbor cells have the greatest coupling capacitance with the victim cell, as they likely have a large effective facing area to, and a small distance from, the victim cell compared to other surrounding cells. Second, we observe that the coupling coefficient grows as the feature size decreases [23, 31]. As NAND flash memory

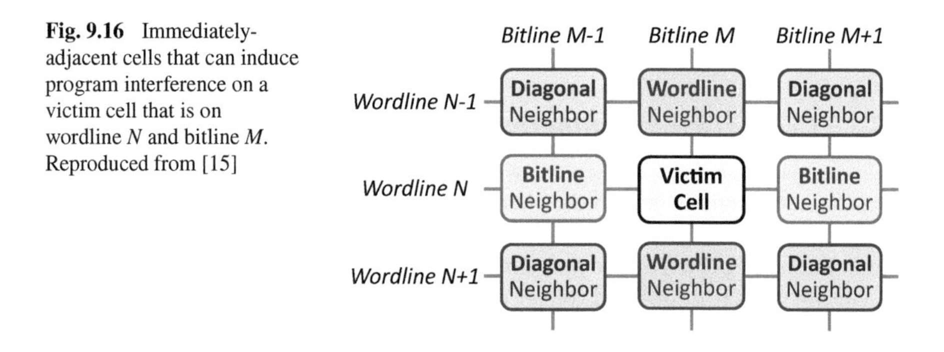

Fig. 9.16 Immediately-adjacent cells that can induce program interference on a victim cell that is on wordline *N* and bitline *M*. Reproduced from [15]

Table 9.2 Coupling coefficients for immediately-adjacent cells

Process technology	Wordline neighbor	Bitline neighbor	Diagonal neighbor
2y-nm	0.060	0.032	0.012
1x-nm	0.110	0.055	0.020

process technology scales down to smaller feature sizes, cells become smaller and get closer to each other, which increases the effective capacitance between them. As a result, at smaller feature sizes, it is easier for an immediately-adjacent cell to induce program interference on a victim cell. We conclude that (1) the program interference an immediately-adjacent cell induces on a victim cell is primarily determined by the distance between the cells and the immediately-adjacent cell's effective area facing the victim cell; and (2) the wordline neighbor cell causes the highest such interference, based on empirical measurements.

Due to the order of program operations performed in NAND flash memory, many immediately-adjacent cells do *not* end up inducing interference after a victim cell is fully programmed (i.e., once the victim cell is at its target voltage). In modern all-bitline NAND flash memory, all flash cells on the same wordline are programmed at the same time, and wordlines are fully programmed sequentially (i.e., the cells on wordline i are fully programmed before the cells on wordline $i + 1$). As a result, an immediately-adjacent cell on the wordline below the victim cell or on the same wordline as the victim cell does *not* induce program interference on a fully-programmed victim cell. Therefore, the major source of program interference on a fully-programmed victim cell is the programming of the wordline immediately above it.

Figure 9.17 shows how the threshold voltage distribution of a victim cell shifts when different values are programmed onto its immediately-adjacent cells in the wordline above the victim cell for MLC NAND flash, when one-shot programming

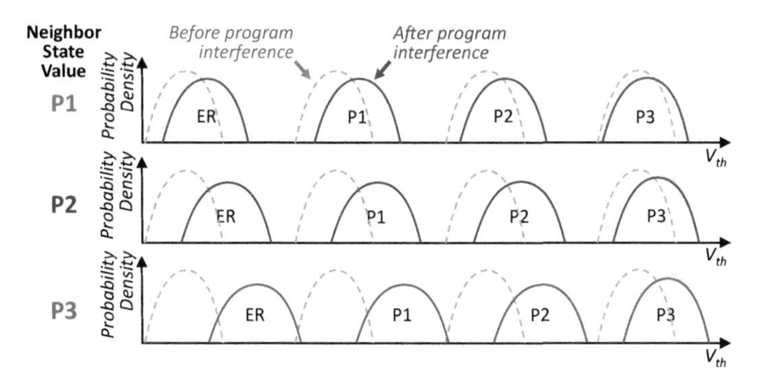

Fig. 9.17 Impact of cell-to-cell program interference on a victim cell during one-shot programming, depending on the value its neighboring cell is programmed to. Reproduced from [15]

is used. The amount by which the victim cell distribution shifts is directly correlated with the number of programming step-pulses applied to the immediately-adjacent cell. That is, when an immediately-adjacent cell is programmed to a higher-voltage state (which requires more step-pulses for programming), the victim cell distribution shifts further to the right [23]. When an immediately-adjacent cell is set to the ER state, no step-pulses are applied, as an unprogrammed cell is already in the ER state. Thus, no interference takes place. Note that the amount by which a fully-programmed victim cell distribution shifts is different when two-step programming is used, as a fully-programmed cell experiences interference from only one of the two programming steps of a neighboring wordline [17].

More findings on the nature of cell-to-cell program interference and the impact of cell-to-cell program interference on NAND flash memory errors and lifetime can be found in our prior work [14, 17, 23, 31].

9.3.4 Data Retention Errors

Retention errors are caused by charge leakage over time after a flash cell is programmed, and are the dominant source of flash memory errors, as demonstrated previously [19, 22, 29, 30, 182, 256]. As flash memory process technology scales to smaller feature sizes, the capacitance of a flash cell, and the number of electrons stored on it, decreases. State-of-the-art (i.e., 1x-nm) MLC flash memory cells can store only ~100 electrons [294]. Gaining or losing several electrons on a cell can significantly change the cell's voltage level and eventually alter its state. Charge leakage is caused by the unavoidable trapping of charge in the tunnel oxide [22, 150]. The amount of trapped charge increases with the electrical stress induced by repeated program and erase operations, which degrade the insulating property of the oxide.

Two failure mechanisms of the tunnel oxide lead to retention loss. *Trap-assisted tunneling* (TAT) occurs because the trapped charge forms an electrical tunnel, which exacerbates the weak tunneling current, SILC (see Sect. 9.2.4). As a result of this TAT effect, the electrons present in the floating gate (FG) leak away much faster through the intrinsic electric field. Hence, the threshold voltage of the flash cell decreases over time. As the flash cell wears out with increasing P/E cycles, the amount of trapped charge also increases [22, 150], and so does the TAT effect. At high P/E cycles, the amount of trapped charge is large enough to form percolation paths that significantly hamper the insulating properties of the gate dielectric [22, 60], resulting in retention failure. *Charge detrapping*, where charge previously trapped in the tunnel oxide is freed spontaneously, can also occur over time [22, 60, 150, 285]. The charge polarity can be either negative (i.e., electrons) or positive (i.e., holes). Hence, charge detrapping can either decrease or increase the threshold voltage of a flash cell, depending on the polarity of the detrapped charge.

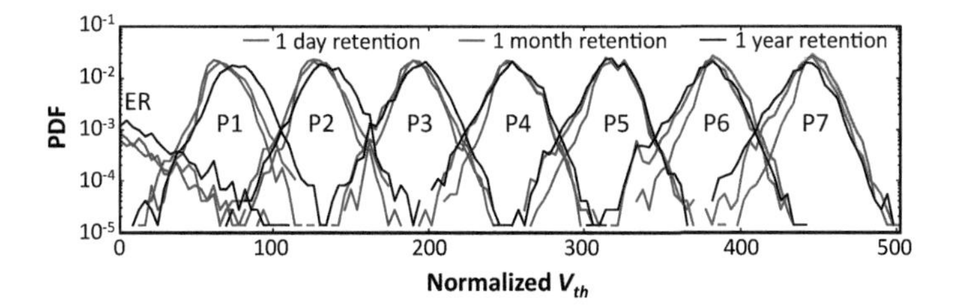

Fig. 9.18 Threshold voltage distribution for TLC NAND flash memory after one day, one month, and one year of retention time. Reproduced from [15]

Figure 9.18 illustrates how the voltage distribution shifts for data we program into TLC NAND flash, as the data sits untouched over a period of one day, one month, and one year. The mean and standard deviation are provided in Table 9.6 in the Appendix (which includes data for other retention ages as well). These results are obtained from real flash memory chips we tested. We distill three major findings from these results, which are similar to our previously reported findings for retention behavior on MLC NAND flash memory [22].

First, as the *retention age* (i.e., the length of time after programming) of the data increases, the threshold voltage distributions of the higher-voltage states shift to lower voltages, while the threshold voltage distributions of the lower-voltage states shift to higher voltages. As the intrinsic electric field strength is higher for the cells in higher-voltage states, TAT is the dominant failure mechanism for these cells, which can *only* decrease the threshold voltage, as the resulting SILC can flow only in the direction of the intrinsic electric field generated by the electrons in the FG. Cells at the lowest-voltage states, where the intrinsic electric field strength is low, do not experience high TAT, and instead contain many *holes* (i.e., positive charge) that leak away as the retention age grows, leading to increase in threshold voltage.

Second, the threshold voltage distribution of each state becomes wider with retention age. Charge detrapping can cause cells to shift in either direction (i.e., toward lower or higher voltages), contributing to the widening of the distribution. The rate at which TAT occurs can also vary from cell to cell, as a result of process variation, which further widens the distribution.

Third, the threshold voltage distributions of higher-voltage states shift by a larger amount than the distributions of lower-voltage states. This is again a result of TAT. Cells at higher-voltage states have greater intrinsic electric field intensity, which leads to larger SILC. A cell where the SILC is larger experiences a greater drop in its threshold voltage than a cell where the SILC is smaller.

More findings on the nature of data retention and the impact of data retention behavior on NAND flash memory errors and lifetime can be found in our prior work [14, 19, 22, 29, 30].

9.3.5 Read Disturb Errors

Read disturb is a phenomenon in NAND flash memory where reading data from a flash cell can cause the threshold voltages of other (unread) cells in the same block to shift to a higher value [19, 21, 58, 75, 182, 206, 254]. While a single threshold voltage shift is small, such shifts can accumulate over time, eventually becoming large enough to alter the state of some cells and hence generate *read disturb errors*.

The failure mechanism of a read disturb error is similar to the mechanism of a normal program operation. A program operation applies a high programming voltage (e.g., +15 V) to the cell to change the cell's threshold voltage to the desired range. Similarly, a read operation applies a *high pass-through voltage* (e.g., +6 V) to *all other cells* that share the same bitline with the cell that is being read. Although the pass-through voltage is not as high as the programming voltage, it still generates a *weak programming effect* on the cells it is applied to [21], which can unintentionally change these cells' threshold voltages.

Figure 9.19 shows how read disturb errors impact threshold voltage distributions in real TLC NAND flash memory chips. We use blocks that have endured 2,000 P/E cycles, and we experimentally study the impact of read disturb on a single wordline in each block. We then read from a second wordline in the same block 1, 10 and 100 K times to induce different levels of read disturb. The mean and standard deviation of each distribution are provided in Table 9.7 in the Appendix. We derive three major findings from these results, which are similar to our previous findings for read disturb behavior in MLC NAND flash memory [21].

First, as the read disturb count increases, the threshold voltages increase (i.e., the voltage distribution shifts to the right). In particular, we find that the distribution shifts are greater for lower-voltage states, indicating that read disturb impacts cells in the ER and P1 states the most. This is because we apply the same pass-through voltage (V_{pass}) to *all* unread cells during a read operation, *regardless* of the threshold voltages of the cells. A lower threshold voltage on a cell induces a larger voltage difference ($V_{pass} - V_{th}$) through the tunnel oxide layer of the cell, and in turn generates a stronger tunneling current, making the cell more vulnerable to read disturb (as described in detail in our prior work [21]).

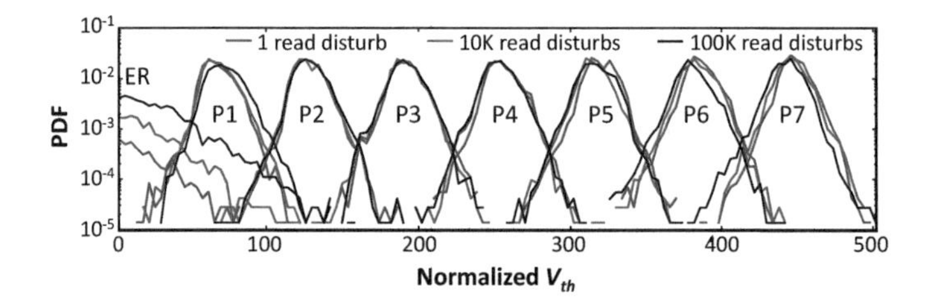

Fig. 9.19 Threshold voltage distribution for TLC NAND flash memory after 1, 10 and 100 K read disturb operations. Reproduced from [15]

Second, cells whose threshold voltages are closer to the point at which the voltage distributions of the ER and P1 states intersect are more vulnerable to read disturb errors. This is because process variation causes different cells to have different degrees of vulnerability to read disturb. We find that cells that are *prone* to read disturb end up at the right tail of the threshold voltage distribution of the ER state, as these cells' threshold voltages increase more rapidly, and that cells that are relatively *resistant* to read disturb end up at the left tail of the threshold voltage distribution of the P1 state, as their threshold voltages increase more slowly. We can exploit this divergent behavior of cells that end up at the left and right distribution tails to perform error recovery in the event of an uncorrectable error, as we discuss in Sect. 9.5.4.

Third, unlike with the other states, the threshold voltages of the cells at the left tail of the highest-voltage state (P7) in TLC NAND flash memory actually *decreases* as the read disturb count increases. This occurs for two reasons: (1) applying V_{pass} causes electrons to move from the floating gate to the control gate for a cell at high voltage (i.e., a cell containing a large number of electrons), thus *reducing* its threshold voltage [21, 289]; and (2) some retention time elapses while we sweep the voltages during our read disturb experiments, inducing trap-assisted tunneling (see Sect. 9.3.4) and leading to retention errors that decrease the voltage.

More findings on the nature of read disturb and the impact of read disturb on NAND flash memory errors and lifetime can be found in our prior work [21].

9.3.6 Large-Scale Studies on SSD Errors

The error characterization studies we have discussed so far examine the susceptibility of real NAND flash memory devices to specific error sources, by conducting controlled experiments on individual flash devices in controlled environments. To examine the *aggregate* effect of these error sources on flash devices that operate in the field, several recent studies have analyzed the reliability of SSDs deployed at a large scale (e.g., hundreds of thousands of SSDs) in production data centers [174, 199, 233]. Unlike the controlled low-level error characterization studies discussed in Sect. 9.3.1 through 9.3.5, these large-scale studies analyze the observed errors and error rates in an *uncontrolled* manner, i.e., based on real data center workloads operating at field conditions (as opposed to carefully controlling access patterns and operating conditions). As such, these large-scale studies can study flash memory behavior and reliability using only a black-box approach, where they are able to access only the registers used by the SSD to record select statistics. Because of this, their conclusions are usually correlational in nature, as opposed to identifying the underlying causes behind the observations. On the other hand, these studies incorporate the effects of a real system, including the system software stack and real workloads [174] and real operational conditions in data centers, on the flash memory devices, which is not present in the controlled small-scale studies.

These recent large-scale studies have made a number of observations across large sets of SSDs employed in the data centers of large internet companies:

Facebook [174], Google [233], and Microsoft [199]. We highlight six key observations from these studies about the *SSD failure rate*, which is the fraction of SSDs that have experienced at least one uncorrectable error.

First, the number of uncorrectable errors observed varies significantly for each SSD. Figure 9.20 shows the distribution of uncorrectable errors per SSD across a large set of SSDs used by Facebook. The distributions are grouped into six different *platforms* that are deployed in Facebook's data center.[2] For every platform, we observe that the top 10% of SSDs, when sorted by their uncorrectable error count, account for over 80% of the total uncorrectable errors observed across all SSDs for that platform. We find that the distribution of uncorrectable errors across all SSDs belonging to a platform follows a Weibull distribution, which we show using a solid black line in Fig. 9.20.

Second, the SSD failure rate does *not* increase monotonically with the P/E cycle count. Instead, we observe several *distinct* periods of reliability, as illustrated pictorially and abstractly in Fig. 9.21, which is based on data obtained from analyzing errors in SSDs used in Facebook's data centers [174]. The failure rate increases when the SSDs are relatively new (shown as the *early detection* period in Fig. 9.21), as the SSD controller identifies unreliable NAND flash cells during the initial read and write operations to the devices and removes them from the address space (see Sect. 9.1.3.9). As the SSDs are used more, they enter the *early failure* period, where failures are less likely to occur. When the SSDs approach the end of their lifetime (*useful life/wearout* in the figure), the failure rate increases again, as more cells become unreliable due to wearout. Figure 9.22 shows how the measured failure rate changes as more writes are performed to the SSDs (i.e., how real data collected

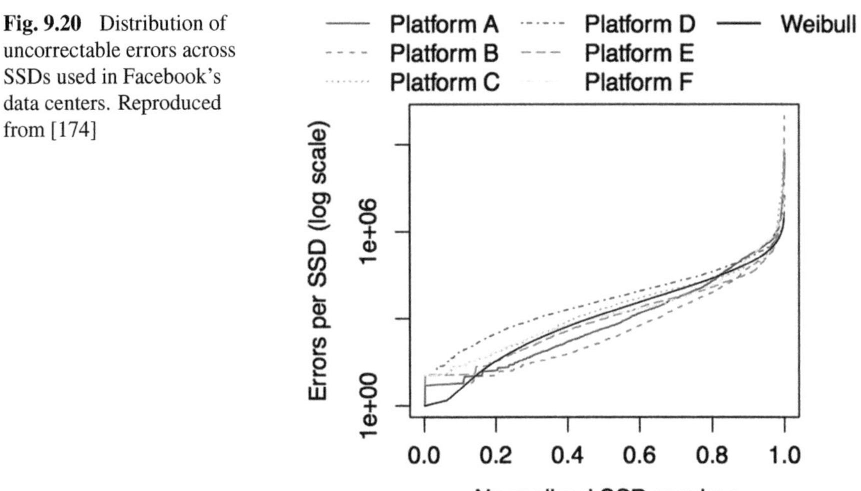

Fig. 9.20 Distribution of uncorrectable errors across SSDs used in Facebook's data centers. Reproduced from [174]

[2]Each platform has a different combination of SSDs, host controller interfaces, and workloads. The six platforms are described in detail in [174].

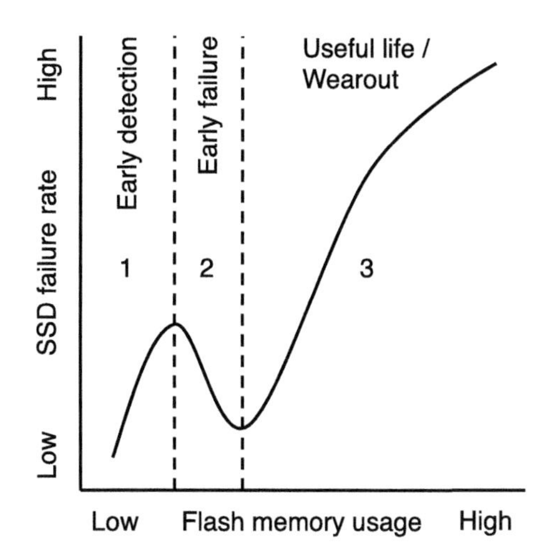

Fig. 9.21 Pictorial and abstract depiction of the pattern of SSD failure rates observed in real SSDs operating in a modern data center. An SSD fails at different rates during distinct periods throughout the SSD lifetime. Reproduced from [174]

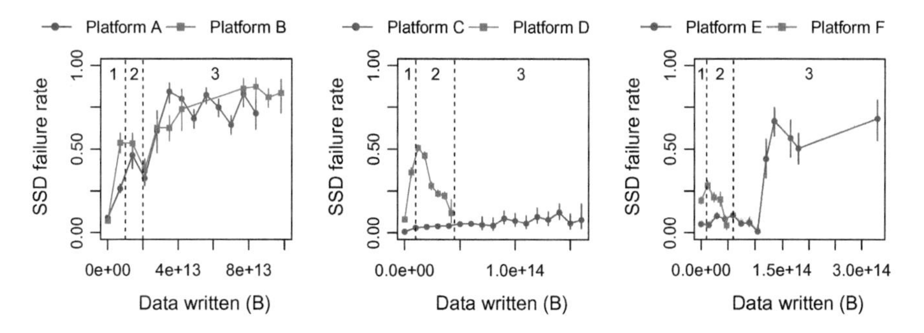

Fig. 9.22 SSD failure rate versus the amount of data written to the SSD. The three periods of failure rates, shown pictorially and abstractly in Fig. 9.21, are annotated on each graph: (1) early detection, (2) early failure, and (3) useful life/wearout. Reproduced from [174]

from Facebook's SSDs corresponds to the pictorial depiction in Fig. 9.21) for the same six platforms shown in Fig. 9.20. We observe that the failure rates in each platform exhibit the distinct periods that are illustrated in Fig. 9.21. For example, let us consider the SSDs in Platforms A and B, which have more data written to their cells than SSDs in other platforms. We observe from Fig. 9.22 that for SSDs in Platform A, there is an 81.7% increase from the failure rate during the early detection period to the failure rate during the wearout period [174].

Third, the raw bit error rate grows with the age of the device even if the P/E cycle count is held constant, indicating that mechanisms such as silicon aging likely contribute to the error rate [199].

Fourth, the observed failure rate of SSDs has been noted to be significantly higher than the failure rates specified by the manufacturers [233].

Fifth, higher operating temperatures can lead to higher failure rates, but modern SSDs employ throttling techniques that reduce the access rates to the underlying flash chips, which can greatly reduce the negative reliability impact of higher temperatures [174]. For example, Fig. 9.23 shows the SSD failure rate as the SSD operating temperature varies, for SSDs from the same six platforms shown in Fig. 9.20 [174]. We observe that at an operating temperature range of 30–40 °C, SSDs either (1) have similar failure rates across the different temperatures, or (2) experience slight increases in the failure rate as the temperature increases. As the temperature increases beyond 40 °C, the SSDs fall into three categories: (1) temperature-sensitive with increasing failure rate (Platforms A and B), (2) less temperature-sensitive (Platforms C and E), and (3) temperature-sensitive with decreasing failure rate (Platforms D and F). There are two factors that affect the temperature sensitivity of each platform: (1) some, but not all, of the platforms employ techniques to throttle SSD activity at high operating temperatures to reduce the failure rate (e.g., Platform D); and (2) the platform configuration (e.g., the number of SSDs in each machine, system airflow) can shorten or prolong the effects of higher operating temperatures.

Sixth, while SSD failure rates are higher than specified by the manufacturers, the overall occurrence of *uncorrectable* errors is lower than expected [174] because (1) effective bad block management policies (see Sect. 9.1.3.9) are implemented in SSD controllers; and (2) certain types of error sources, such as read disturb [174, 199] and incomplete erase operations [199], have yet to become a major source of uncorrectable errors at the system level.

9.4 Error Mitigation

Several different types of errors can occur in NAND flash memory, as we described in Sect. 9.3. As NAND flash memory continues to scale to smaller technology nodes, the magnitude of these errors has been increasing [168, 212, 294]. This, in turn,

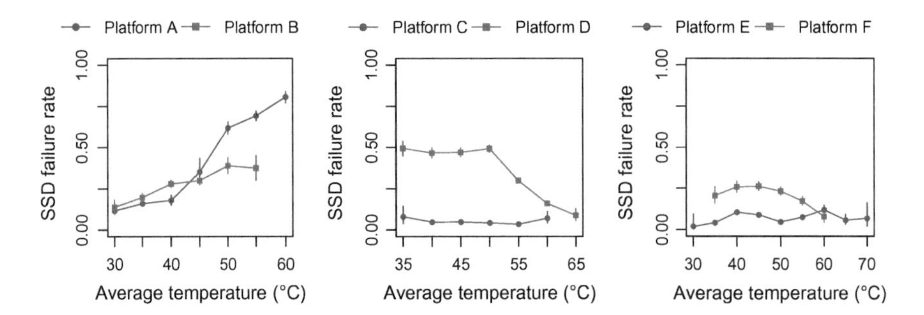

Fig. 9.23 SSD failure rate versus operating temperature. Reproduced from [174]

uses up the limited error correction capability of ECC more rapidly than in past flash memory generations and shortens the lifetime of modern SSDs. To overcome the decrease in lifetime, a number of error mitigation techniques have been designed. These techniques exploit intrinsic properties of the different types of errors to reduce the rate at which they lead to raw bit errors. In this section, we discuss how the flash controller mitigates each of the error types via various proposed error mitigation mechanisms. Table 9.3 shows the techniques we overview and which errors (from Sect. 9.3) they mitigate.

9.4.1 Shadow Program Sequencing

As discussed in Sect. 9.3.3, cell-to-cell program interference is a function of the distance between the cells of the wordline that is being programmed and the cells of the victim wordline. The impact of program interference is greatest on a victim wordline when either of the victim's immediately-adjacent wordlines is programmed (e.g.,

Table 9.3 List of different types of errors mitigated by various NAND flash error mitigation mechanisms

Mitigation mechanism	Error type				
	P/E cycling [19, 20, 162] (Sect. 9.3.1)	Program [17, 162, 212] (Sect. 9.3.2)	Cell-to-cell interference [19, 23, 31, 151] (Sect. 9.3.3)	Data retention [19, 22, 29, 30, 182] (Sect. 9.3.4)	Read disturb [19, 21, 75, 182] (Sect. 9.3.5)
Shadow program sequencing [17, 23] (Sect. 9.4.1)			X		
Neighbor-cell assisted error [31] (Sect. 9.4.2)			X		
Refresh [29, 30, 185, 205] (Sect. 9.4.3)				X	X
Read-retry [20, 70, 287] (Sect. 9.4.4)	X			X	X
Voltage optimization [21, 22, 106] (Sect. 9.4.5)	X			X	X
Hot data management [81, 82, 161] (Sect. 9.4.6)	X	X	X	X	X
Adaptive error mitigation [28, 44, 86, 272, 275] (Sect. 9.4.7)	X	X	X	X	X

if we program WL1 in Fig. 9.8, WL0 and WL2 experience the greatest amount of interference). Early MLC flash memories used one-shot programming, where both the LSB and MSB pages of a wordline are programmed at the same time. As flash memory scaled to smaller process technologies, one-shot programming resulted in much larger amounts of cell-to-cell program interference. As a result, manufacturers introduced two-step programming for MLC NAND flash (see Sect. 9.2.4), where the SSD controller writes values of the two pages within a wordline in two independent steps.

The SSD controller minimizes the interference that occurs during two-step programming by using *shadow program sequencing* [17, 23, 207] to determine the order that data is written to different pages in a block. If we program the LSB and MSB pages of the same wordline back to back, as shown in Fig. 9.24a, both programming steps induce interference on a *fully-programmed wordline* (i.e., a wordline where both the LSB and MSB pages are already written). For example, if the controller programs both pages of WL1 back to back, shown as bold page programming operations in Fig. 9.24a, the program operations induce a high amount of interference on WL0, which is fully programmed. The key idea of shadow program sequencing is to ensure that a fully-programmed wordline experiences interference minimally, i.e., *only* during MSB page programming (and *not* during LSB page programming). In shadow program sequencing, we assign a unique page number to each page within a block, as shown in Fig. 9.24b. The LSB page of wordline i is numbered page $2i - 1$, and the MSB page is numbered page $2i + 2$. The only exceptions to the numbering are the LSB page of wordline 0 (page 0) and the MSB page of the last wordline n (page $2n + 1$). Two-step programming writes to pages in *increasing* order of page number inside a block [17, 23, 207], such that a *fully-programmed* wordline experiences interference only from the MSB page programming of the wordline directly above it, shown as the bold page programming operation in Fig. 9.24b. With this programming order/sequence, the LSB page of the wordline above, and both pages of the wordline below, do *not* cause interference to fully-programmed data [17, 23, 207], as these two pages are programmed *before* programming the MSB page of the given wordline. Foggy-fine programming in TLC NAND flash (see Sect. 9.2.4) uses a similar ordering to reduce cell-to-cell program interference, as shown in Fig. 9.24c.

Shadow program sequencing is an effective solution to minimize cell-to-cell program interference on fully-programmed wordlines during two-step programming, and is employed in commercial SSDs today.

9.4.2 Neighbor-Cell Assisted Error Correction

The threshold voltage shift that occurs due to program interference is highly correlated with the values stored in the cells of the *immediately-adjacent wordlines*, as we discussed in Sect. 9.3.3. Due to this correlation, knowing the value programmed

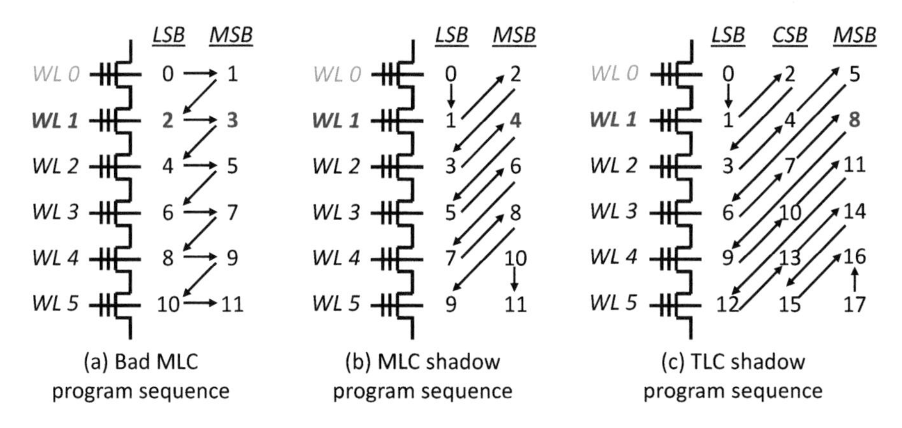

Fig. 9.24 Order in which the pages of each wordline (WL) are programmed using **a** a bad programming sequence, and using shadow sequencing for **b** MLC and **c** TLC NAND flash. The bold page programming operations for WL1 induce cell-to-cell program interference when WL0 is fully programmed. Reproduced from [15]

in the immediately-adjacent cell (i.e., a *neighbor cell*) makes it easier to correctly determine the value stored in the flash cell that is being read [31]. We describe a recently proposed error correction method that takes advantage of this observation, called *neighbor-cell-assisted error correction* (NAC). The key idea of NAC is to use the data values stored in the cells of the immediately-adjacent wordline to determine a better set of read reference voltages for the wordline that is being read. Doing so leads to a more accurate identification of the logical data value that is being read, as the data in the immediately-adjacent wordline was *partially responsible* for shifting the threshold voltage of the cells in the wordline that is being read when the immediately-adjacent wordline was programmed.

Figure 9.25 shows an operational example of NAC that is applied to eight bitlines (BL) of an MLC flash wordline. The SSD controller first reads a flash page from a wordline using the standard read reference voltages (step 1 in Fig. 9.25). The bit values read from the wordline are then buffered in the controller. If there are no errors uncorrectable by ECC, the read was successful, and nothing else is done. However, if there are errors that are *uncorrectable* by ECC, we assume that the threshold voltage distribution of the page shifted due to cell-to-cell program interference, triggering further correction. In this case, NAC reads the LSB and MSB pages of the wordline *immediately above* the requested page (i.e., the *adjacent* wordline that was programmed *after* the requested page) to classify the cells of the requested page (step 2). NAC then identifies the cells adjacent to (i.e., connected to the same bitline as) the ER cells (i.e., cells in the immediately above wordline that are in the

		BL0	BL1	BL2	BL3	BL4	BL5	BL6	BL7
	Originally-programmed value	11	00	01	10	11	00	01	00
	1. Read (using V_{opt}) with *errors*	01	00	00	00	11	10	00	01
N	2. Read **adjacent wordline**	**P2**	ER	**P2**	ER	P1	P3	P1	ER
A	3. *Correct cells* adjacent to ER	01	00	00	***10***	11	10	00	***00***
C	4. *Correct cells* adjacent to P1	01	00	00	10	11	10	***01***	00

Fig. 9.25 Overview of neighbor-cell-assisted error correction (NAC). Reproduced from [15]

ER state), such as the cells on BL1, BL3, and BL7 in Fig. 9.25. NAC rereads these cells using read reference voltages that *compensate for* the threshold voltage shift caused by programming the adjacent cell to the ER state (step 3). If ECC can correct the remaining errors, the controller returns the corrected page to the host. If ECC fails again, the process is repeated using a different set of read reference voltages for cells that are adjacent to the P1 cells (step 4). If ECC continues to fail, the process is repeated for cells that are adjacent to P2 and P3 cells (steps 5 and 6, respectively, which are not shown in the figure) until either ECC is able to correct the page or all possible adjacent values are exhausted.

NAC extends the lifetime of an SSD by reducing the number of errors that need to be corrected using the limited correction capability of ECC. With the use of experimental data collected from real MLC NAND flash memory chips, we show that NAC extends the NAND flash memory lifetime by 33% [31]. Our previous work [31] provides a detailed description of NAC, including a theoretical treatment of why it works and a practical implementation that minimizes the number of reads performed, even in the case when the neighboring wordline itself has errors.

9.4.3 Refresh Mechanisms

As we see in Fig. 9.12, during the time period after a flash page is programmed, retention (Sect. 9.3.4) and read disturb (Sect. 9.3.5) can cause an increasing number of raw bit errors to accumulate over time. This is particularly problematic for a page that is not updated frequently. Due to the limited error correction capability, the accumulation of these errors can potentially lead to data loss for a page with a *high retention age* (i.e., a page that has not been programmed for a long time). To avoid data loss, *refresh mechanisms* have been proposed, where the stored data is periodically read, corrected, and reprogrammed, in order to eliminate the retention and read disturb errors that have accumulated prior to this periodic read/correction/reprogramming (i.e., refresh). The concept of refresh in flash memory is thus conceptually similar to the refresh mechanisms found in DRAM [35, 104, 157, 158]. By performing refresh and limiting the number of retention and read disturb errors that can accumulate, the lifetime of the SSD increases significantly. In this section, we describe three types of refresh mechanisms used in modern SSDs: remapping-based refresh, in-place refresh, and read reclaim.

Remapping-Based Refresh. Flash cells must first be erased before they can be reprogrammed, due to the fact the programming a cell via ISPP can only increase the charge level of the cell but not reduce it (Sect. 9.2.4). The key idea of *remapping-based refresh* is to periodically read data from each valid flash block, correct any data errors, and *remap the data to a different physical location*, in order to prevent the data from accumulating too many retention errors [14, 29, 30, 185, 205]. During each refresh interval, a block with valid data that needs to be refreshed is selected. The valid data in the selected block is read out page by page and moved to the SSD controller. The ECC engine in the SSD controller corrects the errors in the read data, including retention errors that have accumulated since the last refresh. A new block is then selected from the free list (see Sect. 9.1.3.2), the error-free data is programmed to a page within the new block, and the logical address is remapped to point to the newly-programmed physical page. By reducing the accumulation of retention and read disturb errors, remapping-based refresh increases SSD lifetime by an average of 9x for a variety of disk workloads [29, 30].

Prior work proposes extensions to the basic remapping-based refresh approach. One work, *refresh SSDs*, proposes a refresh scheduling algorithm based on an earliest deadline first policy to guarantee that all data is refreshed in time [185]. The *quasi-nonvolatile SSD* proposes to use remapping-based refresh to choose between improving flash endurance and reducing the flash programming latency (by using larger ISPP step-pulses) [205]. In the quasi-nonvolatile SSD, refresh requests are deprioritized, scheduled at idle times, and can be interrupted after refreshing any page within a block, to minimize the delays that refresh can cause for the response time of pending workload requests to the SSD. A refresh operation can also be triggered proactively based on the data read latency observed for a page, which is indicative of how many errors the page has experienced [24]. Triggering refresh *proactively* based on the observed read latency (as opposed to doing so *periodically*) improves SSD latency and throughput [24]. Whenever the read latency for a page within a block exceeds a fixed threshold, the valid data in the block is refreshed, i.e., remapped to a new block [24].

In-place Refresh. A major drawback of remapping-based refresh is that it performs *additional writes* to the NAND flash memory, accelerating wearout. To reduce the wearout overhead of refresh, we propose *in-place refresh* [14, 29, 30]. As data sits unmodified in the SSD, data retention errors dominate [19, 30, 256], leading to charge loss and causing the threshold voltage distribution to shift to the left, as we showed in Sect. 9.3.4. The key idea of in-place refresh is to incrementally replenish the lost charge of each page *at its current location*, i.e., in place, without the need for remapping.

Figure 9.26 shows a high-level overview of in-place refresh for a wordline. The SSD controller first reads all of the pages in the wordline (❶ in Fig. 9.26). The

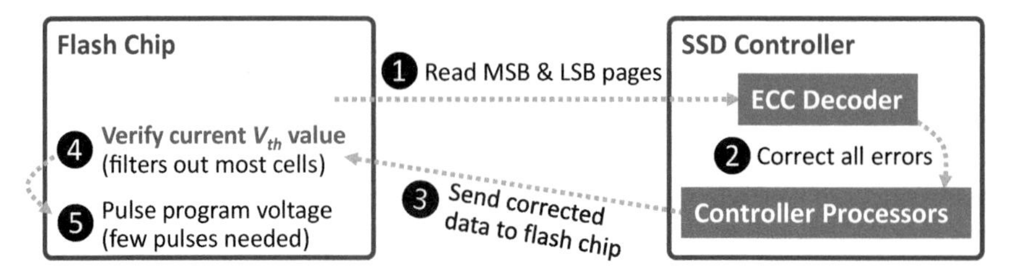

Fig. 9.26 Overview of in-place refresh mechanism for MLC NAND flash memory. Reproduced from [15]

controller invokes the ECC decoder to correct the errors within each page (❷), and sends the corrected data back to the flash chips (❸). In-place refresh then invokes a modified version of the ISPP mechanism (see Sect. 9.2.4), which we call *Verify-ISPP* (V-ISPP), to compensate for retention errors by restoring the charge that was lost. In V-ISPP, we first verify the voltage currently programmed in a flash cell (❹). If the current voltage of the cell is *lower* than the target threshold voltage of the state that the cell should be in, V-ISPP pulses the programming voltage in steps, gradually injecting charge into the cell until the cell returns to the target threshold voltage (❺). If the current voltage of the cell is *higher* than the target threshold voltage, V-ISPP inhibits the programming pulses to the cell.

When the controller invokes in-place refresh, it is unable to use shadow program sequencing (Sect. 9.4.1), as all of the pages within the wordline have already been programmed. However, unlike traditional ISPP, V-ISPP does not introduce a high amount of cell-to-cell program interference (Sect. 9.3.3) for two reasons. First, V-ISPP programs *only* those cells that have retention errors, which typically account for less than 1% of the total number of cells in a wordline selected for refresh [29]. Second, for the small number of cells that are selected to be refreshed, their threshold voltage is usually only slightly lower than the target threshold voltage, which means that only a few programming pulses need to be applied. As cell-to-cell interference is linearly correlated with the threshold voltage change to immediately-adjacent cells [23, 31], the small voltage change on these in-place refreshed cells leads to only a small interference effect.

One issue with in-place refresh is that it is unable to correct retention errors for cells in lower-voltage states. Retention errors cause the threshold voltage of a cell in a lower-voltage state to *increase* (e.g., see Sect. 9.3.4, ER and P1 states in Fig. 9.18), but V-ISPP *cannot* decrease the threshold voltage of a cell. To achieve a balance between the wearout overhead due to remapping-based refresh and errors that increase the threshold voltage due to in-place refresh, we propose *hybrid in-place refresh* [14, 29, 30]. The key idea is to use in-place refresh when the number of program errors (caused due to reprogramming) is within the correction capability of ECC, but to use remapping-based refresh if the number of program errors is too large to tolerate. To accomplish this, the controller tracks the number of *right-shift errors* (i.e., errors that move a cell to a higher-voltage state) [29, 30]. If the number of right-shift errors remains under a certain threshold, the controller performs in-place refresh; otherwise, it performs remapping-based refresh. Such a hybrid in-place refresh mechanism increases SSD lifetime by an average of 31x for a variety of disk workloads [29, 30].

Read Reclaim to Reduce Read Disturb Errors. We can also mitigate read disturb errors using an idea similar to remapping-based refresh, known as *read reclaim*. The key idea of read reclaim is to remap the data in a block to a new flash block, if the block has experienced a high number of reads [81, 82, 127]. To bound the number of read disturb errors, some flash vendors specify a maximum number of tolerable reads for a flash block, at which point read reclaim rewrites the data to a new block (just as is done for remapping- based refresh).

Adaptive Refresh and Read Reclaim Mechanisms. For the refresh and read reclaim mechanisms discussed above, the SSD controller can (1) invoke the mechanisms at fixed regular intervals; or (2) *adapt* the rate at which it invokes the mechanisms, based on various conditions that impact the rate at which data retention and read disturb errors occur. By adapting the mechanisms based on the current conditions of the SSD, the controller can reduce the overhead of performing refresh or read reclaim. The controller can adaptively adjust the rate that the mechanisms are invoked based on (1) the wearout (i.e., the current P/E cycle count) of the NAND flash memory [29, 30]; or (2) the temperature of the SSD [19, 22].

As we discuss in Sect. 9.3.4, for data with a given retention age, the number of retention errors grows as the P/E cycle count increases. Exploiting this P/E cycle dependent behavior of retention time, the SSD controller can perform refresh less frequently (e.g., once every year) when the P/E cycle count is low, and more frequently (e.g., once every week) when the P/E cycle count is high, as proposed and described in our prior works [29, 30]. Similarly, for data with a given read disturb count, as the P/E cycle count increases, the number of read disturb errors increases as well [21]. As a result, the SSD controller can perform read reclaim less frequently (i.e., it increases the maximum number of tolerable reads per block before read reclaim is triggered) when the P/E cycle count is low, and more frequently when the P/E cycle count is high.

Prior works demonstrate that for a given retention time, the number of data retention errors increases as the NAND flash memory's operating temperature increases [19, 22]. To compensate for the increased number of retention errors at high temperature, a state-of-the-art SSD controller adapts the rate at which it triggers refresh. The SSD contains sensors that monitor the current environmental temperature every few milliseconds [174, 269]. The controller then uses the Arrhenius equation [4, 185, 282] to estimate the rate at which retention errors accumulate at the current temperature of the SSD. Based on the error rate estimate, the controller decides if it needs to increase the rate at which it triggers refresh to ensure that the data is not lost.

By employing adaptive refresh and/or read reclaim mechanisms, the SSD controller can successfully reduce the mechanism overheads while effectively mitigating the larger number of data retention errors that occur under various conditions.

9.4.4 Read-Retry

In earlier generations of NAND flash memory, the read reference voltage values were fixed at design time [20, 182]. However, several types of errors cause the threshold voltage distribution to shift, as shown in Fig. 9.13. To compensate for threshold voltage distribution shifts, a mechanism called *read-retry* has been implemented in modern flash memories (typically those below 30 nm for planar flash [20, 70, 241, 287]).

The read-retry mechanism allows the read reference voltages to dynamically adjust to changes in distributions. During read-retry, the SSD controller first reads the data out of NAND flash memory with the default read reference voltage. It then sends the data for error correction. If ECC successfully corrects the errors in the data, the read operation succeeds. Otherwise, the SSD controller reads the memory again with a *different* read reference voltage. The controller repeats these steps until it either successfully reads the data using a certain set of read reference voltages or is unable to correctly read the data using all of the read reference voltages that are available to the mechanism.

While read-retry is widely implemented today, it can significantly increase the overall read operation latency due to the multiple read attempts it causes [22]. Mechanisms have been proposed to reduce the number of read-retry attempts while taking advantage of the effective capability of read-retry for reducing read errors, and read-retry has also been used to enable mitigation mechanisms for various other types of errors, as we describe in Sect. 9.4.5. As a result, read-retry is an essential mechanism in modern SSDs to mitigate read errors (i.e., errors that manifest themselves during a read operation).

9.4.5 Voltage Optimization

Many raw bit errors in NAND flash memory are affected by the various voltages used within the memory to enable reading of values. We give two examples. First, a suboptimal *read reference voltage* can lead to a large number of read errors (Sect. 9.3), especially after the threshold voltage distribution shifts. Second, as we saw in Sect. 9.3.5, the *pass-through voltage* can have a significant effect on the number of read disturb errors that occur. As a result, optimizing these voltages such that they minimize the total number of errors that are induced can greatly mitigate error counts. In this section, we discuss mechanisms that can discover and employ the optimal[3] read reference and pass-through voltages.

Optimizing Read Reference Voltages Using Disparity-Based Approximation and Sampling. As we discussed in Sect. 9.4.4, when the threshold voltage distribution

[3]Or, more precisely, near-optimal, if the read-retry steps are too coarse grained to find the optimal voltage.

shifts, it is important to move the read reference voltage to the point where the number of read errors is minimized. After the shift occurs and the threshold voltage distribution of each state widens, the distributions of different states may overlap with each other, causing many of the cells within the overlapping regions to be misread. The number of errors due to misread cells can be minimized by setting the read reference voltage to be exactly at the point where the distributions of two neighboring states intersect, which we call the *optimal read reference voltage* (V_{opt}) [22, 23, 31, 162, 206], illustrated in Fig. 9.27. Once the optimal read reference voltage is applied, the raw bit error rate is minimized, improving the reliability of the device.

One approach to finding V_{opt} is to adaptively learn and apply the optimal read reference voltage for each flash block through sampling [22, 45, 56, 280]. The key idea is to periodically (1) use *disparity* information (i.e., the ratio of 1s to 0s in the data) to attempt to find a read reference voltage for which the error rate is lower than the ECC correction capability; and to (2) use *sampling* to efficiently tune the read reference voltage to its optimal value to reduce the read operation latency. Prior characterization of real NAND flash memory [22, 206] found that the value of V_{opt} does *not* shift greatly over a short period of time (e.g., a day), and that all pages within a block experience *similar* amounts of threshold voltage shifts, as they have the same amount of wearout and are programmed around the same time [22, 206]. Therefore, we can invoke our V_{opt} learning mechanism periodically (e.g., daily) to efficiently tune the *initial read reference voltage* (i.e., the first read reference voltage used when the controller invokes the read-retry mechanism, described in Sect. 9.4.4) for each flash block, ensuring that the initial voltage used by read-retry stays close to V_{opt} even as the threshold voltage distribution shifts.

The SSD controller searches for V_{opt} by counting the number of errors that need to be corrected by ECC during a read. However, there may be times where the initial read reference voltage ($V_{initial}$) is set to a value at which the number of errors during a read exceeds the ECC correction capability, such as the raw bit error rate for $V_{initial}$ in Fig. 9.27 (right). When the ECC correction capability is exceeded, the SSD controller

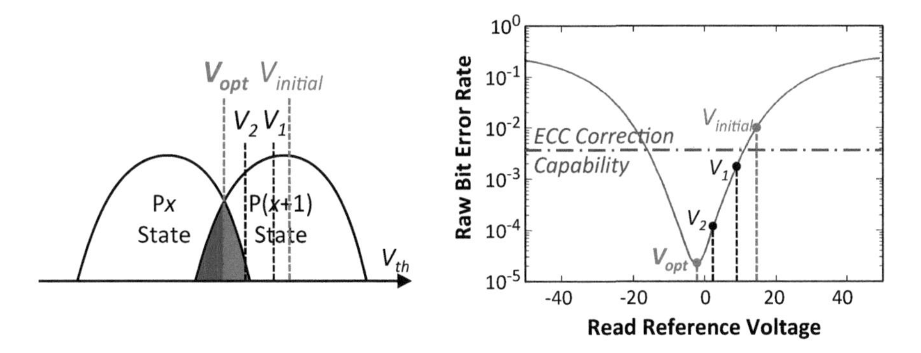

Fig. 9.27 Finding the optimal read reference voltage after the threshold voltage distributions overlap (left), and raw bit error rate as a function of the selected read reference voltage (right). Reproduced from [15]

is unable to count how many errors exist in the raw data. The SSD controller uses *disparity-based read reference voltage approximation* [45, 56, 280] for each flash block to try to bring $V_{initial}$ to a region where the number of errors does not exceed the ECC correction capability. Disparity-based read reference voltage approximation takes advantage of data scrambling. Recall from Sect. 9.1.3.6 that to minimize data value dependencies for the error rate, the SSD controller scrambles the data written to the SSD to probabilistically ensure that an equal number of 0s to 1s exist in the flash memory cells. The key idea of disparity-based read reference voltage approximation is to find the read reference voltages that result in approximately 50% of the cells reading out bit value 0, and the other 50% of the cells reading out bit value 1. To achieve this, the SSD controller employs a binary search algorithm, which tracks the ratio of 0s to 1s for each read reference voltage it tries. The binary search tests various read reference voltage values, using the ratios of previously tested voltages to narrow down the range where the read reference voltage can have an equal ratio of 0s to 1s. The binary search algorithm continues narrowing down the range until it finds a read reference voltage that satisfies the ratio.

The usage of the binary search algorithm depends on the type of NAND flash memory used within the SSD. For SLC NAND flash, the controller searches for only a single read reference voltage. For MLC NAND flash, there are three read reference voltages: the LSB is determined using V_b, and the MSB is determined using both V_a and V_c (see Sect. 9.2.3). Figure 9.28 illustrates the search procedure for MLC NAND flash. First, the controller uses binary search to find V_b, choosing a voltage that reads the LSB of 50% of the cells as data value 0 (step 1 in Fig. 9.28). For the MSB, the controller uses the discovered V_b value to help search for V_a and V_c. Due to scrambling, cells should be equally distributed across each of the four voltage states. The controller uses binary search to set V_a such that 25% of the cells are in the ER state, by ensuring that half of the cells *to the left of* V_b are read with an MSB of 0 (step 2). Likewise, the controller uses binary search to set V_c such that 25% of the cells are in the P3 state, by ensuring that half of the cells *to the right of* V_b are read with an MSB of 0 (step 3). This procedure is extended in a similar way to approximate the voltages for TLC NAND flash.

If disparity-based approximation finds a value for $V_{initial}$ where the number of errors during a read can be counted by the SSD controller, the controller invokes *sampling-based adaptive V_{opt} discovery* [22] to minimize the error count, and thus reduce the read latency. Sampling-based adaptive V_{opt} discovery learns and records V_{opt} for the *last-programmed page* in each block. We sample only the last-programmed page because it is the page with the lowest data retention age in the flash block. As retention errors cause the higher-voltage states to shift to the left (i.e., to lower voltages), the last-programmed page usually provides an *upper bound* of V_{opt} for the entire block.

During sampling-based adaptive V_{opt} discovery, the SSD controller first reads the last-programmed page using $V_{initial}$, and attempts to correct the errors in the raw data read from the page. Next, it records the number of raw bit errors as the current lowest error count N_{ERR}, and sets the applied read reference voltage (V_{ref}) as $V_{initial}$. Since V_{opt} typically decreases over retention age, the controller first attempts to lower

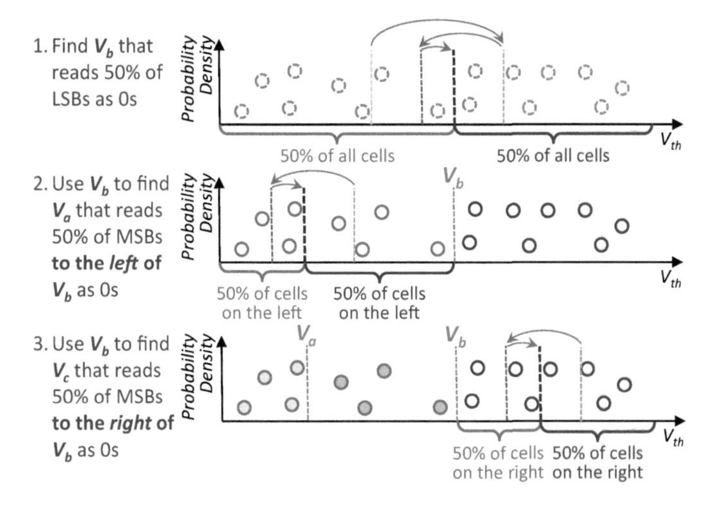

Fig. 9.28 Disparity-based read reference voltage approximation to find $V_{initial}$ for MLC NAND flash memory. Each circle represents a cell, where a dashed border indicates that the LSB is undetermined, a solid border indicates that the LSB is known, a hollow circle indicates that the MSB is unknown, and a filled circle indicates that the MSB is known. Reproduced from [15]

the read reference voltage for the last-programmed page, decreasing the voltage to $V_{ref} - \Delta V$ and reading the page. If the number of corrected errors in the new read is less than or equal to the old N_{ERR}, the controller updates N_{ERR} and V_{ref} with the new values. The controller continues to lower the read reference voltage until the number of corrected errors in the data is greater than the old N_{ERR} or the lowest possible read reference voltage is reached. Since the optimal threshold voltage might increase in rare cases, the controller also tests increasing the read reference voltage. It increases the voltage to $V_{ref} + \Delta V$ and reads the last-programmed page to see if N_{ERR} decreases. Again, it repeats increasing V_{ref} until the number of corrected errors in the data is greater than the old N_{ERR} or the highest possible read reference voltage is reached. The controller sets the initial read reference voltage of the block as the value of V_{ref} at the end of this process so that the next time an uncorrectable error occurs, read-retry starts at a $V_{initial}$ that is hopefully closer to the optimal read reference voltage (V_{opt}).

During the course of the day, as more retention errors (the dominant source of errors on already-programmed blocks) accumulate, the threshold voltage distribution shifts to the left (i.e., voltages decrease), and our initial read reference voltage (i.e., $V_{initial}$) is now an upper bound for the read-retry voltages. Therefore, whenever read-retry is invoked, the controller now needs to only decrease the read reference voltages (as opposed to traditional read-retry, which tries *both* lower and higher voltages [22]). Sampling-based adaptive V_{opt} discovery improves the *endurance* (i.e., the number of P/E cycles before the ECC correction capability is exceeded) of the NAND flash memory by 64% and reduces error correction latency by 10% [22], and is employed in some modern SSDs today.

Other Approaches to Optimizing Read Reference Voltages. One drawback of the sampling-based adaptive technique is that it requires time and storage overhead to find and record the per-block initial voltages. To avoid this, the SSD controller can employ an accurate *online threshold voltage distribution model* [14, 20, 162], which can efficiently track and predict the shift in the distribution over time. The model represents the threshold voltage distribution of each state as a probability density function (PDF), and the controller can use the model to calculate the intersection of the different PDFs. The controller uses the PDF in place of the threshold voltage sampling, determining V_{opt} by calculating the intersection of the distribution of each state in the model. The endurance improvement from our state-of-the-art model-based V_{opt} estimation technique [162] is within 2% of the improvement from an ideal V_{opt} identification mechanism [162]. An online threshold voltage distribution model can be used for a number of other purposes, such as estimating the future growth in the raw bit error rate and improving error correction [162].

Other prior work examines adapting read reference voltages based on P/E cycle count, retention age, or read disturb. In one such work, the controller periodically learns read reference voltages by testing three read reference voltages on six pages per block, which the work demonstrates to be sufficiently accurate [206]. Similarly, error correction using LDPC soft decoding (see Sect. 9.5.2.2) requires reading the same page using multiple sets of read reference voltages to provide fine-grained information on the probability of each cell representing a bit value 0 or a bit value 1. Another prior work optimizes the read reference voltages to increase the ECC correction capability without increasing the coding rate [266].

Optimizing Pass-Through Voltage to Reduce Read Disturb Errors. As we discussed in Sect. 9.3.5, the vulnerability of a cell to read disturb is directly correlated with the voltage difference ($V_{pass} - V_{th}$) through the cell oxide [21]. Traditionally, a single V_{pass} value is used *globally* for the entire flash memory, and the value of V_{pass} must be higher than *all* potential threshold voltages within the chip to ensure that unread cells along a bitline are turned on during a read operation (see Sect. 9.2.3). To reduce the impact of read disturb, we can tune V_{pass} to reduce the size of the voltage difference ($V_{pass} - V_{th}$). However, it is difficult to reduce V_{pass} *globally*, as any cell with a value of $V_{th} > V_{pass}$ introduces an error during a read operation (which we call a *pass-through error*).

We propose a mechanism that can dynamically lower V_{pass} while ensuring that it can correct any new pass-through errors introduced. The key idea of the mechanism is to lower V_{pass} only for those blocks where ECC has enough leftover error correction capability (see Sect. 9.1.3.7) to correct the newly introduced pass-through errors. When the retention age of the data within a block is low, we find that the raw bit error rate of the block is much lower than the rate for the block when the retention age is high, as the number of data retention and read disturb errors remains low at low retention age [21, 82]. As a result, a block with a low retention age has significant *unused* ECC correction capability, which we can use to correct the pass-through errors we introduce when we lower V_{pass}, as shown in Fig. 9.29. Thus, when a block has a low retention age, the controller lowers V_{pass} aggressively, making it

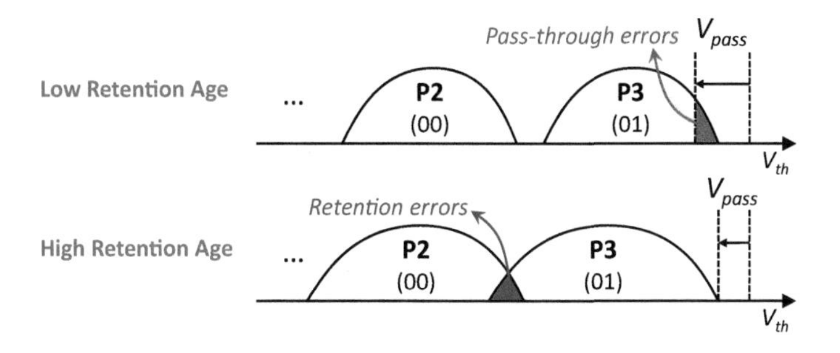

Fig. 9.29 Dynamic pass-through voltage tuning at different retention ages. Reproduced from [15]

much less likely for read disturbs to induce an uncorrectable error. When a block has a high retention age, the controller also lowers V_{pass}, but does not reduce the voltage aggressively, since the limited ECC correction capability now needs to correct retention errors, and might not have enough unused correction capability to correct many new pass-through errors. By reducing V_{pass} aggressively when a block has a low retention age, we can extend the time before the ECC correction capability is exhausted, improving the flash lifetime.

Our read disturb mitigation mechanism [21] learns the minimum pass-through voltage for each block, such that all data within the block can be read correctly with ECC. Our learning mechanism works online and is triggered periodically (e.g., daily). The mechanism is implemented in the controller, and has two components. It first finds the size of the ECC margin M (i.e., the unused correction capability) that can be exploited to tolerate additional read errors for each block. Once it knows the available margin M, our mechanism calibrates V_{pass} on a per-block basis to find the lowest value of V_{pass} that introduces no more than M additional raw errors (i.e., there are no more than M cells where $V_{th} > V_{pass}$). Our findings on MLC NAND flash memory show that the mechanism can improve flash endurance by an average of 21% for a variety of disk workloads [21].

Programming and Erase Voltages. Prior work also examines tuning the programming and erase voltages to extend flash endurance [106]. By decreasing the two voltages when the P/E cycle count is low, the accumulated wearout for each program or erase operation is reduced, which, in turn, increases the overall flash endurance. Decreasing the programming voltage, however, comes at the cost of increasing the time required to perform ISPP, which, in turn, increases the overall SSD write latency [106].

9.4.6 Hot Data Management

The data stored in different locations of an SSD can be accessed by the host at different rates. For example, we find that across a wide range of disk workloads, almost 100% of the write operations target less than 1% of the pages within an SSD [161], as shown in Fig. 9.30. These pages exhibit high temporal write locality, and are called *write-hot* pages. Likewise, pages with a high amount of temporal read locality (i.e., pages that are accessed by a large fraction of the read operations) are called *read-hot* pages. A number of issues can arise when an SSD does not distinguish between write-hot pages and *write-cold* pages (i.e., pages with low temporal write locality), or between read-hot pages and *read-cold* pages (i.e., pages with low temporal read locality). For example, if write-hot pages and write-cold pages are stored within the same block, refresh mechanisms (which operate at the block level; see Sect. 9.4.3) *cannot* avoid refreshes to pages that were overwritten recently. This increases not only the energy consumption but also the write amplification due to remapping-based refresh [161]. Likewise, if read-hot and read-cold pages are stored within the same block, read-cold pages are unnecessarily exposed to a high number of read disturb errors [81, 82]. *Hot data management* refers to a set of mechanisms that can identify and exploit write-hot or read-hot pages in the SSD. The key idea common to such mechanisms is to apply special SSD management policies by placing hot pages and cold pages into *separate* flash blocks.

A state-of-the-art hot data management mechanism is *write-hotness aware refresh management* (WARM) [161], which efficiently identifies write-hot pages and uses this information to carefully place pages within blocks. WARM aims to ensure that every block in the NAND flash memory contains either *only* write-hot pages or *only* write-cold pages. A small pool of blocks in the SSD are designated to exclusively store the small amount of write-hot data (as shown in Fig. 9.30). This block-level

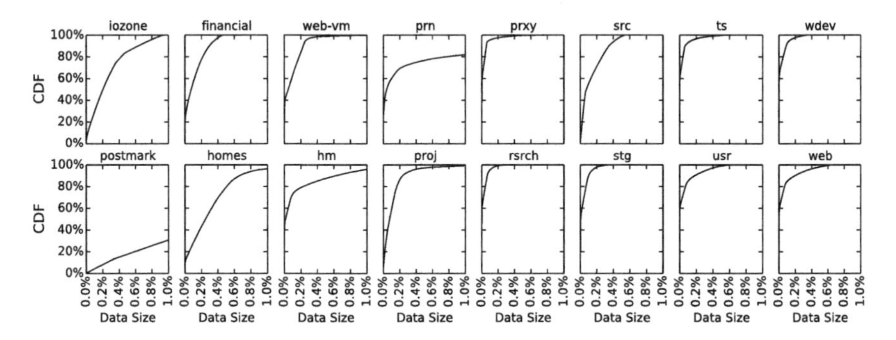

Fig. 9.30 Cumulative distribution function of the fraction of writes performed by a workload to NAND flash memory pages, for 16 evaluated workloads. For every workload except *postmark*, over 95% of all writes performed by the workload are destined for less than 1.0% of the workload's pages. Total data footprint of each workload is 217.6 GB, i.e., 1.0% on the x-axis represents 2.176 GB of data. Reproduced from [161]

segregation between write-hot pages and write-cold pages allows WARM to apply separate specialized management policies based on the write-hotness of the pages in each block.

Two examples of policies for write-hot blocks in WARM are the write-hotness-aware refresh policy (see Sect. 9.4.3 for baseline refresh policies) and the write-hotness-aware garbage collection algorithm (see Sect. 9.1.3.3). In write-hotness-aware refresh, since write-hot data is overwritten more frequently than the refresh interval, the SSD controller skips refresh operations to the write-hot blocks. As the retention time for write-hot data never exceeds the refresh interval, performing refresh to this data does *not* reduce the error rate. By skipping refresh for write-hot data, WARM reduces the total number of writes performed on the SSD, which in turn increases the SSD lifetime, without introducing uncorrectable errors. In write-hotness-aware garbage collection, the SSD controller performs *oldest-block-first* garbage collection. WARM sizes the pool of write-hot blocks such that when a write-hot block becomes the oldest block in the pool of write-hot blocks, all of the data that was in the block is likely to already have been overwritten. As a result, all of the pages within the oldest write-hot block is likely to be invalid, and the block can be erased without the need to migrate any remaining valid pages to a new block. By always selecting the oldest block in the pool of write-hot blocks for garbage collection, the write-hotness-aware garbage collection algorithm (1) does not spend time searching for a block to select (as traditional garbage collection algorithms do), and (2) rarely needs to migrate pages from the selected block. Both of these lead to a reduction in the performance overhead of garbage collection.

WARM continues to use the traditional controller policies (i.e., the policies described in Sect. 9.1.3) and refresh mechanisms for the write-cold blocks. WARM reduces fragmentation within write-cold blocks (i.e., each write-cold block is likely to have few, if any, invalid pages), because each page within the block does *not* get updated frequently by the application. Due to the write-hotness-aware policies and reduced fragmentation, WARM reduces write amplification significantly, which translates to an average lifetime improvement of 21% over an SSD that employs a state-of-the-art refresh mechanism [29] (see *Adaptive Refresh and Read Reclaim Mechanisms* in Sect. 9.4.3), across a wide variety of disk workloads [161].

Another work [265] proposes to reuse the correctly functioning flash pages within bad blocks (see Sect. 9.1.3.9) to store write-cold data. This technique increases the total number of usable blocks available for overprovisioning, and extends flash lifetime by delaying the point at which each flash chip reaches the upper limit of bad blocks it can tolerate.

RedFTL identifies and replicates read-hot pages across multiple flash blocks, allowing the controller to evenly distribute read requests to these pages across the replicas [81]. Other works reduce the number of read reclaims (see Sect. 9.4.3) that need to be performed by mapping read-hot data to particular flash blocks and lowering the maximum possible threshold voltage for such blocks [26, 82]. By lowering the maximum possible threshold voltage for these blocks, the SSD controller can use a lower V_{pass} value (see Sect. 9.4.5) on the blocks without introducing any additional errors during a read operation. To lower the maximum threshold voltage in these blocks, the width of the voltage window for each voltage state is decreased, and each voltage window shifts to the left [26, 82]. Another work applies stronger ECC encodings to *only* read-hot blocks based on the total read count of the block, in order to increase SSD endurance without significantly reducing the amount of overprovisioning [25] (see Sect. 9.1.4 for a discussion on the tradeoff between ECC strength and overprovisioning).

9.4.7 Adaptive Error Mitigation Mechanisms

Due to the many different factors that contribute to raw bit errors, error rates in NAND flash memory can be highly variable. Adaptive error mitigation mechanisms are capable of adapting error tolerance capability to the error rate. They provide stronger error tolerance capability when the error rate is higher, improving flash lifetime significantly. When the error rate is low, adaptive error mitigation techniques reduce error tolerance capability to lower the cost of the error mitigation techniques. In this section, we examine two types of adaptive techniques: (1) multi-rate ECC and (2) dynamic cell levels.

Multi-rate ECC. Some works propose to employ multiple ECC algorithms in the SSD controller [28, 44, 86, 95, 275]. Recall from Sect. 9.1.4 that there is a tradeoff between ECC strength (i.e., the coding rate; see Sect. 9.1.3.7) and overprovisioning, as a codeword (which contains a data chunk *and* its corresponding ECC information) uses more bits when stronger ECC is employed. The key idea of multi-rate ECC is to employ a weaker codeword (i.e., one that uses fewer bits for ECC) when the SSD is relatively new and has a smaller number of raw bit errors, and to use the saved SSD space to provide additional overprovisioning, as shown in Fig. 9.31.

Let us assume that the controller contains a configurable ECC engine that can support n different types of ECC codewords, which we call ECC_i. Figure 9.31 shows an example of multi-rate ECC that uses four ECC engines, where ECC_1 provides the weakest protection but has the smallest codeword, while ECC_4 provides the strongest protection with the largest codeword. We need to ensure that the NAND flash memory has enough space to fit the largest codewords, e.g., those for ECC_4 in Fig. 9.31.

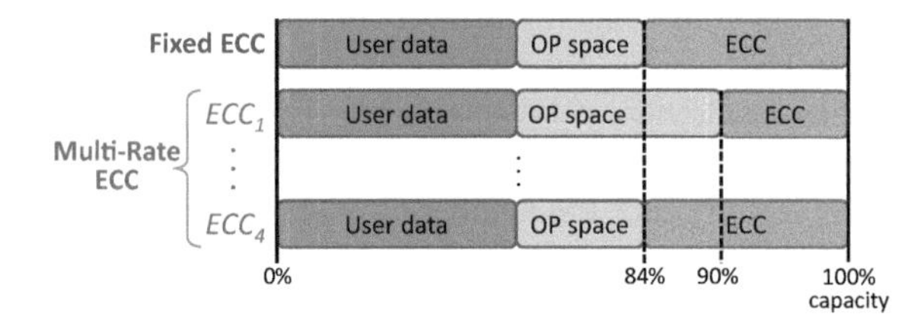

Fig. 9.31 Comparison of space used for user data, overprovisioning, and ECC between a fixed ECC and a multi-rate ECC mechanism. Reproduced from [15]

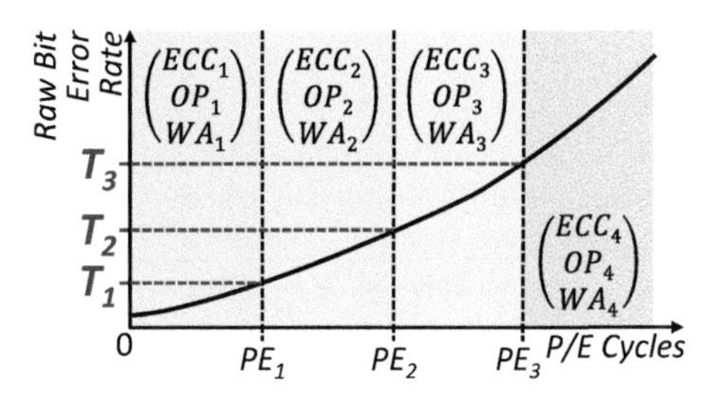

Fig. 9.32 Illustration of how multi-rate ECC switches to different ECC codewords (i.e., ECC_i) as the RBER grows. OP_i is the overprovisioning factor used for engine ECC_i, and WA_i is the resulting write amplification value. Reproduced from [15]

Initially, when the raw bit error rate (RBER) is low, the controller employs ECC_1, as shown in Fig. 9.32. The smaller codeword size for ECC_1 provides additional space for overprovisioning, as shown in Fig. 9.31, and thus reduces the effects of write amplification. Multi-rate ECC works on an interval-by-interval basis. Every interval (in this case, a predefined number of P/E cycles), the controller measures the RBER. When the RBER exceeds the threshold set for transitioning from a weaker ECC to a stronger ECC, the controller switches to the stronger ECC. For example, when the SSD exceeds the first RBER threshold for switching (T_1 in Fig. 9.32), the controller starts switching from ECC_1 to ECC_2. When switching between ECC engines, the controller uses the ECC_1 engine to decode data the next time the data is read out, and stores a new codeword using the ECC_2 engine. This process is repeated during the lifetime of flash memory for each stronger engine ECC_i, where each engine has a corresponding threshold that triggers switching [28, 44, 86], as shown in Fig. 9.32.

Multi-rate ECC allows the same maximum P/E cycle count for each block as if ECC_n was used throughout the lifetime of the SSD, but reduces write amplification and improves performance during the periods where the lower strength engines are employed, by providing additional overprovisioning (see Sect. 9.1.4) during those times. As the lower-strength engines use smaller codewords (e.g., ECC_1 vs. ECC_4 in Fig. 9.31), the resulting free space can instead be employed to further increase the amount of overprovisioning within the NAND flash memory, which in turn increases the total lifetime of the SSD. We compute the lifetime improvement by modifying (9.4) (Sect. 9.1.4) to account for each engine, as follows:

$$\text{Lifetime} = \sum_{i=1}^{n} \frac{\text{PEC}_i \times (1 + \text{OP}_i)}{365 \times \text{DWPD} \times \text{WA}_i \times R_{compress}} \tag{9.9}$$

In (9.9), WA_i and OP_i are the write amplification and overprovisioning factor for ECC_i, and PEC_i is the number of P/E cycles that ECC_i is used for. Manufacturers can set parameters to maximize SSD lifetime in (9.9), by optimizing the values of WA_i and OP_i.

Figure 9.33 shows the lifetime improvements for a four-engine multi-rate ECC, with the coding rates for the four ECC engines (ECC_1–ECC_4) set to 0.90, 0.88, 0.86,

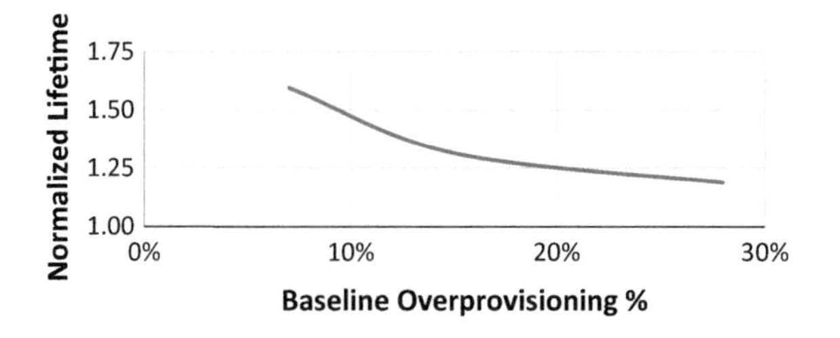

Fig. 9.33 Lifetime improvements of using multi-rate ECC over using a fixed ECC coding rate. Reproduced from [15]

and 0.84 (recall that a *lower* coding rate provides stronger protection; see Sect. 9.1.4), over a fixed ECC engine that employs a coding rate of 0.84. We see that the lifetime improvements of using multi-rate ECC are: (1) significant, with a 31.2% increase if the baseline NAND flash memory has 15% overprovisioning; and (2) greater when the SSD initially has a smaller amount of overprovisioning.

Dynamic Cell Levels. A major reason that errors occur in NAND flash memory is because the threshold voltage distribution of each state overlaps more with those of neighboring states as the distributions widen over time. Distribution overlaps are a greater problem when more states are encoded within the same voltage range. Hence, TLC flash has a much lower endurance than MLC, and MLC has a much lower endurance than SLC (assuming the same process technology node). If we can increase the margins between the states' threshold voltage distributions, the amount of overlap can be reduced significantly, which in turn reduces the number of errors.

Prior work proposes to increase margins by *dynamically* reducing the number of bits stored within a cell, e.g., by going from three bits that encode eight states (TLC) to two bits that encode four states (equivalent to MLC), or to one bit that encodes two states (equivalent to SLC) [26, 272]. Recall that TLC uses the ER state and states P1–P7, which are spaced out approximately equally. When we *downgrade* a flash block (i.e., reduce the number of states its cells can represent) from eight states to four, the cells in the block now employ only the ER state and states P3, P5, and P7. As we can see from Fig. 9.34, this provides large margins between states P3, P5, and P7, and provides an even larger margin between ER and P3. The SSD controller maintains a list of all of the blocks that have been downgraded. For each read operation, the SSD controller checks if the target block is in the downgraded block list, and uses this information to interpret the data that it reads out from the wordline of the block.

A cell can be downgraded to reduce various types of errors (e.g., wearout, read disturb). To reduce wearout, a cell is downgraded when it has high wearout. To reduce read disturb, a cell can be downgraded if it stores *read-hot* data (i.e., the most frequently read data in the SSD). By using fewer states for a block that holds read-hot data, we can reduce the impact of read disturb because it becomes harder for the read disturb mechanism to affect the distributions enough for them to overlap. As an optimization, the SSD controller can employ various hot-cold data partitioning mechanisms (e.g., [25, 26, 81, 161]) to keep read-hot data in specially designated

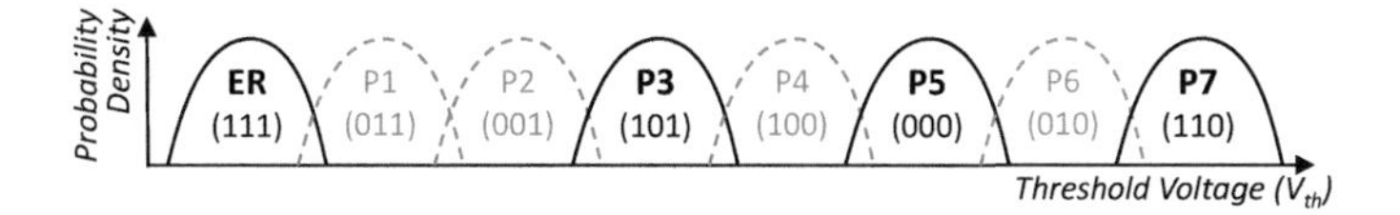

Fig. 9.34 States used when a TLC cell (with 8 states) is downgraded to an MLC cell (with 4 states). Reproduced from [15]

blocks [25, 26, 81, 82], allowing the controller to reduce the size of the downgraded block list and isolate the impact of read disturb from *read-cold* (i.e., infrequently read) data.

Another approach to dynamically increasing the distribution margins is to perform program and erase operations more slowly when the SSD write request throughput is low [26, 106]. Slower program/erase operations allow the final voltage of a cell to be programmed more precisely, and reduce the amount of oxide degradation that occurs during programming. As a result, the distribution of each state is initially much narrower, and subsequent widening of the distributions results in much lower overlap for a given P/E cycle count. This technique improves the SSD lifetime by an average of 61.2% for a variety of disk workloads [106]. Unfortunately, the slower program/erase operations come at the cost of higher SSD latency, and are thus not applied during periods of high write traffic. One way to mitigate the impact of the higher write latency is to perform slower program/erase operations only during garbage collection, which ensures that the higher latency occurs only when the SSD is idle [26]. As a result, read and write requests from the host do not experience any additional delays.

9.5 Error Correction and Data Recovery Techniques

Now that we have described a variety of error mitigation mechanisms that can target various types of error sources, we turn our attention to the error correction flow that is employed in modern SSDs as well as *data recovery techniques* that can be employed when the error correction flow fails to produce correct data. In this section, we briefly overview the major error correction steps an SSD performs when reading data. We first discuss two ECC encodings that are typically used by modern SSDs: Bose–Chaudhuri–Hocquenghem (BCH) codes [10, 92, 153, 243] and low-density parity-check (LDPC) codes [72, 73, 167, 243] (Sect. 9.5.1). Next, we go through example error correction flows for an SSD that uses either BCH codes or LDPC codes (Sect. 9.5.2). Then, we compare the error correction strength (i.e., the number of errors that ECC can correct) when we employ BCH codes or LDPC codes in an SSD (Sect. 9.5.3). Finally, we discuss techniques that can rescue data from an SSD when the BCH/LDPC decoding fails to correct all errors (Sect. 9.5.4).

9.5.1 Error-Correcting Codes Used in SSDs

Modern SSDs typically employ one of two types of ECC. Bose–Chaudhuri–Hocquenghem (BCH) codes allow for the correction of multiple bit errors [10, 92, 153, 243], and are used to correct the errors observed during a *single* read from the NAND flash memory [153]. Low-density parity-check (LDPC) codes employ information accumulated over *multiple* read operations to determine the likelihood of each cell containing a bit value 1 or a bit value 0 [72, 73, 167, 243], providing stronger protection at the cost of greater decoding latency and storage overhead [266, 298]. Next, we describe the basics of BCH and LDPC codes.

9.5.1.1 Bose–Chaudhuri–Hocquenghem (BCH) Codes

BCH codes [10, 92, 153, 243] have been widely used in modern SSDs during the past decade due to their ability to detect and correct multi-bit errors while keeping the latency and hardware cost of encoding and decoding low [42, 153, 170, 179]. For SSDs, BCH codes are designed to be *systematic*, which means that the original data message is embedded *verbatim* within the codeword. Within an n-bit codeword (see Sect. 9.1.3.7), error-correcting codes use the first k bits of the codeword, called *data bits*, to hold the data message bits, and the remaining $(n - k)$ bits, called *check bits*, to hold error correction information that protects the data bits. BCH codes are designed to *guarantee* that they correct up to a certain number of raw bit errors (e.g., t error bits) within each codeword, which depends on the values chosen for n and k. A stronger error correction strength (i.e., a larger t) requires more redundant check bits (i.e., $(n - k)$) or a longer codeword length (i.e., n).

A BCH code [10, 92, 153, 243] is a linear block code that consists of check bits generated by an algorithm. The codeword generation algorithm ensures that the check bits are selected such that the check bits can be used during a parity check to detect and correct up to t bit errors in the codeword. A BCH code is defined by (1) a generator matrix G, which informs the generation algorithm of how to generate each check bit using the data bits; and (2) a parity check matrix H, which can be applied to the codeword to detect if any errors exist. In order for a BCH code to guarantee that it can correct t errors within each codeword, the minimum separation d (i.e., the Hamming distance) between valid codewords must be at least $d = 2t + 1$ [243].

BCH Encoding. The codeword generation algorithm encodes a k-bit data message m into an n-bit BCH codeword c, by computing the dot product of m and the generator matrix G (i.e., $c = m \cdot G$). G is defined within a finite Galois field $GF(2^d) = \{0, \alpha^0, \alpha^1, \ldots, \alpha^{2^d-1}\}$, where α is a *primitive element* of the field and d is a positive integer [64]. An SSD manufacturer constructs G from a set of polynomials $g_1(x), g_2(x), \ldots g_{2t}(x)$, where $g_i(\alpha^i) = 0$. Each polynomial generates a *parity bit*, which is used during decoding to determine if any errors were introduced. The i-th row of G encodes the i-th polynomial $g_i(x)$. When decoding, the codeword c can be viewed as a polynomial $c(x)$. Since $c(x)$ is generated by $g_i(x)$ which has a root

α^i, α^i should also be a root of $c(x)$. The parity check matrix H is constructed such that cH^t calculates $c(\alpha_i)$. Thus, the element in the i-th row and j-th column of H is $H_{ij} = \alpha^{(j-1)(i+1)}$. This allows the decoder to use H to quickly determine if any of the parity bits do not match, which indicates that there are errors within the codeword. BCH codes in SSDs are typically designed to be *systematic*, which guarantees that a verbatim copy of the data message is embedded within the codeword. To form a systematic BCH code, the generator matrix and the parity check matrix are transformed such that they contain the identity matrix.

BCH Decoding. When the SSD controller is servicing a read request, it must extract the data bits (i.e., the k-bit data message m) from the BCH codeword that is stored in the NAND flash memory chips. Once the controller retrieves the codeword, which we call r, from NAND flash memory, it sends r to a BCH decoder. The decoder performs five steps, as illustrated in Fig. 9.35, which correct the retrieved codeword r to obtain the originally-written codeword c, and then extract the data message m from c. In Step 1, the decoder uses *syndrome calculation* to detect if any errors exist within the retrieved codeword r. If no errors are detected, the decoder uses the retrieved codeword as the original codeword, c, and skips to Step 5. Otherwise, the decoder continues on to correct the errors and recover c. In Step 2, the decoder uses the syndromes from Step 1 to construct an *error location polynomial*, which encodes the locations of each detected bit error within r. In Step 3, the decoder extracts the specific location of each detected bit error from the error location polynomial. In Step 4, the decoder corrects each detected bit error in the retrieved codeword r to recover the original codeword c. In Step 5, the decoder extracts the data message from the original codeword c. We describe the algorithms most commonly used by BCH decoders in SSDs [48, 153, 160] for each step in detail below.

Step 1—Syndrome Calculation: To determine whether the retrieved codeword r contains any errors, the decoder computes the *syndrome vector*, S, which indicates how many of the parity check polynomials no longer match with the parity bits originally computed during encoding. The i-th syndrome, S_i, is set to one if parity bit i does *not* match its corresponding polynomial, and to zero otherwise. To calculate

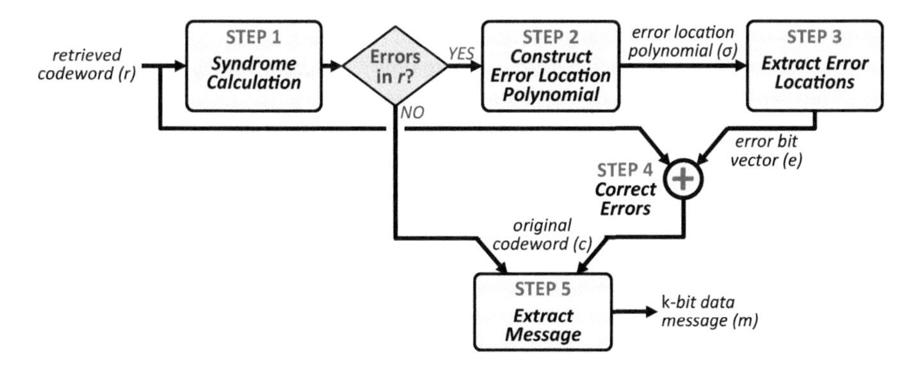

Fig. 9.35 BCH decoding steps

S, the decoder calculates the dot product of r and the parity check matrix H (i.e., $S = r \cdot H$). If every syndrome in S is set to 0, the decoder does not detect any errors within the codeword, and skips to Step 5. Otherwise, the decoder proceeds to Step 2.

Step 2—Constructing the Error Location Polynomial: A state-of-the-art BCH decoder uses the Berlekamp–Massey algorithm [8, 42, 171, 230] to construct an error location polynomial, $\sigma(x)$, whose roots encode the error locations of the codeword:

$$\sigma(x) = 1 + \sigma_1 \cdot x + \sigma_2 \cdot x^2 + \cdots + \sigma_b \cdot x^b \tag{9.10}$$

In (9.10), b is the number of raw bit errors in the codeword.

The polynomial is constructed using an iterative process. Since b is not known initially, the algorithm initially assumes that $b = 0$ (i.e., $\sigma(x) = 1$). Then, it updates $\sigma(x)$ by adding a *correction term* to the equation in each iteration, until $\sigma(x)$ successfully encodes all of the errors that were detected during syndrome calculation. In each iteration, a new correction term is calculated using both the syndromes from Step 1 and the $\sigma(x)$ equations from prior iterations of the algorithm, as long as these prior values of $\sigma(x)$ satisfy certain conditions. This algorithm successfully finds $\sigma(x)$ after $n = (t + b)/2$ iterations, where t is the maximum number of bit errors correctable by the BCH code [64].

Note that (1) the highest order of the polynomial, b, is directly correlated with the number of errors in the codeword; (2) the number of iterations, n, is also proportional to the number of errors; (3) each iteration is compute-intensive, as it involves several multiply and add operations; and (4) this algorithm cannot be parallelized across iterations, as the computation in each iteration is dependent on the previous ones.

Step 3—Extracting Bit Error Locations from the Error Polynomial: A state-of-the-art decoder applies the Chien search [46, 243] on the error location polynomial to find the location of *all* raw bit errors that have been detected during Step 1 in the retrieved codeword r. Each bit error location is encoded with a known function f [230]. The error polynomial from Step 2 is constructed such that if the i-th bit of the codeword has an error, the error location polynomial $\sigma(f(i)) = 0$; otherwise, if the i-th bit does *not* have an error, $\sigma(f(i)) \neq 0$. The Chien search simply uses trial-and-error (i.e., tests if $\sigma(f(i))$ is zero), testing each bit in the codeword starting at bit 0. As the decoder needs to correct only the first k bits of the codeword that contain the data message m, the Chien search needs to evaluate only k different values of $\sigma(f(i))$. The algorithm builds a bit vector e, which is the same length as the retrieved codeword r, where the i-th bit of e is set to one if bit i of r contains a bit error, and is set to zero if bit i of r does *not* contain an error, or if $i \geq k$ (since there is no need to correct the parity bits).

Note that (1) the calculation of $\sigma(f(i))$ is compute-intensive, but can be parallelized because the calculation of each bit i is independent of the other bits, and (2) the complexity of Step 3 is linearly correlated with the number of detected errors in the codeword.

Step 4—Correcting the Bit Errors: The decoder corrects each detected bit error location by flipping the bit at that location in the retrieved codeword r. This simply involves XORing r with the error vector e created in Step 3. After the errors are corrected, the decoder now has the estimated value of the originally-written codeword c (i.e., $c = r \oplus e$). The decoded version of c is only an *estimate* of the original codeword, since if r contains more bit errors than the maximum number of errors (t) that the BCH can correct, there may be some uncorrectable errors that were *not* detected during syndrome calculation (Step 1). In such cases, the decoder cannot *guarantee* that it has determined the actual original codeword. In a modern SSD, the bit error rate of a codeword *after* BCH correction is expected to be less than 10^{-15} [105].

Step 5—Extracting the Message from the Codeword: As we discuss above, during BCH codeword encoding, the generator matrix G contains the identity matrix, to ensure that the k-bit message m is embedded verbatim into the codeword c. Therefore, the decoder recovers m by simply truncating the last $(n-k)$ bits from the n-bit codeword c.

BCH Decoder Latency Analysis. We can model the latency of the state-of-the-art BCH decoder (T^{dec}_{BCH}) that we described above as:

$$T^{dec}_{BCH} = T_{Syndrome} + N \cdot T_{Berlekamp} + \frac{k}{p} \cdot T_{Chien} \tag{9.11}$$

In (9.11), $T_{Syndrome}$ is the latency for calculating the syndrome, which is determined by the size of the parity check matrix H; $T_{Berlekamp}$ is the latency of one iteration of the Berlekamp–Massey algorithm; N is the total number of iterations that the Berlekamp–Massey algorithm performs; T_{Chien} is the latency for deciding whether or not a single bit location contains an error, using the Chien search; k is the length of the data message m; and p is the number of bits that are processed in parallel in Step 3. In this equation, $T_{Syndrome}$, $T_{Berlekamp}$, k, and p are constants for a BCH decoder implementation, while N and T_{Chien} are proportional to the raw bit error count of the codeword. Note that Steps 4 and 5 can typically be implemented such that they take less than one clock cycle in modern hardware, and thus their latencies are *not* included in (9.11).

9.5.1.2 Low-Density Parity-Check (LDPC) Codes

LDPC codes [72, 73, 167, 243] are now used widely in modern SSDs, as LDPC codes provide a stronger error correction capability than BCH codes, albeit at a greater storage cost [266, 298]. LDPC codes are one type of *capacity-approaching codes*, which are error-correcting codes that come close to the *Shannon limit*, i.e., the maximum number of data message bits (k_{max}) that can be delivered without errors for a certain codeword size (n) under a given error rate [239, 240]. Unlike BCH codes, LDPC codes cannot *guarantee* that they will correct a minimum number of raw bit errors. Instead, a good LDPC code guarantees that the *failure rate* (i.e., the fraction of all reads where the LDPC code cannot successfully correct the data) is less than a target rate for a given number of bit errors. Like BCH codes, LDPC codes for SSDs are designed to be *systematic*, i.e., to contain the data message verbatim within the codeword.

An LDPC code [72, 73, 167, 243] is a linear code that, like a BCH code, consists of check bits generated by an algorithm. For an LDPC code, these check bits are used to form a bipartite graph, where one side of the graph contains nodes that represent each bit in the codeword, and the other side of the graph contains nodes that represent the parity check equations used to generate each parity bit. When a codeword containing errors is retrieved from memory, an LDPC decoder applies *belief propagation* [218] to iteratively identify the bits within the codeword that are *most likely* to contain a bit error.

An LDPC code is defined using a binary parity check matrix H, where H is very sparse (i.e., there are few ones in the matrix). Figure 9.36a shows an example H matrix for a seven-bit codeword c (see Sect. 9.1.3.7). For an n-bit codeword that encodes a k-bit data message, H is sized to be an $(n-k) \times n$ matrix. Within the matrix, each *row* represents a *parity check equation*, while each *column* represents one of the seven bits in the codeword. As our example matrix has three rows, this means that our error correction uses three parity check equations (denoted as f). A bit value 1 in row i, column j indicates that parity check equation f_i contains bit c_j. Each parity check equation XORs all of the codeword bits in the equation to see whether the output is zero. For example, parity check equation f_1 from the H matrix in Fig. 9.36a is:

$$f_1 = c_1 \oplus c_2 \oplus c_4 \oplus c_5 = 0 \tag{9.12}$$

This means that c is a valid codeword only if $H \cdot c^T = 0$, where c^T is the transpose matrix of the codeword c.

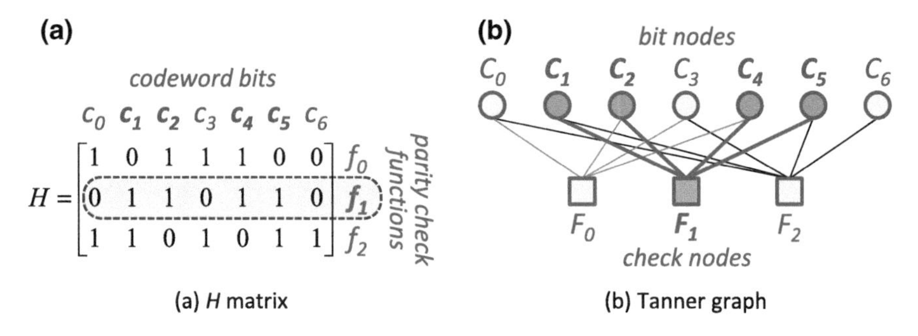

Fig. 9.36 Example LDPC code for a seven-bit codeword with a four-bit data message (stored in bits c_0, c_1, c_2, and c_3) and three parity check equations (i.e., $n = 7$, $k = 4$), represented as **a** an H matrix and **b** a Tanner graph

In order to perform belief propagation, H can be represented using a *Tanner graph* [258]. A Tanner graph is a bipartite graph that contains *check nodes*, which represent the parity check equations, and *bit nodes*, which represent the bits in the codeword. An edge connects a check node F_i to a bit node C_j only if parity check equation f_i contains bit c_j. Figure 9.36b shows the Tanner graph that corresponds to the H matrix in Fig. 9.36a. For example, since parity check equation f_1 uses codeword bits c_1, c_2, c_4, and c_5, the F_1 check node in Fig. 9.36b is connected to bit nodes C_1, C_2, C_4, and C_5.

LDPC Encoding. As was the case with BCH, the LDPC codeword generation algorithm encodes a k-bit data message m into an n-bit LDPC codeword c by computing the dot product of m and a generator matrix G (i.e., $c = m \cdot G$). For an LDPC code, the generator matrix is designed to (1) preserve m verbatim within the codeword, and (2) generate the parity bits for each parity check equation in H. Thus, G is defined using the parity check matrix H. With linear algebra based transformations, H can be expressed in the form $H = [A, I_{(n-k)}]$, where H is composed of A, an $(n-k) \times k$ binary matrix, and $I_{(n-k)}$, an $(n-k) \times (n-k)$ identity matrix [110]. The generator matrix G can then be created using the composition $G = [I_k, A^T]$, where A^T is the transpose matrix of A.

LDPC Decoding. When the SSD controller is servicing a read request, it must extract the k-bit data message from the LDPC codeword r that is stored in NAND flash memory. In an SSD, an LDPC decoder performs multiple *levels* of decoding [64, 263, 298], which correct the retrieved codeword r to obtain the originally-written codeword c and extract the data message m from c. Initially, the decoder performs a single level of *hard decoding*, where it uses the information from a single read operation on the codeword to attempt to correct the codeword bit errors. If the decoder cannot correct all errors using hard decoding, it then initiates the first level of *soft decoding*, where a *second* read operation is performed on the *same* codeword using a *different* set of read reference voltages. The second read provides *additional* information on the *probability* that each bit in the codeword is a zero or a one. An LDPC decoder typically uses multiple levels of soft decoding, where each new level performs an additional read operation to calculate a more accurate probability for each bit value. We discuss multi-level soft decoding in detail in Sect. 9.5.2.2.

For each level, the decoder performs five steps, as illustrated in Fig. 9.37. At each level, the decoder uses two pieces of information to determine which bits are *most likely* to contain errors: (1) the *probability* that each bit in r is a zero or a one, and (2) the parity check equations. In Step 1 (Fig. 9.37), the decoder computes an initial *log likelihood ratio* (LLR) for each bit of the stored codeword. We refer to the initial codeword LLR values as L, where L_j is the LLR value for bit j of the codeword. L_j expresses the likelihood (i.e., *confidence*) that bit j *should be* a zero or a one, based on the current threshold voltage of the NAND flash cell where bit j is stored. The decoder uses L as the initial *LLR message* generated using the bit nodes. An LLR message consists of the LLR values for each bit, which are updated by and communicated between the check nodes and bit nodes during each step of belief

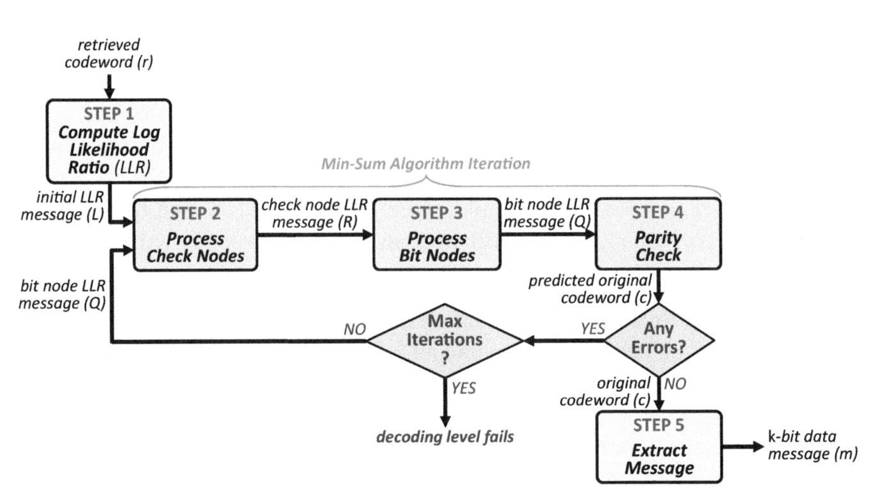

Fig. 9.37 LDPC decoding steps for a single level of hard or soft decoding

propagation.[4] In Steps 2 through 4, the belief propagation algorithm [218] iteratively updates the LLR message, using the Tanner graph to identify those bits that are most likely to be incorrect (i.e., the codeword bits whose (1) bit nodes are connected to the largest number of check nodes that currently contain a parity error, and (2) LLR values indicate low confidence). Several decoding algorithms exist to perform belief propagation for LDPC codes. The most commonly-used belief propagation algorithm is the *min-sum algorithm* [43, 68], a simplified version of the original sum-product algorithm for LDPC [72, 73] with near-equivalent error correction capability [3]. During each iteration of the min-sum algorithm, the decoder identifies a set of codeword bits that likely contain errors and thus need to be flipped. The decoder accomplishes this by (1) having each check node use its parity check information to determine how much the LLR value of each bit should be updated by, using the most recent LLR messages from the bit nodes; (2) having each bit node gather the LLR updates from each bit to generate a new LLR value for the bit, using the most recent LLR messages from the check nodes; and (3) using the parity check equations to see if the values predicted by the new LLR message for each node are correct. The min-sum algorithm terminates under one of two conditions: (1) the predicted bit values after the most recent iteration are all correct, which means that the decoder now has an estimate of the original codeword c, and can advance to Step 5; or (2) the algorithm exceeds a predetermined number of iterations, at which point the decoder moves onto the next decoding level, or returns a decoding failure if the max-

[4]Note that an LLR message is *not* the same as the k-bit data message. The *data message* refers to the actual data stored within the SSD, which, when read, is modeled in information theory as a message that is transmitted across a noisy communication channel. In contrast, an *LLR message* refers to the updated LLR values for each bit of the codeword that are exchanged between the check nodes and the bit nodes during belief propagation. Thus, there is no relationship between a data message and an LLR message.

imum number of decoding levels have been performed. In Step 5, once the errors are corrected, and the decoder has the original codeword c, the decoder extracts the k-bit data message m from the codeword. We describe the steps used by a state-of-the-art decoder in detail below, which uses an optimized version of the min-sum algorithm that can be implemented efficiently in hardware [78, 79].

Step 1—Computing the Log Likelihood Ratio (LLR): The LDPC decoder uses the *probability* (i.e., *likelihood*) that a bit is a zero or a one to identify errors, instead of using the bit values directly. The *log likelihood ratio* (LLR) is the probability that a certain bit is zero, i.e., $P(x = 0|V_{th})$, over the probability that the bit is one, i.e., $P(x = 1|V_{th})$, given a certain threshold voltage range (V_{th}) bounded by two threshold voltage values (i.e., the maximum and the minimum voltage of the threshold voltage range) [266, 298]:

$$\text{LLR} = \log \frac{P(x = 0|V_{th})}{P(x = 1|V_{th})} \tag{9.13}$$

The sign of the LLR value indicates whether the bit is likely to be a zero (when the LLR value is positive) or a one (when the LLR value is negative). A *larger* magnitude (i.e., absolute value) of the LLR value indicates a *greater* confidence that a bit should be zero or one, while an LLR value closer to zero indicates low confidence. The bits whose LLR values have the smallest magnitudes are the ones that are most likely to contain errors.

There are several alternatives for how to compute the LLR values. A common approach for LLR computation is to treat a flash cell as a communication channel, where the channel takes an input program signal (i.e., the target threshold voltage for the cell) and outputs an observed signal (i.e., the current threshold voltage of the cell) [20]. The observed signal differs from the input signal due to the various types of NAND flash memory errors. The communication channel model allows us to break down the threshold voltage of a cell into two components: (1) the expected signal; and (2) the additive signal noise due to errors. By enabling the modeling of these two components separately, the communication channel model allows us to estimate the current threshold voltage distribution of each state [20]. The threshold voltage distributions can be used to predict how likely a cell within a certain voltage region is to belong to a particular voltage state.

One popular variant of the communication channel model assumes that the threshold voltage distribution of each state can be modeled as a Gaussian distribution [20]. If we use the mean observed threshold voltage of each state (denoted as μ) to represent the signal, we find that the P/E cycling noise (i.e., the shift in the distribution of threshold voltages due to the accumulation of charge from repeated programming operations; see Sect. 9.3.1) can be modeled as *additive white Gaussian noise* (AWGN) [20], which is represented by the standard deviation of the distribution (denoted as σ). The closed-form AWGN-based model can be used to determine the LLR value for a cell with threshold voltage y, as follows:

$$\text{LLR}(y) = \frac{\mu_1^2 - \mu_0^2}{2\sigma^2} + \frac{y(\mu_0 - \mu_1)}{\sigma^2} \tag{9.14}$$

where μ_0 and μ_1 are the mean threshold voltages for the distributions of the threshold voltage states for bit value 0 and bit value 1, respectively, and σ is the standard deviation of both distributions (assuming that the standard deviation of each threshold voltage state distribution is equal). Since the SSD controller uses threshold voltage ranges to categorize a flash cell, we can substitute μ_{R_j}, the mean threshold voltage of the threshold voltage range R_j, in place of y in (9.14).

The AWGN-based LLR model in (9.14) provides only an estimate of the LLR, because (1) the actual threshold voltage distributions observed in NAND flash memory are *not* perfectly Gaussian in nature [20, 162]; (2) the controller uses the mean voltage of the threshold voltage range to *approximate* the actual threshold voltage of a cell; and (3) the standard deviations of each threshold voltage state distribution are *not* perfectly equal (see Tables 9.5, 9.6 and 9.7 in the Appendix). A number of methods have been proposed to improve upon the AWGN-based LLR estimate by: (1) using nonlinear transformations to convert the AWGN-based LLR into a more accurate LLR value [278]; (2) scaling and rounding the AWGN-based LLR to compensate for the estimation error [277]; (3) initially using the AWGN-based LLR to read the data, and, if the read fails, using the ECC information from the failed read attempt to optimize the LLR and to perform the read again with the optimized LLR [57]; and (4) using online and offline training to empirically determine the LLR values under a wide range of conditions (e.g., P/E cycle count, retention time, read disturb count) [279]. The SSD controller can either compute the LLR values at runtime, or statically store precomputed LLR values in a table.

Once the decoder calculates the LLR values for each bit of the codeword, which we call the initial LLR message L, the decoder starts the first iteration of the min-sum algorithm (Steps 2–4 below).

Step 2—Check Node Processing: In every iteration of the min-sum algorithm, each check node i (see Fig. 9.36) generates a revised check node LLR message R_{ij} to send to each bit node j (see Fig. 9.36) that is connected to check node i. The decoder computes R_{ij} as:

$$R_{ij} = \delta_{ij}\kappa_{ij} \tag{9.15}$$

where δ_{ij} is the sign of the LLR message, and κ_{ij} is the magnitude of the LLR message. The decoder determines the values of both δ_{ij} and κ_{ij} using the bit node LLR message Q'_{ji}. At a high level, each check node collects LLR values sent from each bit node (Q'_{ji}), and then determines how much each bit's LLR value should be adjusted using the parity information available at the check node. These LLR value updates are then bundled together into the LLR message R_{ij}. During the first iteration of the min-sum algorithm, the decoder sets $Q'_{ji} = L_j$, the initial LLR value from Step 1. In subsequent iterations, the decoder uses the value of Q'_{ji} that was generated in Step 3 of the *previous* iteration. The decoder calculates δ_{ij}, the sign of the check node LLR message, as:

$$\delta_{ij} = \prod_{J} \text{sgn}(Q'_{Ji}) \tag{9.16}$$

where J represents all bit nodes connected to check node i *except* for bit node j. The sign of a bit node indicates whether the value of a bit is predicted to be a zero (if the sign is positive) or a one (if the sign is negative). The decoder calculates κ_{ij}, the magnitude of the check node LLR message, as:

$$\kappa_{ij} = \min_{J} |Q'_{Ji}| \tag{9.17}$$

In essence, the smaller the magnitude of Q'_{ji} is, the more uncertain we are about whether the bit should be a zero or a one. At each check node, the decoder updates the LLR value of each bit node j, adjusting the LLR by the smallest value of Q' for any of the other bits connected to the check node (i.e., the LLR value of the most uncertain bit aside from bit j).

Step 3—Bit Node Processing: Once each check node generates the LLR messages for each bit node, we combine the LLR messages received by each bit node to update the LLR value of the bit. The decoder first generates the LLR messages to be used by the check nodes in the next iteration of the min-sum algorithm. The decoder calculates the bit node LLR message Q_{ji} to send from bit node j to check node i as follows:

$$Q_{ji} = L_j + \sum_{I} R_{Ij} \tag{9.18}$$

where I represents all check nodes connected to bit node j *except* for check node i, and L_j is the original LLR value for bit j generated in Step 1. In essence, for each check node, the bit node LLR message combines the LLR messages from the *other check nodes* to ensure that all of the LLR value updates are propagated globally across all of the check nodes.

Step 4—Parity Check: After the bit node processing is complete, the decoder uses the revised LLR information to predict the value of each bit. For bit node j, the predicted bit value P_j is calculated as:

$$P_j = L_j + \sum_{i} R_{ij} \tag{9.19}$$

where i represents *all* check nodes connected to bit node j, *including* check node i, and L_j is the original LLR value for bit j generated in Step 1. If P_j is positive, bit j of the original codeword c is predicted to be a zero; otherwise, bit j is predicted to be a one. Once the predicted values have been computed for all bits of c, the H matrix is used to check the parity, by computing $H \cdot c^T$. If $H \cdot c^T = 0$, then the predicted bit values are correct, the min-sum algorithm terminates, and the decoder goes to Step 5. Otherwise, at least one bit is still incorrect, and the decoder goes back to Step 2 to perform the next iteration of the min-sum algorithm. In the next iteration, the min-sum algorithm uses the updated LLR values from the current iteration to identify the next set of bits that are most likely incorrect and need to be flipped.

The current decoding level fails to correct the data when the decoder cannot determine the correct codeword bit values after a predetermined number of min-sum algorithm iterations. If the decoder has more soft decoding levels left to perform, it advances to the next soft decoding level. For the new level, the SSD controller performs an additional read operation using a different set of read reference voltages than the ones it used for the prior decoding levels. The decoder then goes back to Step 1 to generate the new LLR information, using the output of *all* of the read operations performed for each decoding level so far. We discuss how the number of decoding levels and the read reference voltages are determined, as well as what happens if *all* soft decoding levels fail, in Sect. 9.5.2.2.

Step 5—Extracting the Message from the Codeword: As we discuss above, during LDPC codeword encoding, the generator matrix G contains the identity matrix, to ensure that the codeword c includes a verbatim version of m. Therefore, the decoder recovers the k-bit data message m by simply truncating the last $(n - k)$ bits from the n-bit codeword c.

9.5.2 Error Correction Flow

For both BCH and LDPC codes, the SSD controller performs several stages of error correction to retrieve the data, known as the *error correction flow*. The error correction flow is invoked when the SSD performs a read operation. The SSD starts the read operation by using the initial read reference voltages ($V_{initial}$; see Sect. 9.4.5) to read the raw data stored within a page of NAND flash memory into the controller. Once the raw data is read, the controller starts error correction.

Algorithm 1 Example BCH/LDPC Error Correction Procedure

```
First Stage: BCH/LDPC Hard Decoding

Controller gets stored V_initial values to use as V_ref
Flash chips read page using V_ref
ECC decoder decodes BCH/LDPC
if ECC succeeds then
  Controller sends data to host; exit algorithm
else if number of stage iterations not exceeded then
  Controller invokes V_ref optimization to find new V_ref;
             repeats first stage
end

Second Stage (BCH only): NAC

Controller reads immediately-adjacent wordline W
while ECC fails and all possible voltage states for
             adjacent wordline not yet tried do
  Controller goes to next neighbor voltage state V
  Controller sets V_ref based on neighbor voltage state V
  Flash chips read page using V_ref
  Controller corrects cells adjacent to W's cells that
             were programmed to V
ECC decoder decodes BCH
if ECC succeeds then
    Controller sends data to host; exit algorithm
    end
end

Second Stage (LDPC only): Level X LDPC Soft Decoding

while ECC fails and X < maximum level N do
  Controller selects optimal value of V^X_ref
  Flash chips do read-retry using V^X_ref
  Controller recomputes LLR^R0_X to LLR^RX_X
  ECC decoder decodes LDPC
  if ECC succeeds then
    Controller sends data to host; exit algorithm
else
    Controller goes to soft decoding level X + 1
    end
end

Third Stage: Superpage-Level Parity Recovery

Flash chips read all other pages in the superpage
Controller XORs all other pages in the superpage
if data extraction succeeds then
    Controller sends data to host
else
    Controller reports uncorrectable error
end
```

Algorithm 1 lists the three stages of an example error correction flow, which can be used to decode either BCH codes or LDPC codes. In the first stage, the ECC engine performs *hard decoding* on the raw data. In hard decoding, the ECC engine uses only the *hard* bit value information (i.e., either a 1 or a 0) read for a cell using a *single* set of read reference voltages. If the first stage succeeds (i.e., the controller detects that the error rate of the data after correction is lower than a predetermined threshold), the flow finishes. If the first stage fails, then the flow moves on to the second stage of error correction. The second stage differs significantly for BCH and for LDPC, which we discuss below. If the second stage succeeds, the flow terminates; otherwise, the flow moves to the third stage of error correction. In the third stage, the controller tries to correct the errors using the more expensive superpage-level parity recovery (see Sect. 9.1.3.10). The steps for superpage-level parity recovery are shown in the third stage of Algorithm 1. If the data can be extracted successfully from the other pages in the superpage, the data from the target page can be recovered. Whenever data is successfully decoded or recovered, the data is sent to the host (and it is also reprogrammed into a new physical page to ensure that the *corrected* data values are stored for the logical page). Otherwise, the SSD controller reports an uncorrectable error to the host.

Figure 9.38 compares the error correction flow with BCH codes to the flow with LDPC codes. Next, we discuss the flows used with both BCH codes (Sect. 9.5.2.1) and LDPC codes (Sect. 9.5.2.2).

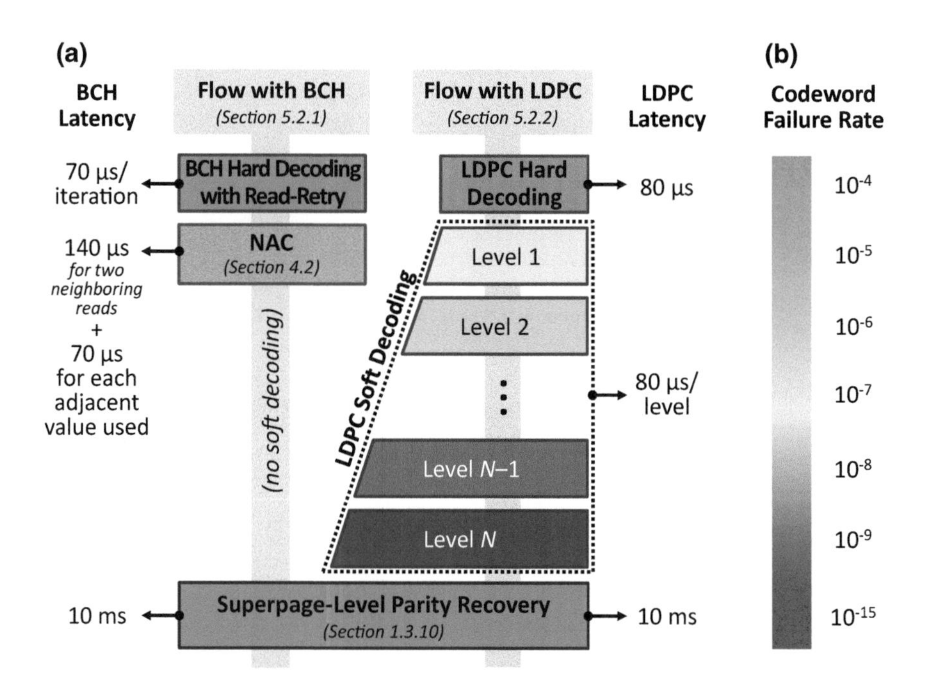

Fig. 9.38 **a** Example error correction flow using BCH codes and LDPC codes, with average latency of each BCH/LDPC stage. **b** The corresponding codeword failure rate for each LDPC stage. Adapted from [15]

9.5.2.1 Flow Stages for BCH Codes

An example flow of the stages for BCH decoding is shown on the left-hand side of Fig. 9.38a. In the first stage, the ECC engine performs BCH hard decoding on the raw data, which reports the total number of bit errors in the data. If the data cannot be corrected by the implemented BCH codes, many controllers invoke read-retry (Sect. 9.4.4) or read reference voltage optimization (Sect. 9.4.5) to find a new set of read reference voltages (V_{ref}) that lower the raw bit error rate of the data from the error rate when using $V_{initial}$. The controller uses the new V_{ref} values to read the data again, and then repeats the BCH decoding. We discuss the algorithm used to perform decoding for BCH codes in Sect. 9.5.1.1.

If the controller exhausts the maximum number of read attempts (specified as a parameter in the controller), it employs correction techniques such as neighbor-cell-assisted correction (NAC; see Sect. 9.4.2) to further reduce the error rate, as shown in the second BCH stage of Algorithm 1. If NAC cannot successfully read the data, the controller then tries to correct the errors using the more expensive superpage-level parity recovery (see Sect. 9.1.3.10).

9.5.2.2 Flow Stages for LDPC Codes

An example flow of the stages for LDPC decoding is shown on the right-hand side of Fig. 9.38a. LDPC decoding consists of three major steps. First, the SSD controller performs LDPC hard decoding, where the controller reads the data using the optimal read reference voltages. The process for LDPC hard decoding is similar to that of BCH hard decoding (as shown in the first stage of Algorithm 1), but does not typically invoke read-retry if the first read attempt fails. Second, if LDPC hard decoding cannot correct all of the errors, the controller uses LDPC *soft decoding* to decode the data (which we describe in detail below). Third, if LDPC soft decoding also cannot correct all of the errors, the controller invokes superpage-level parity. We discuss the algorithm used to perform hard and soft decoding for LDPC codes in Sect. 9.5.1.2.

Soft Decoding. Unlike BCH codes, which require the invocation of expensive superpage-level parity recovery immediately if the hard decoding attempts (i.e., BCH hard decoding with read-retry or NAC) fail to return correct data, LDPC decoding fails more gracefully: it can perform multiple levels of *soft decoding* (shown in the second stage of Algorithm 1) after hard decoding fails before invoking superpage-level parity recovery [266, 298]. The key idea of soft decoding is to use *soft* information for each cell (i.e., the *probability* that the cell contains a 1 or a 0) obtained from *multiple* reads of the cell via the use of different sets of read reference voltages [64, 72, 73, 167, 243, 298]. Soft information is typically represented by the *log likelihood ratio* (LLR; see Sect. 9.5.1.2).

Every additional level of soft decoding (i.e., the use of a new set of read reference voltages, which we call V_{ref}^{X} for level X) increases the strength of the error correction, as the level *adds* new information about the cell (as opposed to hard decoding, where a new decoding step simply *replaces* prior information about the cell). The new read reference voltages, unlike the ones used for hard decoding, are optimized such that the amount of useful information (or *mutual information*) provided to the LDPC decoder is maximized [266]. Thus, the use of soft decoding reduces the frequency at which superpage-level parity needs to be invoked.

Figure 9.39 illustrates the read reference voltages used during LDPC hard decoding and during the first two levels of LDPC soft decoding. At each level, a new read reference voltage is applied, which divides an existing threshold voltage range into two ranges. Based on the bit values read using the various read reference voltages, the SSD controller bins each cell into a certain V_{th} range, and sends the bin categorization of all the cells to the LDPC decoder. For each cell, the decoder applies an LLR value, precomputed by the SSD manufacturer, which corresponds to the cell's bin and decodes the data. For example, as shown in the bottom of Fig. 9.39, the three read reference voltages in Level 2 soft decoding form four threshold voltage ranges (i.e., R0–R3). Each of these ranges corresponds to a different LLR value (i.e., LLR_2^{R0} to LLR_2^{R3}, where LLR_i^{Rj} is the LLR value for range R_j in soft decoding level i). Compared with hard decoding (shown at the top of Fig. 9.39), which has only two LLR values, Level 2 soft decoding provides more accurate information to the decoder, and thus has stronger error correction capability.

Determining the Number of Soft Decoding Levels. If the final level of soft decoding, i.e., level N in Fig. 9.38a, fails, the controller attempts to read the data using superpage-level parity (see Sect. 9.1.3.10). The number of levels used for soft decoding depends on the improved reliability that each additional level provides, taking into account the latency of performing additional decoding. Figure 9.38b shows a

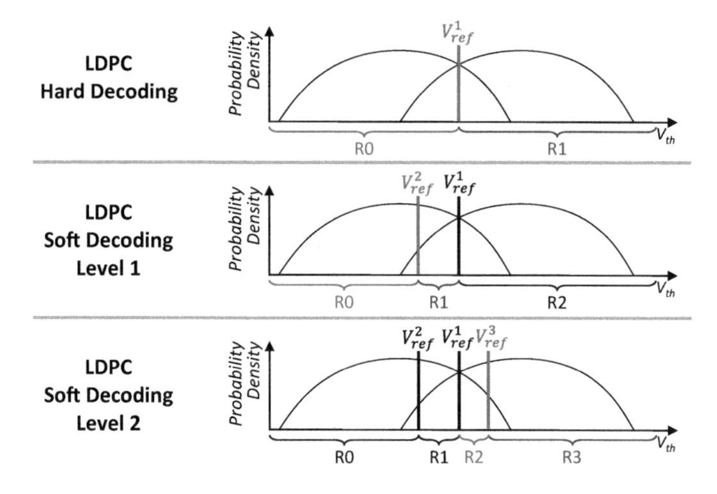

Fig. 9.39 LDPC hard decoding and the first two levels of LDPC soft decoding, showing the V_{ref} value added at each level, and the resulting threshold voltage ranges (R0–R3) used for flash cell categorization. Adapted from [15]

rough estimation of the average latency and the codeword failure rate for each stage. There is a tradeoff between the number of levels employed for soft decoding and the expected read latency. For a smaller number of levels, the additional reliability can be worth the latency penalty. For example, while a five-level soft decoding step requires up to 480 μs, it effectively reduces the codeword failure rate by five orders of magnitude. This not only improves overall reliability, but also reduces the frequency of triggering expensive superpage-level parity recovery, which can take around 10 ms [86]. However, manufacturers limit the number of levels, as the benefit of employing an additional soft decoding level (which requires more read operations) becomes smaller due to diminishing returns in the number of additional errors corrected.

9.5.3 BCH and LDPC Error Correction Strength

BCH and LDPC codes provide different strengths of error correction. While LDPC codes can offer a stronger error correction capability, soft LDPC decoding can lead to a greater latency for error correction. Figure 9.40 compares the error correction strength of BCH codes, hard LDPC codes, and soft LDPC codes [85]. The x-axis shows the raw bit error rate (RBER) of the data being corrected, and the y-axis shows the *uncorrectable bit error rate* (UBER), or the error rate after correction, once the error correction code has been applied. The UBER is defined as the ECC codeword (see Sect. 9.1.3.7) failure rate divided by the codeword length [103]. To ensure a fair comparison, we choose a similar codeword length for both BCH and LDPC codes, and use a similar coding rate (0.935 for BCH, and 0.936 for LDPC) [85]. We make two observations from Fig. 9.40.

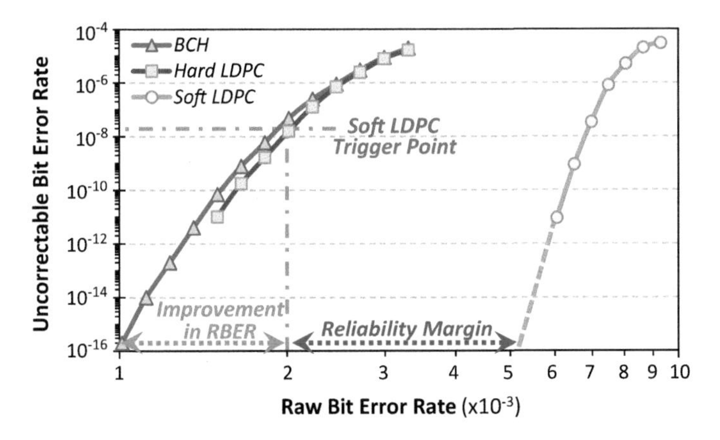

Fig. 9.40 Raw bit error rate versus uncorrectable bit error rate for BCH codes, hard LDPC codes, and soft LDPC codes. Reproduced from [15]

First, we observe that the error correction strength of the hard LDPC code is similar to that of the BCH codes. Thus, on its own, hard LDPC does not provide a significant advantage over BCH codes, as it provides an equivalent degree of error correction with similar latency (i.e., one read operation). Second, we observe that soft LDPC decoding provides a significant advantage in error correction capability. Contemporary SSD manufacturers target a UBER of 10^{-16} [103]. The example BCH code with a coding rate of 0.935 can successfully correct data with an RBER of 1.0×10^{-3} while remaining within the target UBER. The example LDPC code with a coding rate of 0.936 is more successful with soft decoding, and can correct data with an RBER as high as 5.0×10^{-3} while remaining within the target UBER, based on the error rate extrapolation shown in Fig. 9.40. While soft LDPC can tolerate up to five times the raw bit errors as BCH, this comes at a cost of latency (not shown on the graph), as soft LDPC can require several additional read operations after hard LDPC decoding fails, while BCH requires only the original read.

To understand the benefit of LDPC codes over BCH codes, we need to consider the combined effect of hard LDPC decoding and soft LDPC decoding. As discussed in Sect. 9.5.2.2, soft LDPC decoding is invoked *only when hard LDPC decoding fails*. To balance error correction strength with read performance, SSD manufacturers can require that the hard LDPC failure rate cannot exceed a certain threshold, and that the overall read latency (which includes the error correction time) cannot exceed a certain target [85, 86]. For example, to limit the impact of error correction on read performance, a manufacturer can require 99.99% of the error correction operations to be completed after a single read. To meet our example requirement, the hard LDPC failure rate should not be greater than 10^{-4} (i.e., 99.99%), which corresponds to an RBER of 2.0×10^{-3} and a UBER of 10^{-8} (shown as *Soft LDPC Trigger Point* in Fig. 9.40). For only the data that contains one or more failed codewords, soft LDPC is invoked (i.e., soft LDPC is invoked only 0.01% of the time). For our example LDPC code with a coding rate of 0.936, soft LDPC decoding is able to correct these codewords: for an RBER of 2.0×10^{-3}, using soft LDPC results in a UBER well below 10^{-16}, as shown in Fig. 9.40.

To gauge the combined effectiveness of hard and soft LDPC codes, we calculate the overhead of using the combined LDPC decoding over using BCH decoding. If 0.01% of the codeword corrections fail, we can assume that in the worst case, each failed codeword resides in a different flash page. As the failure of a single codeword in a flash page causes soft LDPC to be invoked for the entire flash page, our assumption maximizes the number of flash pages that require soft LDPC decoding. For an SSD with four codewords per flash page, our assumption results in up to 0.04% of the data reads requiring soft LDPC decoding. Assuming that the example soft LDPC decoding requires seven additional reads, this corresponds to 0.28% more reads when using combined hard and soft LDPC over BCH codes. Thus, with a 0.28% overhead in the number of reads performed, the combined hard and soft LDPC decoding provides twice the error correction strength of BCH codes (shown as *Improvement in RBER* in Fig. 9.40).

In our example, the lifetime of an SSD is limited by both the UBER and whether more than 0.01% of the codeword corrections invoke soft LDPC, to ensure that the overhead of error correction does not significantly increase the read latency [85]. In this case, when the lifetime of the SSD ends, we can still read out the data correctly from the SSD, albeit at an increased read latency. This is because even though we capped the SSD lifetime to an RBER of 2.0×10^{-3} in our example shown in Fig. 9.40, soft LDPC is able to correct data with an RBER as high as 5.0×10^{-3} while still maintaining an acceptable UBER (10^{-16}) based on the error rate extrapolation shown. Thus, LDPC codes have a margin, which we call the *reliability margin* and show in Fig. 9.40. This reliability margin enables us to trade off lifetime with read latency.

We conclude that with a combination of hard and soft LDPC decoding, an SSD can offer a significant improvement in error correction strength over using BCH codes.

9.5.4 SSD Data Recovery

When the number of errors in data exceeds the ECC correction capability and the error correction techniques in Sects. 9.5.2.1 and 9.5.2.2 are unable to correct the read data, then data loss can occur. At this point, the SSD is considered to have reached the end of its lifetime. In order to avoid such data loss and *recover* (or, *rescue*) the data from the SSD, we can harness our understanding of data retention and read disturb behavior. The SSD controller can employ two conceptually similar mechanisms, *Retention Failure Recovery* (RFR) [22] and *Read Disturb Recovery* (RDR) [21], to undo errors that were introduced into the data as a result of data retention and read disturb, respectively. The key idea of both of these mechanisms is to exploit the wide variation of different flash cells in their susceptibility to data retention loss and read disturbance effects, respectively, in order to correct some of the errors *without* the assistance of ECC so that the remaining error count falls within the ECC error correction capability.

When a flash page read fails (i.e., uncorrectable errors exist), RFR and RDR record the current threshold voltages of each cell in the page using the read-retry mechanism (see Sect. 9.4.4), and identify the cells that are *susceptible* to generating errors due to retention and read disturb (i.e., cells that lie at the tails of the threshold voltage distributions of each state, where the distributions overlap with each other), respectively. We observe that some flash cells are more likely to be affected by retention leakage and read disturb than others, as a result of process variation [21, 22]. We call these cells retention/read disturb *prone*, while cells that are less likely to be affected are called retention/read disturb *resistant*. RFR and RDR classify the susceptible cells as retention/read disturb prone or resistant by inducing *even more* retention and read disturb on the failed flash page, and then recording the new threshold voltages of the susceptible cells. We classify the susceptible cells by observing the magnitude of the threshold voltage shift due to the additional retention/read disturb induction.

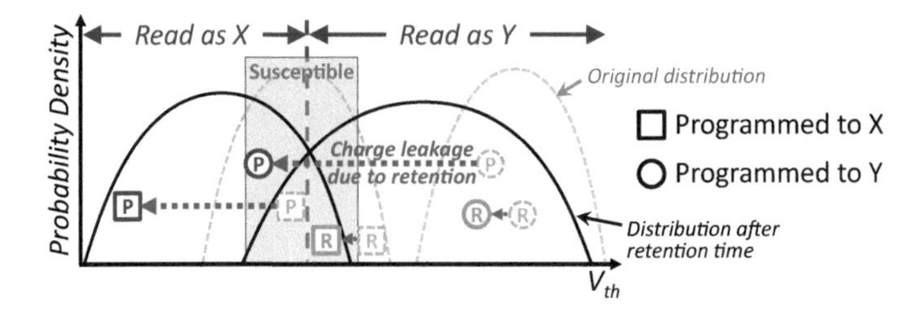

Fig. 9.41 Some retention-prone (P) and retention-resistant (R) cells are incorrectly read after charge leakage due to retention time. RFR identifies and corrects the incorrectly read cells based on their leakage behavior. Reproduced from [15]

Figure 9.41 shows how the threshold voltage of a retention-prone cell (i.e., a *fast-leaking* cell, labeled P in the figure) decreases over time (i.e., the cell shifts to the left) due to retention leakage, while the threshold voltage of a retention- resistant cell (i.e., a *slow-leaking* cell, labeled R in the figure) does not change significantly over time. Retention Failure Recovery (RFR) uses this classification of retention-prone versus retention-resistant cells to correct the data from the failed page *without* the assistance of ECC. Without loss of generality, let us assume that we are studying susceptible cells near the intersection of two threshold voltage distributions X and Y, where Y contains higher voltages than X. Figure 9.41 highlights the region of cells considered susceptible by RFR using a box, labeled *Susceptible*. A susceptible cell within the box that is retention prone likely belongs to distribution Y, as a retention-prone cell shifts rapidly to a lower voltage (see the circled cell labeled P within the *susceptible* region in the figure). A retention-resistant cell in the same *susceptible* region likely belongs to distribution X (see the boxed cell labeled R within the *susceptible* region in the figure).

Similarly, Read Disturb Recovery (RDR) uses the classification of read disturb prone versus read disturb resistant cells to correct data. For RDR, disturb-prone cells shift more rapidly to higher voltages, and are thus likely to belong to distribution X, while disturb-resistant cells shift little and are thus likely to belong to distribution Y. Both RFR and RDR correct the bit errors for the susceptible cells based on such *expected* behavior, reducing the number of errors that ECC needs to correct.

RFR and RDR are highly effective at reducing the error rate of failed pages, reducing the raw bit error rate by 50% and 36%, respectively, as shown in our prior works [21, 22], where more detailed information and analyses can be found.

9.6 Emerging Reliability Issues for 3D NAND Flash Memory

While the demand for NAND flash memory capacity continues to grow, manufacturers have found it increasingly difficult to rely on manufacturing process technology scaling to achieve increased capacity [210]. Due to a combination of limitations in manufacturing process technology and the increasing reliability issues as manufacturers move to smaller process technology nodes, planar (i.e., 2D) NAND flash scaling has become difficult for manufacturers to sustain. This has led manufacturers to seek alternative approaches to increase NAND flash memory capacity.

Recently, manufacturers have begun to produce SSDs that contain *three-dimensional* (3D) NAND flash memory [98, 112, 177, 178, 210, 292]. In 3D NAND flash memory, *multiple layers* of flash cells are stacked vertically to increase the density and to improve the scalability of the memory [292]. In order to achieve this stacking, manufacturers have changed a number of underlying properties of the flash memory design. In this section, we examine these changes, and discuss how they affect the reliability of the flash memory devices. In Sect. 9.6.1, we discuss the flash memory cell design commonly used in contemporary 3D NAND flash memory, and how these cells are organized across the multiple layers. In Sect. 9.6.2, we discuss how the reliability of 3D NAND flash memory compares to the reliability of the planar NAND flash memory that we have studied so far in this work. Table 9.4 summarizes the differences observed in 3D NAND flash memory reliability. In Sect. 9.6.3, we briefly discuss error mitigation mechanisms that cater to emerging reliability issues in 3D NAND flash memory.

9.6.1 3D NAND Flash Design and Operation

As we discuss in Sect. 9.2.1, NAND flash memory stores data as the threshold voltage of each flash cell. In planar NAND flash memory, we achieve this using a floating-gate transistor as a flash cell, as shown in Fig. 9.6. The floating-gate transistor stores charge in the floating gate of the cell, which consists of a conductive material. The floating gate is surrounded on both sides by an oxide layer. When high

Table 9.4 Changes in behavior of different types of errors in 3D NAND flash memory, compared to planar (i.e., two-dimensional) NAND flash memory. See Sect. 9.6.2 for a detailed discussion

Error type	Change in 3D versus Planar
P/E Cycling (Sect. 9.3.1)	3D is *less susceptible*, due to current use of charge trap transistors for flash cells
Program (Sect. 9.3.2)	3D is *less susceptible for now*, due to use of one-shot programming (see Sect. 9.2.4)
Cell-to-cell interference (Sect. 9.3.3)	3D is *less susceptible for now*, due to larger manufacturing process technology
Data retention (Sect. 9.3.4)	3D is *more susceptible*, due to early retention loss
Read disturb (Sect. 9.3.5)	3D is *less susceptible for now*, due to larger manufacturing process technology

voltage is applied to the control gate of the transistor, charge can migrate through the oxide layers into the floating gate due to Fowler-Nordheim (FN) tunneling [69] (see Sect. 9.2.4).

Most manufacturers use a *charge trap transistor* [65, 268] as the flash cell in 3D NAND flash memories, instead of using a floating-gate transistor. Figure 9.42 shows the cross section of a charge trap transistor. Unlike a floating-gate transistor, which stores data in the form of charge within a *conductive* material, a charge trap transistor stores data as charge within an *insulating* material, known as the *charge trap*. In a 3D circuit, the charge trap wraps around a cylindrical transistor substrate, which contains the source (labeled *S* in Fig. 9.42) and drain (labeled *D* in the figure), and a control gate wraps around the charge trap. This arrangement allows the channel between the source and drain to form *vertically* within the transistor. As is the case with a floating-gate transistor, a tunnel oxide layer exists between the charge trap and the substrate, and a gate oxide layer exists between the charge trap and the control gate.

Despite the change in cell structure, the mechanism for transferring charge into and out of the charge trap is similar to the mechanism for transferring charge into and out of the floating gate. In 3D NAND flash memory, the charge trap transistor typically employs FN tunneling to change the threshold voltage of the charge trap [115, 210].[5] When high voltage is applied to the control gate, electrons are injected into the charge trap from the substrate. As this behavior is similar to how electrons are injected into a floating gate, read, program, and erase operations remain the same for both planar and 3D NAND flash memory.

Figure 9.43 shows how multiple charge trap transistors are physically organized within 3D NAND flash memory to form flash blocks, wordlines, and bitlines (see Sect. 9.2.2). As mentioned above, the channel within a charge trap transistor forms vertically, as opposed to the horizontal channel that forms within a floating-gate transistor. The vertical orientation of the channel allows us to stack multiple transistors *on top of each other* (i.e., along the z-axis) within the chip, using 3D-stacked circuit integration. The vertically-connected channels form one bitline of a flash block in

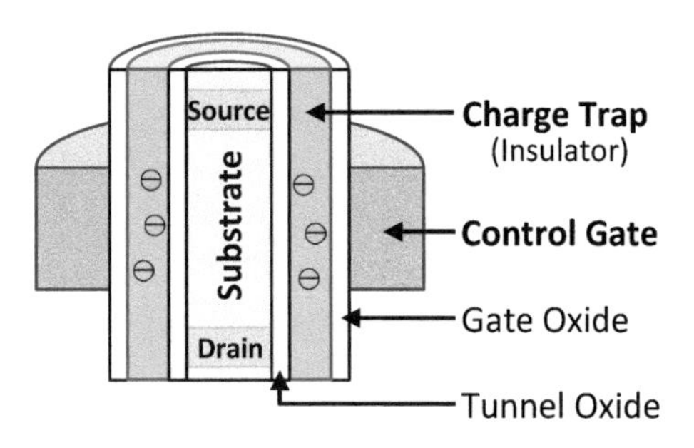

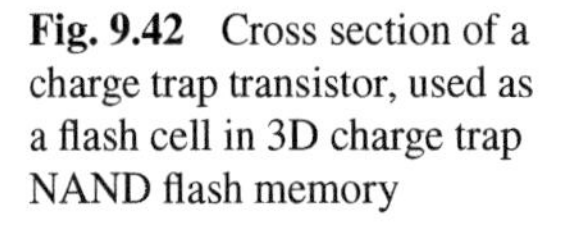
Fig. 9.42 Cross section of a charge trap transistor, used as a flash cell in 3D charge trap NAND flash memory

[5]Note that *not* all charge trap transistors rely on FN tunneling. Charge trap transistors used for NOR flash memory change their threshold voltage using *channel hot electron injection*, also known as *hot carrier injection* [166].

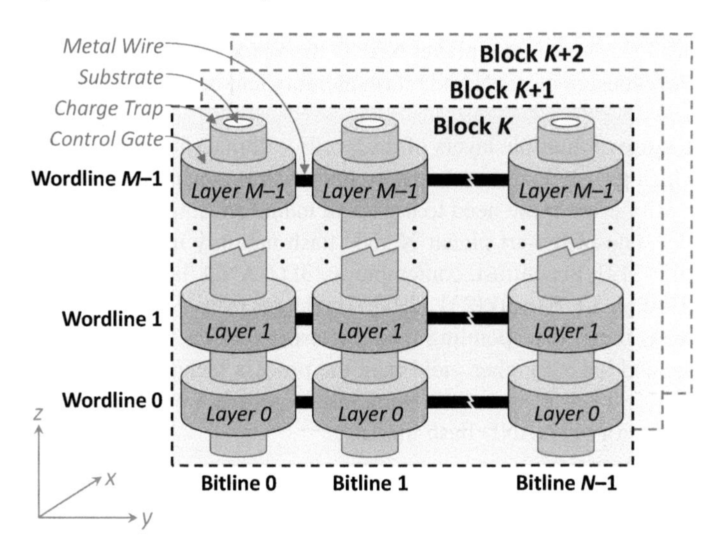

Fig. 9.43 Organization of flash cells in an M-layer 3D charge trap NAND flash memory chip, where each block consists of M wordlines and N bitlines

3D NAND flash memory. Unlike in planar NAND flash memory, where only the substrates of flash cells on the same bitline are connected together, flash cells along the same bitline in 3D NAND flash memory share a common substrate and a common insulator (i.e., charge trap). The FN tunneling induced by the control gate of the transistor forms a tunnel only in a local region of the insulator, and, thus, electrons are injected only into that local region. Due to the strong insulating properties of the material used for the insulator, different regions of a single insulator can have different voltages. This means that each region of the insulator can store a different data value, and thus, the data of *multiple* 3D NAND flash memory cells can be stored reliably in a *single* insulator. This is because the FN tunneling induced by the control gate of the transistor forms a tunnel only in a *local* region of the insulator, and, thus, electrons are injected only into that local region.

Each cell along a bitline belongs to a different *layer* of the flash memory chip. Thus, a bitline crosses *all* of the layers within the chip. Contemporary 3D NAND flash memory contains 24–96 layers [66, 112, 122, 210, 260, 292]. Along the y-axis, the control gates of cells within a *single layer* are connected together to form one wordline of a flash block. As we show in Fig. 9.43, a block in 3D NAND flash memory consists of all of the flash cells within the same y-z plane (i.e., all cells that have the same coordinate along the x-axis). Note that, while not depicted in Fig. 9.43, each bitline within a 3D NAND flash block includes a sense amplifier and two selection transistors used to select the bitline (i.e., the SSL and GSL transistors; see Sect. 9.2.2). The sense amplifier and selection transistors are connected in series with the charge trap transistors that belong to the same bitline, in a similar manner to the connections shown for a planar NAND flash block in Fig. 9.8. More detail on the circuit-level design of 3D NAND flash memory can be found in [102, 115, 137, 255].

Due to the use of multiple layers of flash cells within a single NAND flash memory chip, which greatly increases capacity per unit area, manufacturers can achieve high cell density *without* the need to use small manufacturing process technologies. For example, state-of-the-art planar NAND flash memory uses the 15–19 nm feature size [162, 212]. In contrast, contemporary 3D NAND flash memory uses larger feature sizes (e.g., 30–50 nm) [231, 292]. The larger feature sizes reduce manufacturing costs, as their corresponding manufacturing process technologies are much more mature and have a higher yield than the process technologies used for small feature sizes. As we discuss in Sect. 9.6.2, the larger feature size also has an effect on the reliability of 3D NAND flash memory.

9.6.2 Errors in 3D NAND Flash Memory

While the high-level behavior of 3D NAND flash memory is similar to the behavior of 2D planar NAND flash memory, there are a number of differences between the reliability of 3D NAND flash and planar NAND flash, which we summarize in Table 9.4. There are two reasons for the differences in reliability: (1) the use of charge trap transistors instead of floating-gate transistors, and (2) moving to a larger manufacturing process technology. We categorize the changes based on the reason for the change below.

Effects of Charge Trap Transistors. Compared to the reliability issues discussed in Sect. 9.3 for planar NAND flash memory, the use of charge trap transistors introduces two key differences: (1) *early retention loss* [47, 183, 292, 301], and (2) a *reduction in P/E cycling errors* [210, 292, 301].

First, early retention loss refers to the rapid leaking of electrons from a flash cell soon after the cell is programmed [47, 292]. Early retention loss occurs in 3D NAND flash memory because charge can now migrate out of the charge trap in *three* dimensions. In planar NAND flash memory, charge leakage due to retention occurs across the tunnel oxide, which occupies two dimensions (see Sect. 9.3.4). In 3D NAND flash memory, charge can leak across *both* the tunnel oxide *and* the insulator that is used for the charge trap, i.e., across *three* dimensions. The additional charge leakage takes place for only a few seconds after cell programming. After a few seconds have passed, the impact of leakage through the charge trap decreases, and the long-term cell retention behavior is similar to that of flash cells in planar NAND flash memory [47, 183, 292].

Second, P/E cycling errors (see Sect. 9.3.1) reduce with 3D NAND flash memory because the tunneling oxide in charge trap transistors is *less* susceptible to breakdown than the oxide in floating-gate transistors during high-voltage operation [183, 292]. As a result, the oxide is less likely to contain trapped electrons once a cell is erased, which in turn makes it less likely that the cell is subsequently programmed to an incorrect threshold voltage. One benefit of the reduction in P/E cycling errors is that the endurance (i.e., the maximum P/E cycle count) for a 3D flash memory cell has increased by more than an order of magnitude [211, 213].

Effects of Larger Manufacturing Process Technologies. Due to the use of larger manufacturing process technologies for 3D NAND flash memory, many of the errors that we observe in 2D planar NAND flash (see Sect. 9.3) are not as prevalent in 3D NAND flash memory. For example, while read disturb is a prominent source of errors at small feature sizes (e.g., 20–24 nm), its effects are small at larger feature sizes [21, 301]. Likewise, there are much fewer errors due to cell-to-cell program interference (see Sect. 9.3.3) in 3D NAND flash memory, as the physical distance between neighboring cells is much larger due to the increased feature size. As a result, both cell-to-cell program interference and read disturb are *currently* not major issues in 3D NAND flash memory reliability [210, 213, 292, 301].

One advantage of the lower cell-to-cell program interference is that 3D NAND flash memory uses the older *one-shot programming* algorithm [211, 213, 293] (see Sect. 9.2.4). In planar NAND flash memory, one-shot programming was replaced by two-step programming (for MLC) and foggy-fine programming (for TLC) in order to reduce the impact of cell-to-cell program interference on fully-programmed cells (as we describe in Sect. 9.2.4). The lower interference in 3D NAND flash memory makes two-step and foggy-fine programming unnecessary. As a result, none of the cells in 3D NAND flash memory are partially-programmed, significantly reducing the number of program errors (see Sect. 9.3.2) that occur [213, 301].

Unlike the effects on reliability due to the use of a charge trap transistor, which are likely longer-term, the effects on reliability due to the use of larger manufacturing process technologies are expected to be shorter-term. As manufacturers seek to further increase the density of 3D NAND flash memory, they will reach an upper limit for the number of layers that can be integrated within a 3D-stacked flash memory chip, which is currently projected to be in the range of 300–512 layers [139, 152]. At that point, manufacturers will once again need to scale down the chip to *smaller* manufacturing process technologies [292], which, in turn, will reintroduce high amounts of read disturb and cell-to-cell program interference (just as it happened for planar NAND flash memory [21, 23, 31, 133, 209]).

9.6.3 Changes in Error Mitigation for 3D NAND Flash Memory

Due to the reduction in a number of sources of errors, fewer error mitigation mechanisms are currently needed for 3D NAND flash memory. For example, because the number of errors introduced by cell-to-cell program interference is currently low, manufacturers have *reverted* to using one-shot programming (see Sect. 9.2.4) for 3D NAND flash [211, 213, 293]. As a result of the currently small effect of read disturb errors, mitigation and recovery mechanisms for read disturb (e.g., pass-through voltage optimization in Sect. 9.4.5, Read Disturb Recovery in Sect. 9.5.4) may not be needed, for the time being. We expect that once 3D NAND flash memory begins to scale down to smaller manufacturing process technologies, approaching the current feature sizes used for planar NAND flash memory, there will be a significant need for 3D NAND flash memory to use many, if not all, of the error mitigation mechanisms we discuss in Sect. 9.4.

To our knowledge, no mechanisms have been designed yet to reduce the impact of early retention loss,[6] which is a new error mechanism in 3D NAND flash memory. This is in part due to the reduced overall impact of retention errors in 3D NAND flash memory compared to planar NAND flash memory [47, 301], since a larger cell contains a greater number of electrons than a smaller cell at the same threshold voltage. As a result, existing refresh mechanisms (see Sect. 9.4.3) can be used to tolerate errors introduced by early retention loss with little modification. However, as 3D NAND flash memory scales into future smaller technology nodes, the early retention loss problem may require new mitigation techniques.

At the time of writing, only a few rigorous studies examine error characteristics of and error mitigation techniques for 3D NAND flash memories. An example of such a study is by Luo et al. [164], which (1) examines the *self-recovery effect* in 3D NAND flash memory, where the damage caused by wearout due to P/E cycling (see Sect. 9.3.1) can be repaired by *detrapping* electrons that are *inadvertently* trapped in flash cells; (2) examines how the operating temperature of 3D NAND flash memory affects the raw bit error rate; (3) comprehensively models the impact of wearout, data retention, self-recovery, and temperature on 3D NAND flash reliability; and (4) proposes a new technique to mitigate errors in 3D NAND flash memory using this comprehensive model. Other such studies (1) may expose additional sources of errors that have not yet been observed, and that may be unique to 3D NAND flash memory; and (2) can enable a solid understanding of current error mechanisms in 3D NAND flash memory so that appropriate specialized mitigation mechanisms can be developed. We expect that future works will experimentally examine such sources of errors, and will potentially introduce novel mitigation mechanisms for these errors. Thus, the field (both academia and industry) is currently in much need of rigorous experimental characterization and analysis of 3D NAND flash memory devices (see footnote 6).

9.7 Similar Errors in Other Memory Technologies

As we discussed in Sect. 9.3, there are five major sources of errors in flash-memory-based SSDs. Many of these error sources can also be found in other types of memory and storage technologies. In this section, we take a brief look at the major reliability issues that exist within DRAM and in emerging nonvolatile memories. In particular,

[6]One such work will be presented at the June 2018 Sigmetrics conference [301] as this chapter is being sent to print.

we focus on DRAM in our discussion, as modern SSD controllers have access to dedicated DRAM of considerable capacity (e.g., 1 GB for every 1 TB of SSD capacity), which exists within the SSD package (see Sect. 9.1). Major sources of errors in DRAM include data retention, cell-to-cell interference, and read disturb. There is a wide body of work on mitigation mechanisms for the DRAM and emerging memory technology errors we describe in this section, but we explicitly discuss only a select number of them here, since a full treatment of such mechanisms is out of the scope of this current chapter.

9.7.1 Cell-to-Cell Interference Errors in DRAM

One similarity between the capacitive DRAM cell and the floating-gate cell in NAND flash memory is that they are both vulnerable to cell-to-cell interference. In DRAM, one important way in which cell-to-cell interference exhibits itself is the data-dependent retention behavior, where the retention time of a DRAM cell is dependent on the values written to *nearby* DRAM cells [116–119, 157, 214]. This phenomenon is called *data pattern dependence* (DPD) [157]. Data pattern dependence in DRAM is similar to the data-dependent nature of program interference that exists in NAND flash memory (see Sect. 9.3.3). Within DRAM, data dependence occurs as a result of parasitic capacitance coupling (between DRAM cells). Due to this coupling, the amount of charge stored in one cell's capacitor can inadvertently affect the amount of charge stored in an adjacent cell's capacitor [116–119, 157, 214]. As DRAM cells become smaller with technology scaling, cell-to-cell interference worsens because parasitic capacitance coupling between cells increases [116, 157]. More findings on cell-to-cell interference and the data-dependent nature of cell retention times in DRAM, along with experimental data obtained from modern DRAM chips, can be found in our prior works [34, 116–119, 157, 214, 223].

9.7.2 Data Retention Errors in DRAM

DRAM uses the charge within a capacitor to represent one bit of data. Much like the floating gate within NAND flash memory, charge leaks from the DRAM capacitor over time, leading to data retention issues. Charge leakage in DRAM, if left unmitigated, can lead to much more rapid data loss than the leakage observed in a NAND flash cell. While leakage from a NAND flash cell typically leads to data loss after several days to years of retention time (see Sect. 9.3.4), leakage from a DRAM cell leads to data loss after a retention time on the order of *milliseconds* to *seconds* [157].

The retention time of a DRAM cell depends upon several factors, including (1) manufacturing process variation and (2) temperature [157]. Manufacturing process variation affects the amount of current that leaks from each DRAM cell's capacitor and access transistor [157]. As a result, the retention time of the cells within a

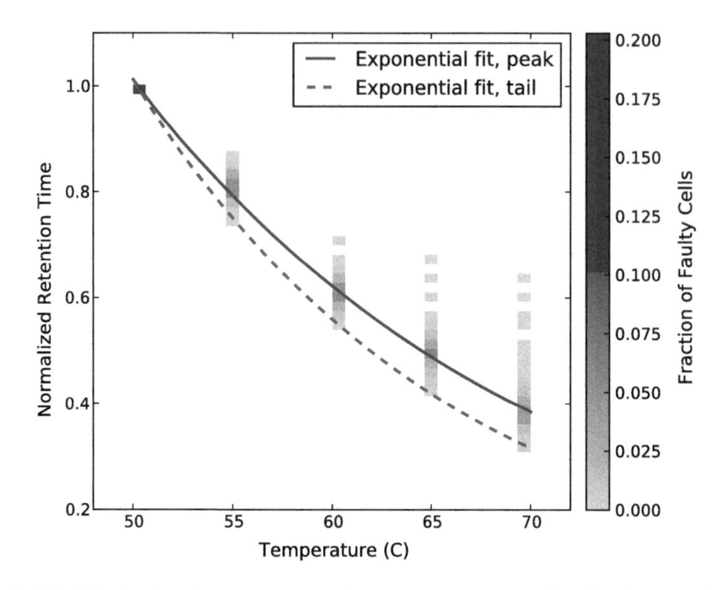

Fig. 9.44 DRAM retention time versus operating temperature, normalized to the retention time of each DRAM cell at 50 °C. Reproduced from [157]

single DRAM chip vary significantly, resulting in *strong cells* that have high retention times and *weak cells* that have low retention times within each chip. The operating temperature affects the rate at which charge leaks from the capacitor. As the operating temperature increases, the retention time of a DRAM cell decreases exponentially [83, 157]. Figure 9.44 shows the change in retention time as we vary the operating temperature, as measured from real DRAM chips [157]. In Fig. 9.44, we normalize the retention time of each cell to its retention time at an operating temperature of 50 °C. As the number of cells is large, we group the normalized retention times into bins, and plot the density of each bin. We draw two exponential-fit curves: (1) the *peak* curve, which is drawn through the most populous bin at each temperature measured; and (2) the *tail* curve, which is drawn through the lowest non-zero bin for each temperature measured. Figure 9.44 provides us with three major conclusions about the relationship between DRAM cell retention time and temperature. First, both of the exponential-fit curves fit well, which confirms the exponential decrease in retention time as the operating temperature increases in modern DRAM devices. Second, the retention times of different DRAM cells are affected very differently by changes in temperature. Third, the variation in retention time across cells increases greatly as temperature increases. More analysis of factors that affect DRAM retention times can be found in our recent works [116–119, 157, 214, 223].

Due to the rapid charge leakage from DRAM cells, a DRAM controller periodically refreshes all DRAM cells in place [35, 104, 116, 157, 158, 214, 223] (similar to the techniques discussed in Sect. 9.4.3, but at a much smaller time scale). DRAM standards require a DRAM cell to be refreshed once every 64 ms [104].

As the density of DRAM continues to increase over successive product generations (e.g., by 128x between 1999 and 2017 [34, 37]), enabled by the scaling of DRAM to smaller manufacturing process technology nodes [169], the performance and energy overheads required to refresh an entire DRAM module have grown significantly [35, 158]. It is expected that the refresh problem will get worse and limit DRAM density scaling, as described in a recent work by Samsung and Intel [114] and by our group [158]. Refresh operations in DRAM cause both (1) performance loss and (2) energy waste, both of which together lead to a difficult technology scaling challenge. Refresh operations degrade performance due to three major reasons. First, refresh operations increase the memory latency, as a request to a DRAM bank that is refreshing must wait for the refresh latency before it can be serviced. Second, they reduce the amount of bank-level parallelism available to requests, as a DRAM bank cannot service requests during refresh. Third, they decrease the row buffer hit rate, as a refresh operation causes all open rows in a bank to be closed. When a DRAM chip scales to a greater capacity, there are more DRAM rows that need to be refreshed. As Fig. 9.45a shows, the amount of time spent on each refresh operation scales linearly with the capacity of the DRAM chip. The additional time spent on refresh causes the DRAM data throughput loss due to refresh to become more severe in denser DRAM chips, as shown in Fig. 9.45b. For a chip with a density of 64 Gbit, nearly 50% of the data throughput is lost due to the high amount of time spent on refreshing all of the rows in the chip. The increased refresh time also increases the effect of refresh on power consumption. As we observe from Fig. 9.45c, the fraction of DRAM power spent on refresh is expected to be the dominant component of the total DRAM power consumption, as DRAM chip capacity scales to become larger. For a chip with a density of 64 Gbit, nearly 50% of the DRAM chip power is spent on refresh operations. Thus, refresh poses a clear challenge to DRAM scalability.

To combat the growing performance and energy overheads of refresh, two classes of techniques have been developed. The first class of techniques reduce the *frequency* of refresh operations without sacrificing the reliability of data stored in DRAM (e.g., [6, 101, 116, 118, 119, 158, 214, 223, 264]). Various experimental studies of real

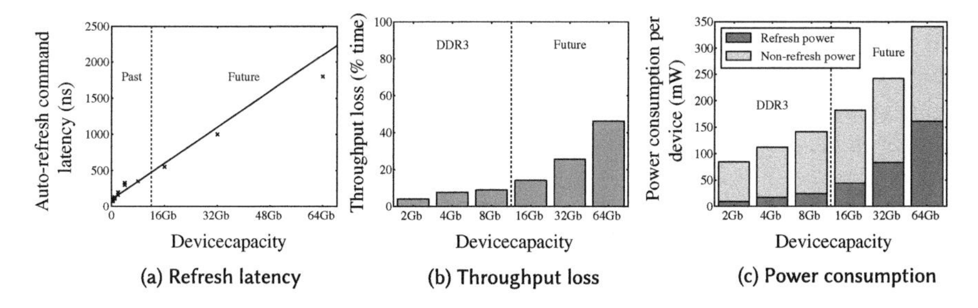

Fig. 9.45 Negative performance and power consumption effects of refresh in contemporary and future DRAM devices. We expect that as the capacity of each DRAM chip increases, **a** the refresh latency, **b** the DRAM throughput lost during refresh operations, and **c** the power consumed by refresh will all increase. Reproduced from [158]

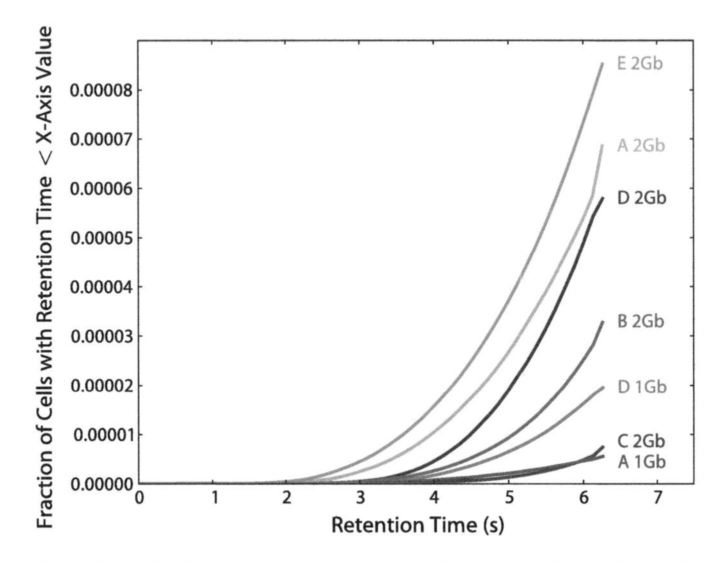

Fig. 9.46 Cumulative distribution of the number of cells in a DRAM module with a retention time less than the value on the x-axis, plotted for seven different DRAM modules. Reproduced from [157]

DRAM chips (e.g., [87, 116, 117, 126, 149, 157, 158, 214, 223]) have studied the data retention time of DRAM cells in modern chips. Figure 9.46 shows the retention time measured from seven different real DRAM modules (by manufacturers A, B, C, D, and E) at an operating temperature of 45 °C, as a cumulative distribution (CDF) of the fraction of cells that have a retention time less than the x-axis value [157]. We observe from the figure that even for the DRAM module whose cells have the worst retention time (i.e., the CDF is the highest), fewer than only 0.001% of the total cells have a retention time smaller than 3 s at 45 °C. As shown in Fig. 9.44, the retention time decreases exponentially as the temperature increases. We can extrapolate our observations from Fig. 9.46 to the worst-case operating conditions by using the tail curve from Fig. 9.44. DRAM standards specify that the operating temperature of DRAM should not exceed 85 °C [104]. Using the tail curve, we find that a retention time of 3 s at 45 °C is equivalent to a retention time of 246 ms at the worst-case temperature of 85 °C. Thus, the vast majority of DRAM cells can retain data without loss for much longer than the 64 ms retention time specified by DRAM standards. The other experimental studies of DRAM chips have validated this observation as well [87, 116, 117, 126, 149, 158, 214, 223].

A number of works take advantage of this variability in data retention time behavior across DRAM cells, by introducing heterogeneous refresh rates, i.e., different refresh rates for different DRAM rows. Thus, these works can reduce the frequency at which the vast majority of DRAM rows within a module are refreshed (e.g., [6, 101, 116, 118, 157, 158, 214, 223, 264]). For example, the key idea of RAIDR [158] is to refresh the *strong* DRAM rows (i.e., those rows that can retain data for much longer than the minimum 64 ms retention time in the DDR4 standard [104]) less frequently, and refresh the *weak* DRAM rows (i.e., those rows that can retain data only for the minimum retention time) more frequently. The major challenge in such works is how to accurately identify the retention time of each DRAM row. To solve this challenge, many recent works examine (online) DRAM retention time profiling techniques [116, 117, 119, 157, 214, 223].

The second class of techniques reduce the interference caused by refresh requests on demand requests (e.g., [35, 190, 249]). These works either change the scheduling order of refresh requests [35, 190, 249] or slightly modify the DRAM architecture to enable the servicing of refresh and demand requests in parallel [35].

One critical challenge in developing techniques to reduce refresh overheads is that it is getting significantly more difficult to determine the minimum retention time of a DRAM cell, as we have shown experimentally on modern DRAM chips [116, 117, 157, 214, 223]. Thus, determining the correct rate at which to refresh DRAM cells has become more difficult, as also indicated by industry [114]. This is due to two major phenomena, both of which get worse (i.e., become more prominent) with manufacturing process technology scaling. The first phenomenon is *variable retention time* (VRT), where the retention time of some DRAM cells can change drastically over time, due to a memoryless random process that results in very fast charge loss via a phenomenon called *trap-assisted gate-induced drain leakage* [157, 223, 228, 286]. VRT, as far as we know, is very difficult to test for, because there seems to be no way of determining that a cell exhibits VRT until that cell is observed to exhibit VRT, and the time scale of a cell exhibiting VRT does not seem to be bounded, based on the current experimental data on modern DRAM devices [157, 214]. The second phenomenon is *data pattern dependence* (DPD), which we discuss in Sect. 9.7.1. Both of these phenomena greatly complicate the accurate determination of minimum data retention time of DRAM cells. Therefore, data retention in DRAM continues to be a vulnerability that can greatly affect DRAM technology scaling (and thus performance and energy consumption) as well as the reliability and security of current and future DRAM generations.

More findings on the nature of DRAM data retention and associated errors, as well as relevant experimental data from modern DRAM chips, can be found in our prior works [34, 35, 87, 116–119, 149, 157, 158, 193, 214, 223].

9.7.3 Read Disturb Errors in DRAM

Commodity DRAM chips that are sold and used in the field today exhibit read disturb errors [134], also called *RowHammer*-induced errors [193], which are *conceptually* similar to the read disturb errors found in NAND flash memory (see Sect. 9.3.5). Repeatedly accessing the same row in DRAM can cause bit flips in data stored in adjacent DRAM rows. In order to access data within DRAM, the row of cells corresponding to the requested address must be *activated* (i.e., opened for read and write operations). This row must be *precharged* (i.e., closed) when another row in the same

DRAM bank needs to be activated. Through experimental studies on a large number of real DRAM chips, we show that when a DRAM row is activated and precharged repeatedly (i.e., *hammered*) enough times within a DRAM refresh interval, one or more bits in physically-adjacent DRAM rows can be flipped to the wrong value [134].

We tested 129 DRAM modules manufactured by three major manufacturers (A, B, and C) between 2008 and 2014, using an FPGA-based experimental DRAM testing infrastructure [87] (more detail on our experimental setup, along with a list of all modules and their characteristics, can be found in our original RowHammer paper [134]). Figure 9.47 shows the rate of RowHammer errors that we found, with the 129 modules that we tested categorized based on their manufacturing date. We find that 110 of our tested modules exhibit RowHammer errors, with the earliest such module dating back to 2010. In particular, we find that *all* of the modules manufactured in 2012–2013 that we tested are vulnerable to RowHammer. Like with many NAND flash memory error mechanisms, especially read disturb, RowHammer is a recent phenomenon that especially affects DRAM chips manufactured with more advanced manufacturing process technology generations.

Figure 9.48 shows the distribution of the number of rows (plotted in log scale on the y-axis) within a DRAM module that flip the number of bits along the x-axis, as measured for example DRAM modules from three different DRAM manufacturers [134]. We make two observations from the figure. First, the number of bits flipped when we hammer a row (known as the *aggressor row*) can vary significantly within a module. Second, each module has a different distribution of the number of rows. Despite these differences, we find that this DRAM failure mode affects more than 80% of the DRAM chips we tested [134]. As indicated above, this read disturb error mechanism in DRAM is popularly called RowHammer [193].

Various recent works show that RowHammer can be maliciously exploited by user-level software programs to (1) induce errors in existing DRAM modules [134, 193] and (2) launch attacks to compromise the security of various systems [11, 13, 76, 77, 193, 227, 235, 236, 262, 281]. For example, by exploiting the RowHammer read disturb mechanism, a user-level program can gain kernel-level privileges

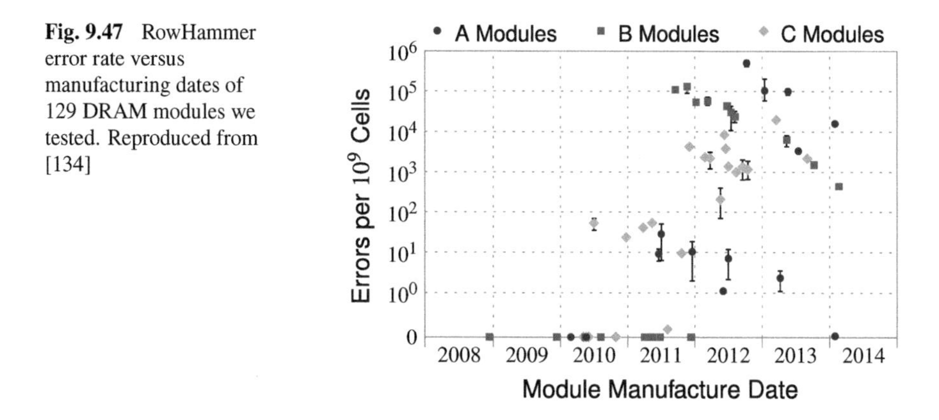

Fig. 9.47 RowHammer error rate versus manufacturing dates of 129 DRAM modules we tested. Reproduced from [134]

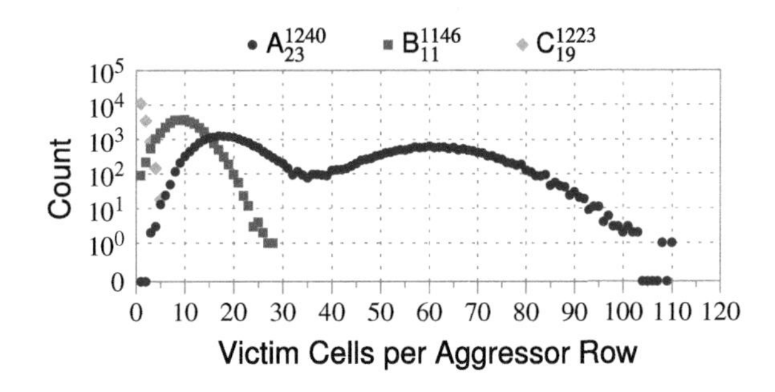

Fig. 9.48 Number of victim cells (i.e., number of bit errors) when an aggressor row is repeatedly activated, for three representative DRAM modules from three major manufacturers. We label the modules in the format X_n^{yyww}, where X is the manufacturer (A, B, or C), $yyww$ is the manufacture year (yy) and week of the year (ww), and n is the number of the selected module. Reproduced from [134]

on real laptop systems [235, 236], take over a server vulnerable to RowHammer [77], take over a victim virtual machine running on the same system [11], and take over a mobile device [262]. Thus, the RowHammer read disturb mechanism is a prime (and perhaps the first) example of how a circuit-level failure mechanism in DRAM can cause a practical and widespread system security vulnerability. We believe similar (yet likely more difficult to exploit) vulnerabilities exist in MLC NAND flash memory as well, as described in our recent work [17].

Note that various solutions to RowHammer exist [128, 134, 193], but we do not discuss them in detail here. Our recent work [193] provides a comprehensive overview. A very promising proposal is to modify either the memory controller or the DRAM chip such that it probabilistically refreshes the physically-adjacent rows of a recently-activated row, with very low probability. This solution is called *Probabilistic Adjacent Row Activation* (PARA) [134]. Our prior work shows that this low-cost, low-complexity solution, which does not require any storage overhead, greatly closes the RowHammer vulnerability [134].

The RowHammer effect in DRAM worsens as the manufacturing process scales down to smaller node sizes [134, 193]. More findings on RowHammer, along with extensive experimental data from real DRAM devices, can be found in our prior works [128, 134, 193].

9.7.4 Large-Scale DRAM Error Studies

Like flash memory, DRAM is employed in a wide range of computing systems, at scale. Thus, there is a similar need to study the aggregate behavior of errors observed in a large number of DRAM chips deployed in the field. Akin to the large-scale flash memory SSD reliability studies discussed in Sect. 9.3.6, a number of experimental

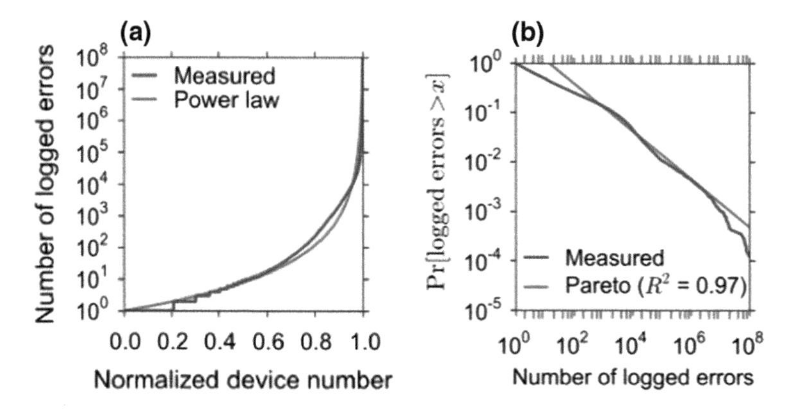

Fig. 9.49 Distribution of memory errors among servers with errors (**a**), which resembles a power law distribution. Memory errors follow a Pareto distribution among servers with errors (**b**). Reproduced from [175]

studies characterize the reliability of DRAM at large scale in the field (e.g., [96, 175, 234, 246, 247]). We highlight three notable results from these studies.

First, as we saw for large-scale studies of SSDs (see Sect. 9.3.6), the number of errors observed varies significantly for each DRAM module [175]. Figure 9.49a shows the distribution of correctable errors across the *entire fleet* of servers at Facebook over a fourteen-month period, omitting the servers that did not exhibit any correctable DRAM errors. The x-axis shows the normalized device number, with devices sorted based on the number of errors they experienced in a month. As we saw in the case of SSDs, a small number of servers accounts for the majority of errors. As we see from Fig. 9.49a, the top 1% of servers account for 97.8% of all observed correctable DRAM errors. The distribution of the number of errors among servers follows a power law model. We show the probability density distribution of correctable errors in Fig. 9.49b, which indicates that the distribution of errors across servers follows a Pareto distribution, with a decreasing hazard rate [175]. This means that a server that has experienced more errors in the past is likely to experience more errors in the future.

Second, unlike SSDs, DRAM does *not* seem to show any clearly discernible trend where higher utilization and age lead to a greater raw bit error rate [175].

Third, the increase in the density of DRAM chips with technology scaling leads to higher error rates [175]. The latter is illustrated in Fig. 9.50, which shows how different DRAM chip densities are related to device failure rate. We can see that there is a clear trend of increasing failure rate with increasing chip density. We find that the failure rate increases because despite small improvements in the reliability of an *individual* cell, the quadratic increase in the number of cells per chip greatly increases the probability of observing a single error in the whole chip [175].

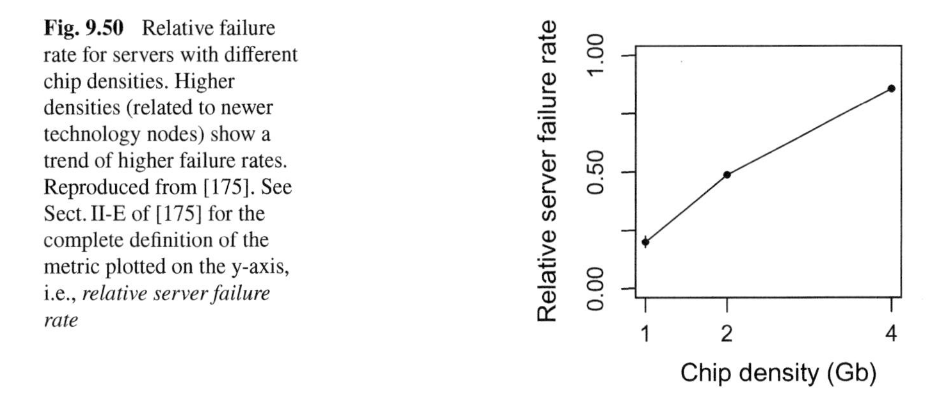

Fig. 9.50 Relative failure rate for servers with different chip densities. Higher densities (related to newer technology nodes) show a trend of higher failure rates. Reproduced from [175]. See Sect. II-E of [175] for the complete definition of the metric plotted on the y-axis, i.e., *relative server failure rate*

9.7.5 Latency-Related Errors in DRAM

Various experimental studies examine the tradeoff between DRAM reliability and latency [33, 34, 37, 38, 87, 125, 144, 146, 149]. These works perform extensive experimental studies on real DRAM chips to identify the effect of (1) temperature, (2) supply voltage, and (3) manufacturing process variation that exists in DRAM on the latency and reliability characteristics of different DRAM cells and chips. The temperature, supply voltage, and manufacturing process variation all dictate the amount of time that each cell needs to safely complete its operations. Several of our works [37, 38, 146, 149] examine how one can *reliably* exploit different effects of variation to improve DRAM performance or energy consumption.

Adaptive-Latency DRAM (AL-DRAM) [149] shows that significant variation exists in the access latency of (1) different DRAM modules, as a result of manufacturing process variation; and (2) the same DRAM module over time, as a result of varying operating temperature, since at low temperatures DRAM can be accessed faster. The key idea of AL-DRAM is to adapt the DRAM latency to the operating temperature and the DRAM module that is being accessed. Experimental results show that AL-DRAM can reduce DRAM read latency by 32.7% and write latency by 55.1%, averaged across 115 DRAM modules operating at 55 °C [149].

Voltron [38] identifies the relationship between the DRAM supply voltage and access latency variation. Voltron uses this relationship to identify the combination of voltage and access latency that minimizes system-level energy consumption without exceeding a user-specified threshold for the maximum acceptable performance loss. For example, at an average performance loss of only 1.8%, Voltron reduces the DRAM energy consumption by 10.5%, which translates to a reduction in the overall system energy consumption of 7.3%, averaged over seven memory-intensive quad-core workloads [38].

Flexible-Latency DRAM (FLY-DRAM) [37] captures access latency variation across DRAM cells *within* a single DRAM chip due to manufacturing process variation. For example, Fig. 9.51 shows how the bit error rate (BER) changes if we reduce

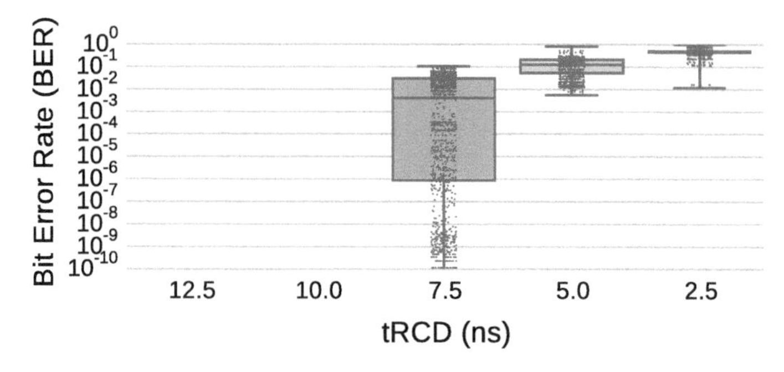

Fig. 9.51 Bit error rates of tested DRAM modules as we reduce the DRAM access latency (i.e., the t_{RCD} timing parameter). Reproduced from [37]

one of the timing parameters used to control the DRAM access latency below the minimum value specified by the manufacturer [37]. We use an FPGA-based experimental DRAM testing infrastructure [87] to measure the BER of 30 real DRAM modules, over a total of 7500 rounds of tests, as we lower the t_{RCD} timing parameter (i.e., how long it takes to open a DRAM row) below its standard value of 13.125 ns.[7] In this figure, we use a box plot to summarize the bit error rate measured during each round. For each box, the bottom, middle, and top lines indicate the 25th, 50th, and 75th percentile of the population. The ends of the whiskers indicate the minimum and maximum BER of all modules for a given t_{RCD} value. Each round of BER measurement is represented as a single point overlaid upon the box. From the figure, we make three observations. First, the BER decreases exponentially as we reduce t_{RCD}. Second, there are no errors when t_{RCD} is at 12.5 ns or at 10.0 ns, indicating that manufacturers provide a significant latency *guardband* to provide additional protection against process variation. Third, the BER variation across different models becomes smaller as t_{RCD} decreases. The reliability of a module operating at $t_{RCD} = 7.5$ ns varies significantly based on the DRAM manufacturer and model. This variation occurs because the number of DRAM cells that experience an error within a DRAM chip varies significantly from module to module. Yet, the BER variation across different modules operating at $t_{RCD} = 2.5$ ns is much smaller, as most modules fail when the latency is reduced so significantly.

From other experiments that we describe in our FLY-DRAM paper [37], we find that there is spatial locality in the slower cells, resulting in *fast regions* (i.e., regions where all DRAM cells can operate at significantly-reduced access latency without experiencing errors) and *slow regions* (i.e., regions where *some* of the DRAM cells *cannot* operate at significantly-reduced access latency without experiencing errors) within each chip. To take advantage of this heterogeneity in the reliable access latency of DRAM cells within a chip, FLY-DRAM (1) categorizes the cells into fast and slow regions; and (2) lowers the overall DRAM latency by accessing fast

[7]More detail on our experimental setup, along with a list of all modules and their characteristics, can be found in our original FLY-DRAM paper [37].

regions with a lower latency. FLY-DRAM lowers the timing parameters used for the fast region by as much as 42.8% [37]. FLY-DRAM improves system performance for a wide variety of real workloads, with the average improvement for an eight-core system ranging between 13.3 and 19.5%, depending on the amount of variation that exists in each module [37].

Design-Induced Variation-Aware DRAM (DIVA-DRAM) [146] identifies the latency variation within a single DRAM chip that occurs due to the architectural design of the chip. For example, a cell that is further away from the row decoder requires a longer access time than a cell that is close to the row decoder. Similarly, a cell that is farther away from the wordline driver requires a larger access time than a cell that is close to the wordline driver. DIVA-DRAM uses design-induced variation to reduce the access latency to different parts of the chip. One can further reduce latency by sacrificing some amount of reliability and performing error correction to fix the resulting errors [146]. Experimental results show that DIVA-DRAM can reduce DRAM read latency by 40.0% and write latency by 60.5% [146]. In an eight-core system running a wide variety of real workloads, DIVA-DRAM improves system performance by an average of 13.8% [146].

More information about the errors caused by reduced latency and reduced voltage operation in DRAM chips and the tradeoff between reliability and latency and voltage can be found in our prior works [34, 37, 38, 87, 144, 146, 149, 165].

9.7.6 *Error Correction in DRAM*

In order to protect the data stored within DRAM from various types of errors, some (but not all) DRAM modules employ ECC [165]. The ECC employed within DRAM is much weaker than the ECC employed in SSDs (see Sect. 9.5) for various reasons. First, DRAM has a much lower access latency, and error correction mechanisms should be designed to ensure that DRAM access latency does not increase significantly. Second, the error rate of a DRAM chip tends to be lower than that of a flash memory chip. Third, the granularity of access is much smaller in a DRAM chip than in a flash memory chip, and hence sophisticated error correction can come at a high cost. The most common ECC algorithm used in commodity DRAM modules is *SECDED* (single error correction, double error detection) [165]. Another ECC algorithm available for some commodity DRAM modules is *Chipkill*, which can tolerate the failure of an *entire* DRAM chip within a module [61] at the expense of higher storage overhead and higher latency. For both SECDED and Chipkill, the ECC information is stored on one or more extra chips within the DRAM module, and, on a read request, this information is sent alongside the data to the memory controller, which performs the error detection and correction.

As DRAM scales to smaller technology nodes, its error rate continues to increase [114, 134, 169, 175, 192, 193, 196]. Effects like read disturb [134], cell-to-cell interference [116–119, 157, 214], and variable retention time [116, 157, 214, 223] become more severe [114, 134, 192, 193, 196]. As a result, there is an increasing need for (1) employing ECC algorithms in *all* DRAM chips/modules; (2) developing more sophisticated and efficient ECC algorithms for DRAM chips/modules; and (3) developing error-specific mechanisms for error correction. To this end, recent work follows various directions. First, in-DRAM ECC, where correction is performed within the DRAM module itself (as opposed to in the controller), is proposed [114]. One work shows how exposing this in-DRAM ECC information to the memory controller can provide Chipkill-like error protection at much lower overhead than the traditional Chipkill mechanism [198]. Second, various works explore and develop stronger ECC algorithms for DRAM (e.g., [123, 124, 270]), and explore how to make ECC more efficient based on the current DRAM error rate (e.g., [2, 49, 61, 146, 261]). Third, recent work shows how the cost of ECC protection can be reduced by (1) exploiting *heterogeneous reliability memory* [165], where different portions of DRAM use different strengths of error protection based on the error tolerance of different applications and different types of data [159, 165], and (2) using the additional DRAM capacity that is otherwise used for ECC to improve system performance when reliability is not as important for the given application and/or data [163].

Many of these works that propose error mitigation mechanisms for DRAM do *not* distinguish between the characteristics of different types of errors. We believe that, in addition to providing sophisticated and efficient ECC mechanisms in DRAM, there is also significant value in and opportunity for exploring *specialized* error mitigation mechanisms that are *customized for different error types*, just as it is done for flash memory (as we discussed in Sect. 9.4). One such example of a specialized error mitigation mechanism is targeted to fix the RowHammer read disturb mechanism, and is called *Probabilistic Adjacent Row Activation* (PARA) [134, 193], as we discussed earlier. Recall that the key idea of PARA is to refresh the rows that are physically adjacent to an activated row, with a very low probability. PARA is shown to be very effective in fixing the RowHammer problem at no storage cost and at very low performance overhead [134]. PARA is a specialized yet very effective solution for fixing a specific error mechanism that is important and prevalent in modern DRAM devices.

9.7.7 Errors in Emerging Nonvolatile Memory Technologies

DRAM operations are several orders of magnitude faster than SSD operations, but DRAM has two major disadvantages. First, DRAM offers orders of magnitude less storage density than NAND-flash-memory-based SSDs. Second, DRAM is volatile (i.e., the stored data is lost on a power outage). Emerging nonvolatile memories, such as *phase-change memory* (PCM) [140–142, 224, 274, 290, 299], *spin-transfer torque magnetic RAM* (STT-RAM or STT-MRAM) [138, 197], *metal-oxide resistive RAM* (RRAM) [273], and *memristors* [53, 248], are expected to bridge the gap between DRAM and SSDs, providing DRAM-like access latency and energy, and at the same time SSD-like large capacity and nonvolatility (and hence SSD-like data persistence). These technologies are also expected to be used as part of *hybrid memory systems* (also called *heterogeneous memory systems*), where one part of the memory consists of DRAM modules and another part consists of modules of emerging technologies [41, 50, 51, 109, 155, 173, 176, 220, 224–226, 290, 291, 295, 296]. PCM-based devices are expected to have a limited lifetime, as PCM can only endure a certain number of writes [140, 224, 274], similar to the P/E cycling errors in NAND-flash-memory-based SSDs (though PCM's write endurance is higher than that of SSDs). PCM suffers from (1) *resistance drift* [97, 221, 274], where the resistance used to represent the value becomes higher over time (and eventually can introduce a bit error), similar to how charge leakage in NAND flash memory and DRAM lead to retention errors over time; and (2) *write disturb* [108], where the heat generated during the programming of one PCM cell dissipates into neighboring cells and can change the value that is stored within the neighboring cells. STT-RAM suffers from (1) *retention failures*, where the value stored for a single bit (as the magnetic orientation of the layer that stores the bit) can flip over time; and (2) *read disturb* (a conceptually different phenomenon from the read disturb in DRAM and flash memory), where reading a bit in STT-RAM can inadvertently induce a write to that same bit [197]. Due to the nascent nature of emerging nonvolatile memory technologies and the lack of availability of large-capacity devices built with them, extensive and dependable experimental studies have yet to be conducted on the reliability of real PCM, STT-RAM, RRAM, and memristor chips. However, we believe that error mechanisms conceptually or abstractly similar to those we discussed in this paper for flash memory and DRAM are likely to be prevalent in emerging technologies as well (as supported by some recent studies [5, 108, 120, 197, 244, 245, 297]), albeit with different underlying mechanisms and error rates.

9.8 Conclusion

We provide a survey of the fundamentals of and recent research in NAND-flash-memory-based SSD reliability. As the underlying NAND flash memory within SSDs scales to increase storage density, we find that the rate at which raw bit errors occur in the memory increases significantly, which in turn reduces the lifetime of the SSD. We describe the prevalent error mechanisms that affect NAND flash memory, and examine how they behave in modern NAND flash memory chips. To compensate for the increased raw bit error rate with technology scaling, a wide range of error mitigation and data recovery mechanisms have been proposed. These techniques effectively undo some of the SSD lifetime reductions that occur due to flash memory scaling. We describe the state-of-the-art techniques for error mitigation and data recovery, and discuss their benefits. Even though our focus is on MLC and TLC NAND flash memories, for which we provide data from real flash chips, we believe that these techniques will be applicable to emerging 3D NAND flash memory technology as well, especially when the process technology scales to smaller nodes. Thus, we hope the tutorial presented in this work on fundamentals and recent research not only enables practitioners to get acquainted with flash memory errors and how they are mitigated, but also helps inform future directions in NAND flash memory and SSD development as well as system design using flash memory. We believe future is bright for system-level approaches that codesign system and memory [192, 193, 196] to enhance overall scaling of platforms, and we hope that the many examples of this approach presented in this tutorial inspire researchers and developers to enhance future computing platforms via such system-memory codesign.

Acknowledgements The authors would like to thank Rino Micheloni for his helpful feedback on earlier drafts of the chapter. They would also like to thank Seagate for their continued dedicated support. Special thanks also goes to our research group SAFARI's industrial sponsors over the past six years, especially Facebook, Google, Huawei, Intel, Samsung, Seagate, VMware. This work was also partially supported by ETH Zürich, the Intel Science and Technology Center for Cloud Computing, the Data Storage Systems Center at Carnegie Mellon University, and NSF grants 1212962 and 1320531. An earlier, shorter version of this book chapter appears on arxiv.org [15] and in the *Proceedings of the IEEE* [16].

Appendix: TLC Threshold Voltage Distribution Data

See Tables 9.5, 9.6 and 9.7.

Table 9.5 Normalized mean (top) and standard deviation (bottom) values for threshold voltage distribution of each voltage state at various P/E cycle counts (Sect. 9.3.1)

P/E cycles	ER	P1	P2	P3	P4	P5	P6	P7
0	−110.0	65.9	127.4	191.6	254.9	318.4	384.8	448.3
200	−110.4	66.6	128.3	192.8	255.5	319.3	385.0	448.6
400	−105.0	66.0	127.3	191.7	254.5	318.2	383.9	447.7
1,000	−99.9	66.5	127.1	191.7	254.8	318.1	384.4	447.8
2,000	−92.7	66.6	128.1	191.9	254.9	318.3	384.3	448.1
3,000	−84.1	68.3	128.2	193.1	255.7	319.2	385.4	449.1
P/E cycles	ER	P1	P2	P3	P4	P5	P6	P7
0	45.9	9.0	9.4	8.9	8.8	8.9	9.3	8.5
200	46.2	9.2	9.8	9.0	8.8	9.0	9.1	8.5
400	46.4	9.2	9.5	9.1	8.8	8.8	9.0	8.6
1,000	47.3	9.5	9.4	9.1	9.3	8.9	9.4	8.8
2,000	48.2	9.7	9.7	9.4	9.3	9.1	9.5	9.1
3,000	49.4	10.2	10.2	9.6	9.7	9.5	9.8	9.4

Table 9.6 Normalized mean (top) and standard deviation (bottom) values for threshold voltage distribution of each voltage state at various data retention times (Sect. 9.3.4)

Time	ER	P1	P2	P3	P4	P5	P6	P7
1 day	−92.7	66.6	128.1	191.9	254.9	318.3	384.3	448.1
1 week	−86.7	67.5	128.1	191.4	253.8	316.5	381.8	444.9
1 month	−84.4	68.6	128.7	191.6	253.5	315.8	380.9	443.6
3 months	−75.6	72.8	131.6	193.3	254.3	315.7	380.2	442.2
1 year	−69.4	76.6	134.2	195.2	255.3	316.0	379.6	440.8
Time	ER	P1	P2	P3	P4	P5	P6	P7
1 day	48.2	9.7	9.7	9.4	9.3	9.1	9.5	9.1
1 week	46.4	10.7	10.8	10.5	10.6	10.3	10.6	10.6
1 month	46.8	11.3	11.2	11.0	10.9	10.8	11.2	11.1
3 months	45.9	12.0	11.8	11.5	11.4	11.4	11.7	11.7
1 year	45.9	12.8	12.4	12.0	12.0	11.9	12.3	12.4

Table 9.7 Normalized mean (top) and standard deviation (bottom) values for threshold voltage distribution of each voltage state at various read disturb counts (Sect. 9.3.5)

Read disturbs	ER	P1	P2	P3	P4	P5	P6	P7
1	−84.2	66.2	126.3	191.5	253.7	316.8	384.3	448.0
1,000	−76.1	66.7	126.6	191.5	253.6	316.4	383.8	447.5
10,000	−57.0	67.9	127.0	191.5	253.3	315.7	382.9	445.7
50,000	−33.4	69.9	128.0	191.9	253.3	315.4	382.0	444.1
100,000	−20.4	71.6	128.8	192.1	253.3	315.0	381.1	443.0
Read disturbs	ER	P1	P2	P3	P4	P5	P6	P7
1	48.2	9.7	9.7	9.4	9.3	9.1	9.5	9.1
1,000	47.4	10.7	10.8	10.5	10.6	10.3	10.6	10.6
10,000	46.3	12.0	11.7	11.4	11.4	11.4	11.7	11.7
50,000	46.1	12.3	12.1	11.7	11.6	11.7	12.0	12.4
100,000	45.9	12.8	12.4	12.0	12.0	11.9	12.3	12.4

References

1. N. Agrawal, V. Prabhakaran, T. Wobber, J.D. Davis, M. Manasse, R. Panigrahy, Design tradeoffs for SSD performance, in *USENIX ATC* (2008)
2. A.R. Alameldeen, I. Wagner, Z. Chisthi, W. Wu, C. Wilkerson, S.-L. Lu, Energy-efficient cache design using variable-strength error-correcting codes, in *ISCA* (2011)
3. A. Anastasopoulos, A comparison between the sum-product and the min-sum iterative detection algorithms based on density evolution, in *GLOBECOM* (2001)
4. S.A. Arrhenius, *Über die dissociationswärme und den einfluß der temperatur auf den dissociationsgrad der elektrolytae.* Z. Phys. Chem. (1889)

5. A. Athmanathan, M. Stanisavljevic, N. Papandreou, H. Pozidis, E. Eleftheriou, *Multilevel-cell phase-change memory: a viable technology*. J. Emerg. Sel. Top. Circuits Syst. (2016)
6. S. Baek, S. Cho, R. Melhem, *Refresh now and then*. IEEE Trans. Comput. (2014)
7. F.M. Benelli, How to extend 2D-TLC endurance to 3,000 P/E cycles. in *Flash Memory Summit* (2015)
8. E.R. Berlekamp, Nonbinary BCH decoding, in *ISIT* (1967)
9. R. Bez, E. Camerlenghi, A. Modelli, A. Visconti, Introduction to flash memory. Proc. IEEE, April 2003
10. R.C. Bose, D.K. Ray-Chaudhuri, *On a class of error correcting binary group codes*. Inf. Control (1960)
11. E. Bosman, K. Razavi, H. Bos, C. Guiffrida, Dedup est machina: memory deduplication as an advanced exploitation vector, in *SP* (2016)
12. J.E. Brewer, M. Gill, *Nonvolatile Memory Technologies with Emphasis on Flash: A Comprehensive Guide to Understanding and Using NVM Devices* (Wiley, Hoboken, NJ, USA, 2008)
13. W. Burleson, O. Mutlu, and M. Tiwari, Who is the major threat to tomorrow's security? you, the hardware designer, in *DAC* (2016)
14. Y. Cai, NAND flash memory: characterization, analysis, modelling, and mechanisms. Ph.D. Dissertation, Carnegie Mellon University, 2012
15. Y. Cai, S. Ghose, E. F. Haratsch, Y. Luo, O. Mutlu, Error characterization, mitigation, and recovery in flash memory based solid-state drives (2017), arXiv:1706.08642
16. Y. Cai, S. Ghose, E.F. Haratsch, Y. Luo, O. Mutlu, Error characterization, mitigation, and recovery in flash-memory-based solid-state drives. Proc. IEEE, Sept 2017
17. Y. Cai, S. Ghose, Y. Luo, K. Mai, O. Mutlu, E.F. Haratsch, Vulnerabilities in MLC NAND flash memory programming: experimental analysis, exploits, and mitigation techniques, in *HPCA* (2017)
18. Y. Cai, E.F. Haratsch, M. McCartney, K. Mai, FPGA-based solid-state drive prototyping platform, in *FCCM* (2011)
19. Y. Cai, E.F. Haratsch, O. Mutlu, K. Mai, Error patterns in MLC NAND flash memory: measurement, characterization, and analysis, in *DATE* (2012)
20. Y. Cai, E.F. Haratsch, O. Mutlu, K. Mai, Threshold voltage distribution in MLC NAND flash memory: characterization, analysis, and modeling, in *DATE* (2013)
21. Y. Cai, Y. Luo, S. Ghose, E.F. Haratsch, K. Mai, O. Mutlu, Read disturb errors in MLC NAND flash memory: characterization, mitigation, and recovery, in *DSN* (2015)
22. Y. Cai, Y. Luo, E.F. Haratsch, K. Mai, O. Mutlu, Data retention in MLC NAND flash memory: characterization, optimization, and recovery, in *HPCA* (2015)
23. Y. Cai, O. Mutlu, E.F. Haratsch, K. Mai, Program interference in MLC NAND flash memory: characterization, modeling, and mitigation, in *ICCD* (2013)
24. Y. Cai, Y. Wu, N. Chen, E.F. Haratsch, Z. Chen, Systems and methods for latency based data recycling in a solid state memory system, U.S. Patent 9,424,179 (2016)
25. Y. Cai, Y. Wu, E.F. Haratsch, Hot-read data aggregation and code selection, U.S. Patent Application 14/192,110 (2015)
26. Y. Cai, Y. Wu, E.F. Haratsch, System to control a width of a programming threshold voltage distribution width when writing hot-read data, U.S. Patent 9,218,885 (2015)
27. Y. Cai, Y. Wu, E.F. Haratsch, Data recovery once ECC fails to correct the data, U.S. Patent 9,323,607 (2016)
28. Y. Cai, Y. Wu, E.F. Haratsch, Error correction code (ECC) selection using probability density functions of error correction capability in storage controllers with multiple error correction codes, U.S. Patent 9,419,655 (2016)
29. Y. Cai, G. Yalcin, O. Mutlu, E.F. Haratsch, A. Cristal, O. Unsal, K. Mai, Flash correct and refresh: retention aware management for increased lifetime, in *ICCD* (2012)
30. Y. Cai, G. Yalcin, O. Mutlu, E.F. Haratsch, A. Cristal, O. Unsal, K. Mai, Error analysis and retention-aware error management for NAND flash memory. Intel. Technol. J. (2013)

31. Y. Cai, G. Yalcin, O. Mutlu, E.F. Haratsch, O. Unsal, A. Cristal, K. Mai, Neighbor cell assisted error correction in MLC NAND flash memories, in *SIGMETRICS* (2014)
32. J. Cha, S. Kang, Data randomization scheme for endurance enhancement and interference mitigation of multilevel flash memory devices. ETRI J. (2013)
33. K. Chandrasekar, S. Goossens, C. Weis, M. Koedam, B. Akesson, N. Wehn, K. Goossens, Exploiting expendable process-margins in DRAMs for run-time performance optimization, in *DATE* (2014)
34. K.K. Chang, Understanding and improving the latency of DRAM-based memory systems. Ph.D. Dissertation, Carnegie Mellon University, 2017
35. K.K. Chang, D. Lee, Z. Chishti, A.R. Alameldeen, C. Wilkerson, Y. Kim, O. Mutlu, Improving DRAM performance by parallelizing refreshes with accesses, in *HPCA* (2014)
36. K.K. Chang, P.J. Nair, S. Ghose, D. Lee, M.K. Qureshi, O. Mutlu, Low-cost inter-linked subarrays (LISA): enabling fast inter-subarray data movement in DRAM, in *HPCA* (2016)
37. K.K. Chang, A. Kashyap, H. Hassan, S. Ghose, K. Hsieh, D. Lee, T. Li, G. Pekhimenko, S. Khan, O. Mutlu, Understanding latency variation in modern DRAM chips: experimental characterization, analysis, and optimization, in *SIGMETRICS* (2016)
38. K.K. Chang, A.G. Yaglikci, A. Agrawal, N. Chatterjee, S. Ghose, A. Kashyap, H. Hassan, D. Lee, M. O'Connor, O. Mutlu, Understanding reduced-voltage operation in modern DRAM devices: experimental characterization, analysis, and mechanisms, in *SIGMETRICS* (2017)
39. L.-P. Chang, On efficient wear leveling for large-scale flash-memory storage systems, in *SAC* (2007)
40. L.-P. Chang, T.-W. Kuo, S.-W. Lo, *Real-time garbage collection for flash-memory storage systems of real-time embedded systems* (ACM Trans. Embed. Comput, Syst, 2004)
41. N. Chatterjee, M. Shevgoor, R. Balasubramonian, A. Davis, Z. Fang, R. Illikkal, R. Iyer, Leveraging heterogeneity in DRAM main memories to accelerate critical word access, in *MICRO* (2012)
42. C.-L. Chen, High-speed decoding of BCH codes (Corresp.) IEEE Trans. Inf. Theory (1981)
43. J. Chen, M.P.C. Fossorier, *Near optimum universal belief propagation based decoding of low-density parity check codes* (IEEE Trans, Commun, 2002)
44. T.-H. Chen, Y.-Y. Hsiao, Y.-T. Hsing, C.-W. Wu, An adaptive-rate error correction scheme for nand flash memory, in *VTS* (2009)
45. Z. Chen, E.F. Haratsch, S. Sankaranarayanan, Y. Wu, Estimating read reference voltage based on disparity and derivative metrics, U.S. Patent 9,417,797 (2016)
46. R.T. Chien, *Cyclic decoding procedures for the Bose-Chaudhuri-Hocquenghem codes* IEEE Trans. Inf. Theory (1964)
47. B. Choi et al., Comprehensive evaluation of early retention (fast charge loss within a few seconds) characteristics in tube-type 3-D nand flash memory, in *VLSIT* (2016)
48. H. Choi, W. Liu, W. Sung, *VLSI implementation of BCH error correction for multilevel cell NAND flash memory* (IEEE Trans. Very Large Scale Integr, Syst, 2009)
49. C. Chou, P. Nair, M.K. Qureshi, Reducing refresh power in mobile devices with morphable ECC, in *DSN* (2015)
50. C.-C. Chou, A. Jaleel, M.K. Qureshi, CAMEO: a two-level memory organization with capacity of main memory and flexibility of hardware-managed cache, in *MICRO* (2014)
51. C.-C. Chou, A. Jaleel, M.K. Qureshi, BEAR: techniques for mitigating bandwidth bloat in gigascale DRAM caches, in *ISCA* (2015)
52. S. Choudhuri, T. Givargis, Deterministic service guarantees for NAND flash using partial block cleaning, in *CODES + ISSS* (2008)
53. L. Chua, *Memristor–the missing circuit element* (IEEE Trans, Circuit Theory, 1971)
54. T.-S. Chung, D.-J. Park, S. Park, D.-H.L.S.-W. Lee, H.-J. Song, A survey of flash translation layer. J. Syst. Archit. (2009)
55. R. Codandaramane, Securing the SSDs—NVMe controller encryption, in *Flash Memory Summit* (2016)
56. E.T. Cohen, Zero-one balance management in a solid-state disk controller, U.S. Patent 8,839,073 (2014)

57. E.T. Cohen, Y. Cai, E.F. Haratsch, Y. Wu, Method to dynamically update LLRs in an SSD drive and/or controller, U.S. Patent 9,329,935 (2015)
58. J. Cooke, The inconvenient truths of NAND flash memory, in *Flash Memory Summit* (2007)
59. J. Daemen, V. Rijmen, *The Design of Rijndael* (Germany, New York, NY, USA, Springer, Berlin, Heidelberg, 2002)
60. R. Degraeve et al., *Analytical percolation model for predicting anomalous charge loss in flash memories* (IEEE Trans. Electron, Devices, 2004)
61. T.J. Dell, *A white paper on the benefits of chipkill-correct ECC for PC server main memory* IBM Microelectron. Division Tech. Rep. (1997)
62. P. Desnoyers, Analytic modeling of SSD write performance, in *SYSTOR* (2012)
63. C. Dirik, B. Jacob, The performance of PC solid-state disks (SSDs) as a function of bandwidth, concurrency, device architecture, and system organization, in *ISCA* (2009)
64. L. Dolecek, Making error correcting codes work for flash memory, in *Flash Memory Summit* (2014)
65. B. Eitan, Non-volatile semiconductor memory cell utilizing asymmetrical charge trapping, U.S. Patent 5,768,192 (1998)
66. J. Elliott, J. Jeong, *Advancements in SSDs and 3D NAND reshaping storage market* Keynote Present. in *Flash Memory Summit* (2017)
67. Facebook, Inc., Flashcache. https://github.com/facebookarchive/flashcache
68. M.P.C. Fossorier, M. Mihaljević, H. Imai, *Reduced complexity iterative decoding of low-density parity check codes based on belief propagation* (IEEE Trans, Commun, 1999)
69. R.H. Fowler, L. Nordheim, Electron emission in intense electric fields Proc. R. Soc. A (1928)
70. A. Fukami, S. Ghose, Y. Luo, Y. Cai, O. Mutlu, *Improving the reliability of chip-off forensic analysis of NAND flash memory devices* (Digit, Investig, 2017)
71. E. Gal, S. Toledo, *Algorithms and data structures for flash memories* (ACM Comput, Surv, 2005)
72. R.G. Gallager, *Low-density parity-check codes* (IRE Trans. Inf, Theory, 1962)
73. R.G. Gallager, *Low-Density Parity-Check Codes* (MIT Press, Cambridge, MA, USA, 1963)
74. S. Ghose, H. Lee, J.F. Martínez, Improving memory scheduling via processor-side load criticality information, in *ISCA* (2013)
75. L.M. Grupp, A.M. Caulfield, J. Coburn, S. Swanson, E. Yaakobi, P.H. Siegel, J.K. Wolf, Characterizing flash memory: anomalies, observations, and applications, in *MICRO* (2009)
76. D. Gruss, M. Lipp, M. Schwarz, D. Genkin, J. Juffinger, S. O'Connell, W. Schoechl, Y. Yarom, Another flip in the wall of Rowhammer defenses (2017), arXiv:1710.00551
77. D. Gruss, C. Maurice, S. Mangard, Rowhammer.js: a remote software-induced fault attack in javascript, in *DIMVA* (2016)
78. K. Gunnam, LDPC decoding: VLSI architectures and implementations, in *Flash Memory Summit* (2014)
79. K.K. Gunnam, G.S. Choi, M.B. Yeary, M. Atiquzzaman, VLSI architectures for layered decoding for irregular LDPC codes of WiMax, in *ICC* (2007)
80. A. Gupta, Y. Kim, B. Urgaonkar, DFTL: a flash translation layer employing demand-based selective caching of page-level address mappings, in *ASPLOS* (2009)
81. K. Ha, J. Jeong, J. Kim, A read-disturb management technique for high-density NAND flash memory, in *APSys* (2013)
82. K. Ha, J. Jeong, J. Kim, *An integrated approach for managing read disturbs in high-density NAND flash memory* (IEEE Trans. Comput.-Aided Des. Integr, Circuits Syst, 2016)
83. T. Hamamoto, S. Sugiura, S. Sawada, *On the retention time distribution of dynamic random access memory (DRAM)* (IEEE Trans. Electron, Devices, 1998)
84. L. Han, Y. Ryu, K. Yim, CATA: a garbage collection scheme for flash memory file systems, in *UIC* (2006)
85. E.F. Haratsch, Controller concepts for 1y/1z nm and 3D NAND flash, in *Flash Memory Summit* (2015)
86. E.F. Haratsch, Media management for high density NAND flash memories, in *Flash Memory Summit* (2016)

87. H. Hassan, N. Vijaykumar, S. Khan, S. Ghose, K. Chang, G. Pekhimenko, D. Lee, O. Ergin, O. Mutlu, SoftMC: a flexible and practical open-source infrastructure for enabling experimental DRAM studies, in *HPCA* (2017)
88. H. Hassan, G. Pekhimenko, N. Vijaykumar, V. Seshadri, D. Lee, O. Ergin, O. Mutlu, Charge-Cache: reducing DRAM latency by exploiting row access locality, in *HPCA* (2016)
89. J. Haswell, SSD architectures to ensure security and performance, in *Flash Memory Summit* (2016)
90. J. He, S. Kannan, A.C. Arpaci-Dusseau, R.H. Arpaci-Dusseau, The unwritten contract of solid state drives, in *EuroSys* (2017)
91. J. Ho, B. Chester, The iPhone 7 and iPhone 7 Plus review: iterating on a flagship, in *AnandTech* (2016)
92. A. Hocquenghem, *Codes Correcteurs d'Erreurs*. Chiffres (1959)
93. X.-Y. Hu, E. Eleftheriou, R. Haas, I. Iliadis, R. Pletka, Write amplification analysis in flash-based solid state drives, in *SYSTOR* (2009)
94. Y. Hu, H. Jiang, D. Feng, L. Tian, H. Luo, S. Zhang, Performance impact and interplay of SSD parallelism through advanced commands, allocation strategy and data granularity, in *ICS* (2011)
95. P. Huang, P. Subedi, X. He, S. He, K. Zhou, FlexECC: partially relaxing ECC of MLC SSD for better cache performance, in *USENIX ATC* (2014)
96. A. Hwang, I. Stefanovici, B. Schroeder, Cosmic rays don't strike twice: understanding the nature of DRAM errors and the implications for system design, in *ASPLOS* (2012)
97. D. Ielmini, A.L. Lacaita, D. Mantegazza, *Recovery and drift dynamics of resistance and threshold voltages in phase-change memories* (IEEE Trans. Electron, Devices, 2007)
98. J. Im et al., A 128Gb 3b/Cell V-NAND flash memory with 1Gb/s I/O rate, in *ISSCC* (2015)
99. Intel Corp., *Serial ATA Advanced Host Controller Interface (AHCI) 1.3.1* (2012)
100. E. Ipek, O. Mutlu, J. F. Martínez, R. Caruana, Self-optimizing memory controllers: a reinforcement learning approach, in *ISCA* (2008)
101. C. Isen, L. John, ESKIMO—Energy savings using semantic knowledge of inconsequential memory occupancy for DRAM subsystem, in *MICRO* (2009)
102. J. Jang et al., Vertical cell array using TCAT (terabit cell array transistor) technology for ultra high density NAND flash memory, in *VLSIT* (2009)
103. JEDEC Solid State Technology Assn., *Solid-State Drive (SSD) Requirements and Endurance Test Method* (Publication JEP218, 2010)
104. JEDEC Solid State Technology Assn., *DDR4 SDRAM Standard* (Publication JESD79-4A, 2013)
105. JEDEC Solid State Technology Assn., *Failure Mechanisms and Models for Semiconductor Devices* (Publication JEP122H, 2016)
106. J. Jeong, S.S. Hahn, S. Lee, J. Kim, Lifetime improvement of NAND flash-based storage systems using dynamic program and erase scaling, in *FAST* (2014)
107. S. Jeong, K. Lee, S. Lee, S. Son, Y. Won, I/O stack optimization for smartphones, in *USENIX ATC* (2013)
108. L. Jiang, Y. Zhang, J. Yang, Mitigating write disturbance in super-dense phase change memories, in *DSN* (2014)
109. X. Jiang, N. Madan, L. Zhao, M. Upton, R. Iyer, S. Makineni, D. Newell, D. Solihin, R. Balasubramonian, CHOP: adaptive filter-based DRAM caching for CMP server platforms, in *HPCA* (2010)
110. S.J. Johnson, Introducing low-density parity-check codes, http://sigpromu.org/sarah/SJohnsonLDPCintro.pdf
111. D. Kahng, S.M. Sze, A floating gate and its application to memory devices. Bell Syst. Tech. J. (1967)
112. D. Kang et al., 7.1 256Gb 3b/cell V-NAND flash memory with 48 stacked WL layers, in *ISSCC* (2016)
113. J.-U. Kang, H. Jo, J.-S. Kim, J. Lee, A superblock-based flash translation layer for NAND flash memory, in *EMSOFT* (2006)

114. U. Kang, H.-S. Yu, C. Park, H. Zheng, J. Halbert, K. Bains, S. Jang, J. Choi, Co-architecting controllers and DRAM to enhance DRAM process scaling, in *Memory Forum* (2014)
115. R. Katsumata et al., Pipe-Shaped BiCS flash memory with 16 stacked layers and multi-level-cell operation for ultra high density storage devices, in *VLSIT* (2009)
116. S. Khan, D. Lee, Y. Kim, A. Alameldeen, C. Wilkerson, O. Mutlu, The efficacy of error mitigation techniques for DRAM retention failures: a comparative experimental study, in *SIGMETRICS* (2014)
117. S. Khan, D. Lee, O. Mutlu, PARBOR: an efficient system-level technique to detect data-dependent failures in DRAM, in *DSN* (2016)
118. S. Khan, C. Wilkerson, D. Lee, A.R. Alameldeen, O. Mutlu, *A case for memory content-based detection and mitigation of data-dependent failures in DRAM* (IEEE Comput. Archit, Lett, 2016)
119. S. Khan, C. Wilkerson, Z. Wang, A.R. Alameldeen, D. Lee, O. Mutlu, Detecting and mitigating data-dependent DRAM failures by exploiting current memory content, in *MICRO* (2017)
120. W.-S. Khwa et al., A resistance-drift compensation scheme to reduce MLC PCM raw BER by over 100 × for storage-class memory applications, in *ISSCC* (2016)
121. C. Kim et al., *A 21 nm high performance 64 Gb MLC NAND flash memory with 400 MB/s asynchronous toggle DDR interface* (IEEE J, Solid-State Circuits, 2012)
122. C. Kim et al., A 512 Gb 3b/Cell 64-Stacked WL 3D V-NAND flash memory, in *ISSCC* (2017)
123. J. Kim, M. Sullivan, M. Erez, Bamboo ECC: strong, safe, and flexible codes for reliable computer memory, in *HPCA* (2015)
124. J. Kim, M. Sullivan, S.-L. Gong, M. Erez, Frugal ECC: efficient and versatile memory error protection through fine-grained compression, in *SC* (2015)
125. J.S. Kim, M. Patel, H. Hassan, O. Mutlu, The DRAM latency PUF: quickly evaluating physical unclonable functions by exploiting the latency–reliability tradeoff in modern DRAM devices, in *HPCA* (2018)
126. K. Kim, J. Lee, *A new investigation of data retention time in truly nanoscaled DRAMs* (IEEE Electron, Device Lett, 2009)
127. N. Kim, J.-H. Jang, Nonvolatile memory device, method of operating nonvolatile memory device and memory system including nonvolatile memory device. U.S. Patent 8,203,881 (2012)
128. Y. Kim, Architectural techniques to enhance DRAM scaling. Ph.D. Dissertation, Carnegie Mellon Univ., 2015
129. Y. Kim, D. Han, O. Mutlu, M. Harchol-Balter, ATLAS: a scalable and high-performance scheduling algorithm for multiple memory controllers, in *HPCA* (2010)
130. Y. Kim, O. Mutlu, *"Memory Systems," in Computing Handbook*, 3rd edn. (CRC Press, Boca Raton, FL, USA, 2014)
131. Y. Kim, V. Seshadri, D. Lee, J. Liu, O. Mutlu, A case for exploiting subarray-level parallelism (SALP) in DRAM, in *ISCA* (2012)
132. Y. Kim, W. Yang, O. Mutlu, *Ramulator: a fast and extensible DRAM simulator* (IEEE Comput. Archit, Lett, 2016)
133. Y.S. Kim, D.J. Lee, C.K. Lee, H.K. Choi, S.S. Kim, J.H. Song, D.H. Song, J.-H. Choi, K.-D. Suh, C. Chung, New scaling limitation of the floating gate cell in NAND flash memory, in *IRPS* (2010)
134. Y. Kim, R. Daly, J. Kim, C. Fallin, J. H. Lee, D. Lee, C. Wilkerson, K. Lai, O. Mutlu, "Flipping bits in memory without accessing them: an experimental study of DRAM disturbance errors, in *ISCA* (2014)
135. Y. Kim, M. Papamichael, O. Mutlu, M. Harchol-Balter, Thread cluster memory scheduling: exploiting differences in memory access behavior, in *MICRO* (2010)
136. Y. Koh, NAND flash scaling beyond 20 nm, in *IMW* (2009)
137. Y. Komori, M. Kido, M. Kito, R. Katsumata, Y. Fukuzumi, H. Tanaka, Y. Nagata, M. Ishiduki, H. Aochi, A. Nitayama, Disturbless flash memory due to high boost efficiency on BiCS structure and optimal memory film stack for ultra high density storage device, in *IEDM* (2008)

138. E. Kültürsay, M. Kandemir, A. Sivasubramaniam, O. Mutlu, Evaluating STT-RAM as an energy-efficient main memory alternative, in *ISPASS* (2013)
139. M. LaPedus, *How to make 3D NAND* (Semicond, Eng, 2016)
140. B.C. Lee, E. Ipek, O. Mutlu, D. Burger, Architecting phase change memory as a scalable DRAM alternative, in *ISCA* (2009)
141. B.C. Lee, E. Ipek, O. Mutlu, D. Burger, Phase change memory architecture and the quest for scalability. ACM Commun. (2010)
142. B.C. Lee, P. Zhou, J. Yang, Y. Zhang, B. Zhao, E. Ipek, O. Mutlu, D. Burger, Phase-change technology and the future of main memory. IEEE Micro (2010)
143. C.J. Lee, V. Narasiman, O. Mutlu, Y.N. Patt, Improving memory bank-level parallelism in the presence of prefetching, in *MICRO* (2009)
144. D. Lee, Reducing DRAM energy at low cost by exploiting heterogeneity. Ph.D. Dissertation, Carnegie Mellon University, 2016
145. D. Lee, S. Ghose, G. Pekhimenko, S. Khan, O. Mutlu, Simultaneous multi-layer access: improving 3D-stacked memory bandwidth at low cost. ACM TACO (2016)
146. D. Lee, S. Khan, L. Subramanian, S. Ghose, R. Ausavarungnirun, G. Pekhimenko, V. Seshadri, O. Mutlu, Design-induced latency variation in modern DRAM chips: characterization, analysis, and latency reduction mechanisms, in *SIGMETRICS* (2017)
147. D. Lee, Y. Kim, V. Seshadri, J. Liu, L. Subramanian, O. Mutlu, Tiered-Latency DRAM: a low latency and low cost DRAM architecture, in *HPCA* (2013)
148. D. Lee, L. Subramanian, R. Ausavarungnirun, J. Choi, and O. Mutlu, Decoupled direct memory access: isolating CPU and IO traffic by leveraging a dual-data-port DRAM, in *PACT* (2015)
149. D. Lee, Y. Kim, G. Pekhimenko, S. Khan, V. Seshadri, K. Chang, O. Mutlu, Adaptive-latency DRAM: optimizing DRAM timing for the common-case, in *HPCA* (2015)
150. J.-D. Lee, J.-H. Choi, D. Park, K. Kim, Degradation of tunnel oxide by FN current stress and its effects on data retention characteristics of 90 nm NAND flash memory cells, in *IRPS* (2003)
151. J.-D. Lee, S.-H. Hur, J.-D. Choi, *Effects of floating-gate interference on NAND flash memory cell operation* (IEEE Electron, Device Lett, 2002)
152. S.-Y. Lee, Limitations of 3D NAND scaling. EE Times (2017)
153. Y. Lee, H. Yoo, I. Yoo, I.-C. Park, 6.4 Gb/s Multi-threaded BCH encoder and decoder for multi-channel SSD controllers, in *ISSCC* (2012)
154. J. Li, K. Zhao, X. Zhang, J. Ma, M. Zhao, T. Zhang, How much can data compressibility help to improve NAND flash memory lifetime? in *FAST* (2015)
155. Y. Li, S. Ghose, J. Choi, J. Sun, H. Wang, O. Mutlu, Utility-based hybrid memory management, in *CLUSTER* (2017)
156. Y. Li, C. Hsu, K. Oowada, Non-volatile memory and method with improved first pass programming, U.S. Patent 8,811,091 (2014)
157. J. Liu, B. Jaiyen, Y. Kim, C. Wilkerson, O. Mutlu, An experimental study of data retention behavior in modern DRAM devices: implications for retention time profiling mechanisms, in *ISCA* (2013)
158. J. Liu, B. Jaiyen, R. Veras, O. Mutlu, RAIDR: retention-aware intelligent DRAM refresh, in *ISCA* (2012)
159. S. Liu, K. Pattabiraman, T. Moscibroda, B. Zorn, Flikker: saving DRAM refresh-power through critical data partitioning, in *ASPLOS* (2011)
160. W. Liu, J. Rho, W. Sung, Low-power high-throughput BCH error correction VLSI design for multi-level cell NAND flash memories, in *SIPS* (2006)
161. Y. Luo, Y. Cai, S. Ghose, J. Choi, O. Mutlu, WARM: improving NAND flash memory lifetime with write-hotness aware retention management, in *MSST* (2015)
162. Y. Luo, S. Ghose, Y. Cai, E.F. Haratsch, O. Mutlu, *Enabling accurate and practical online flash channel modeling for modern MLC NAND flash memory* (IEEE J. Sel, Areas Commun, 2016)

163. Y. Luo, S. Ghose, T. Li, S. Govindan, B. Sharma, B. Kelly, B. Kelly, A. Boroumand, O. Mutlu, Using ECC DRAM to adaptively increase memory capacity (2017), arXiv:1706.08870
164. Y. Luo, S. Ghose, Y. Cai, E. F. Haratsch, O. Mutlu, HeatWatch: improving 3D NAND flash memory device reliability by exploiting self-recovery and temperature awareness, in *HPCA* (2018)
165. Y. Luo, S. Govindan, B. Sharma, M. Santaniello, J. Meza, A. Kansal, J. Liu, B. Khessib, K. Vaid, O. Mutlu, Characterizing application memory error vulnerability to optimize datacenter cost via heterogeneous-reliability memory, in *DSN* (2014)
166. S. Luryi, A. Kastalsky, A.C. Gossard, R.H. Hendel, *Charge injection transistor based on real-space hot-electron transfer* (IEEE Trans. Electron, Devices, 1984)
167. D.J.C. MacKay, R.M. Neal, *Near Shannon limit performance of low density parity check codes* (IET Electron, Lett, 1997)
168. A. Maislos, A new era in embedded flash memory, in *Flash Memory Summit* (2011)
169. J.A. Mandelman, R.H. Dennard, G.B. Bronner, J.K. DeBrosse, R. Divakaruni, Y. Li, C.J. Radens, *Challenges and future directions for the scaling of dynamic random-access memory (DRAM)* (IBM J. Res, Develop, 2002)
170. A. Marelli, R. Micheloni, *BCH and LDPC error correction codes for NAND flash memories, in 3D Flash Memories* (Springer, Dordrecht, Netherlands, 2016)
171. J.L. Massey, *Shift-register synthesis and BCH decoding* (IEEE Trans. Inf, Theory, 1969)
172. F. Masuoka, M. Momodomi, Y. Iwata, R. Shirota, New ultra high density EPROM and flash EEPROM with NAND structure cell, in *IEDM* (1987)
173. J. Meza, Y. Luo, S. Khan, J. Zhao, Y. Xie, O. Mutlu, A case for efficient hardware-software cooperative management of storage and memory, in *WEED* (2013)
174. J. Meza, Q. Wu, S. Kumar, O. Mutlu, A large-scale study of flash memory errors in the field, in *SIGMETRICS* (2015)
175. J. Meza, Q. Wu, S. Kumar, O. Mutlu, Revisiting memory errors in large-scale production data centers: analysis and modeling of new trends from the field, in *DSN* (2015)
176. J. Meza, J. Chang, H. Yoon, O. Mutlu, P. Ranganathan, *Enabling efficient and scalable hybrid memories using fine-granularity DRAM cache management* (IEEE Comput. Archit, Lett, 2012)
177. R. Micheloni (ed.), *3D Flash Memories* (Netherlands, Springer, Netherlands, Dordrecht, 2016)
178. R. Micheloni, S. Aritome, L. Crippa, Array architectures for 3-D NAND flash memories. Proc. IEEE (2017)
179. R. Micheloni et al., A 4Gb 2b/Cell NAND flash memory with embedded 5b BCH ECC for 36 MB/s system read throughput, in *ISSCC* (2006)
180. Micron Technology, Inc., *Memory Management in NAND Flash Arrays*, Tech Note TN-29-28, 2005
181. Micron Technology, Inc., *Bad Block Management in NAND Flash Memory*, Tech Note TN-29-59, 2011
182. N. Mielke, T. Marquart, N. Wu, J. Kessenich, H. Belgal, E. Schares, F. Trivedi, E. Goodness, L.R. Nevill, Bit error rate in NAND flash memories, in *IRPS* (2008)
183. K. Mizoguchi, T. Takahashi, S. Aritome, K. Takeuchi, Data-retention characteristics comparison of 2D and 3D TLC NAND flash memories, in *IMW* (2017)
184. V. Mohan, Modeling the physical characteristics of NAND flash memory. Ph.D. Dissertation, University of Virginia, 2010
185. V. Mohan, S. Sankar, S. Gurumurthi, W. Redmond, ReFresh SSDs: enabling high endurance, low cost flash in datacenters. Technical Report No. CS-2012-05 (University of Virginia, 2012)
186. V. Mohan, T. Siddiqua, S. Gurumurthi, M.R. Stan, How I learned to stop worrying and love flash endurance, in *HotStorage* (2010)
187. M. Momodomi, F. Masuoka, R. Shirota, Y. Itoh, K. Ohuchi, R. Kirisawa, Electrically erasable programmable read-only memory with NAND cell structure, U.S. Patent 4,959,812 (1988)
188. T. Moscibroda, O. Mutlu, Memory performance attacks: denial of memory service in multi-core systems, in *USENIX Security* (2007)

189. T. Moscibroda, O. Mutlu, Distributed order scheduling and its application to multi-core DRAM controllers, in *PODC* (2008)
190. J. Mukundan, H. Hunter, K.-H. Kim, J. Stuecheli, J.F. Martínez, Understanding and mitigating refresh overheads in high-density DDR4 DRAM systems, in *ISCA* (2013)
191. S.P. Muralidhara, L. Subramanian, O. Mutlu, M. Kandemir, T. Moscibroda, Reducing memory interference in multicore systems via application-aware memory channel partitioning, in *MICRO* (2011)
192. O. Mutlu, Memory scaling: a systems architecture perspective, in *IMW* (2013)
193. O. Mutlu, The Rowhammer problem and other issues we may face as memory becomes denser, in *DATE* (2017)
194. O. Mutlu, T. Moscibroda, Stall-time fair memory access scheduling for chip multiprocessors, in *MICRO* (2007)
195. O. Mutlu, T. Moscibroda, Parallelism-aware batch scheduling: enhancing both performance and fairness of shared DRAM systems, in *ISCA* (2008)
196. O. Mutlu, L. Subramanian, Research problems and opportunities in memory systems, *SUPERFRI* (2014)
197. H. Naeimi, C. Augustine, A. Raychowdhury, S.-L. Lu, J. Tschanz, STT-RAM scaling and retention failure. Intel Technol. J. (2013)
198. P.J. Nair, V. Sridharan, M.K. Qureshi, XED: exposing on-die error detection information for strong memory reliability, in *ISCA* (2016)
199. I. Narayanan, D. Wang, M. Jeon, B. Sharma, L. Caulfield, A. Sivasubramaniam, B. Cutler, J. Liu, B. Khessib, K. Vaid, SSD failures in datacenters: What? When? and Why? in *SYSTOR* (2016)
200. K. Naruke, S. Taguchi, M. Wada, Stress induced leakage current limiting to scale down EEPROM tunnel oxide thickness, in *IEDM* (1988)
201. National Inst. of Standards and Technology, *Specification for the Advanced Encryption Standard (AES)*, FIPS Publication 197, 2001
202. NVM Express, Inc., *NVM Express Specification, Revision 1.3*, 2017
203. S. Ohshima Y. Tanaka, New 3D flash technologies offer both low cost and low power solutions, in *Flash Memory Summit* (2016)
204. Openmoko, NAND Bad Blocks, http://wiki.openmoko.org/wiki/NAND_bad_blocks (2012)
205. Y. Pan, G. Dong, Q. Wu, T. Zhang, Quasi-nonvolatile SSD: trading flash memory nonvolatility to improve storage system performance for enterprise applications, in *HPCA* (2012)
206. N. Papandreou, T. Parnell, H. Pozidis, T. Mittelholzer, E. Eleftheriou, C. Camp, T. Griffin, G. Tressler, A. Walls, Using adaptive read voltage thresholds to enhance the reliability of MLC NAND flash memory systems, in *GLSVLSI* (2014)
207. J. Park, J. Jeong, S. Lee, Y. Song, J. Kim, Improving performance and lifetime of NAND storage systems using relaxed program sequence, in *DAC* (2016)
208. K.-T. Park et al., A 7MB/s 64Gb 3-Bit/Cell DDR NAND Flash Memory in 20nm-node technology, in *ISSCC* (2011)
209. K.-T. Park, M. Kang, D. Kim, S.-W. Hwang, B.Y. Choi, Y.-T. Lee, C. Kim, K. Kim, *A zeroing cell-to-cell interference page architecture with temporary LSB storing and parallel MSB program scheme for MLC NAND flash memories* (IEEE J, Solid-State Circuits, 2008)
210. K. Park et al., Three-dimensional 128 Gb MLC vertical NAND flash memory with 24-WL stacked layers and 50 MB/s high-speed programming. J. Solid-State Circuits (2015)
211. T. Parnell, NAND flash basics and error characteristics: why do we need smart controllers? in *Flash Memory Summit* (2016)
212. T. Parnell, N. Papandreou, T. Mittelholzer, H. Pozidis, Modelling of the threshold voltage distributions of sub-20nm NAND flash memory, in *GLOBECOM* (2014)
213. T. Parnell, R. Pletka, NAND flash basics and error characteristics, in *Flash Memory Summit* (2017)
214. M. Patel, J.S. Kim, O. Mutlu, The reach profiler (REAPER): enabling the mitigation of DRAM retention failures via profiling at aggressive conditions, in *ISCA* (2017)

215. D.A. Patterson, G. Gibson, R.H. Katz, A case for redundant arrays of inexpensive disks (RAID), in *SIGMOD* (1988)
216. P. Pavan, R. Bez, P. Olivo, E. Zanoni, Flash memory cells–an overview. Proc. IEEE (1997)
217. PCI-SIG, *PCI Express Base Specification Revision 3.1a*, 2015
218. J. Pearl, Reverend bayes on inference engines: a distributed hierarchical approach, in *AAAI* (1982)
219. W.W. Peterson, D.T. Brown, *Cyclic codes for error detection* (Proc, IRE, 1961)
220. S. Phadke, S. Narayanasamy, MLP aware heterogeneous memory system, in *DATE* (2011)
221. A. Pirovano, A.L. Lacaita, F. Pellizzer, S.A. Kostylev, A. Benvenuti, R. Bez, *Low-field amorphous state resistance and threshold voltage drift in chalcogenide materials* (IEEE Trans, Electron Devices, 2004)
222. Z. Qin, Y. Wang, D. Liu, Z. Shao, Y. Guan, MNFTL: an efficient flash translation layer for MLC NAND flash memory storage systems, in *DAC* (2011)
223. M. Qureshi, D.H. Kim, S. Khan, P. Nair, O. Mutlu, AVATAR: a variable-retention-time (VRT) aware refresh for DRAM systems, in *DSN* (2015)
224. M.K. Qureshi, V. Srinivasan, J.A. Rivers, Scalable high performance main memory system using phase-change memory technology, in *ISCA* (2009)
225. M.K. Qureshi, G.H. Loh, Fundamental latency trade-off in architecting DRAM caches: outperforming impractical SRAM-tags with a simple and practical design, in *MICRO* (2012)
226. L.E. Ramos, E. Gorbatov, R. Bianchini, Page placement in hybrid memory systems, in *ICS* (2011)
227. K. Razavi, B. Gras, E. Bosman, B. Preneel, C. Guiffrida, H. Bos, Flip Feng Shui: hammering a needle in the software stack, in *USENIX Security* (2016)
228. P.J. Restle, J.W. Park, B.F. Lloyd, DRAM variable retention time, in *IEDM* (1992)
229. D. Rollins, *A Comparison of Client and Enterprise SSD Data Path Protection* (Micron Technology, Inc., 2011)
230. W. Ryan, S. Lin, *Channel Codes: Classical and Modern* (Cambridge University Press, Cambridge, UK, 2009)
231. Samsung Electronics Co., Ltd., Samsung V-NAND Technology (2014), http://www.samsung.com/us/business/oem-solutions/pdfs/V-NAND_technology_WP.pdf
232. Samsung Electronics Co., Ltd., *Samsung SSD 960 PRO M.2 Data Sheet Rev. 1.1* (2017)
233. B. Schroeder, R. Lagisetty, A. Merchant, Flash reliability in production: the expected and the unexpected, in *FAST* (2016)
234. B. Schroeder, E. Pinheiro, W.-D. Weber, DRAM errors in the wild: a large-scale field study, in *SIGMETRICS* (2009)
235. M. Seaborn, T. Dullien, *Exploiting the DRAM Rowhammer Bug to Gain Kernel Privileges*, (Google Project Zero Blog, 2015)
236. M. Seaborn, T. Dullien, Exploiting the DRAM Rowhammer bug to gain kernel privileges, in *BlackHat* (2015)
237. Seagate Technology LLC, *Enterprise Performance 15K HDD Data Sheet* (2016)
238. Serial ATA International Organization, *Serial ATA Revision 3.3 Specification* (2016)
239. C.E. Shannon, A mathematical theory of communication. Bell Syst. Tech. J. (July 1948)
240. C.E. Shannon, A mathematical theory of communication. Bell Syst. Tech. J. (Oct 1948)
241. H. Shim et al., Highly reliable 26nm 64Gb MLC E2NAND (embedded-ECC and enhanced-efficiency) flash memory with MSP (memory signal processing) controller, in *VLSIT* (2011)
242. S.-H. Shin et al., A new 3-bit programming algorithm using SLC-to-TLC migration for 8 MB/s high performance TLC NAND flash memory, in *VLSIC* (2012)
243. L. Shu, D.J. Costello, *Error Control Coding*, 2nd edn. (Prentice-Hall, Englewood Cliffs, NJ, USA, 2004)
244. S. Sills, S. Yasuda, A. Calderoni, C. Cardon, J. Strand, K. Aratani, N. Ramaswamy, Challenges for high-density 16Gb ReRAM with 27nm technology, in *VLSIC* (2015)
245. S. Sills, S. Yasuda, J. Strand, A. Calderoni, K. Aratani, A. Johnson, N. Ramaswamy, A copper ReRAM cell for storage class memory applications, in *VLSIT* (2014)

246. V. Sridharan, J. Stearley, N. DeBardeleben, S. Blanchard, S. Gurumurthi, Feng Shui of supercomputer memory: positional effects in DRAM and SRAM faults, in *SC* (2013)
247. V. Sridharan, N. DeBardeleben, S. Blanchard, K.B. Ferreira, J. Stearley, J. Shalf, S. Gurumurthi, Memory errors in modern systems: the good, the bad, and the ugly, in *ASPLOS* (2015)
248. D.B. Strukov, G.S. Snider, D.R. Stewart, R.S. Williams, The missing memristor found. Nature (2008)
249. J. Stuecheli, D. Kaseridis, H.C. Hunter, L.K. John, Elastic refresh: techniques to mitigate refresh penalties in high density memory, in *MICRO* (2010)
250. L. Subramanian, D. Lee, V. Seshadri, H. Rastogi, O. Mutlu, The blacklisting memory scheduler: achieving high performance and fairness at low cost, in *ICCD* (2014)
251. L. Subramanian, D. Lee, V. Seshadri, H. Rastogi, O. Mutlu, *BLISS: balancing performance, fairness and complexity in memory access scheduling* (IEEE Trans. Parallel Distrib, Syst, 2016)
252. L. Subramanian, V. Seshadri, Y. Kim, B. Jaiyen, O. Mutlu, MISE: providing performance predictability and improving fairness in shared main memory systems, in *HPCA* (2013)
253. K.-D. Suh et al., A 3.3V 32 Mb NAND Flash memory with incremental step pulse programming scheme. IEEE J. Solid-State Circuits (1995)
254. K. Takeuchi, S. Satoh, T. Tanaka, K.-I. Imamiya, K. Sakui, *A negative Vth cell architecture for highly scalable, excellently noise-immune, and highly reliable NAND flash memories* (IEEE J, Solid-State Circuits, 1999)
255. H. Tanaka et al., Bit cost scalable technology with punch and plug process for ultra high density flash memory, in *VLSIT* (2007)
256. S. Tanakamaru, C. Hung, A. Esumi, M. Ito, K. Li, K. Takeuchi, 95%-lower-BER 43%-lower-power intelligent solid-state drive (SSD) with asymmetric coding and stripe pattern elimination algorithm, in *ISSCC* (2011)
257. L. Tang, Q. Huang, W. Lloyd, S. Kumar, K. Li, RIPQ: advanced photo caching on flash for facebook, in *FAST* (2015)
258. R. Tanner, *A recursive approach to low complexity codes* (IEEE Trans. Inf, Theory, 1981)
259. Techman Electronics Co., Techman XC100 NVMe SSD, White Paper v1.0, 2016
260. Toshiba Corp., 3D Flash Memory: Scalable, High Density Storage for Large Capacity Applications (2017), http://www.toshiba.com/taec/adinfo/technologymoves/3d-flash.jsp
261. A.N. Udipi, N. Muralimanohar, R. Balasubramonian, A. Davis, N.P. Jouppi, LOT-ECC: localized and tiered reliability mechanisms for commodity memory systems, in *ISCA* (2012)
262. V. van der Veen, Y. Fratanonio, M. Lindorfer, D. Gruss, C. Maurice, G. Vigna, H. Bos, K. Razavi, C. Guiffrida, Drammer: deterministic Rowhammer attacks on mobile platforms, in *CCS* (2016)
263. N. Varnica, LDPC decoding: VLSI architectures and implementations—module 1: LDPC decoding, in *Flash Memory Summit* (2013)
264. R.K. Venkatesan, S. Herr, E. Rotenberg, Retention-aware placement in DRAM (RAPID): software methods for quasi-non-volatile DRAM, in *HPCA* (2006)
265. C. Wang, W.-F. Wong, Extending the lifetime of NAND flash memory by salvaging bad blocks, in *DATE* (2012)
266. J. Wang, K. Vakilinia, T.-Y. Chen, T. Courtade, G. Dong, T. Zhang, H. Shankar, R. Weselk, *Enhanced precision through multiple reads for LDPC decoding in flash memories* (IEEE J. Sel, Areas Commun, 2014)
267. W. Wang, T. Xie, D. Zhou, Understanding the impact of threshold voltage on MLC flash memory performance and reliability, in *ICS* (2014)
268. H.A.R. Wegener, A.J. Lincoln, H.C. Pao, M.R. O'Connell, R.E. Oleksiak, H. Lawrence, The variable threshold transistor, a new electrically-alterable, non-destructive read-only storage device, in *IEDM* (1967)
269. J. Werner, A look under the hood at some unique SSD features, in *Flash Memory Summit* (2010)
270. C. Wilkerson, A. R. Alameldeen, Z. Chishti, W. Wu, D. Somasekhar, S.-L. Lu, Reducing cache power with low-cost, multi-bit error-correcting codes, in *ISCA* (2010)

271. M. Willett, Encrypted SSDs: self-encryption versus software solutions, in *Flash Memory Summit* 2015
272. E.H. Wilson, M. Jung, M.T. Kandemir, Zombie NAND: resurrecting dead NAND flash for improved SSD longevity, in *MASCOTS* (2014)
273. H.-S.P. Wong, H.-Y. Lee, S. Yu, Y.-S. Chen, Y. Wu, P.-S. Chen, B. Lee, F.T. Chen, M.-J. Tsai, *Metal-Oxide RRAM* (Proc, IEEE, 2012)
274. H.-S.P. Wong, S. Raoux, S. Kim, J. Liang, J.P. Reifenberg, B. Rajendran, M. Asheghi, K.E. Goodson, *Phase change memory* (Proc, IEEE, 2010)
275. G. Wu, X. He, N. Xie, T. Zhang, DiffECC: improving SSD read performance using differentiated error correction coding schemes, in *MASCOTS* (2010)
276. G. Wu, X. He, Reducing SSD read latency via NAND flash program and erase suspension, in *FAST* (2012)
277. Y. Wu, Y. Cai, E.F. Haratsch, Fixed point conversion of LLR values based on correlation, U.S. Patent 9,582,361 (2017)
278. Y. Wu, Y. Cai, E.F. Haratsch, Systems and methods for soft data utilization in a solid state memory system, U.S. Patent 9,201,729 (2017)
279. Y. Wu, Z. Chen, Y. Cai, E.F. Haratsch, Method of erase state handling in flash channel tracking, U.S. Patent 9,213,599 (2015)
280. Y. Wu, E.T. Cohen, Optimization of read thresholds for non-volatile memory, U.S. Patent 9,595,320 (2015)
281. Y. Xiao, X. Zhang, Y. Zhang, R. Teodorescu, One bit flips, one cloud flops: cross-VM Rowhammer attacks and privilege escalation, in *USENIX Security* (2016)
282. M. Xu, M. Li, C. Tan, *Extended Arrhenius law of time-to-breakdown of ultrathin gate oxides* (Appl. Phys, Lett, 2003)
283. Q. Xu, H. Siyamwala, M. Ghosh, M. Awasthi, T. Suri, Z. Guz, A. Shayesteh, V. Balakrishnan, Performance characterization of hyperscale applications on NVMe SSDs, in *SIGMETRICS*, (2015)
284. Q. Xu, H. Siyamwala, M. Ghosh, T. Suri, M. Awasthi, Z. Guz, A. Shayesteh, V. Balakrishnan, Performance analysis of NVMe SSDs and their implication on real world databases, in *SYSTOR* (2015)
285. R.-I. Yamada, Y. Mori, Y. Okuyama, J. Yugami, T. Nishimoto, H. Kume, Analysis of detrap current due to oxide traps to improve flash memory retention, in *IRPS* (2000)
286. D.S. Yaney, C.Y. Lu, R.A. Kohler, M.J. Kelly, J.T. Nelson, A meta-stable leakage phenomenon in DRAM charge storage—variable hold time, in *IEDM* (1987)
287. J. Yang, High-Efficiency SSD for reliable data storage systems, in *Flash Memory Summit* (2011)
288. M.-C. Yang, Y.-M. Chang, C.-W. Tsao, P.-C. Huang, Y.-H. Chang, T.-W. Kuo, Garbage collection and wear leveling for flash memory: past and future, in *SMARTCOMP* (2014)
289. N.N. Yang, C. Avila, S. Sprouse, A. Bauche, Systems and methods for read disturb management in non-volatile memory, U.S. Patent 9,245,637 (2015)
290. H. Yoon, J. Meza, N. Muralimanohar, N.P. Jouppi, O. Mutlu, Efficient data mapping and buffering techniques for multi-level cell phase-change memories. ACM TACO (2014)
291. H. Yoon, J. Meza, R. Ausavarungnirun, R. Harding, O. Mutlu, Row buffer locality aware caching policies for hybrid memories, in *ICCD* (2012)
292. J. H. Yoon, 3D NAND technology: implications to enterprise storage applications, in *Flash Memory Summit* (2015)
293. J. H. Yoon, R. Godse, G. Tressler, H. Hunter, 3D-NAND scaling and 3D-SCM—implications to enterprise storage, in *Flash Memory Summit* (2017)
294. J.H. Yoon, G.A. Tressler, Advanced flash technology status, scaling trends and implications to enterprise SSD technology enablement, in *Flash Memory Summit* (2012)
295. X. Yu, C.J. Hughes, N. Satish, O. Mutlu, S. Devadas, Banshee: bandwidth-efficient DRAM caching via software/hardware cooperation, in *MICRO* (2017)
296. W. Zhang, T. Li, Exploring phase change memory and 3D die-stacking for power/thermal friendly, fast and durable memory architectures, in *PACT* (2009)

297. Z. Zhang, W. Xiao, N. Park, D.J. Lilja, Memory module-level testing and error behaviors for phase change memory, in *ICCD* (2012)
298. K. Zhao, W. Zhao, H. Sun, X. Zhang, N. Zheng, T. Zhang, LDPC-in-SSD: making advanced error correction codes work effectively in solid state drives, in *FAST* (2013)
299. P. Zhou, B. Zhao, J. Yang, Y. Zhang, A durable and energy efficient main memory using phase change memory technology, in *ISCA* (2009)
300. A. Zuck, S. Toledo, D. Sotnikov, D. Harnik, Compression and SSDs: where and how? in *INFLOW* (2014)
301. Y. Luo, S. Ghose, Y. Cai, E.F. Haratsch, O. Mutlu, Improving 3D NAND flash memory lifetime by tolerating early retention loss and process variation, in *SIGMETRICS* (2018)

Chapter 10
Efficient Wear Leveling in NAND Flash Memory

Yuan-Hao Chang and Li-Pin Chang

Abstract In the recent years, flash storage devices such as solid-state drives (SSDs) and flash cards have become a popular choice for the replacement of hard disk drives, especially in the applications of mobile computing devices and consumer electronics. However, the physical constraints of flash memory pose a lifetime limitation on these storage devices. New technologies for ultra-high density flash memory such as multilevel-cell (MLC) flash further degrade flash endurance and worsen this lifetime concern. As a result, flash storage devices may experience a unexpectedly short lifespan, especially when accessing these devices with high frequencies. In order to enhance the endurance of flash storage device, various wear leveling algorithms are proposed to evenly erase blocks of the flash memory so as to prevent wearing out any block excessively. In this chapter, various existing wear leveling algorithms are investigated to point out their design issues and potential problems. Based on this investigation, two efficient wear leveling algorithms (i.e., the evenness-aware algorithm and dual-pool algorithm) are presented to solve the problems of the existing algorithms with the considerations of the limited computing power and memory space in flash storage devices. The evenness-aware algorithm maintains a bit array to keep track of the distribution of block erases to prevent any cold data from staying in any block for a long period of time. The dual-pool algorithm maintains one hot pool and one cold pool to maintain the blocks that store hot data and cold data, respectively, and the excessively erased blocks in the hot pool are exchanged with the rarely erased blocks in the cold pool to prevent any block from being erased excessively. In this chapter, a series of explanations and analyses shows that these two wear leveling algorithms could evenly distribute block erases to the whole flash memory to enhance the endurance of flash memory.

Y.-H. Chang (✉)
Academia Sinica, Institute of Information Science, Taipei, Taiwan
e-mail: johnson@iis.sinica.edu.tw

L.-P. Chang
Department of Computer Science, National Chiao-Tung University, Hsinchu, Taiwan
e-mail: lpchang@cs.nctu.edu.tw

R. Micheloni et al. (eds.), *Inside Solid State Drives (SSDs)*,
Springer Series in Advanced Microelectronics 37,
https://doi.org/10.1007/978-981-13-0599-3_10

10.1 Introduction

NAND flash memory has been widely adopted in various mobile embedded applications, due to its non-volatility, shock-resistance, low-power consumption, and low cost. It is widely adopted in various storage systems, and its applications have grown much beyond its original designs. The two popular NAND flash memory designs are single-level-cell (SLC) flash memory and multi-level-cell (MLC) flash memory. Each SLC flash-memory cell can accommodate 1-bit information while each $MLC_{\times n}$ flash-memory cell can contain n-bit information. As n increases, the endurance of each block in MLC flash memory decreases substantially.[1] In recent years, Well-known examples are flash-memory cache of hard drives (known as TurboMemory) [13, 40, 48], fast booting devices (for Microsoft Windows Visa), and solid-state disks (SSD) (for the replacement of hard drives).

As the low-cost MLC flash-memory designs are gaining market momentum [11], the endurance of flash memory is an even more challenging problem. For example, the endurance of an $MLC_{\times 2}$ flash-memory block is only 10,000 (or 5,000) erase cycles whereas that of its SLC flash memory counterpart is 100,000 erase cycles [35, 41]. As the number of bits of information per cell would keep increasing for MLC in the near future, the endurance of a block might also get worse, such as few thousand or even hundred erase cycles. This underlines the endurance issue of flash memory. However, improving endurance is problematic because flash-memory designs allow little compromise between system performance and cost, especially for low-cost flash storage devices. Such developments reveal the limitations of flash memory, especially in terms of endurance.

A NAND flash storage device or storage system, e.g., a solid-state disk (SSD) and flash cards, may be associated with multiple chips. Each chip is composed of one or more sub-chips or dies. Each sub-chip might have multiple planes. Each plane is organized in terms of blocks that are the basic unit for erase operations. A block is further divided into a fixed number of pages and can only endure limited erase cycles. A page (that is the unit of read and write operations) consists of a user area and a spare area, where the user area is for data storage, and the spare area stores house-keeping information such as the corresponding logical block addresses (LBAs), status flags, and error correction codes (ECCs). When a page is written with data, it is no longer available unless it is erased. This is called the "write-once property". As a result, "out-place updates" are adopted so that data are usually updated over free pages. Pages that contain the latest copy of data (i.e., valid data) are considered as live (or valid) pages, and pages with old versions (i.e., invalid data) are dead (or invalid) pages. Therefore, address translation is needed to map logical addresses of data to their physical addresses, and "garbage collection" is needed to reclaim dead pages. Because each block has a limited number of erase

[1]In this chapter, we consider NAND flash memory, which is the most widely adopted flash memory in storage-system designs.

cycles, "wear-leveling" is needed to evenly erase blocks so as to prevent wearing out some blocks excessively.

Engineers and researchers have recently become concerned with how long flash storage devices can withstand daily use when they are adopted in applications with high access frequencies. The host systems, e.g., smart phones and notebooks, access their secondary storages (such as hard drives and SSDs) with temporal localities [6, 32, 33, 46]. Frequently updated data and rarely updated data coexist under such workloads. When reclaiming free space, block erases are always directed to the blocks with few valid data so as to reduce data-copy overheads. Thus, blocks having many static (or immutable) data are rarely chosen for erases, while other blocks are erased many times to circulate frequently updated data. As a result, some blocks are worn out when other blocks remain fresh. The problem of wearing out blocks is a crucial concern for new-generation flash memory, and *wear leveling* is the policy of evenly erasing all flash-memory blocks to keep all the blocks alive as long as possible. Strategies friendly to wear leveling can be adopted in various system layers, including applications, file systems, and firmware. To closely monitor wear in all blocks, the flash management strategies that are usually implemented as firmware implements wear leveling. However, wear leveling is not free, since extra data movement is required. Alleviating wear-leveling overheads is an important task, as wear leveling activities themselves wear flash memory too.

Many excellent wear leveling algorithms have been proposed by academia and industry. Updating data out of place is a simple wear-leveling technique [12, 23, 29, 31, 38]. However, this simple policy is vulnerable in the presence of static data because static data are rarely invalidated and need to be copied out before their residing blocks are erased. In order to reduce live-data-copying overhead, blocks storing a lot of static data rarely participate in the activities of reclaiming free space. Therefore, the key to wear leveling may be to encourage the blocks with static data to participate in block erases. Kim and Lee [20] and Chiang et al. [9] proposed value functions for choosing victim blocks. In their approach, a block receives a high score if it currently has few valid data or its number of accumulated erase cycles is low. Another technique is to erase blocks in favor of reclaiming free space most of the time, but periodically, a block is erased in favor of wear leveling [24, 47]. A typical strategy is to occasionally erase a random block. Wear leveling activities can also be completely detached from free-space reclaiming. Hot-cold swapping [6, 10, 17, 20, 27] involves swapping data in a frequently erased block with that in an infrequently erased block whenever the wear of all blocks is unbalanced.

These existing approaches share a common idea: encouraging infrequently erased blocks to contribute to erases cycles. Under the workload of most real access patterns, most block erases are contributed by a small fraction of blocks if wear leveling is not used. According to such observations, static wear leveling algorithms are proposed to move static data away from infrequently erased blocks [2, 7, 43]. However, some existing static wear leveling algorithms don't consider the limited computing power or restricted RAM space, while some don't consider the access patterns and data access frequencies [3, 4, 18, 39, 42]. As a result, these existing static wear leveling algorithms either consume too many hardware resources or introduce too many overheads on extra live page copies and block erases. In order to achieve static wear leveling effectively with limited computing power, limited main memory, and limited overheads, two efficient wear leveling algorithms (i.e., the evenness-aware algorithm and dual-pool algorithm) are proposed and presented in this chapter. The *evenness-aware algorithm* [8] maintains a house-keeping data structure, i.e., a bit array, with a cyclic-queue-based scanning procedure to keep track of the distribution of block erases to prevent any static or cold data staying in any block for a long period of time. The objective is to improve the endurance of flash memory with limited overhead and without excessively modifying popular implementations of flash management designs, such as FTL, NFTL, and BL [1, 14, 16, 45]. The *dual-pool algorithm* [5] maintains one hot pool and one cold pool to maintain the blocks that store hot data and cold data, respectively, and the excessively erased blocks in the hot pool are exchanged with the rarely erased blocks in the cold pool to prevent any block being erased excessively. Whenever a block is excessively erased, it is filled with static data. In this way, such blocks stop participating in free-space reclaiming. This strategy helps conserve data movement because the major contributors of block erases are only a small fraction of all blocks. Second, blocks recently involved in wear leveling should be temporarily isolated from wear leveling activities. For example, after static data are written to a block which has been erased many times, the dual-pool algorithm decides how long this block should wait before it can contribute more erase cycles.

The rest of this paper is organized as follows: Sect. 10.2 presents the evenness-aware algorithm with the worst-case analysis. In Sect. 10.3, the dual-pool algorithm is presented with a real case study. Section 10.4 concludes this chapter.

10.2 Evenness-Aware Algorithm

10.2.1 Algorithm Design

10.2.1.1 Overview

The motivation of the evenness-aware algorithm is to prevent static data from staying at any block for a long period of time. It minimizes the maximum erase-count difference between any two blocks, so flash memory lifetime is extended. This algorithm could be implemented as a module. In this algorithm, it maintain a *Block Erasing Table (BET)* that identifies the blocks erased during a given period of time (Sect. 10.2.1.2). The BET is associated with the process *SW Leveler* that is activated by some system parameters for the needs of static wear leveling (Sect. 10.2.1.3). When the SW Leveler runs, it either resets the BET or picks up a block that has not been erased so far (based on the BET information), and triggers the garbage collector to do garbage collection on the block (note that the selection procedure of a block must be performed efficiently and within a limited time). Whenever a block is recycled by the garbage collection, any modification to the address translation is performed as in the original design of a flash management design. The SW Leveler can be implemented as a thread or as a procedure triggered by a timer or the garbage collector based on some preset conditions. Note that, whenever a block is erased, the BET must be updated by a triggering action to the SW Leveler. The design of the BET is scalable to accommodate rapidly increasing flash-memory capacity [34] and the limited RAM space on a controller.

10.2.1.2 Block Erasing Table

The Block Erasing Table (BET) attempts to remember which block has been erased in a pre-determined time frame, referred to as the *resetting interval*, so as to locate blocks of cold data. A BET is a bit array in which each bit corresponds to a set of 2^k contiguous blocks where k is an integer that equals or exceeds 0. Whenever a block is erased by the Cleaner, the SW Leveler is triggered to set the corresponding bit as 1. Initially, the BET is reset to 0 for every bit. As shown in Fig. 10.1, information maintenance is performed in one-to-one and one-to-many modes, and one flag is used to track whether any one of the corresponding 2^k blocks is erased. When $k = 0$, one flag is used for one block (i.e., in the one-to-one mode). The larger the value of k, the greater the chance in the overlooking of blocks of cold data. However, a large value for k could help reduce the RAM space required by a BET controller.

The worst case for a large k value occurs when hot and cold data co-exist in a block set. Fortunately, such a case is eventually resolved when hot data are

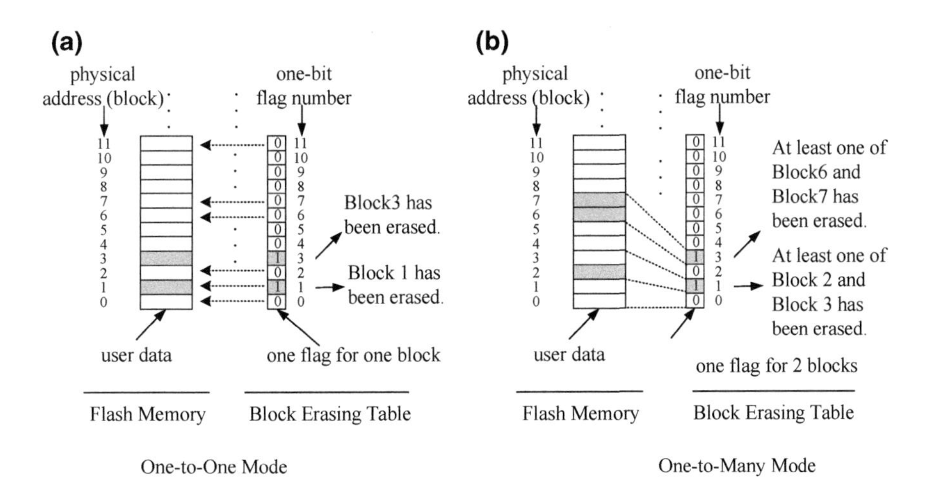

Fig. 10.1 The mapping mechanism between flags and blocks. **a** One-to-One mode. **b** One-to-Many mode

invalidated. As a result, cold data could be moved to other blocks by the SW Leveler (see Sect. 10.2.1.3). The technical problem relies on the tradeoff between the time to resolve such a case (bias in favor of a small k) and the available RAM space for the BET (bias in favor of a large k).

Another technical issue is efficiently rebuilding the BET when a flash-memory storage system is attached. One simple but effective solution is to save the BET in the flash-memory storage system when the system shuts down, and then to reload it from the system when needed. Meanwhile, the whole BET is stored in flash memory and loaded to main memory in an on-demand fashion, so that the required main memory could be minimized. If the system is not properly shut down, we propose loading any existing correct version of the BET when the system is attached. Such a solution is reasonable as long as loss of erase count information is not excessive. Note that the crash resistance of the BET information in the storage system could be provided by the popular dual buffer concept. Scanning of the spare areas of pages when collecting related information should also be avoid because of the potentially huge capacity of a flash-memory storage system.

10.2.1.3 SW Leveler

The SW Leveler consists of two procedures in executing wear leveling: SWL-Procedure and SWL-BETUpdate (please see Algorithms 1 and 2). SWL-BETUpdate is invoked by the garbage collector to update the BET whenever any block is erased by the garbage collector during garbage collection. The SWL-Procedure is invoked whenever static wear leveling is needed. Such a need is tracked by two variables, f_{cnt} and e_{cnt}, which denote the number of 1s in the BET and the total number of block erases performed since the BET was reset, respectively. When the *unevenness level*, i.e., the ratio of e_{cnt} and f_{cnt}, equals or exceeds a given threshold T, SWL-Procedure is invoked to trigger the garbage collector to do garbage collection over selected blocks such that cold data are moved. Note that a high unevenness level reflects the fact that a lot of erases are done on a small portion of the flash memory.

Algorithm 1 shows the algorithm for the SWL-Procedure: the SWL-Procedure simply returns if the BET is just reset (Step 1). When the unevenness level, i.e., $e_{cnt}\ /f_{cnt}$, equals or exceeds a given threshold T, the garbage collector is invoked in each iteration to do garbage collection over a selected set of blocks (Steps 2–15). In each iteration, it is checked up if all of the flags in the BET are set as 1 (Step 3). If so, the BET is reset, and the corresponding variables (i.e., e_{cnt}, f_{cnt}, and f_{index}) are reset (Steps 4–7). The f_{index} is the index in the selection of a block set for static wear leveling and is reset to a randomly selected block set or to a predefined block set, e.g. 0. After the BET is reset, SWL-Procedure simply returns to start the next resetting interval (Step 8). Otherwise, the selection index, i.e., f_{index}, moves to the next block set with a zero-valued flag (Steps 10–12). *Note that the sequential scanning of blocks in the selection of block sets for static wear leveling is very effective in the implementation. We surmise that the design approximates that of an actual random selection policy because cold data can virtually exist in any block in the physical address space of the flash memory.* The SWL-Procedure then invokes the garbage collector to do garbage collection over a selected block set (Step 13) and moves to the next block set (Step 14) for the next iteration. We must point out that f_{cnt} and BET are updated by SWL-BETUpdate because SWL-BETUpdate is invoked by the garbage collector during garbage collection. The loop in static wear leveling ends when the unevenness level drops to a satisfactory value.

Algorithm 1: SWL-Procedure

```
   Input: e_cnt, f_cnt, k, f_index, BET, and T
   Output: null
 1 if f_cnt = 0 then return;
 2 while e_cnt / f_cnt >= T do
       /* size(BET) is the number of flags in the BET.                          */
 3     if f_cnt >= size(BET) then
 4         e_cnt <- 0 ;
 5         f_cnt <- 0 ;
 6         f_index <- RANDOM(0, size(BET) - 1) or 0;
 7         reset all flags in the BET;
 8         return;
 9     end
10     while BET[f_index] = 1 do
11         f_index <- (f_index + 1) mod size(BET)
12     end
13     EraseBlockSet(f_index, k) ;    /* Request the garbage collector to do garbage collection
       over the selected block set. */
14     f_index <- (f_index + 1) mod size(BET) ;
15 end
```

Algorithm 2: SWL-BETUpdate

```
   Input: e_cnt, f_cnt, k, b_index, and BET
   Output: e_cnt, f_cnt and BET are updated based on the erased block address b_index and k in
           the BET mapping.
 1 e_cnt <- e_cnt + 1 ;                                  /* Increase the total erase count. */
   /* Update the BET if needed.                                                            */
 2 if BET[⌊b_index/2^k⌋] = 0 then
 3     BET[⌊b_index/2^k⌋] <- 1 ;
 4     f_cnt <- f_cnt + 1 ;
 5 end
```

The SWL-BETUpdate is as shown in Algorithm 2: Given the address b_{index} of the block erased by the garbage collector, SWL-BETUpdate first increases the number of blocks erased in the resetting interval (Step 1). If the corresponding BET entry is not 1, then the entry is set as 1, and the number of 1s in the BET is increased by one (Steps 2–5). The remaining technical question is how to maintain the values of e_{cnt}, f_{cnt}, and f_{index}. To optimize static wear leveling, e_{cnt}, f_{cnt}, and f_{index} should be saved to flash memory as system parameters and retrieved in the attachment of the flash memory. Notably, these values can tolerate some errors with minor modifications to SWL-Procedure in either the condition in Step 3 or the linear traversal of the BET (Steps 10–12). That is, if the system crashes before their values are saved to flash memory, it simply uses the values previously saved to flash memory.

10.2.2 *Worst-Case Analysis*

10.2.2.1 Worst-Case Model for Extra Overheads

Block recycling overhead is indeed increased by the proposed evenness-aware algorithm. A very minor cause of the increase is the execution of SWL-BETUpdate whenever the garbage collector erases a block, i.e., the value updates of e_{cnt} and f_{cnt} as well as the BET flags (compared to the block erase time, which could be about 1.5 ms over a 1 GB $MLC_{\times 2}$ flash memory [28]). As astute readers might point out, the garbage collector might be triggered more often than before because of wear leveling. That might increase the number of block erases and live-page copyings. The increased overheads caused by extra block erases and extra live-page copyings are apparent in the following worst-case scenario: the flash memory contains blocks of hot data, blocks of static data, and exactly one free block in a resetting interval.

Figure 10.2 shows the worst-case model based on a block-level address translation mechanism. In the block-level address translation mechanism, each LBA is divided into a virtual block address (VBA) and a block offset, and a mapping table is adopted for VBAs and their physical block addresses (PBAs). For each write operation, a free block is allocated to save the data of the remaining valid pages of the original mapped block and the new data of the write operation. Assume there are $(H - 1)$ blocks of hot data and C blocks of static data where the number of blocks in the system is $(H + C)$. The worst-case situation occurs when the C blocks are erased, only due to the evenness-aware algorithm. The worst case occurs when hot data are updated with the same frequency and only to the free block or the blocks of hot data, where $k = 0$. Sections 10.2.2.2 and 10.2.2.3 show the analyses for extra block erases and extra live-page copyings in the worst-case model, respectively.

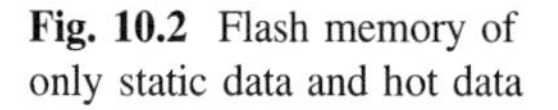
Fig. 10.2 Flash memory of only static data and hot data

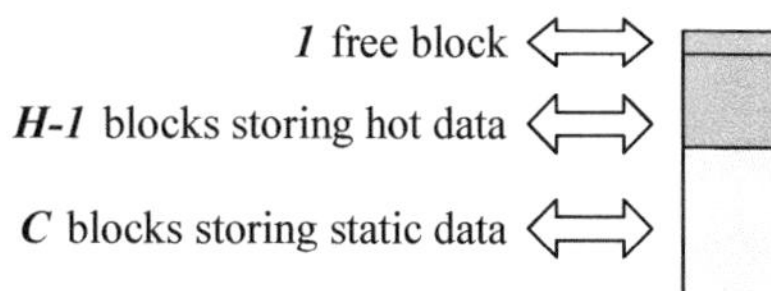

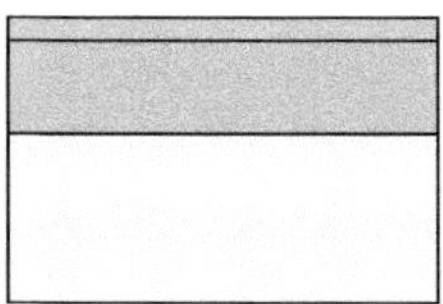

10.2.2.2 Extra Block Erases

When $k = 0$, the BET contains $(H + C)$ bits, i.e., $(H + C)$ 1-bit flags. In each resetting interval, when the updates of hot data result in $(T \times H)$ block erases, SWL-Procedure is activated to recycle one block of cold data for the first time because only H bits of the BET are set, and the unevenness level reaches T (i.e., $(T \times H)/H$). After one block of cold data is recycled by SWL-Procedure, $(H + 1)$ bits of the BET are set, and the number of block erases reaches $(T \times H + 1)$. The unevenness level (i.e., $(T \times H + 1)/(H + 1)$) is then smaller than the threshold T. Thereafter, SWL-Procedure is activated to recycle one block of cold data on all other $(T - 1)$ block erases resulting from hot data updates. Finally, this procedure is repeated C times such that all BET flags are set and the resetting interval ends. Therefore, the resetting interval has $T \times (H + C)$ block erases. For every $T \times (H + C)$ block erases in a resetting interval, SWL-Procedure performs C block erases. Therefore, the increased ratio of block erases (due to static wear leveling) is derived as follows:

$$\frac{C}{T \times (H + C) - C} \approx \frac{C}{T \times (H + C)}, \text{when } T \times (H + C) \gg C.$$

The increased ratio is even worse when C is the dominant part of $(H + C)$ (an earlier study [18] showed that the amount of non-hot data is often several times that of hot data). Table 10.1 shows different increased ratios in extra block erasing for different configurations of H, C, and T. As shown in the table, the increased overhead ratio in extra block erasing is sensitive to the setting of T. Therefore, to avoid excessive triggering of static wear leveling, T must not be set too small.

10.2.2.3 Extra Live-Page Copyings

The extra overheads in live-page copyings due to the static wear leveling mechanism can be explored by the worst-case model. Let N be the number of pages in a block. Suppose that L is the average number of pages copied by the garbage collector when erasing a block of hot data. Thus, in the worst case, totally $(C \times N)$ live-pages are copied when erasing C blocks of static data (due to the evenness-aware algorithm) in a resetting interval, and $(T \times (H + C) - C) \times L$

Table 10.1 The increased ratio of block erases of a 1 GB $MLC_{\times 2}$ flash-memory storage system

H	C	H:C	T	Increased ratio (%)
256	3,840	1:15	10	9.46
2048	2,048	1:1	10	5.03
256	3,840	1:15	100	0.95
2,048	2,048	1:1	100	0.50
256	3,840	1:15	1,000	0.09
2,048	2,048	1:1	1,000	0.05

Table 10.2 The increased ratio in live-page copyings of a 1 GB $MLC_{\times 2}$ flash-memory storage system

H	C	H:C	T	L	$\frac{N}{T \times L}$	Increased ratio (%)
256	3,840	1:15	10	16	0.800	75.72
2,048	2,048	1:1	10	16	0.800	40.02
256	3,840	1:15	10	32	0.400	37.86
2,048	2,048	1:1	10	32	0.400	20.00
256	3,840	1:15	100	16	0.0800	7.57
2,048	2,048	1:1	100	16	0.0800	4.00
256	3,840	1:15	100	32	0.0400	3.79
2,048	2,048	1:1	100	32	0.0400	2.00
256	3,840	1:15	1,000	16	0.0080	0.76
2,048	2,048	1:1	1,000	16	0.0080	0.40
256	3,840	1:15	1,000	32	0.0040	0.38
2,048	2,048	1:1	1,000	32	0.0040	0.20

live-page copyings are performed in the course of regular garbage collection activities in a resetting interval. The increased ratio in live-page copyings, due to static wear leveling, can be derived as follows:

$$\frac{C \times N}{(T \times (H+C) - C) \times L} \approx \frac{C \times N}{T \times L \times (H+C)}, \text{when } T \times (H+C) \gg C.$$

Table 10.2 shows varying increases in the ratios of live-page copyings for different configurations of H, C, T, and L, when $N = 128$. The increased ratio of live-page copyingscan be estimated by $\frac{N}{L}$ times the increased ratio of extra block erases. For example, when $T = 100$, $L = 16$, $N = 128$, and $\frac{H}{C} = \frac{1}{15}$, the increased ratio of block erases is 0.95% (the third row of Table 10.1) and its corresponding increased ratio of live-page copyings is 7.57%, i.e., 0.95% $\times \frac{128}{16}$ (the third row of Table 10.2). As shown in Tables 10.1 and 10.2, the increased ratios of block erases and live-page copyings would be limited with a proper selection of T and other parameters. The increased ratios could be limited to very small percentages of flash management strategies when the evenness-aware algorithm is supported.

10.3 Dual-Pool Algorithm

10.3.1 Algorithm Design

10.3.1.1 Algorithm Concept

This section introduces the basic concepts of the dual-pool algorithm. Let write requests arriving at the flash storage device be ordered by their arrival times. Let the temperature of a piece of data be inversely proportional to the number of requests between the two most recent writes to that data. A piece of data is *hot* if its temperature is higher than the average temperature of all data. Otherwise, the data is *cold* or *non-hot*. A block is referred to as a *young(/old) block* if its erase-cycle count is smaller(/larger) than the average erase-cycle count of all blocks.

We say that a block contributes or accumulates erase cycles if garbage collection erases this block to reclaim free space. Garbage collection avoids erasing a block having many valid data. If a block has more cold data than other blocks, then it will stop contributing erase cycles. This is because cold data remains valid in the block for a long time. Conversely, if a block has many hot data, then it can accumulate erase cycles faster than other blocks. This is because hot data are invalidated faster than cold data, and the block can become a victim of garbage collection before other blocks. After the block is erased, it can again be written with many hot data, because writes to hot data arrive more frequently than writes to cold data. Thus, this block is again erased and is written with many hot data.

The dual-pool algorithm monitors the erase-cycle count of each block. If an old block's erase-cycle count is larger than that of a young block by a predefined threshold, wear leveling activities are triggered. Cold data are moved to the old block to prevent it from being erased by garbage collection. This strategy is referred to as *cold-data migration*. After this, the old block should stop accumulating erase cycles. Compared to encouraging young blocks to contribute erase cycles, this strategy reduces data-movement overhead. This is because only a small fraction of blocks are worn into old blocks, while the majority are young blocks. Right after cold data are written to an old block, the old block still has a large erase-cycle count. If we are not aware that the old block has been involved in cold-data migration, we may again write some other cold data to the old block. This pointlessly reduces the block's lifetime. Similarly, after a young block is involved in cold-data migration, cold data previously stored in the block are removed. At this point, the young block has no cold data, even though its erase-cycle count is small. So, right after a block is involved in cold-data migration, it should be protected from immediate re-involvement. This strategy is called *block protection*. The protection of an old block is no longer required when other blocks become older than it. The protection of a young block expires when it is worn into an old block.

The access patterns from the host to the flash storage devices can change periodically. For example, a user application in the host may finish using some files and then begin accessing other files. These application-level behaviors can change the frequency with which a piece of data is updated, and thus cold data can change into hot data. Consider an old block written with cold data for cold-data migration. The old block is then protected against cold-data migration. Now suppose that the cold data in the old block happens to become hot. The protected old block will again start participating in garbage collection, and continues to age without interruption from wear leveling because its protection cannot expire. Now consider a young block under protection. The block should accumulate erase cycles. If the young block happens to be written with many cold data, then it stops contributing erase cycles. The young block attracts no attention from wear leveling because its protection cannot expire. This dilemma highlights the special cases that must be carefully considered by block protection.

10.3.1.2 The Dual-Pool Algorithm: A Basic Form

The dual-pool algorithm, as implied by its name, uses a *hot pool* and a *cold pool.* A pool is merely a logical aggregation of blocks. Initially, a block arbitrarily joins one of these two pools. Note that the dual-pool algorithm is not to write cold data to blocks in the cold pool. Instead, it migrates blocks storing cold data to the cold pool.

The dual-pool algorithm uses priority queues to sort blocks in terms of different wearing information. The following section defines some symbols for ease of presentation: Let C and H denote the cold pool and the hot pool, respectively. Each element in C and H is a block. Let U be a collection of all blocks. $C \cap H = \emptyset$ and $C \cup H = U$ are invariants. Let Q_P^w be a priority queue that prioritizes all blocks in pool P in terms of wearing information w. The larger the value of w is, the higher the priority is. Each element Q_P^w in w corresponds to a block. For block b, let function $ec(b)$ present its erase-cycle count. In priority queue Q_P^w, $M\left(Q_P^w\right)$ is the element with the highest priority and $m\left(Q_P^w\right)$ is the element with the lowest priority. $M\left(Q_P^w\right)$ and $m\left(Q_P^w\right)$ are referred to as the largest queue head and the smallest queue head, respectively. For example, $m\left(Q_C^{ec}\right)$ denotes the block with the smallest erase-cycle count of all the blocks in the cold pool.

The dual-pool algorithm adopts a user-configurable parameter **TH** to direct how even the wear of blocks is to be pursued. The smaller the value of *TH* is, the more aggressive the wear-leveling activities would be. Table 10.3 summarizes the symbol definitions, and the following section defines cold-data migration (CDM for short): **Cold-Data Migration (CDM)**: Upon the completion of block erase, check the following condition:

$$ec\left(M\left(Q_H^{ec}\right)\right) - ec\left(m\left(Q_C^{ec}\right)\right) > TH.$$

Table 10.3 A summary of symbols used in the dual-pool algorithm

Symbol	Definition
C	The cold pool, a collection of blocks
H	The hot pool, a collection of blocks
U	A collection of all blocks. $C \cap H = \emptyset$ and $C \cup H = U$
Q_P^w	A priority queue that sorts blocks in pool P in terms of information w
$M\left(Q_P^w\right)$	The element with the largest priority in Q_P^w
$m\left(Q_P^w\right)$	The element with the smallest priority in Q_P^w
$ec(b)$	The erase-cycle count of block b
$rec(b)$	The recent erase-cycle count of block b
TH	The threshold parameter for wear leveling

If this condition is true, then the largest erase-cycle count of the blocks in the hot pool is larger than the smallest count of the blocks in the cold pool by *TH*. Perform the following procedure:

Step 1.	Copy data from $m(Q_C^{ec})$ to $M(Q_H^{ec})$
Step 2.	Erase $m(Q_C^{ec})$; $ec(m(Q_C^{ec})) \leftarrow ec(m(Q_C^{ec})) + 1$
Step 3.	$C \leftarrow C \cup \{M(Q_H^{ec})\}$; $H \leftarrow H \backslash \{M(Q_H^{ec})\}$
Step 4.	$H \leftarrow H \cup \{m(Q_C^{ec})\}$; $C \leftarrow C \backslash \{m(Q_C^{ec})\}$

Because cold-data migration checks the condition immediately after a block is erased, block $ec(M(Q_H^{ec}))$ must be the most-recently erased block if the condition is true. Whenever $ec(M(Q_H^{ec})) - ec(m(Q_C^{ec}))$ is found larger than *TH*, it is deduced that, on the one hand, block $m(Q_C^{ec})$ has not been erased for a long time because of the storing of many cold data. On the other hand, garbage collection had erased block $M(Q_H^{ec})$ many times, because this block infrequently stores cold data. Next, migrate cold data from block $m(Q_C^{ec})$ to block $M(Q_H^{ec})$. Step 1 moves data from block $m(Q_C^{ec})$ to block $M(Q_H^{ec})$ to complete cold-data migration. After this move, block $M(Q_H^{ec})$ can stop being erased by garbage collection. Step 2 erases block $m(Q_C^{ec})$ and increases the block's erase-cycle count. This erase does not affect the pool membership of block $m(Q_C^{ec})$.

Step 3 moves block $M(Q_H^{ec})$ to the cold pool, and Step 4 moves block $m(Q_C^{ec})$ to the hot pool. These steps swap the two blocks' pool memberships, and enable block protection. When the young block (previously $m(Q_C^{ec})$) joins the hot pool, it may be younger than many blocks in the hot pool. That is because most of the blocks in the hot pool are old. The young block is then protected, because cold-data migration is not interested in a young block in the hot pool. Analogously, when the old block (previously block $M(Q_H^{ec})$) migrates to the cold pool, it may be older than many blocks in the cold pool. The old block in the cold pool is then protected, as cold-data migration is concerned with the youngest block in the cold pool.

The young block in the hot pool (previously $m(Q_C^{ec})$) starts accumulating erase cycles. When the block is worn into the oldest in the hot pool, it will again participate in cold-data migration. On the other hand, the old block in the cold pool (previously $M(Q_C^{ec})$) now stops being erased. When the block becomes the youngest in the cold pool, it is again ready for cold-data migration.

10.3.1.3 Pool Adjustment

The cold pool collects blocks that store cold data. However, the cold pool may also contain blocks that have no cold data. This may be because all the blocks' pool memberships were arbitrarily decided in the very beginning, as all blocks' erase-cycle counts are initially zero. Another possible cause is that applications in the host may change their data-access behaviors. These changes can turn a piece of cold data into hot data.

Garbage collection selects erase victims based on how many invalid data a block has, regardless the block's pool membership. If a block has no cold data, it will continue participating in garbage collection even if it is in the cold pool. In this case, the block's erase-cycle count increases without interruption from wear leveling. This is because cold-data migration always involves the youngest block in the cold pool. Similarly, if a block in the hot pool has many cold data, garbage collection avoids erasing this block. The block cannot be erased into the oldest block in the hot pool, and cannot attract attention from wear leveling.

To deal with this problem, the dual-pool algorithm introduces two operations, cold-pool adjustment (CPA for short) and hot-pool adjustment (HPA for short). These two operations identify and correct any improper pool membership in the blocks. Specifically, blocks' pool membership is adjusted according to how frequently they have been erased since their last involvement in cold-data migration. Hot-pool adjustment removes the blocks that do not accumulate erase cycles from the hot pool. Cold-pool adjustment removes the blocks that actively contribute erase cycles from the cold pool. To enable these operations to function, new block-wearing information (i.e., the *recent erase-cycle count*) is introduced. A block's recent erase-cycle count is initially zero. It increases as along with the erase-cycle count, but reset to zero whenever the block is involved in cold-data migration. Thus, cold-data migration includes a new step:

(CDM) Step 5. $rec(M(Q_H^{ec}))\leftarrow 0; rec(m(Q_C^{ec}))\leftarrow 0$

The hot-pool adjustment and cold-pool adjustment operations also require new priority queues and queue heads, which are summarized in Table 10.4. Let function *rec*() return the recent erase-cycle count of a block. The hot-pool adjustment and cold-pool adjustment are then as follows:

Cold-Pool Adjustment (CPA): Upon completion of block erase, check the following condition:

$$rec(M(Q_C^{rec})) - rec(m(Q_H^{rec})) > TH.$$

Table 10.4 A summary of the five queue heads used by the dual-pool algorithm

Queue heads	Belongs to	Used in
$M(Q_H^{ec})$	The hot pool	Cold-data migration and hot-pool adjustment
$m(Q_H^{ec})$	The hot pool	Hot-pool adjustment
$m(Q_H^{rec})$	The hot pool	Cold-pool adjustment
$m(Q_C^{ec})$	The cold pool	Cold-data migration
$M(Q_C^{rec})$	The cold pool	Cold-pool adjustment

If it holds, then the largest recent erase-cycle count of the blocks in the cold pool is larger than the smallest count of the blocks in the hot pool by *TH*. Perform the following steps:

Step 1. $H \leftarrow H \cup \{M(Q_C^{rec})\}$; $C \leftarrow C \backslash \{M(Q_C^{rec})\}$

If a block has a large recent erase-cycle count, then the block has contributed many erase cycles since the last time it was involved in cold-data migration. Cold-pool adjustment evicts such a block from the cold pool. This is because the last attempt to stop the block from being erased was not successful, or the block did not have cold data in the very beginning.

Hot-Pool Adjustment (HPA): Upon completion of block erase, check the following condition:

$$ec(M(Q_H^{ec})) - ec(m(Q_H^{ec})) > 2 \times TH.$$

If this condition holds, then in the hot pool the smallest erase-cycle count is smaller than the largest count by 2 × *TH*. Perform the following steps:

Step 1. $C \leftarrow C \cup \{m(Q_H^{ec})\}; H \leftarrow H \backslash \{m(Q_H^{ec})\}$

Whether or not a block should be written with cold data for wear leveling depends on the size of its erase-cycle count. If a block in the hot pool accumulates erase cycles more slowly than other blocks, then the block contains cold data, and the hot-pool adjustment operations removes this block from the hot pool. *Readers may question that why 2 × TH is in this condition.* It is to prevent hot-pool adjustment from conflicting with cold-data migration: when cold-data migration moves a young block from the cold pool to the hot pool, the young block's erase-cycle count is already smaller than the oldest block in the hot pool by TH (see the condition for cold-data migration). To prevent hot-pool adjustment from immediately bouncing the young block back to the cold pool, the condition of hot-pool adjustment allows additional TH cycles (2 × *TH* in total).

In the worst case, every time after cold-data migration writes cold data to an old block and moves this block to the cold pool, the cold data become hot. Cold-pool adjustment can identify this old block and move it to the hot pool, after the block contributes *TH* more cycles of erase operations. Right after this, cold-data migration makes another attempt to write cold data to the block. So in this worst case, the dual-pool algorithm guarantees to involve this old block every other *TH* erase operations to this block.

10.3.1.4 Algorithm Demonstration

This section presents an example demonstrating how the dual-pool algorithm accomplishes wear leveling.

In Fig. 10.3, there are six flash-memory blocks, labeled from PBA 0 to PBA 5. The threshold parameter *TH* is 16. In the illustration, each block corresponds to two

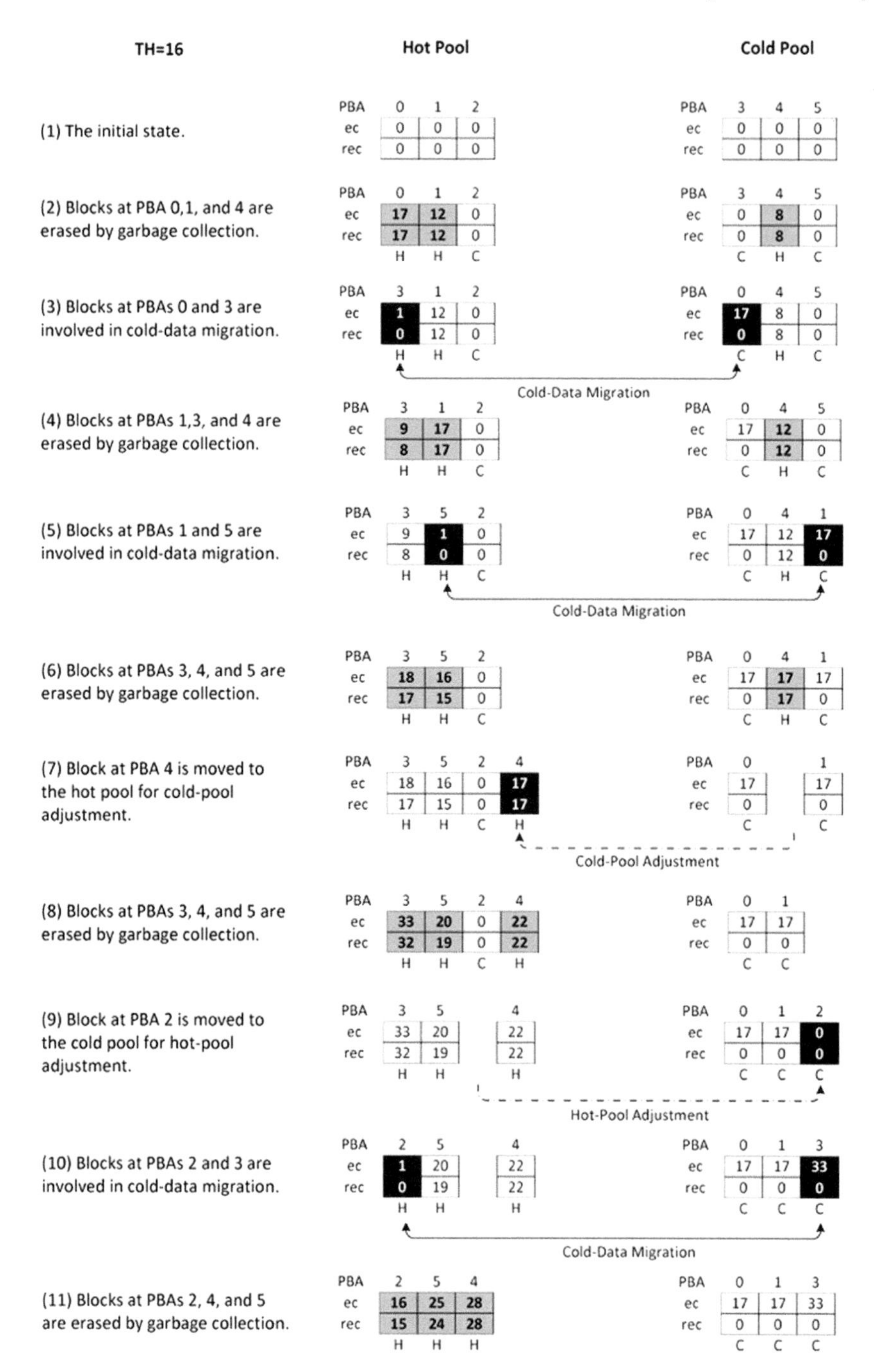

Fig. 10.3 A scenario of the dual-pool algorithm. There are six flash-memory blocks, labeled from PBA 0 to PBA 5. Each block is associated with an erase-cycle count (ec), a recent erase-cycle count (rec), and the attribute of its data (hot or cold)

boxes, which indicate the block's erase-cycle count (ec) and recent erase-cycle count (rec). If a block currently stores cold data, then "C" appears under the block's boxes, and "H" otherwise. The example includes 11 steps. At each step, a block's boxes are shaded in gray if the block has been erased by garbage collection since the last step. A block's boxes are indicated black if it is currently involved in wear leveling. The following discussion refers to a block at PBA x as Block x, where x can be from 0 to 5.

In Step 1, the first three blocks join the hot pool and the rest join the cold pool. Step 2 shows that Blocks 0, 1, and 4 start accumulating erase cycles because they store no cold data. At this point, the largest erase-cycle count in the hot pool and the smallest erase-cycle count in the cold pool are 17 and 0, respectively. As this difference is greater than TH = 16, cold-data migration is triggered. Step 3 shows that the cold data in Block 3 are moved to Block 0, and the pool memberships are switched for both blocks. Notice that a block's wearing information sticks together with that block during cold-data migration. In Step 4, garbage collection erases Blocks 1, 3, and 4 because they had no cold data since Step 3.

Block 0, an old block previously involved in cold-data migration, is written with cold data and stops accumulating erase cycles since Step 3. Even though Block 0 is the oldest among all the blocks in the cold pool, it is now protected against cold-data migration because it is not youngest in the cold pool. In Step 5, cold-data migration is triggered by Blocks 1 and 5, and cold data are migrated from Blocks 5 to 1. In Step 6, Blocks 3–5 contribute some more erase cycles since Step 5. Note that after two cold-data migrations, Blocks 0 and 1, which were previously the contributors of erase cycles in Step 2, now store cold data in the cold pool and are no longer being erased.

In Step 6, Block 4 in the cold pool stores no cold data. In Step 7, it is evicted from the cold pool by cold-pool adjustment, because the difference between Block 4 s recent erase-cycle count and the smallest recent erase-cycle count in the hot pool (i.e., that of Block 2) is greater than TH = 16. In Step 8, Blocks 3–5 keep accumulating erase cycles, and have done so since Step 5. In Step 9, hot-pool adjustment is triggered because the difference between the erase-cycle counts of Blocks 2 and 3 is greater than $2 \times \text{TH} = 32$. Hot-pool adjustment moves Block 2 to the cold pool. Right after Step 9, cold-data migration for Blocks 2 and 3 occurs in Step 10. In Step 11, garbage collection erases some more blocks. At this point, the wear of all blocks is considered even, with respect to TH = 16.

10.3.2 Case Study: An SSD Implementation of the Dual-Pool Algorithm

10.3.2.1 The Firmware and Disk Emulation

The SSD platform in this study is the FreeScale M68KIT912UF32 development kit [15, 25]. This platform integrates an MC9S12UF32 SoC (referred to as the SSD controller hereafter), various flash-memory interfaces, and a USB interface. The controller contains a 16-bit MCU M68HCS12, 3 KB of RAM, 32 KB of EEPROM, a USB 2.0 interface controller, various flash-memory host controllers, and a DMA engine with an 1.5 KB buffer. The MCU is normally rated at 33 MHz. The NAND flash considered in this study is a 128 MB SmartMedia card (abbreviated as SM card hereafter). SM cards have the same appearance as bare NAND-flash chips in terms of physical characteristics. The block size and the page size of the SM card are 16 KB and 512 bytes, respectively, and it has a block endurance of 100 K erase cycles. Readers may notice that its geometry is finer than that of mainstream NAND flash memory [37]. However, the design and implementation of the proposed algorithm is independent of the block size and the page size.

An SSD presents itself to the host system as a logical disk,[2] so ordinary disk-based file systems (such as FAT and NTFS) are compatible with SSDs. The flash-translation layer (FTL), which is a part of SSD firmware, performs disk emulation [21, 22, 26, 44]. Basically, FTL implements a mapping scheme, an update policy, and a garbage-collection policy. For ease of presentation, this section introduces some necessary terms and assumptions: Let a disk be addressed in terms of disk sectors, each of which is as large as a flash-memory page. A *physical block* refers to a flash-memory block. Let the entire disk space be partitioned in terms of *logical blocks*, each of which is as large as a physical block. LBAs and PBAs are abbreviations of logical-block addresses and physical-block addresses, respectively. Let a *physical segment* be a group of contiguous physical blocks, and a *logical segment* be a group of contiguous logical blocks.

The FTL needs logical-to-physical translation because data in flash memory are updated out of place. However, a solid-state-disk controller cannot afford the space overhead of the RAM-resident data structures for this translation. To save RAM-space requirements, the FTL adopts a two-level mapping scheme. The fist level maps eight logical segments to eight physical segments. This first-level mapping has a one-to-one correspondence. The first level uses a RAM-resident segment translation table ("segment L2P table" for short). This table is indexed by logical-segment numbers, and each table entry represents a physical-segment number. As the first level maps a logical segment to a physical segment, the second level uses a RAM-resident block translation table ("block L2P table" for short) to map the 1,000 logical blocks in the logical block to the 1,024 physical blocks in the physical segment. This table is indexed by logical-block addresses and each table entry represents a physical-block address. Each physical segment has $1,024-1,000=24$ unmapped physical blocks, which are spare blocks for garbage collection and bad-block retirement. Thus, the SSD has a total volume of 8 * 1,000 = 8,000 logical blocks, while the SM card has 8 * 1,024 = 8, 192 physical blocks.

[2]A logical disk is also referred to as a logical unit (i.e., LUN) [30].

The FTL sequentially writes all sectors of a logical block to the physical block mapped to this logical block, because the smallest granularity for address translation is one block. To translate an LBA into a PBA, first divide the LBA by 1,000. The quotient and the remainder are the logical-segment number and the logical-block offset, respectively. Looking up the segment L2P table and the block L2P table generates a physical segment number and a physical-block offset, respectively. The final PBA is calculated by adding the physical-block offset to the physical-segment number multiplied by 1,024.

For this FTL, there are two types of sector write operations: a write no larger than 4 KB (i.e., eight 512-byte disk sectors) and a write larger than 4 KB. A write larger than 4 KB effectively rewrites a logical block with the necessary copy-back operations: Unchanged sector data are copied from the logical block encompassing the written sectors, and combined with the newly written sector data. A spare block is allocated from the physical segment to which the logical block is mapped, and the combined data are then written to the spare block. The block L2P table is then revised to re-map the logical block to the spare block. The old physical block of the invalidated logical block is erased and converted to a spare block. Spare blocks are allocated in a FIFO fashion for fair use.

Writes no larger than 4 KB are handled in a different way. In this case, a separate spare block collects the newly written data. This spare block is referred to as a log block, as it can be seen as a log of small writes. Whenever the log block is full, the logical blocks modified by the writes recorded in the log block must be rewritten with copy-back operations to apply the changes. In this way, rewriting logical blocks is delayed until the log block is full. After rewriting all the involved logical blocks, the physical blocks previously mapped to the logical blocks and the spare blocks can be erased and converted to spare blocks. Note that the 4 KB threshold is an empirical setting, and this study provides no further discussion on it. erase and data copy activities for free-space reclaiming are referred to as *garbage collection*.

Figure 10.4 depicts a scenario of the proposed disk-emulation algorithm involving three logical blocks and five physical blocks. Let each physical block have four pages, and let each page be as large as a disk sector. A write is considered large if it is larger than two sectors. The left upper corner shows the initial state. Let a write be denoted by sector numbers enclosed within a pair of braces. Three small writes {0}, {0}, and {0, 1} arrive in turn. As they are small, they are appended to the free space in the log block at PBA 1 in Step 1. At this point, the log block is full. Step 2 then conducts copy-back operations to gather valid data from blocks at PBAs 0 and 1, and then rewrites the valid data to the block at PBA 3. Step 3 erases the blocks at PBAs 0 and 1. Step 4 revised the block L2P table. In Step 5, the fourth write {5, 6, 7} arrives. This write is large, and therefore requires that a logical block be rewritten. However, the unchanged data of Sector 4 are first copied from the block at PBA 4 to the log block at PBA 0. Step 6 then appends {5, 6, 7} to the log block, and Step 7 erases the block of invalid data. Step 8 then revises the block L2P table. Note that disk emulation is traditionally considered to be an issue independent of wear leveling. Refer to [19, 21, 22, 44] for further discussion on disk-emulation algorithms.

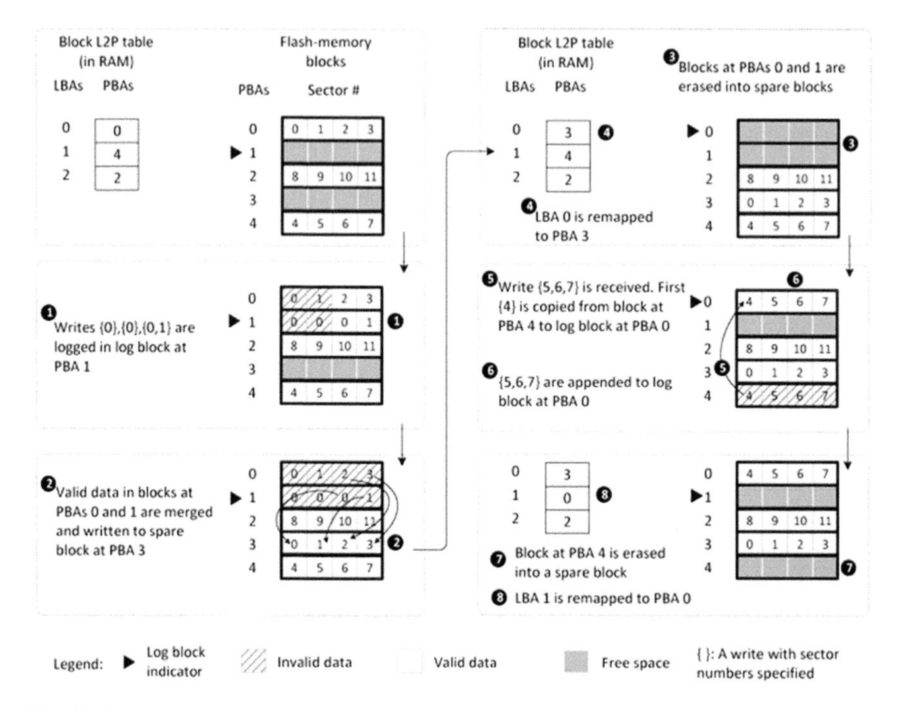

Fig. 10.4 A scenario of our disk-emulation algorithm

The segment L2P table is small enough to be kept in RAM because it has only eight entries. There are eight block L2P tables, one for each pair of a logical segment and a physical segment. As mentioned above, since RAM space is very limited, only two block L2P tables can be cached in RAM. Whenever a block L2P table is needed but is absent from RAM, the least-recently used table in the cache is discarded. The needed table is then constructed by scanning all the physical blocks of the corresponding physical segment. This scanning involves only the spare areas of every physical block's first page, which contain the mapping information.[3]

10.3.2.2 Block-Wearing Information and Priority Queues

The dual-pool algorithm keeps track of every block's wearing information. This includes an erase-cycle count, a recent erase-cycle count, and pool membership. Ideally, this information should be kept in RAM for efficient access. However,

[3]The scanning is read-only and does not affect wear leveling. Previous research has developed excellent methods for reducing the time overhead of this scanning. Refer to [19, 21] for details.

this is not feasible because the SSD controller has only about 1 KB of RAM as working space.

One option is to write a block's wearing information in its spare areas [26]. In this approach, a block's wearing information must be committed to one of its spare areas immediately after the block is erased. Later on, when user data are written, error-correcting codes and mapping information are also written to these spare areas. However, this approach can overwrite a spare area multiple times. This is prohibited by many new NAND flash [36, 37]. One alternative is to exclusively write the wearing information to a spare area, but this spoils the existing data layout in spare areas for disk emulation.

Our approach is to reserve one physical block for writing the wearing information. An on-flash block-wearing information table (“**BWI table**” for short) keeps the blocks' wearing information. A new BWI-table can be written to an arbitrarily allocated spare block, which means that the BWI table is subject to wear leveling. Since the entire flash memory is divided into eight physical segments, each segment has its own BWI table. A BWI table contains 1,024 entries, one for each physical block. Each table entry has 4 bytes, including a 18-bit erase-cycle count, a 13-bit recent erase-cycle count, and 1 bit for pool membership. Note that 13 bits are large enough for a recent erase-cycle count because it is reset upon cold-data migration. A BWI table is 1,024 * 4 = 4 KB large, so one 16-KB physical block can accommodate four revisions of a BWI table. If the block is full, another spare block is allocated for writing the BWI table, and the prior block is discarded for erase.

The on-flash BWI table can be entirely rewritten every time a block's wearing information changes. However, this method considerably increases write traffic to flash memory. Instead, the PBAs of the recently erased blocks are temporarily logged in a RAM buffer. In the current design, this buffer, named the erase-history table (“**EH table**” for short), has eight entries. If the EH table is full, a new version of the BWI table is written to the block reserved for the BWI table to apply the changes. After this, the in-RAM EH table is emptied.

Blocks are sorted in terms of different wearing information, and the dual-pool algorithm must check queue heads every time it is invoked. To scan the on-flash BWI table to find the queue heads is very slow. To reduce the frequency of BWI-table scanning, a small number of queue-head elements can be fetched for later use. For example, for fast access to $M(Q_H^{ec})$, after the BWI table is scanned, the wearing information of the two blocks with the two largest erase-cycles counts in the hot pool can be stored in RAM. An in-RAM queue-head table (“**QH table**” for short) is created for this purpose. The size of the QH table is fixed, and each of the five types of queue heads (shown in Table 10.4) is allocated to two table entries. A QH-table entry consists of a 2-byte PBA and 4-byte block-wearing information. Cold-data migration, hot-pool adjustment, and cold-pool adjustment check the QH table for queue heads. Wear leveling consumes QH-table entries and modifies the wearing information in the entries. A modified table entry is treated as an EH-table entry. The following section discusses when and how a QH table can be refreshed.

10.3.2.3 Segment Check-In/Check-Out

This section shows how the proposed wear-leveling data structures can be integrated into the segmented management scheme for disk emulation.

Disk emulation uses a two-level mapping scheme, as previously mentioned in Sect. 10.3.2.1. The segment L2P table is indexed by logical-segment numbers, has only eight entries, and is always stored in RAM. Second-level mapping manages the physical segments as if they were small pieces of flash memory. Each segment has an in-RAM L2P table, which maps 1,000 logical blocks to 1,024 physical blocks. Only two segments can have their block L2P tables cached in RAM. A segment is cached if its block L2P table is in RAM.

Each of the two cached segment uses an in-RAM EH table and an in-RAM QH table. Whenever a logical block is accessed, the corresponding physical segment is located by the segment L2P table. The dual-pool algorithm then checks if the segment's block L2P table, the EH table, and the QH table are in RAM. If they are absent, the following procedure, named *segment check-in*, is performed to bring them in: The in-RAM block L2P table is constructed by scanning the spare areas of each block's first page containing the mapping information. During scanning, if a block is found storing the on-flash BWI table, then the most up-to-date BWI table in the block is scanned to create the in-RAM QH table. By the end of this segment check-in procedure, the QH table and the block L2P table are ready. The in-RAM EH table is emptied, and the segment is all set for data access.

As the EH table continues to record the PBAs of erased blocks, sooner or later it will become be full. In this case, a new version of the on-flash BWI table should be created to merge the wearing information in the current on-flash BWI table, the in-RAM EH table, and the in-RAM QH table. The QH table is involved because QH-table entries could have been switched to EH-table entries. This merging procedure, called the *BWI-table merge*, is as follows: First the block storing the current BWI table is located. The dual-pool algorithm creates a new BWI table in the same block right after the current BWI table. If there is no free space left, a new spare block is allocated. The four flash-memory pages storing the current BWI table are then copied to the new location. During copying a BWI-table page, the DMA engine first loads one of the four pages from flash memory into the DMA buffer, and then the dual-pool algorithm performs a three-way synchronization that involves the wearing information from the DMA buffer, the QH table, and the EH table. By the end of this merging procedure, the QH table is refreshed to contain new queue-head physical block addresses and their wearing information, and the EH table is emptied.

A segment's in-RAM data structures can also be evicted from RAM to accommodate those of a newly accessed segment. Before a segment vacates RAM space, its EH table and QH table must be merged with the on-flash BWI table. This process is called *segment check-out*. To check out a segment, the BWI-table merge procedure is first performed, and the in-RAM structures of the segment can then be discarded.

Figure 10.5 shows how by wear leveling, disk emulation, and segment operations use the proposed data structures. Step 1 shows that when a segment is checked

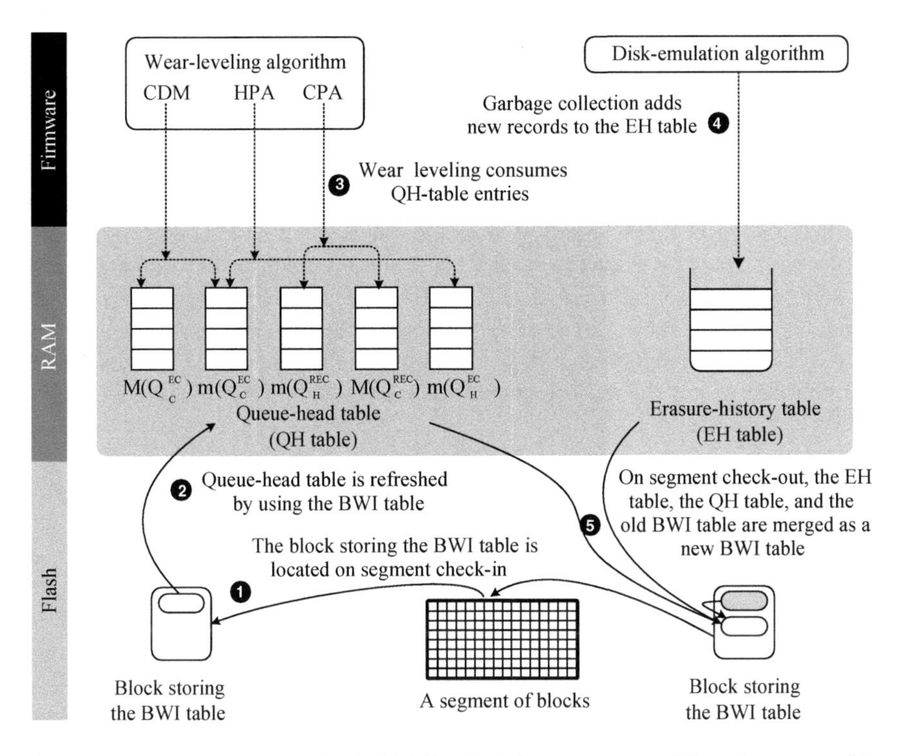

Fig. 10.5 Relationship between the in-RAM/on-flash data structures and how they are used by wear leveling, disk emulation, and segment operations

in, the spare areas of the blocks in that segment are scanned to build the in-RAM block L2P table. This scanning process also locates the block storing the on-flash BWI table. Step 2 refreshes the in-RAM QH table of the segment with information in the on-flash BWI table. Step 3 shows that QH-table entries are consumed by wear leveling. If any block is erased by garbage collection, then a record of the erase is appended to the in-RAM EH table, as shown in Step 4. When the segment is checked out, Step 5 merges the information in the in-RAM QH table, in-RAM EH table, and on-flash old BWI table and writes it to a new BWI table on flash.

10.4 Conclusion

This work addresses a key endurance issue in the deployment of flash memory in various system designs. Unlike the wear leveling algorithms proposed in the previous work, two efficient wear leveling algorithms (i.e., the evenness-aware algorithm and dual-pool algorithm) are presented to solve the problems of the existing algorithms with the considerations of the limited computing power and memory space in flash storage devices. The evenness-aware algorithm proactively moves static or infrequently updated data with an efficient implementation and limited memory-space requirements so as to spread out the wear-leveling actions over the entire physical address space. It proposes an adjustable house-keeping data structure and an efficient wear leveling implementation based on cyclic queue scanning. Its goal is to improve the endurance of flash memory with only limited increases in overhead and without extensive modifications of popular implementation designs. The dual-pool algorithm is to protect a flash-memory block from being worn out if the block is already excessively erased. This goal is accomplished by moving rarely updated data to excessively erased blocks. Because the micro-controllers of flash storage devices are subject to very tight resource budgets, keeping track of wear levels for a large number of blocks is a very challenging task. The dual-pool algorithm keeps only the most frequently accessed data in RAM, while the rest is written to flash memory.

References

1. A. Ban, Flash file system. US Patent 5,404,485, in *M-Systems*, Apr 1995
2. A. Ban, Wear leveling of static areas in flash memory. US Patent 6732221 (2004)
3. A. Ban, R. Hasbaron, Wear leveling of static areas in flash memory, US Patent 6,732,221, in *M-systems*, May 2004
4. A. Ben-Aroya, S. Toledo, Competitive analysis of flash-memory algorithms, in *Proceedings of the 14th Conference on Annual European Symposium* (2006)
5. L.-P. Chang, On efficient wear-leveling for large-scale flash-memory storage systems, in *22nd ACM Symposium on Applied Computing (ACM SAC)*, Mar 2007
6. L.-P. Chang, T.-W. Kuo, Efficient management for large-scale flash-memory stroage systems with resource conservation. ACM Trans. Storage **1**(4), 381–418 (2005)
7. L.-P. Chang, T.-W. Kuo, S.-W. Lo, Real-time garbage collection for flash-memory storage systems of real-time embedded systems. ACM Trans. Embed. Comput. Syst. **3**(4), 837–863 (2004)
8. Y.-H. Chang, J.-W. Hsieh, T.-W. Kuo, Endurance enhancement of flash-memory storage systems: an efficient static wear leveling design, in *DAC'07: Proceedings of the 44th Annual Conference on Design Automation* New York, NY, USA, (ACM, 2007), pp. 212–217
9. M.L. Chiang, P.C.H. Lee, R. Chuan Chang, Using data clustering to improve cleaning performance for flash memory. Softw. Pract. Exp. **29**(3), 267–290 (1999)
10. R.J. Defouw, T. Nguyen, Method and system for improving usable life of memory devices using vector processing. US Patent 7139863 (2006)
11. DRAM market-share games shifting from a knockout to a marathon; 4× nm process and multi-bit/cell as fundamental criteria to judge NAND Flash production competitiveness. Technical report, DRAMeXchange, Apr 2008
12. R.A.R.P. Estakhri, M. Assar, B. Iman, Method of and architecture for controlling system data with automatic wear leveling in a semiconductor non-volatile mass storage memory. US Patent 5835935 (1998)
13. Flash Cache Memory Puts Robson in the Middle. Intel
14. Flash-memory translation layer for NAND flash (NFTL). M-Systems (1998)
15. Freescale Semiconductor. USB Thumb Drive reference design DRM061 (2004)
16. FTL Logger Exchanging Data with FTL Systems. Technical Report, Intel
17. C.J. Gonzalez, K.M. Conley, Automated wear leveling in non-volatile storage systems. US Patent 7120729 (2006)

18. Increasing Flash Solid State Disk Reliability. Technical report, SiliconSystems, Apr 2005
19. J.-U. Kang, H. Jo, J.-S. Kim, J. Lee, A superblock-based flash translation layer for NAND flash memory, in *EMSOFT '06: Proceedings of the 6th ACM and IEEE International Conference on Embedded Software*, New York, NY, USA (ACM, 2006), pp. 161–170
21. H.-J. Kim, S.-G. Lee, An effective flash memory manager for reliable flash memory space management. IEICE Trans. Inf. Syst. **85**(6), 950–964 (2002)
22. J. Kim, J.-M. Kim, S. Noh, S.-L. Min, Y. Cho, A space-efficient flash translation layer for compact flash systems. IEEE Trans. Consum. Electron. **48**(2), 366–375 (2002)
23. S.-W. Lee, D.-J. Park, T.-S. Chung, D.-H. Lee, S. Park, H.-J. Song, A log buffer-based flash translation layer using fully-associative sector translation. Trans. Embed. Comput. Syst. **6**(3), 18 (2007)
24. Micron Technology, *Wear-Leveling Techniques in NAND Flash Devices* (2008)
25. Microsoft, Flash-memory abstraction layer (FAL), in *Windows Embedded CE 6.0 Source Code* (2007)
26. Motorola, Inc., MC9S12UF32 System on a Chip Guide V01.04 (2002)
27. M-Systems, Flash-Memory Translation Layer for NAND Flash (NFTL) (1998)
28. M-Systems. TrufFFS Wear-Leveling Mechanism, Technical Note TN-DOC-017 (2002)
29. NAND08Gx3C2A 8Gbit Multi-level NAND Flash Memory. STMicroelectronics (2005)
30. Numonyx, Wear Leveling in NAND Flash Memories (2008)
31. Open NAND Flash Interface (ONFi), Open NAND Flash Interface Specification Revision 2.1 (2009)
20. K. Perdue, *Wear Leveling* (2008)
32. C. Ruemmler, J. Wilkes, UNIX disk access patterns, in *Usenix Conference* (Winter 1993), pp. 405–420
33. D. Roselli, J.R. Lorch, T.E. Anderson, A comparison of file system workloads, in *Proceedings of the USENIX Annual Technical Conference*, pp. 41–54
34. M. Rosenblum, J.K. Ousterhout, The design and implementation of a log-structured file system. ACM Trans. Comput. Syst. **10**(1) (1992)
35. Samsung Electronics, K9F2808U0B 16 M * 8 Bit NAND Flash Memory Data Sheet (2001)
36. Samsung Electronics Company, K9GAG08U0 M 2G * 8 Bit MLC NAND Flash Memory Data Sheet (Preliminary)
37. Samsung Electronics Company, K9NBG08U5 M 4 Gb * 8 Bit NAND Flash Memory Data Sheet
38. SanDisk Corporation, Sandisk Flash Memory Cards Wear Leveling (2003)
39. D. Shmidt, Technical note: Trueffs wear-leveling mechanism (tn-doc-017). Technical report, M-System (2002)
40. Software Concerns of Implementing a Resident Flash Disk. Intel
41. Spectek, NAND Flash Memory MLC (2003)
42. M. Spivak, S. Toledo, Storing a persistent transactional object heap on flash memory, in *LCTES '06: Proceedings of the 2006 ACM SIGPLAN/SIGBED Conference on Language, Compilers, and Tool Support for Embedded Systems* (2006), pp. 22–33
43. STMicroelectronics, Wear Leveling in Single Level Cell NAND Flash Memories (2006)
44. S.P.D.-H.L.S.-W.L. Tae-Sun Chung, D.-J. Park, H.-J. Song, System software for flash memory: a survey, in *EUC '06: Embedded and Ubiquitous Computing* (2006), pp. 394–404
45. Understanding the Flash Translation Layer (FTL) Specification. Technical report, Intel Corporation (Dec 1998), http://developer.intel.com/
46. W. Vogels, File system usage in windows nt 4.0. SIGOPS Oper. Syst. Rev. **33**(5), 93–109 (1999)
47. D. Woodhouse, Jffs: the journalling flash file system, in *Proceedings of Ottawa Linux Symposium* (2001)
48. M. Wu, W. Zwaenepoel, eNVy: a non-volatile main memory storage system, in *Proceedings of the Sixth International Conference on Architectural Support for Programming Languages and Operating Systems* (1994), pp. 86–97

Chapter 11
BCH Codes for Solid-State-Drives

Alessia Marelli and Rino Micheloni

Abstract Given that the NAND Flash memory is not a very reliable medium, it follows that a Solid State Disk needs some help to achieve a reliability suitable for computing applications: the *Error Correction Code* (ECC). As the NAND technology scales down, ECC becomes a critical design topic. This chapter deals with BCH, the most common ECC in solid state disks. Two main issues arise when an ECC is used inside an SSD. First of all, the ECC should not limit the bandwidth, being the bottleneck of the entire drive: this translates in a hardware implementation that needs to handle multiple devices (channel) in parallel. At the same time, ECC must avoid erroneous corrections when the error correction capability of the code is overcome, i.e. it must have a high detection property. In this chapter the ECC definitions are reviewed, then the BCH code is presented along with the multi-channel topic. Finally, BCH and LDPC detection property are discussed.

11.1 Error Correction Codes Basic Definitions

In 1948 Claude Shannon's article "A Mathematical Theory of Communication" gave birth to the two twin disciplines: information theory and coding theory. The article specifies the meaning of efficient and reliable information and, there, the very well known term "bit" has been used for the first time [1]. Anyway, it was only with Richard Hamming in 1950 that a constructive generating method and the basic parameters of Error Correction Codes (ECC) were defined.

Hamming made his discovery at the Bell Telephone's laboratories during a study on communication on long telephone lines corrupted by lightening and crosstalk. The discovery environment shows how the interest in error-correcting codes has taken shape, since the beginning, outside a purely mathematical field.

A. Marelli (✉) · R. Micheloni
Storage Solutions, Microsemi Corporation, Vimercate, MB, Italy
e-mail: alessiamarelli@gmail.com

R. Micheloni
e-mail: rino.micheloni@ieee.org

R. Micheloni et al. (eds.), *Inside Solid State Drives (SSDs)*,
Springer Series in Advanced Microelectronics 37,
https://doi.org/10.1007/978-981-13-0599-3_11

The codes discovered by Hamming are able to correct only one error, they are simple and widely used in several applications where the probability of error is small and the correction of a single error is considered sufficient.

More powerful codes, such as BCH and Reed-Solomon, were discovered between 1958 and 1960. The first ones were described by Bose and Chaudhuri [2] and through an independent study by Hocquengheim [3]; the second ones were defined by Reed and Solomon a few years later, between 1959 and 1960 [4]. They were immediately used in space missions, and today they are still used in compact discs.

Afterwards, they stopped being of interest for space missions and were replaced by convolutional codes, introduced for the first time by Elias in 1955. Convolutional codes can also be combined with cyclic codes. The study of optimum convolutional codes and the best decoding algorithms continued until 1993 when turbo codes were presented for the first time in the communication environment [5]. In fact, it is in the sector of telecommunications where they have received greater success.

A singular history is that of LDPC (Low Density Parity Checks) codes first discovered in 1962 by Gallager [6], but whose applications are being studied only today [7].

Error correction codes add redundant bits called parity bits to the information data bits so that, on reception, it is possible to detect the errors and to recover the message that has most probably been transmitted.

One of the biggest families in coding theory is block codes [8–10]. Block coding deals with messages of fixed length. Schematically (Fig. 11.1), a block m of k symbols is encoded in a block c of n symbols ($n > k$) and written in a memory. Inside the memory, different sources may generate errors e, so that the block message r is read. The block r is then decoded in d by using the maximum likelihood decoding strategy, so that d is the message that has most probably been written.

A *Code C* is the set of codewords obtained by associating the q^k messages of length k of the space A to q^k words of length n of the space B in a univocal way.

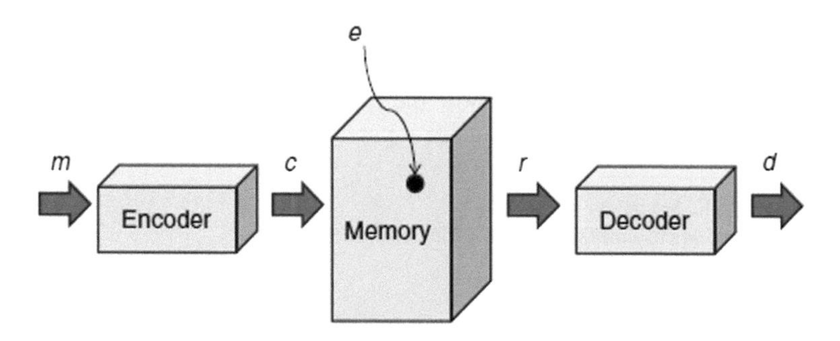

Fig. 11.1 Representation of coding and decoding operations for block codes

A code is defined as *linear* if, given two codewords, also their sum is a codeword. When a code is linear, encoding and decoding can be described with matrix operations.

Definition 11.1.1 G is called *generator matrix of a code* C when all the codewords are obtainable as a combination of the rows of G.

Each code has more than one generator matrix, i.e. all its linear combinations. It follows that each code has infinite equivalent codes, i.e. all those obtained by permutations or linear combinations of the matrix G.

Definition 11.1.2 A set of equations that gives parity positions in terms of data positions is called *parity equations set.*

It is possible to express all these equations as a matrix. The matrix H is called *parity matrix* for a block code.

Therefore, with reference to Fig. 11.1, encoding a data message m consists in multiplying the message m by the code generator matrix G, according to (11.1).

$$c = m \cdot G \tag{11.1}$$

Definition 11.1.3 G is said in *standard form* or in *systematic form* if $G = (I_k, P)$, where I_k is the identity matrix $k \times k$ and P is a matrix $k \times (n - k)$. If G is in standard form then the first k symbols of a word are called information symbols.

Theorem 11.1.1 *If a code $C[n, k]$ has a matrix $G - (I_k, P)$ in standard form, then a parity matrix of C is $H = (-P^T, I_{n-k})$ where P^T is the transpose of P and it is a matrix $(n - k)$ x k and I_{n-k} is the identity matrix $(n - k) \times (n - k)$.*

Systematic codes have the advantage that the data message is visible in the codeword and can be read before decoding. For codes in non-systematic form the message is no more recognizable in the encoded sequence and it is necessary to have the inverse encoding function to recognize the data sequence.

Definition 11.1.4 The code rate is defined as the ratio between the number of information bits and the codeword length. Given a linear code $[n, k]$ the ratio k/n is defined as *code efficiency.*

Definition 11.1.5 It is called *minimum distance* or *Hamming distance* d of a code, the minimum number of different symbols between any two codewords.

We can see that for a linear code the minimum distance is equivalent to the minimum distance between all the codewords and the codeword 0.

Definition 11.1.6 A code has *detection capability* v if it is able to recognize all the messages, containing v errors at the most, as corrupted.

The detection capability is related to the minimum distance as described in (11.2).

$$v = d - 1 \tag{11.2}$$

Definition 11.1.7 A code has *correction capability* t if it is able to correct each combination of a number of errors equal to t at the most. The correction capability is calculated from the minimum distance d by the relation:

$$t = \left[\frac{d-1}{2}\right] \tag{11.3}$$

where the square brackets mean the floor function.

Definition 11.1.8 Given a code $C[n, k]$ A_i represents the number of codeword with weight i. The set $\{A_i\}$ is called weight distribution of the code C and $\{A_i\}$ are called the weights of C.

The code C has a symmetric distribution if (11.4) holds true.

$$A_i = A_{n-1} \quad 0 \leq i \leq \left[\frac{n-1}{2}\right] \tag{11.4}$$

Definition 11.1.9 The dual code C^* of a code $C[n, k]$ is the set of vectors orthogonal to all the codewords of C:

$$C^* = \{v \in (F_q)^n | v \cdot c = 0 \forall c \in C\} \tag{11.5}$$

The weight distribution or the distance of the dual code gives a lot of information on the code itself as the following sections show.

Definition 11.1.10 Given a code C, its dual code C^* and $\{A_i\}$ with $i = \{0, \ldots, n\}$ its weight distribution, we define the weight enumerator of the code C the polynomial

$$W_C(x, y) = \sum_{i=0}^{n} A_i x^{n-i} y^i \in Z[x, y] \tag{11.6}$$

A fundamental theorem that describe the relationship between $W_C(x, y)$ and $W_{C*}(x, y)$ is MacWilliams theorem. Here it will be present only the version for linear binary codes.

Theorem 11.1.2 (MacWilliams equality for binary codes) *Given a linear binary code $C[n, k]$ and C^* its dual code the following equation holds true*

$$W_{C^*}(x, y) = \frac{1}{|C|} W_C(x + y, x - y) \tag{11.7}$$

where $|C| = 2^k$ is the number of words in C. In other words:

$$\sum_{i=0}^{n} A_i^* x^{n-i} y^i = \frac{1}{|C|} \sum_{i=0}^{n} A_i (x + y)^{n-i} (x - y)^i \tag{11.8}$$

This latter equality is called MacWilliams identity.

There is an important operation we can apply to a linear code C called extension. Also in this case, there is a relationship between the original code weights and the weights of its extension.

Definition 11.1.11 Given a code $C[n, k, d]$ we call extension of the code C_E the code obtained from C by adding one more parity bit computed as the logical XOR of all the other bits. The code C_E has only even weighted codeword and

- if d is even C_E is a code $[n + 1, k, d]$
- if d is odd C_E is a code $[n + 1, k, d + 1]$

Given a code C with weight distribution $\{a_i\}$, and $\{A_i\}$ the weight distribution of its extension code C_E, we have

$$A_{2i} = a_{2i} + a_{2i-1} \tag{11.9}$$

with $2 \leq 2i \leq n - 1$.

Given a code $C[n, k, d]$ and C_E its extension: if n is odd and C has a symmetrical weight distribution, then C_E has symmetrical weight distribution.

In a lot of applications there are external factors not subject to error check which determine the length permitted to an error correction code. Non volatile memories, for example, operate on codewords that have a length power of 2.

When the "natural" length of the code is not suitable it is possible to change it with the shortening operation.

Definition 11.1.12 A code $C[n, k]$ is *shortened* into a code $C\,[n - j, k - j]$ by erasing j columns of the parity matrix.

Codes can be combined together in order to improve their correction capabilities. One way to combine them is with concatenation. In this operation we have an inner code (C_{IN}) and an outer code (C_{OUT}) that work together (Fig. 11.2).

Typically, C_{IN} is decoded with a maximum-likelihood approach and C_{OUT} is a block code of length n. In this way the concatenation combines the error probability property of the inner code and the decoding time property of the outer code.

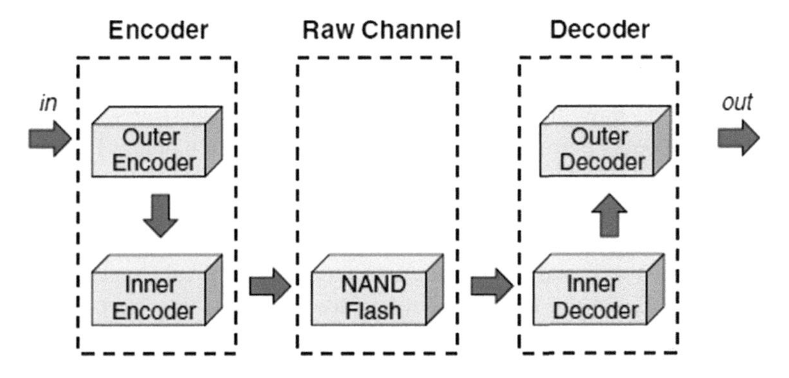

Fig. 11.2 Schematic diagram of code concatenation

As sketched in Fig. 11.2 the message is firstly encoded with the outer code and the resulting codeword is encoded with the inner code. During the decoding phase the message is decoded with the inner code and the result is then decoded with the outer code. From this description it is clear that a key feature is that the inner code must have a good detection property, since we must be sure that the inner code doesn't perform erroneous correction when the correction capability of the code is overcome.

11.2 BCH Codes

BCH codes belong to the family of cyclic codes. These are, perhaps, the most used codes in applications, since they can be implemented by using high-speed shift-register encoders and decoders [11–13].

Definition 11.2.1 A linear code $C[n, k]$ is called *cyclic* if $(x_1,x_2, \ldots,x_n) \in C => (x_n, x_1, \ldots,x_{n-1}) \in C$.

In other words, if we write the vector $a(x) = (a_0, \ldots,a_{n-1})$ as the polynomial $a_0 + a_1x + a_2x^2 + \cdots + a_{n1}x^{n-1}$, the previous definition states that, if $a(x) \in C$, then also the right shift belongs to C.

As seen in the previous section, the distance is a key feature in characterizing a code; in BCH codes the minimum distance can be ensured during construction.

Generally speaking, in order to know the minimum distance for a linear code with generator polynomial $g(x)$, it is necessary to compute the distance between all the possible codewords. BCH codes, by imposing some constraints on the generator polynomial, are able to ensure a "designed distance".

Definition 11.2.2 Let β be an element of $GF(q^m)$. Let b be a non-negative integer. A BCH code with "designed" distance d is generated by the polynomial $g(x)$ of minimal degree that has $d - 1$ consecutive powers of β: β^b, β^{b+1}, ..., β^{b+d-2} as roots. Given Ψ_i the minimal polynomial of β^{b+i} for $0 \leq i < d - 1$, $g(x)$ is computed as:

$$g(x) = LCM\{\psi_0(x), \psi_1(x), \ldots, \psi_{d-2}(x)\} \tag{11.10}$$

and the data protected by the code is $k = n\text{-}deg(g(x))$.

It is possible to show that the designed distance d is at least $2t + 1$, hence the code is able to correct t errors. The number of parity bits for a binary BCH code is less than or equal to mt. Generally, this number is equal to mt; it is less only when the minimum distance is greater than the designed distance the code is constructed with.

If we assume $b = 1$, and β a primitive element of $GF(q^m)$ the code becomes a *narrow-sense* and *primitive* BCH code of length $q^m - 1$ able to correct t errors. We shall now consider primitive BCH codes.

As regards the distance, the important result of Carlitz-Uchiyama inequality is proven.

Theorem 11.2.1 (Carlitz-Uchiyama inequality) *Given a binary BCH code C of length $n = 2^{m-1}$ with designed distance $\delta = 2t + 1$, for the minimum distance of the dual code C^* the following inequality holds true*

$$d^* \geq 2^{m-1} - (t-1)2^{\left[\frac{m}{2}\right]} \tag{11.11}$$

The general decoding structure is represented in Fig. 11.3.

In BCH structure there is only one step to encode a message, while there are three steps to decode a message. Generally, we can state that the decoding is ten times more complex than encoding.

The encoding of a systematic BCH code is performed by multiplying the message m(x) by x^{n-k} and calculating the parity bits as the remainder of the division of this multiplication by the generator polynomial, in accordance with (11.12) *and* (11.13).

$$\frac{m(x) \cdot x^{n-k}}{g(x)} = q(x) + \frac{r(x)}{g(x)} \tag{11.12}$$

$$c(x) = m(x) \cdot x^{n-k} + r(x) \tag{11.13}$$

The structure that implements this division is represented in Fig. 11.4.

The decoding operation follows three fundamental steps, as shown in Fig. 11.3.

- *calculation of the syndromes;*
- *calculation of the coefficients of the error locator polynomial;*
- *calculation of the roots of the error locator polynomial.*

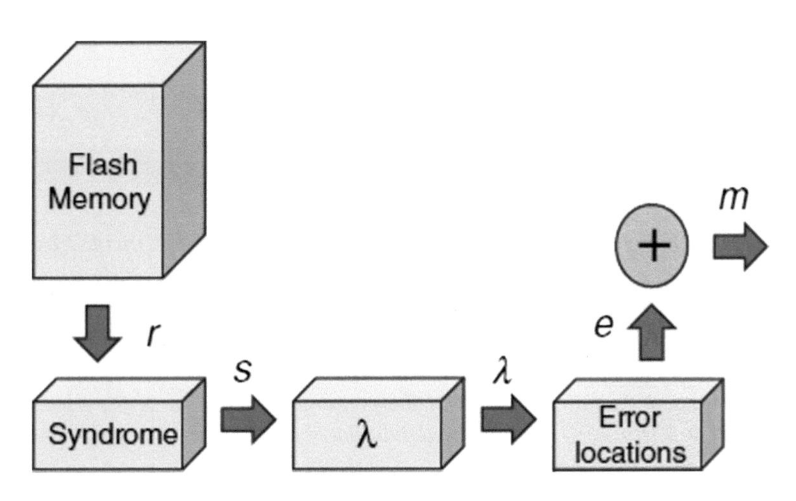

Fig. 11.3 Structure of a binary BCH decoder

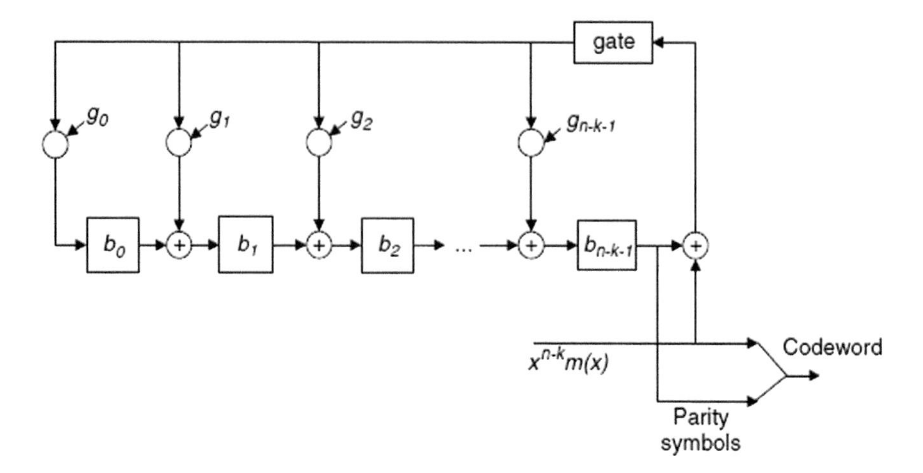

Fig. 11.4 Binary BCH encoder

Errors in the storage media can be represented by a polynomial that has coefficient 1 in correspondence with every error's position:

$$E(x) = E_0 + E_1 x + \cdots + E_{n-1} x^{n-1} \tag{11.14}$$

Observe that, in order for the code to be corrector of t errors, at most t non-null coefficients are allowed in (11.14). *The read vector R(x) is therefore:*

$$R(x) = c(x) + E(x) \tag{11.15}$$

The first decoding step consists in calculating the 2t syndromes for the read message:

$$\frac{R(x)}{\psi_i(x)} = Q_i(x) + \frac{S_i(x)}{\psi_i(x)} \quad \text{with} \quad 1 \leq i \leq 2t \tag{11.16}$$

$$S_i(x) = Q_i(x) \cdot \psi_i(x) + R(x) \quad \text{with} \quad 1 \leq i \leq 2t \tag{11.17}$$

In accordance with (11.16) *and* (11.17), *the received vector is divided by each minimal polynomial Ψ_i forming the generator polynomial, thus getting a quotient $Q_i(x)$ and a remainder $S_i(x)$ called syndrome.*

At this point the 2t syndromes must be evaluated into the elements β, β^2, β^3, ..., β^{2t} whose Ψ_i are the minimal polynomials. With reference to (11.18), *this evaluation is computed as the evaluation of the message received in β, β^2, β^3, ..., β^{2t}, since $\Psi_i(\beta i) = 0$ (for $1 \leq i \leq 2t$) by definition of minimal polynomial.*

$$S_i(\beta^i) = S_i = Q_i(\beta^i) \cdot \psi_i(\beta^i) + R(\beta^i) = R(\beta^i) \tag{11.18}$$

Consequently, the i-th syndrome can be calculated either as the remainder of the division between the received message and the minimal polynomial Ψ_i, then evaluated in β^i, or as the evaluation in β^i of the received message.

Observe that, in case no errors occur, the polynomial received is a codeword: therefore the remainder of the division of (11.16) *is null and all the syndromes are identically null. On the other hand, verifying if the syndromes are identically null is a necessary and sufficient condition to understand if the read message is a codeword or if some errors occurred.*

An useful property described in (11.19) *can be exploited to compute only t syndromes.*

$$S_{2i} = S_i^2 \tag{11.19}$$

The syndromes calculation for a BCH code existing over GF(2^m) involves t structures which contemporarily calculate the remainder of the divisions between the received polynomial and the t minimal polynomials. These structures are very similar to the one depicted in Fig. 11.4.

Once the syndromes are computed, they are used to search the error locator polynomial.

By indicating the error positions with X and the number of errors that occurred with v the following equality holds true:

$$S_i = \sum_{l=1}^{v} X_l^i \tag{11.20}$$

Definition 11.2.3 It is defined *error locator polynomial* $\Lambda(x)$ the polynomial whose roots are the inverse of the error positions.

From the definition we have:

$$\Lambda(x) = \prod_{i=1}^{v} (1 - xX_i) \tag{11.21}$$

Please observe that the degree of the error locator polynomial gives the number of errors that occurred. The degree of $\Lambda(x)$ is t at most, hence, in the case more than t errors occur, the polynomial $\Lambda(x)$ could erroneously indicate t or less errors.

The most used algebraic method to perform this step of the decoding is the Berlekamp-Massey algorithm [14]. The complexity of this algorithm grows in a linear way, enabling the construction of efficient decoders able to correct dozens of errors.

Berlekamp algorithm finds the coefficients of the error locator polynomial in an iterative way. At the i-th step of the algorithm we find a polynomial $\Lambda(x)$ whose coefficients solve the first i equations of (11.20). Then, we test if $\Lambda(x)$ also solves the equation $i + 1$; if not, we calculate the discrepancy term d so that $\Lambda(x) + d$ solves the first $i + 1$ equations. After $2t$ iterations $\Lambda(x)$ is the error locator polynomial.

In the binary case it is possible to perform the Berlekamp algorithm in t iterations. There are a number of different implementations of Berlekamp algorithm [15–17], here below we will explain the one following the diagram of Fig. 11.5.

Equation (11.22) shows the syndrome polynomial and the initial conditions for the algorithm:

$$\begin{aligned} 1+S &= 1+S_1z+S_2z^2+\cdots+S_{2t-1}z^{2t-1} \\ \Lambda^{(0)}(z) &= 1 \quad d^{(0)}=1 \end{aligned} \tag{11.22}$$

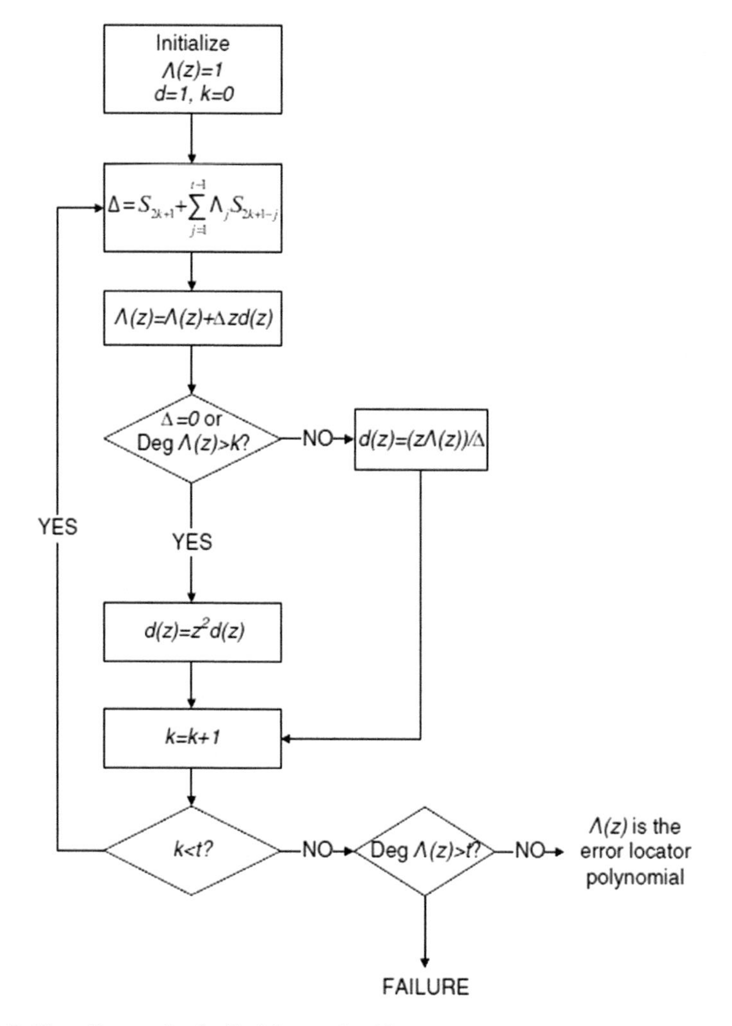

Fig. 11.5 Flow diagram for the Berlekamp algorithm

At the i-th step we proceed as follows:

- if S_{2i+1} is unknown the algorithm is finished;
- otherwise we define $\Delta^{(2i)}$ the coefficient of z^{2i+1} in the product $(1 + S(z))\Lambda^{(2i)}(z)$.

$$\Lambda^{(2i+2)}(z) = \Lambda^{(2i)}(z) + \Delta^{(2i)} \cdot d^{(2i)}(z) \cdot z \tag{11.23}$$

$$d^{(2i+2)}(z) = \begin{cases} z^2 d^{(2i)}(z) & \text{if } \Delta^{(2i)} = 0 \text{ or if } {}^\circ\Lambda^{(2i)}(z) > i \\ \dfrac{z\Lambda^{(2i)}(z)}{\Delta^{(2i)}} & \text{if } \Delta^{(2i)} \neq 0 \text{ or if } {}^\circ\Lambda^{(2i)}(z) \leq i \end{cases} \tag{11.24}$$

The polynomial $\Lambda^{(2t)}(z)$ is the error locator polynomial.

A number of paper have been published to avoid the inversion or to parallelize the structure. It is not the purpose of this chapter to present these paper but they can be found in [15–17].

The last step of the decoding process consists in searching for the roots of the error locator polynomial. If the roots are separate and they are in the field, then it is enough to calculate their inverse to have the error positions. If they are not separate or they are not in the correct field, it means that the word received has a distance from a codeword greater than t. In this case an uncorrectable error pattern occurred and the decoding process fails.

The algorithm used to search the roots, known as Chien algorithm, is a method based on trial and error. Substantially each field element is substituted in the error locator polynomial: if it satisfies the equation it is a root, otherwise the following element is tested. The inverse of the found root indicates an error location.

Recall that the error locator polynomial $\Lambda(x)$ of degree t at the most, for a $BCH[n, k]$, is defined as:

$$\Lambda(x) = 1 + \Lambda_1 x + \cdots + \Lambda_t x^t \tag{11.25}$$

Hence, verifying if a field element α^i satisfies the equation means verifying (11.26):

$$1 + \Lambda_1 \alpha^i + \cdots + \Lambda_t \left(\alpha^i\right)^t = 0 \tag{11.26}$$

If the equation is not satisfied the following element is considered, otherwise αi is a root. In this case the inverse is an error position, i.e. the position 2^{m-1-i} is the erroneous one.

11.3 BCH Decoding Failures

BCH codes are not perfect codes: for this reason it is difficult that a codeword with more than t errors moves in the correction sphere of another codeword. The codewords of BCH codes are well separated one from another and only a number of errors much greater than t could partially overlap their correction spheres.

This is the reason why, when more than t errors occur, most of the time the decoding process fails but erroneous corrections are not performed. It is therefore possible to use an error message showing that more than t errors have occurred.

Suppose we have a message containing more than t errors and see how the decoding proceeds. At the exit of the syndromes calculation block it is not possible to detect if the correction capability has been exceeded; on the contrary, the calculation is completed with success by finding t, apparently valid, syndromes.

The syndromes are transferred to the block that searches for the error locator polynomial. As mentioned, the Berlekamp algorithm is a recursive algorithm that searches for the coefficients of the error locator polynomial using successive approximations, by adding at the i-th iteration a discrepancy term d so that $\Lambda(x) + d$ solves the first $i + 1$ equations. The discrepancy term is a monomial that is added to the error locator polynomial previously found. When the degree of the monomial to be added is greater than t, the correction capability of the code has been exceeded. Recalling that the degree of the error locator polynomial is equal to the number of errors that most likely occurred, we can state that this number is reliable up to t. If a degree higher than t is detected at some point in the algorithm, the decoding terminates with an error message.

Unfortunately it is not granted that, when the correction capability of the code is exceeded, this is what happens. On the contrary, most of the times this does not happen and the error locator polynomial apparently seems a valid one with a degree smaller than or equal to t (most of the times equal to t). Consequently, these coefficients, apparently valid, are loaded into the Chien machine.

When the correction capability of the code is exceeded, the Chien algorithm discloses it, since one of the following cases occurs:

- there are coincident roots;
- a sufficient number of roots is not found. Remember that a number of roots equal to the degree of the error locator polynomial has to be found;
- in case of shortened codes it can also happen that the shortened positions, those ones ideally filled in with 0 s, are recognized as erroneous.

In practical implementations the first condition never happens because, given the implementation of the Chien machine, the same element is never tested more than once.

The second condition is the one that actually occurs in real applications. At the end of the Chien algorithm we verify, through a comparator, if the number of roots found is equal to the degree of the error locator polynomial. If this condition is not satisfied, an error message shows that the correction capability of the code has been exceeded.

Finally, the third condition generally never occurs in shortened codes cases, because the use of an "initialization" constant avoids the testing of shortened positions.

However, remember that if the number of errors is much greater then the correction capability, the received message can be found in a correction sphere of another codeword: in this case the code might not be able to understand if the correction capability has been exceeded and might perform erroneous corrections.

Summarizing, we can state that the BCH decoder can be approximate with an ideal one, since erroneous correction are very unlikely to occur [8], unless the received message really falls in another correction sphere. These cases must be studied based on the algebraic structure of the code and will be presented in the next sections.

11.4 Detection Properties

As explained in the previous section, it is unlikely that the BCH decoding algorithm makes erroneous corrections, i.e. we can approximate it with an ideal decoding. It follows that the erroneous corrections are made only when the received message is located in a correction sphere different from the original codeword.

Definition 11.4.1 Given a binary linear code C able to correct t errors, we call the probability of miscorrection P_{ME} the probability that an ideal bounded distance decoder executes erroneous corrections.

Definition 11.4.2 The weighted probability $P_E(w)$ is the probability of executing erroneous corrections when w errors occurred.

Observe that the probability P_{ME} depends on the code C and on the transmission channel.

Theorem 11.4.1 *The weighted probability $P_E(w)$ is computed as:*

$$P_E(w) = \frac{D_w}{\binom{n}{w}} \tag{11.27}$$

where D_w is the number of decodable words and w is in the range $[t + 1, n]$.

The number of decodable words can be computed as

$$D_w = \sum_{i=0}^{n} a_i \sum_{s=0}^{t} N(i, w; s) \tag{11.28}$$

where N(i,w;s) is the number of words with weight w with a distance s from a word of weight i. This is computed by (11.29)

$$N(i,w;s) = \begin{cases} \binom{n-i}{\frac{s+w-i}{2}}\binom{i}{\frac{s-w+i}{2}} & \text{if } |w-i| \le s \\ 0 & \text{if } |w-i| > s \end{cases} \tag{11.29}$$

Substituting (11.28) *in* (11.27) *we have:*

$$P_E(w) = \frac{\sum_{i=0}^{n} a_i \sum_{s=0}^{t} N(i,w;s)}{\binom{n}{w}} \tag{11.30}$$

P_{ME} is computed based on $P_E(w)$ as described in (11.31)

$$P_{ME} = \sum_{w=t+1}^{n} P_E(w)\phi(w) \tag{11.31}$$

where $\Phi(w)$ is the probability that a word has weight w.

For a binary symmetric channel BSC we have:

$$P_{ME} = \sum_{w=t+1}^{n} D_w p^w (1-p)^{n-w} \tag{11.32}$$

where p is the bit error probability.

It follows that we have to compute the value D_w. This value can be computed according with (11.28). *Unfortunately the weights a_i are unknown for BCH codes and must be estimated.*

11.5 BCH Weight Estimation

There are a number of different theorems that helps in estimating the weight of a BCH code. Here below we will see the major ones and how they behave in comparison with real weights.

First of all, we present a result that establishes a relationship between the weight distribution of a BCH code and the weight distribution of its dual code.

Theorem 11.5.1 *Given C a BCH[n, k, d] code and C_E its extension, C has weight distribution $\{a_i\}$ and C_E has weight distribution $\{A_i\}$. The following equations hold true:*

$$
\begin{aligned}
a_{2i-1} &= \frac{2i}{n} A_{2i} \\
a_{2i} &= \frac{n-2i}{n} A_{2i}
\end{aligned}
\tag{11.33}
$$

Observe that a BCH code has symmetrical weight distribution and the word composed by all 1 is a valid codeword. Moreover, given that BCH has an odd length (i.e $\cdot$ $n = 2^{m-1}$*), also for the extension* BCH_E *the word composed by all 1 is a valid codeword. Finally, observe that the dual code of* C_E *has only even weight codewords.*

One of the most important weight estimation is the Peterson one. It was the first estimation and it is not an upper or a lower bound but an approximation.

Theorem 11.5.2 ([18] Peterson Estimation) *The weight* a_i *of a primitive BCH code of length n and error correction capability t can be approximated as*

$$
a_i \cong \frac{\binom{n}{i}}{(n+1)^t}
\tag{11.34}
$$

In order to have upper bounds, different correction terms are added to (11.34). *In other words, for the extension code* BCH_E *the estimations use the following relationship:*

$$
A_i = \begin{cases} 0 & i \equiv 1 \bmod 2 \\ \frac{\binom{n}{i}}{2^{mt}}(1+E_i) & i \equiv 0 \bmod 2 \end{cases}
\tag{11.35}
$$

In order to compare different estimations a real case is shown. For BCH [255,207,13] the weights w are known. Figure 11.6 *shows the relative errors with respect to the real weights with different estimations for this code. On the x-axis there is the weight w, while on the y-axis we find*

$$
R(w) = \frac{A(w)_{EST} - A(w)_{REAL}}{A(w)_{REAL}}
\tag{11.36}
$$

We distinguish three different behaviors. The first estimation set has a very low error on the first weights but a very high error in the middle.

The following theorems describe the estimations belonging to this set.

Theorem 11.5.3 ([19]) *Given*

$$
t \geq 3, w = n - 2d^*, t < i \leq \frac{n-w}{4}
$$

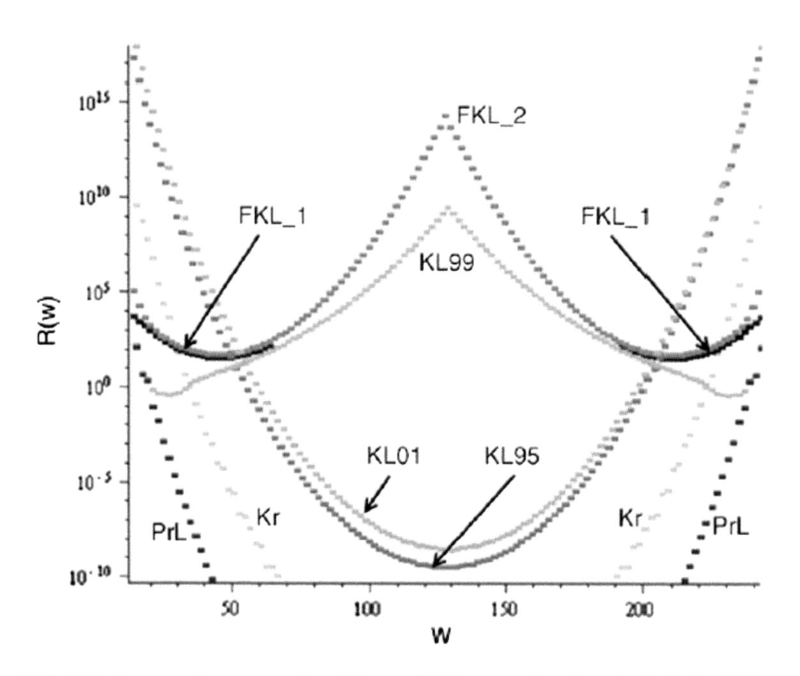

Fig. 11.6 Relative error between real weights and different estimations for BCH[255,207,13]

For cases:

$$\begin{aligned}
i = t+1, \quad & t=3, m \geq 5 \\
& t=4, m \geq 9 \\
& t=5, m \geq 15 \\
i = t+2, \quad & t=4, m \geq 7 \\
& t=5, m \geq 9 \\
& t=6, m \geq 11 \\
& t=7, m \geq 15
\end{aligned}$$

Equation (11.37) *holds true*

$$\frac{E_{2i}}{2} \leq 2^{-m(i-t)} \prod_{h=1}^{2i-1} \left(1 - \frac{h}{n+1}\right)^{-1} \prod_{h=1}^{i} (2h-1)\left(2^{i-1}-1\right) \qquad (11.37)$$

For all other cases (11.38) *holds true*

$$\frac{E_{2i}}{2} \le 2^{-m(i-t)} \prod_{h=1}^{2i-1} \left(1 - \frac{h}{n+1}\right)^{-1} \left\{ \prod_{h=1}^{i} (2h-1) \sum_{h=\lfloor\frac{i-t}{2}\rfloor+1}^{\lfloor\frac{i}{2}\rfloor} \binom{i}{2h} + \prod_{h=1}^{i} (2h-1) \frac{(2i)![2(t-1)]^{2i-2q-2t}}{2^q q!(2i-2q)!} \right\}$$
$$q = \left\lceil i + \tfrac{3}{4} + (t-1)^2 - \sqrt{\tfrac{1}{16} + \left(2i + \tfrac{3}{2}\right)(t-1)^2 + (t-1)^4} \right\rceil \tag{11.38}$$

The proof of this theorem is behind the purpose of this chapter. However, note that this is a very complex equation that estimates very well the weights, but the big drawback is that it can be applied only to some cases. For BCH[255,207,13] sketched in Fig. 11.6 *this estimation (labeled as FKL_1) is applicable only for weight w in the range [13,64] and in the range [256–64, 256]. However, there is another estimation (labeled as FKL_2 in Fig. 11.6) extended to all weights.*

Theorem 11.5.4 ([19]) *For* $t < i \le 2^{m-2}$ *the following inequality holds true:*

$$\left|\frac{E_{2i}}{2}\right| \le 2^{-m(i-t)} \left\{ \sqrt{\frac{8}{\pi}\left(1 - \frac{2i}{n+1}\right)} (2i)^i e^{-ic(i,t-1)} + \frac{2}{\pi}\sqrt{\frac{2i}{2i-1}} \prod_{j=1}^{t} (2j-1) |2(t-1)|^{-2t} \left(t - 1 + \sqrt{2i + (t-1)^2}\right)^{2i} e^{2ib(2i)} \right\}$$
$$c(u,x) = 1 - (\ln 2) H\left(\frac{x}{u}\right) - \frac{5u}{2(n+1)} \tag{11.39}$$
$$H(x) \begin{cases} -x\log_2 x - (1-x)\log_2(1-x) & 0 \le x \le \frac{1}{2} \\ 1 & \frac{1}{2} < x \le 1 \end{cases}$$
$$b(s) = -\frac{1}{2} + \frac{5s}{8(n+1)} + \frac{1}{1 + \sqrt{1 + 4s(n+1)u^{-2}}}$$

As shown in Fig. 11.6 *this estimation is very similar to the previous one for small weights but it has a huge error in the middle.*

Another family of estimations has a big error on the first weights but a very low error in the middle.

Theorem 11.5.5 ([20]) *For a primitive BCH code of length* $n = 2^{m-1}$*, an upper bound for the weight distribution is:*

$$a_i = \frac{\binom{n}{i}}{(n+1)^t}(1 + E_{i^*}) \tag{11.40}$$

*where i * = i + 1 if i is odd and i* = i if i is even and E_i is computed with the following inequality:*

$$|E_i| \leq \frac{n^t \binom{n+1}{\frac{n+1}{2}} \binom{\frac{n+1}{2}}{\frac{i}{2}}}{\binom{n+1}{i} \binom{n+1}{d^*}} \tag{11.41}$$

Another estimation of the correction term E_i is proposed in the following theorem.

Theorem 11.5.6 ([21]) *The correction term E_i can be estimated as:*

$$|E_i| \leq (n+1)^t \sqrt{\frac{(2i(n+1-i)+n+1)(n+1)^t \binom{i}{\frac{i}{2}} \binom{n+1-i}{\frac{n+1-i}{2}}}{2(n-d^*) \binom{n+1}{d^*} \binom{n+1}{i}}} \tag{11.42}$$

As before, the proof of this theorem is behind the purpose of this chapter. However, it is important to state that it is based on the maximization of a specific class of polynomials called Krawtchouk polynomials.

Definition 11.5.1 For every positive integer n we call Krawtchouk polynomial of degree k $P_k(x, n) = P_k(x)$

$$P_k(x,n) = \sum_{j=0}^{k} (-1)^j \binom{x}{j} \binom{n-x}{k-j} \tag{11.43}$$

By using a result of [22] it is possible to prove that:

$$|E_i| \leq \frac{2^{mt}}{\binom{n}{i}} \max_{d^* \leq x \leq 2^{m-1}} |P_i(x)| \tag{11.44}$$

It follows that an upper bound for max $|P_i(x)|$ is an upper bound for E_i.

It is possible to compute exactly this maximum value, with the drawback of a high computational cost. Hence, it is not always possible for all the length and correction capability. In Fig. 11.6 the relative error obtained with the maximization of Krawtchouk polynomials is labeled as "Kr". Finally the last estimation is presented below.

Theorem 11.5.7 ([23]) *Given f(x) and g(x) even function with respect (n + 1)/2 described by* (11.45)

$$f(x) = \sum_{i=0}^{k} f_{2i} P_{2i}(x) \text{ and } g(x) = \sum_{i=0}^{k} g_{2i} P_{2i}(x) \tag{11.45}$$

with the following properties

$$\begin{aligned} & f_{2k} > 0 \\ & g_{2k} > 0 \\ & f(x) \geq 0 \quad \forall x \\ & g(x) \leq 0 \quad d^* \leq x \leq n - d^* \end{aligned} \tag{11.46}$$

It follows that

$$\begin{aligned} & A_{2k}^f \leq A_{2k} \leq A_{2k}^g \\ & A_{2k}^f = \frac{\binom{2^m}{2k}}{2^{mt}} \left(1 - E_{2k}^f\right) = \frac{\frac{f(0)}{2^{mt}} - f(0) - \sum_{i=t+1}^{k-1} f_{2i} A_{2i}}{f_{2k}} \\ & A_{2k}^g = \frac{\binom{2^m}{2k}}{2^{mt}} \left(1 - E_{2k}^g\right) = \frac{\frac{g(0)}{2^{mt}} - g(0) - \sum_{i=t+1}^{k-1} g_{2i} A_{2i}}{g_{2k}} \end{aligned} \tag{11.47}$$

A convenient choice for function f(x) and g(x) is proposed by the authors and is the following

$$\begin{aligned} & f(x) = P_k^2(x) \\ & g(x) = (P_2(x) + C_t) P_{k-1}^2(x) \quad C_t = -P_2\left(d^*\right) \end{aligned} \tag{11.48}$$

Figure 11.7 *shows the relative error with respect the real weights (as described in* (11.36)*) for the estimation made with the maximization of Krawtchouk polynomials, for the best estimation among all the theoretical estimation and for the estimation with linear programming technique.*

As shown in Fig. 11.7*, Krawtchouk estimation is better compared with the minimum among all other theoretical estimations. Anyway, we have the special case of linear programming estimation which shows a quasi-null error.*

A linear programming problem (LP) with N real variables $x_1, \ldots, x_N$ *with M constraints like*

$$\sum_{j=1}^{N} \alpha_{ij} x_j \leq c_i \quad \text{or} \quad \sum_{j=1}^{N} \alpha_{ij} x_j = c_i \tag{11.49}$$

with c_i *and* α_i *positive real variables, can be represented in a matrix form:*

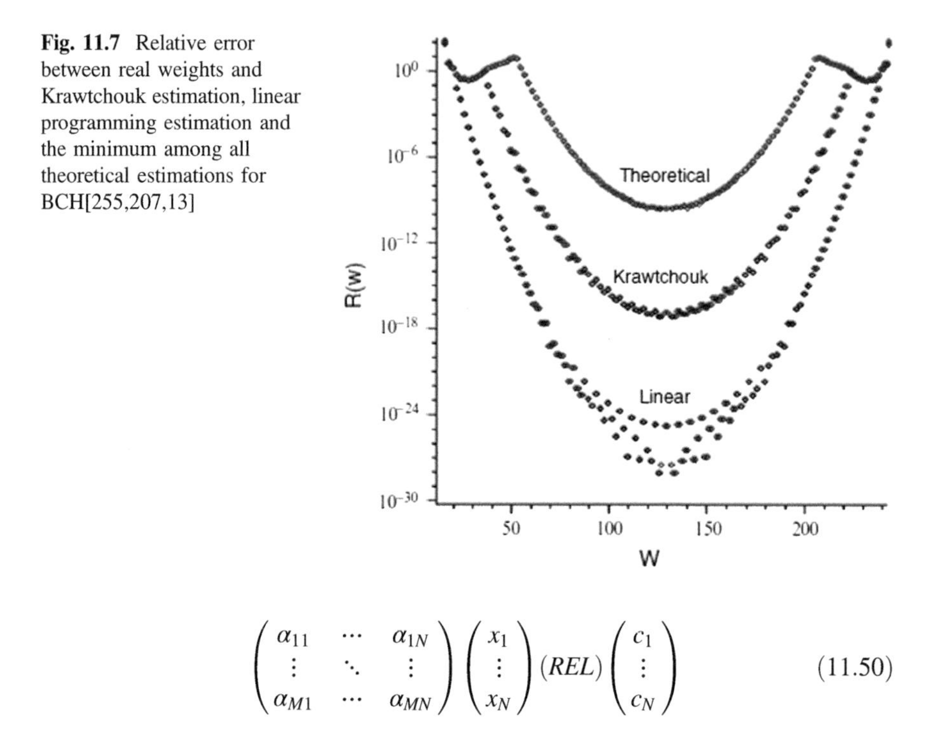

Fig. 11.7 Relative error between real weights and Krawtchouk estimation, linear programming estimation and the minimum among all theoretical estimations for BCH[255,207,13]

$$\begin{pmatrix} \alpha_{11} & \cdots & \alpha_{1N} \\ \vdots & \ddots & \vdots \\ \alpha_{M1} & \cdots & \alpha_{MN} \end{pmatrix} \begin{pmatrix} x_1 \\ \vdots \\ x_N \end{pmatrix} (REL) \begin{pmatrix} c_1 \\ \vdots \\ c_N \end{pmatrix} \tag{11.50}$$

where REL represents the relationship for each components. The purpose is to find a solution x able to maximize or minimize the objective function

$$\sum_{i=1}^{N} o_i x_i \tag{11.51}$$

Linear programming technique is applied to the BCH weight estimation by means of Fujiwara algorithm described in [19, 24].

MacWilliams identity is the objective function we need to maximize, where B_j is the weight distribution for the extension of the dual code:

$$\max \sum_{i=d^*}^{2^m - d^* - 1} P_s(j) B_j \quad s = d^*, \ldots, 2^m - d^* - 1 \tag{11.52}$$

The constraints that we need to add on B_j are the Pless power-moment identities [14]

$$\begin{aligned} &\sum_{j=d*}^{2^m - d^* - 1} \left(\tfrac{2^m}{2} - j\right)^{2l} B_j = 2^{2^m - k} M_{2l} - (1 + B_n)\left(\tfrac{2^m}{2}\right)^{2l} 0 \le l \le t \\ &M_i = 2^{-i} \left(\tfrac{d^i}{dx^i} \cosh^{2^m} x\right)_{x=0} \end{aligned} \tag{11.53}$$

The more constraints we add the easier to find a solution in a fast way. In this case we can add constraints involving the distance of the extension of the dual code as shown in the following example.

Example 1 Let's take the extension of BCH[2048,1992] with distance 12. We can exploit the properties of Reed-Muller codes [9, 10] so that

$$\begin{aligned} & RM(r,m) \subset BCH_E(2^m, 2^{m-r}-1) \\ & RM(7,11) \subset BCH_E(2048,16) \subset BCH_E(2048,12) \\ & RM(7,11) \supset BCH_{E^*}(2048,16) \supset BCH_{E^*}(2048,12) \\ & RM(11-7-1,11) \subset BCH_{E^*}(2048,12) \end{aligned} \tag{11.54}$$

The weights of this Reed-Muller code are non-null and multiple of 2^s with $s=3$. It follows that the weights of the code EBCH*(2048,12) are non-null and multiple of 8 starting from the minimum distance. By using Reed-Muller properties this distance is $d^* \geq 2^{m-r} = 256$, while we obtain $d^* \geq 2^{m-1} - (t-1)\,2^m = 844$ with Carlitz-Uchiyama inequality.

The estimation made with Krawtchouk maximization and Linear Programming technique are the most effective ones even if they are prohibitive for long codes due to computational complexity. Moreover, it's not always possible to find all the constraints and, even if they would be available, it could take a year to find an estimation with today's computers.

11.6 BCH Weight Estimation: Real Cases Analysis

In this part we will analyze different cases to compare different behaviors among estimations.

11.6.1 BCH[255,207,13]

In this case the weights of the code are known, since the code is short and the error correction capability of the code is small, i.e. six errors. Figures 11.6 and 11.7 represent the weight estimation comparison for this code. Observe that the weight estimation depends a lot on the estimation on the first weight.

P_{ME} graph is shown on Fig. 11.8. As it is possible to see the behavior is monotonic increasing. The comparison graph for P_E estimation is shown in Fig. 11.9. In the graph, there is the P_E computed by taking the minimum among all the theoretical estimations, the estimation obtained with linear programming technique and the real one (as we know the real weights).

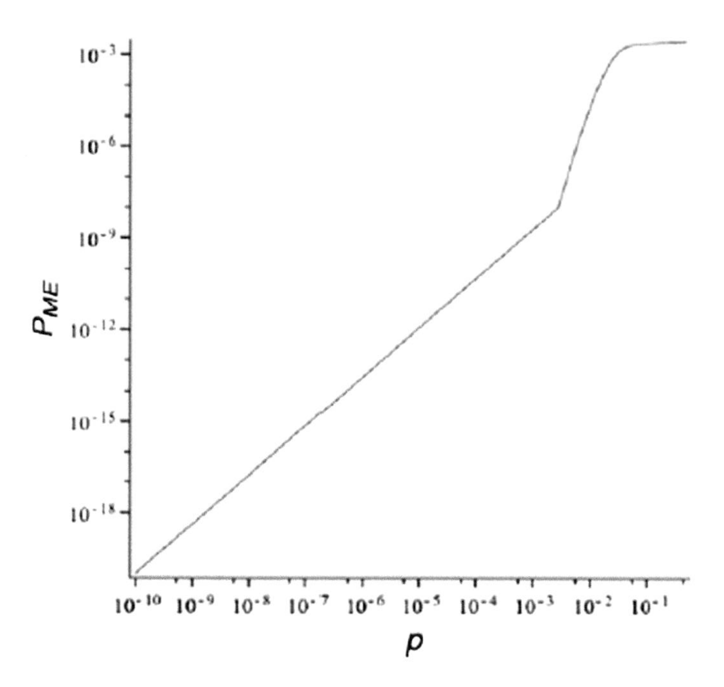

Fig. 11.8 P_{ME} behavior for BCH[255,207,13]

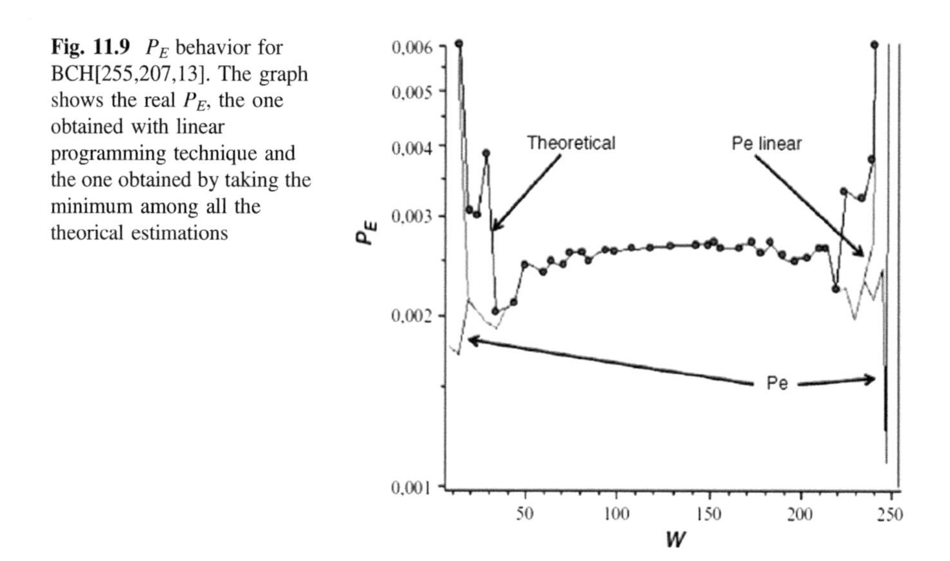

Fig. 11.9 P_E behavior for BCH[255,207,13]. The graph shows the real P_E, the one obtained with linear programming technique and the one obtained by taking the minimum among all the theorical estimations

As it is possible to see, the real P_E is a monotonic increasing function, while all the estimations are very good in the middle but have a bump on the first (and last) weight.

11.6.2 BCH[1023,993,7]

Also in this case the weights are known since the error correction capability of the code is only 3. Figures 11.10 and 11.11 shows the relative error for different weight estimations compared with the real weights and compared with the minimum among all the estimations.

As it possible to see the two figures are quite identical. Figure 11.12 shows *PE* behavior for this code using real weights. Also in this case we note that the behavior is monotonic increasing with a very long floor in the middle.

In Fig. 11.13 P_{ME} is shown. The x-axis represents a probability belonging to the range [0, 1/2] and is divided in 500 subsets.

11.6.3 BCH[4095,3975,21]

This is the first case where we don't know the real weights (Fig. 11.14). The behavior is similar to the previous cases, but a bump on the first weight estimation pops up. This is partially due to the fact that we don't know the minimum distance of the dual code of the extension.

For example, in BCH[255,207,13] case the distance of the code with Carlitz-Uchiyama bound is 48, while the real one is 64. Note that here it is not possible anymore to use Linear Programming technique due to computational complexity.

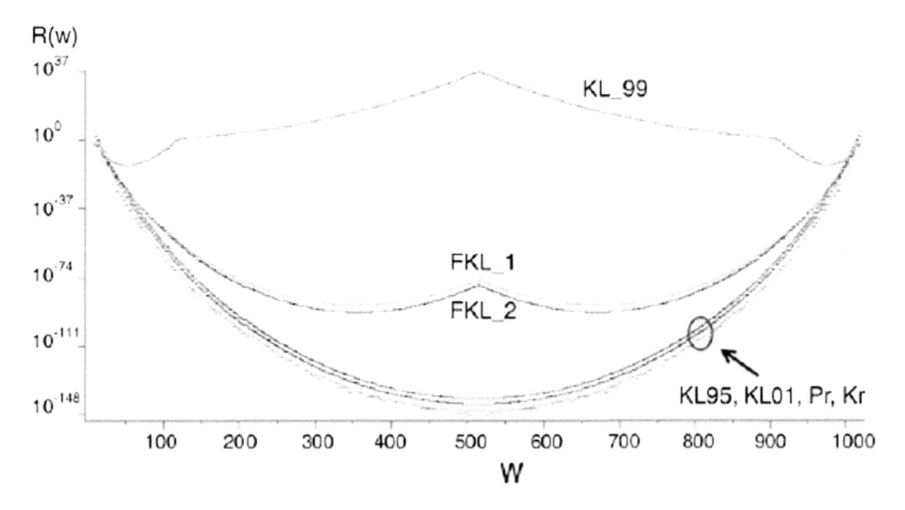

Fig. 11.10 Relative error between real weights and different estimations for BCH[1023,993,7]

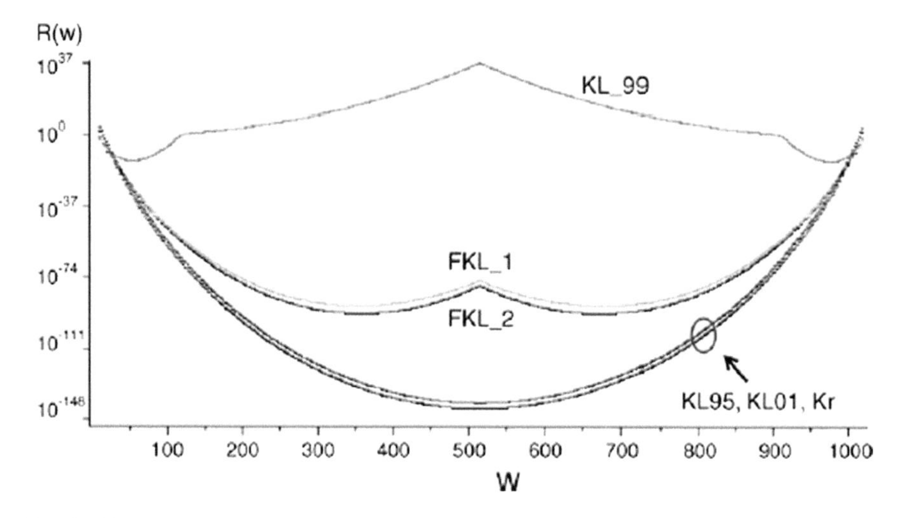

Fig. 11.11 Relative error between different estimations and the minimum among all the different estimations for BCH[1023,993,7]

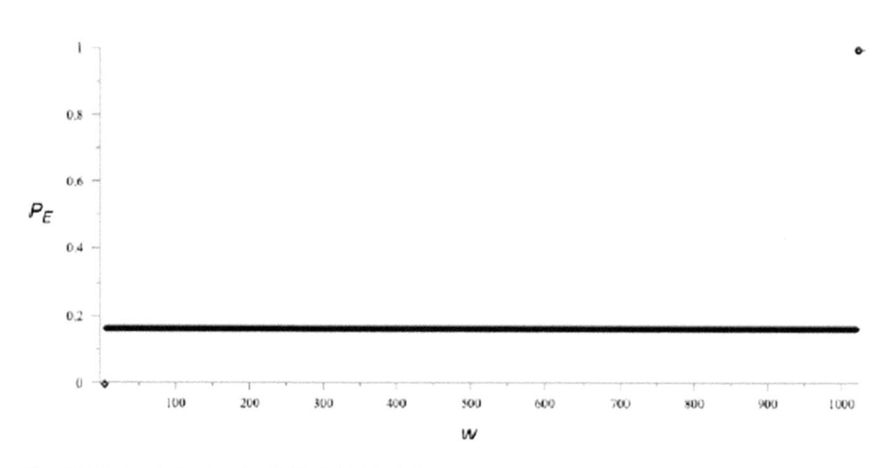

Fig. 11.12 P_E behavior for BCH[1023,993,7]

Figure 11.15 shows P_E behavior: it has the usual long floor, but for the first time we see a bump on the first weights. This behavior is shown also in Fig. 11.16 where we have a zoom on the first weights.

P_{ME} function is not represented here, since it has the same behavior as P_E.

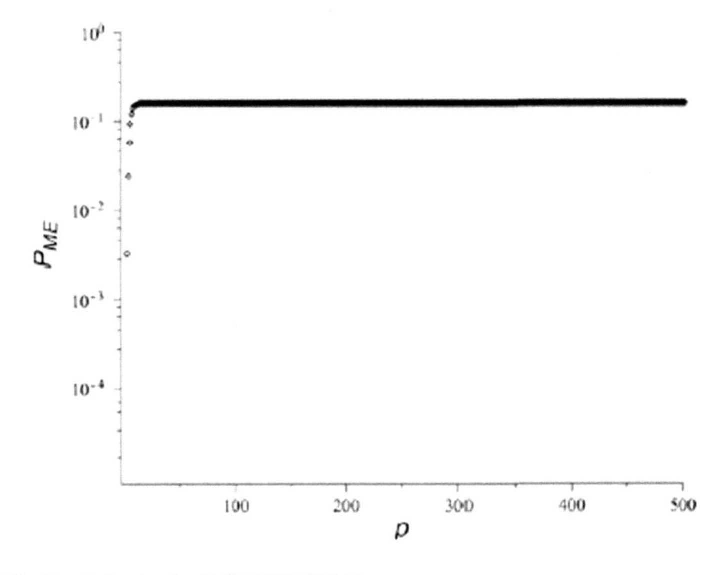

Fig. 11.13 P_{ME} behavior for BCH[1023,993,7]

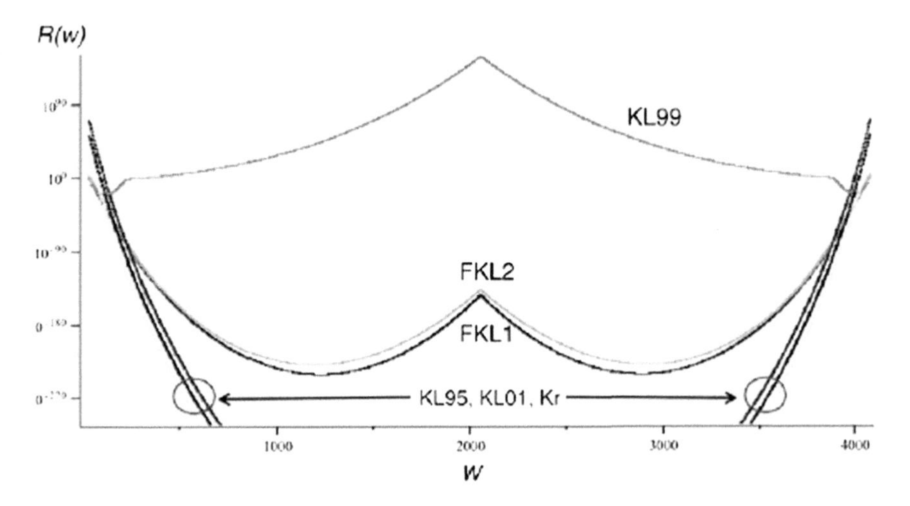

Fig. 11.14 Relative error between different estimations and the minimum among all the different estimations for BCH[4095,3975,21]

11.6.4 BCH[16383,15851,77]

Also in this case the real weights are unknown. Moreover, it is not possible anymore to use the Krawtchouk estimation and the linear programming estimation due to computational complexity. As shown in Fig. 11.17, we have an error on the first

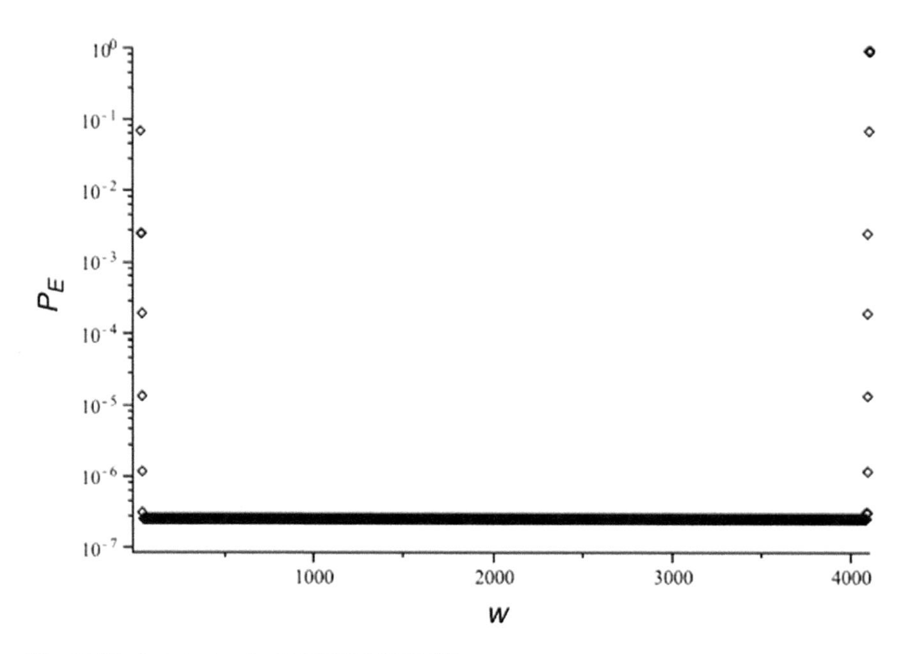

Fig. 11.15 P_E behavior for BCH[4095,3975,21]

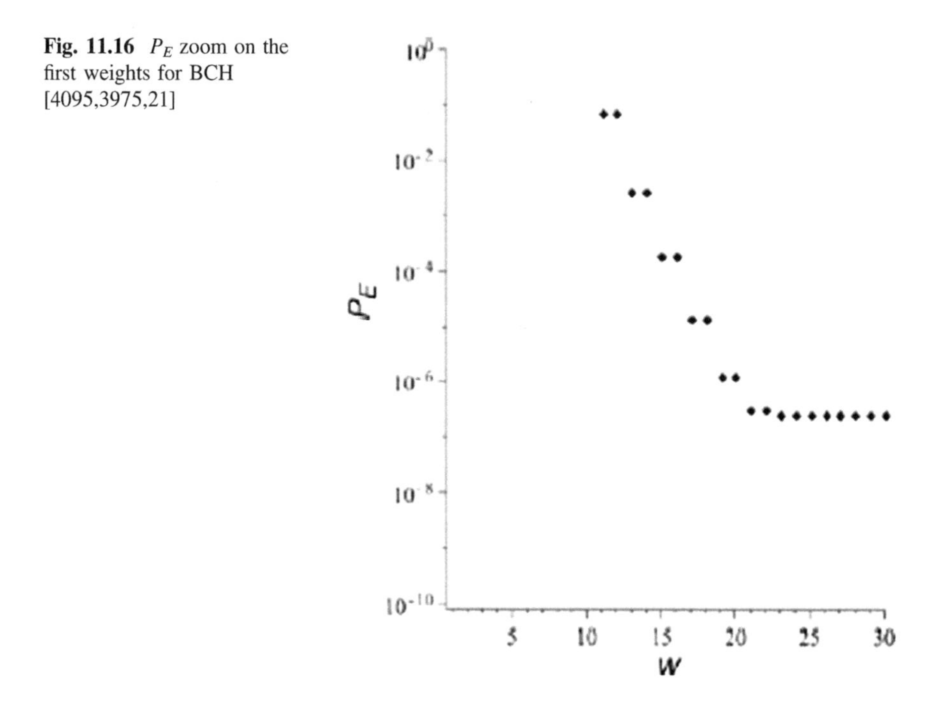

Fig. 11.16 P_E zoom on the first weights for BCH [4095,3975,21]

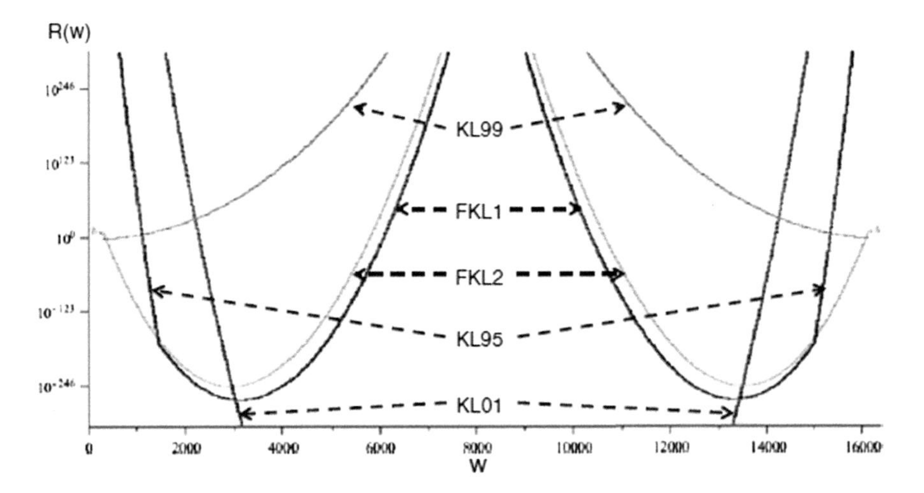

Fig. 11.17 Relative error between different estimations and the minimum among all the different estimations for BCH[16383,15851,77]

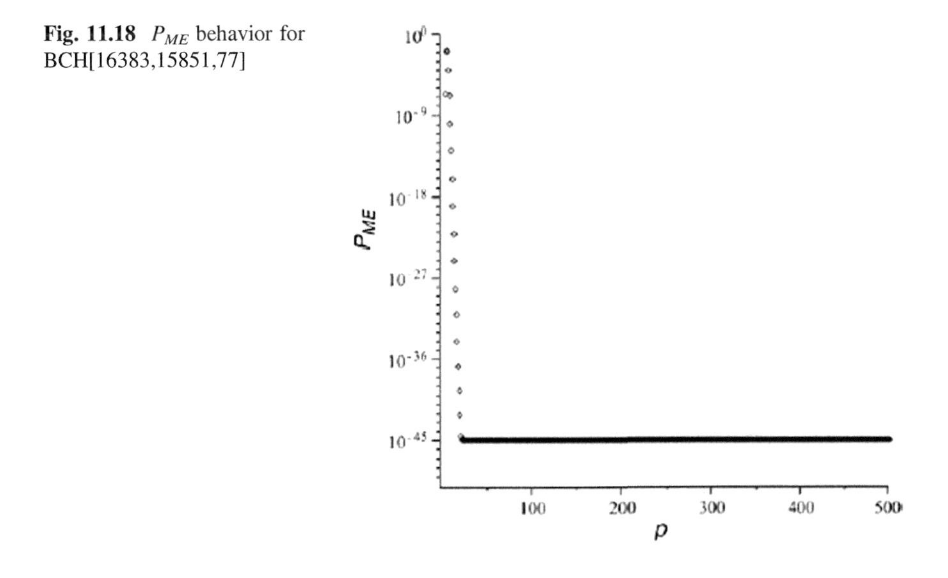

Fig. 11.18 P_{ME} behavior for BCH[16383,15851,77]

weights of hundreds of order of magnitudes. This is mainly due to the poor estimation on the distance of the dual code that gives also a big error on P_E and P_{ME} estimation (Figs. 11.18 and 11.19).

Please observe that, in this case, P_{ME} is almost 1 for low p values!

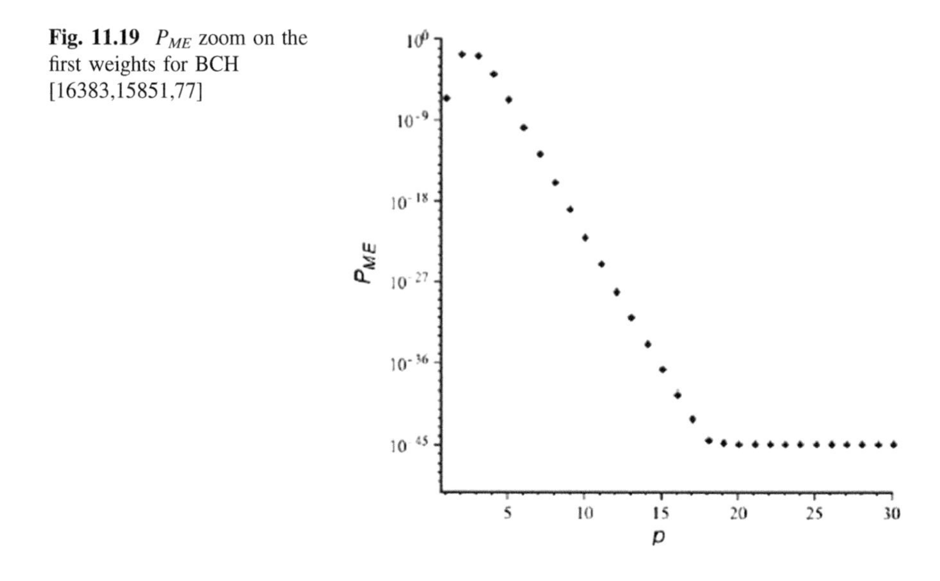

Fig. 11.19 P_{ME} zoom on the first weights for BCH [16383,15851,77]

11.7 BCH Detection Conclusion

As shown in the previous section, the error on the estimation on the first weights has a huge effect on P_E and P_{ME}. In particular a poor estimation shows up as a bump on the first weights that become greater as the code length increases.

One of the best estimation, even if it is not an upper bound, is the Peterson estimation (Theorem 11.5.2). Figures 11.20 and 11.21 shows P_E and P_{ME} behavior for BCH[255,207,13] using Peterson estimation. We can see that we have the monotonic increasing behavior that we expect when real weights are known.

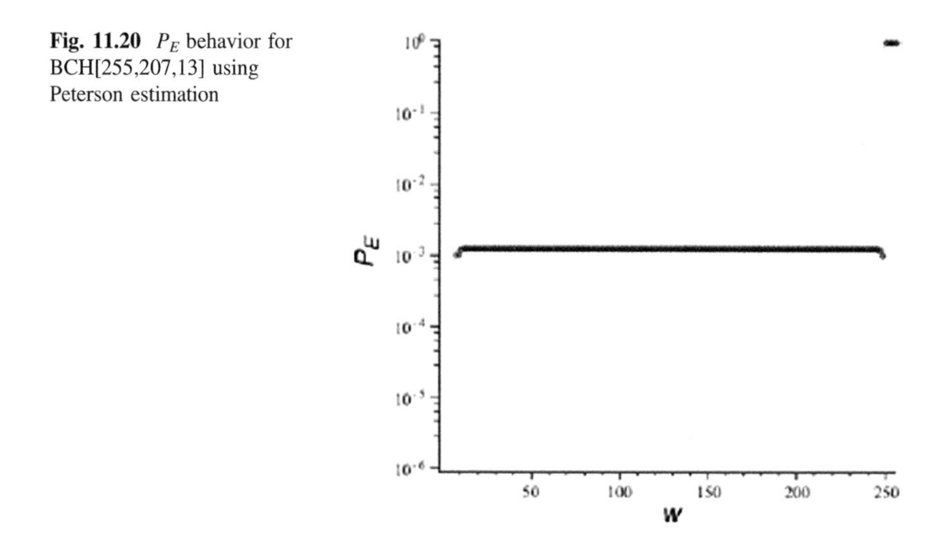

Fig. 11.20 P_E behavior for BCH[255,207,13] using Peterson estimation

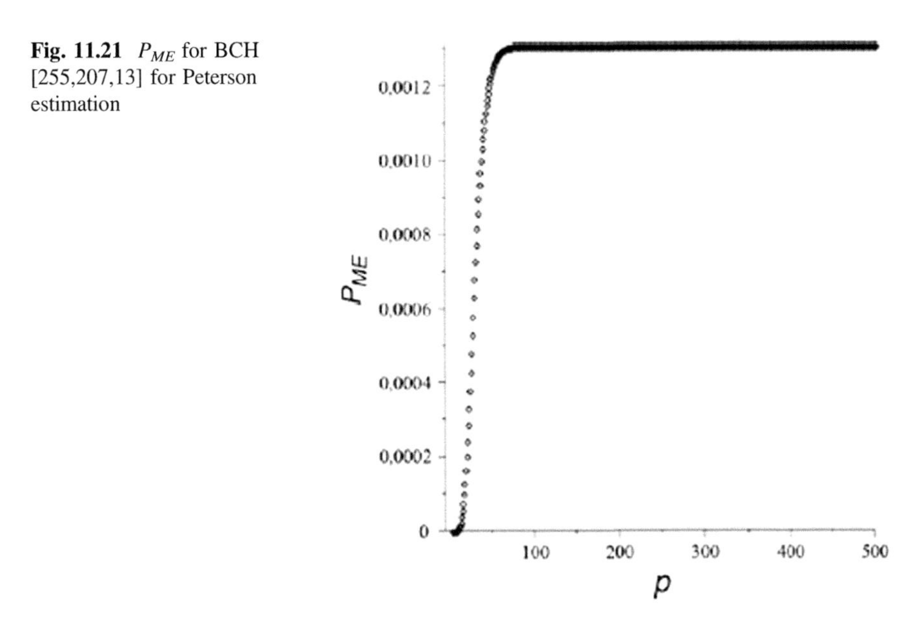

Fig. 11.21 P_{ME} for BCH [255,207,13] for Peterson estimation

Figure 11.22 shows P_{ME} behavior for BCH[16383,15851,77] using Peterson estimation. Recall (Sect. 11.6.4) that here the real weights are unknown and we had a bump at the beginning using upper bound estimations. Instead, by using Peterson estimation a monotonic behavior can be seen.

It follows that the real P_E and P_{ME} profile should be monotonic with a wide floor in the middle. When the code length is high and the code rate is high this floor can be approximate with [25].

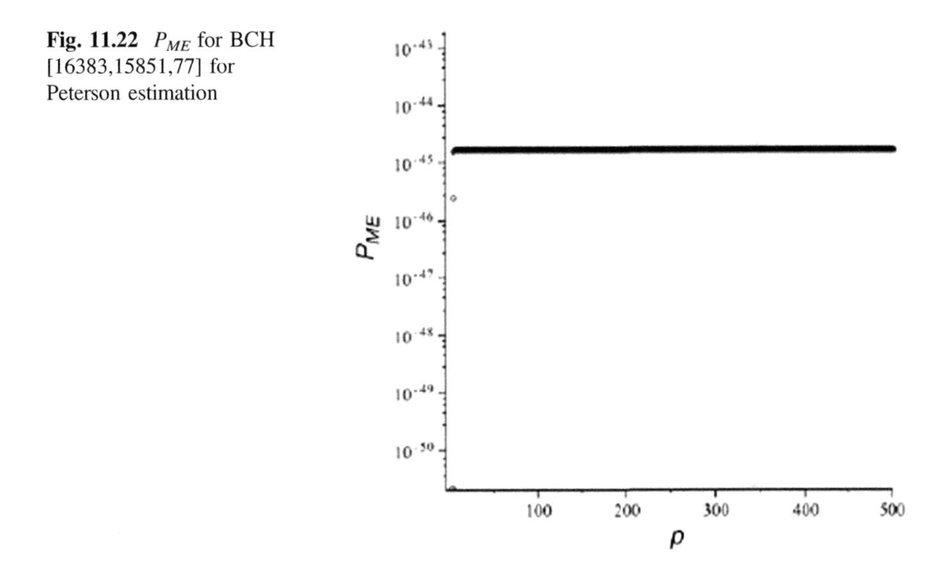

Fig. 11.22 P_{ME} for BCH [16383,15851,77] for Peterson estimation

$$Q = 2^{-(n-k)} \sum_{s=0}^{t} \binom{n}{s} \tag{11.55}$$

Summarizing, we can state that when the length is high, the BCH code has a very good detection properties that made it suitable for the implementation in SSDs. In fact, when a catastrophic error occurs or when the error correction capability of the code is passed, the BCH code declares a decoding failure without attempting erroneous corrections. This is a key point when using BCH code concatenated with another code.

11.8 Multi-channel BCH

Solid State Disks are built with many Flash channels connected to the host through a high-speed interface such as SATA or PCI Express (Chap. 2). In this scenario the performance of the SSD is determined by the ECC needed to overcome the high error-rate. It follows that binary BCH code must have a structure able to handle a number of channels together, without being the performance bottleneck.

It has already been studied [4] how the native serial structure of BCH can be parallelized to work on one byte or dword at a time. In multi-channel architectures, this is not enough and multiple encoding and decoding machines must be implemented. In particular, given raw bit error rate higher than 10^{-4}, the most likely situation is that almost all the pages read in parallel need correction.

Figure 11.23 shows the probability that *n* chunks require correction given a bit error probability. For example, if the bit error probability is 10^{-5}, we have a probability of around 10^{-2} of having 32 error-free chunks, a probability of 10^{-1} that three chunks over 32 require correction, a probability of 10^{-10} that 24 chunks require correction and so on. The highest curve in the graph is the most likely number of correction required: for example, at BER of 10^{-5} 3-err and 6-err are the highest.

If BER is higher than $2 * 10^{-4}$ the most likely number of correction over 32 chunks is 32; in other words every chunk requires correction.

In order to keep up with the bandwidth requirements, the most straightforward solution would be to have one encoder and one decoder per channel. However, this approach is extremely area consuming, especially because of the decoder.

As far as the encoding is concerned, it is very important that data coming from the host are dispatched to the various channels without latency. There are three possible approaches, starting from the less area consuming:

- single encoder shared among all Flash channels [26];
- a pool of encoders;
- one encoder per channel.

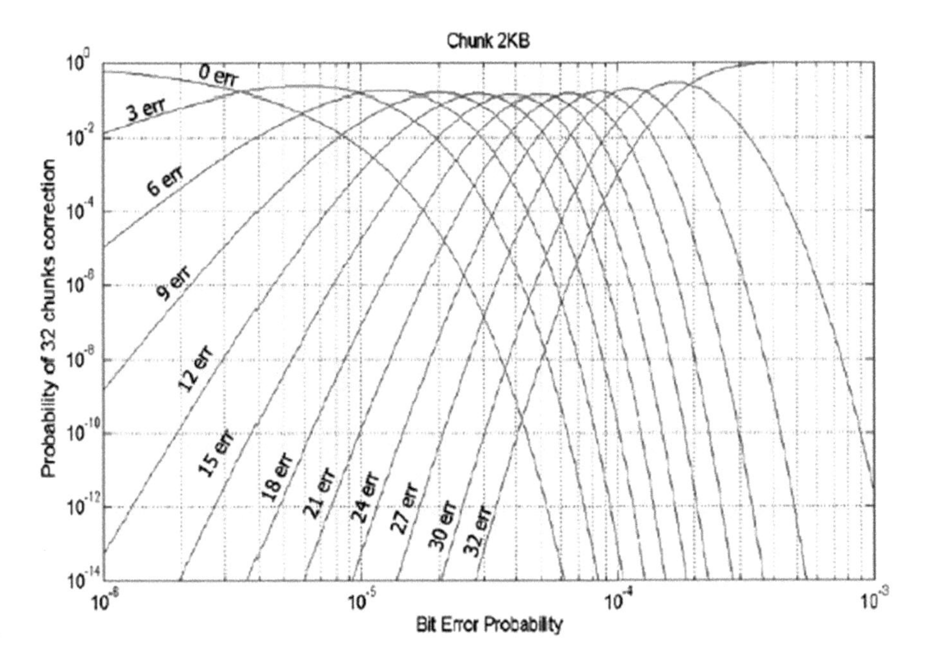

Fig. 11.23 Probability of n chunks over 32 requiring correction

The right hardware choice comes from the tradeoff between silicon area and latency.

The decoding is trickier than encoding since the algorithm is composed of three steps as shown in Fig. 11.3. Please note that null syndromes mean an error-free message: therefore, decoding doesn't need to go through Berlekamp and Chien. This situation is very common when the solid state drive is fresh.

As the reader can notice, Fig. 11.3 shows a pipelined structure. In order to design each decoding step in the correct way, we need to study its latency:

- the input of the syndrome computation is the read codeword of length n. If the t syndrome machines works with a parallelism of b bits, the latency of the syndrome computation is proportional to n/b;
- Berlekamp algorithm takes the t syndromes as input and finds the error locator polynomial coefficients in t iterations. It follows that it has a latency proportional to t;
- Chien search takes the error locator polynomial computed by Berlekamp as input. It substitutes n elements in the polynomial to see if they are roots. If the machine is able to work on c elements at a time, its latency is proportional to n/c. If the degree of the error locator polynomial is v, the Chien machine stops when it finds v roots, without substituting the remaining elements of the field. Hence n/c is the worst case latency.

In order to exploit the pipelined architecture we must have numbers n/b, t, and n/c as similar as possible. However, n is generally much higher than t and there are design constraints on b and c. Hence, we can achieve a balance by adding more machines either to syndrome or Chien.

Let's assume $n1$ HW machines to perform syndrome computation, $n2$ HW machines to execute Berlekamp algorithm and $n3$ HW machines to perform Chien search. We choose $n1$, $n2$ and $n3$ so that (11.56) holds true:

$$\frac{n}{(n1 * b)} \approx \frac{t}{n2} \approx \frac{n}{(n3 * c)} \tag{11.56}$$

The resulting decoding structure is sketched in Fig. 11.24.

Finally we can use probabilistic consideration in choosing the number and the size of Chien machines. The size of the Chien block depends on the parallelism and on the error correction capability. In other words, a machine able to correct x errors is half in area with respect to a machine able to correct $2x$ errors (given the same parallelism). What happens in a real SSD is that the decoding (and so the Chien machine) is always needed but the number of errors that must be corrected is not always t.

Figure 11.25 shows, for a 2112-Byte page, the probability of having to correct only one error, the probability of having to correct two to five errors, and the probability of error (PER) after 5 bits correction as a function of the BER_{in}. For a value of BER_{in} around 10^{-6}, we have that the probability of a single error is equal to $3 * 10^{-2}$ and the probability of two to five errors is equal to $6 * 10^{-4}$ respectively. The probability of a single error is definitely more significant and since the Berlekamp algorithm exactly indicates the number of errors to correct, it may be useful to exploit this information [8]. For example, suppose that from (11.56) we obtain $n3 = 3$. If t is equal to 5 and the area of a Chien machine able to correct one error is 1 U, we obtain an area of $5 * 3 = 15$ U for implementing three machines able to correct five errors. However, from Fig. 11.25 we see that most of the time the correction of only one error is required. It follows that we can implement twomachines able to correct one error and only one complete Chien machine able to correct five errors. The area would be $1 + 1 + 5 = 7$ U with a gain of 6 U at same performances.

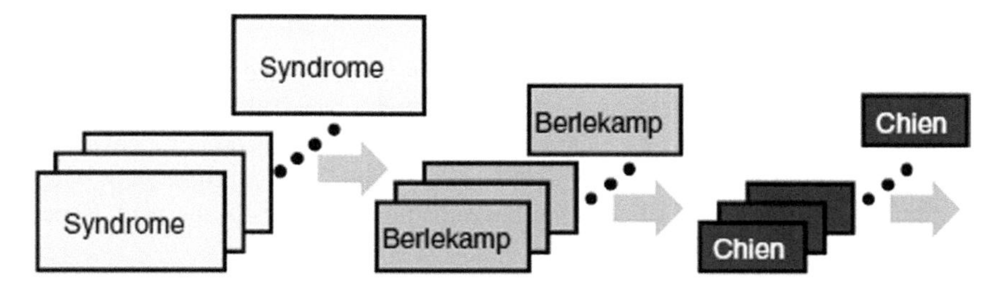

Fig. 11.24 ECC decoding structure for handling multiple channels

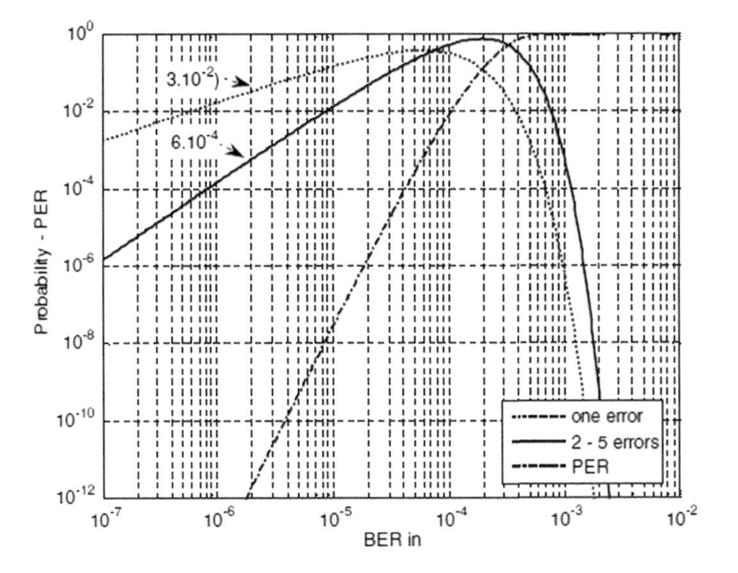

Fig. 11.25 For a 2112-Byte page, representation of single error probability, of two to five error probability and of page error rate using an ECC able to correct five errors

This approach can always be used when the error density function is known. The result is that we can have a pool of Chien machines with different correction capabilities and parallelism. It's Berlekamp machine's task to dispatch the message to the correct Chien machines depending on its degree.

With a good optimization in the number of machines per each step, BCH does not limit the bandwidth between the drive and the host.

As explained in this chapter, multi-channel management and detection properties are the key points to address when developing a BCH engine for Enterprise Class Solid State Disks.

11.9 LDPC False Correction

As already described for BCH codes, false correction probability, also called detection, is a key point for SSD applications. Having a false corrected message is much worse than having an uncorrected message because it is a silent catastrophic event; in other words users will use the message as corrected while it is not, without any suspect of the additional errors introduced.

False correction is part of the code itself and can't be avoided [27]. Figure 11.26 shows the sphere-packing scheme for a bounded distance decoder. In this example there are 4 codewords labeled as CW1, …, CW4 with minimum distance *d* and a correction sphere of radius *t*. Error patterns, represented as squares, can lead to three different cases:

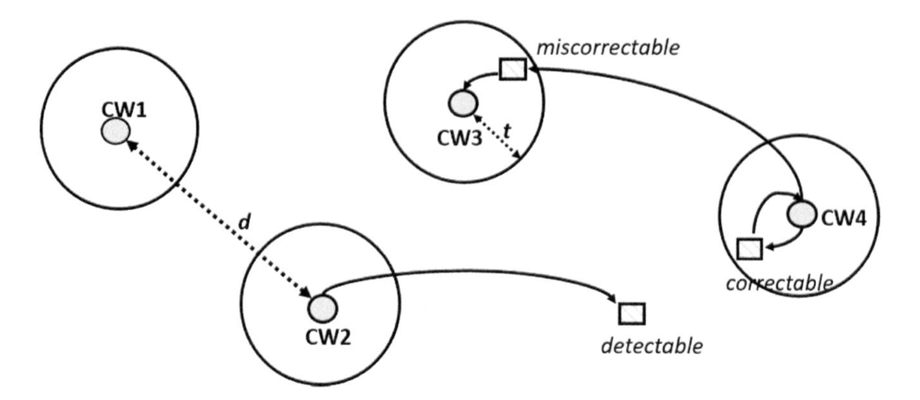

Fig. 11.26 Sphere-packing scheme for a bounded distance decoder

- Correctable error: starting from CW4, less than *t* errors are introduced, so that the erroneous pattern falls in the correction sphere of CW4 and can be corrected.
- Detectable error: starting from CW2, more than *t* errors are introduced, so that the error pattern is outside CW2 correction sphere but does not fall in any correction sphere. In this case the code is able to recognize that errors occurred but is unable to perform any correction.
- Miscorrectable error: starting from CW4, more than *t* errors are introduced, so that the error pattern is outside CW4 correction sphere and falls in CW3 correction sphere. Due to the decoder used, in this case the code will correct the erroneous pattern with CW3, introducing new errors and creating a silent catastrophic event.

Even if the BCH weight estimation was difficult, the issue with LDPC is even worse because the used decoder is not bounded distance: belief propagation is used instead [28, 29]. With reference to Fig. 11.27, we see that it is still true that error patterns can lead to correctable, detectable or miscorrectable error but there isn't the correction sphere concept. In addition to that, this picture represents only hard decoding, soft decoding has a different approach. In the picture we can see that the correction area is different from one codeword to another and is not deterministic. In addition to that, since we are dealing with iterative decoding, the shape of the correction area changes iteration after iteration. It follows that it is not possible to apply weight estimation in this case or to predict the false correction probability with a closed formula. LDPC hard decoding is very powerful because it relies on its sparseness, so that it is very unlikely to fall in another codeword correction region because there are few codewords in the algebraic space.

Figure 11.28 shows the space for soft belief propagation. Here the Hamming distance is replaced by the Euclidean distance. For sake of simplicity, in the picture we have 3 codewords generating a 3D-space. As the reader can see, even if the algebraic space changes, the false decoding probability does not, because we are still using belief propagation decoding algorithm. In addition to that, soft decoding

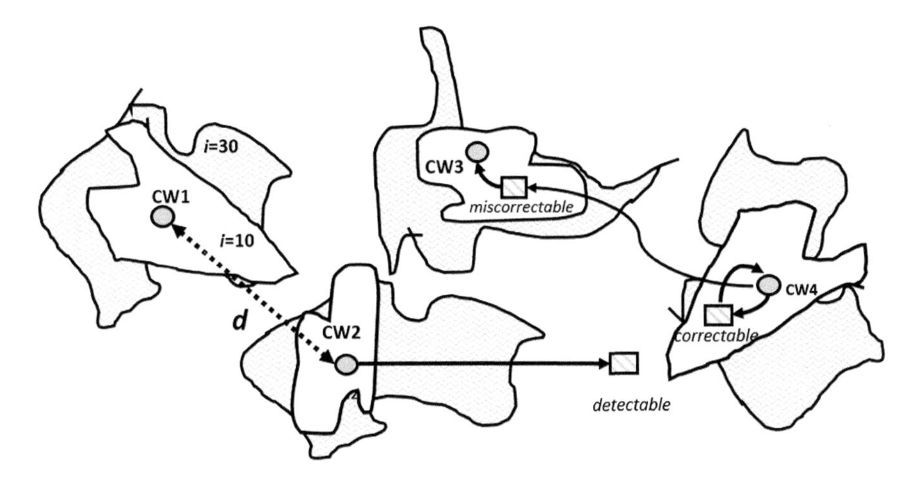

Fig. 11.27 Correction areas for hard LDPC belief propagation decoding algorithm

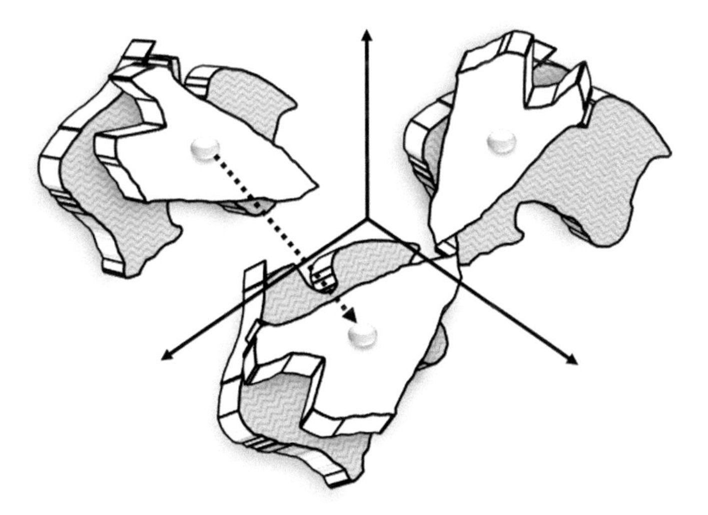

Fig. 11.28 Correction areas for soft LDPC belief propagation decoding algorithm

probability can be higher in the soft case than in the hard one, and can not be deducted from the latter one.

Due to the fact that it is not possible to predict this probability in the LDPC case, right now the only viable solution is simulation. Generally speaking, performance in simulation is evaluated on a known codeword, e.g. all0 codeword, so that even when the decoder ends with a success, there is a comparison between the decoded codeword and the known input codeword [27]. If we call P_{DEC} the probability that

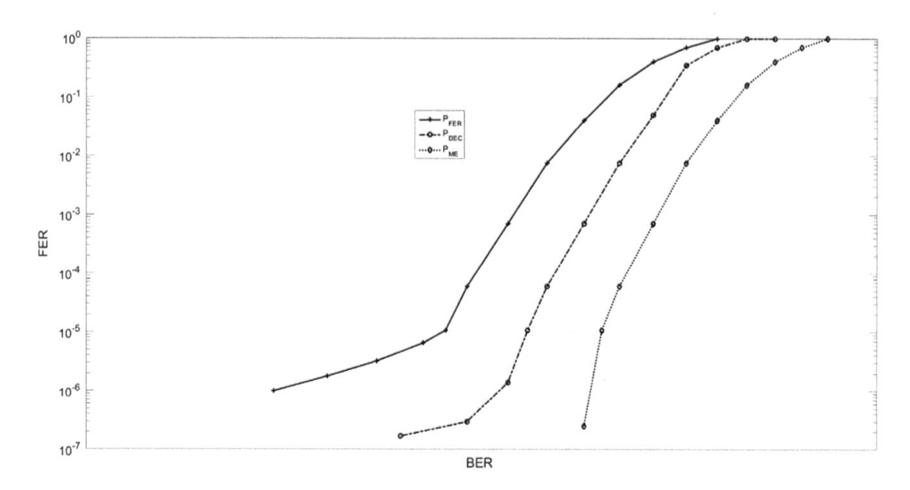

Fig. 11.29 FER versus BER curve for LDPC code

the decoder is unable to recover the input codeword and P_{ME} the probability of having a miscorrection we get that

$$P_{FER} = P_{ME} + P_{DEC} \tag{11.57}$$

as shown in Fig. 11.29. It follows that P_{FER} is an upper bound for P_{ME}.

When the upper bound of P_{FER} is not enough for the P_{ME} requirent, and lower values are needed, LDPC codes can be concatenated with CRC.

Figure 11.30 shows the concatenation scheme for LDPC and CRC. Messages are first encoded with CRC code; the message with CRC parity appended is then encoded with LDPC. On the decoder side, read message is first decoded with LDPC and then with CRC. In this way if the LDPC has performed some erroneous correction, CRC will detect it.

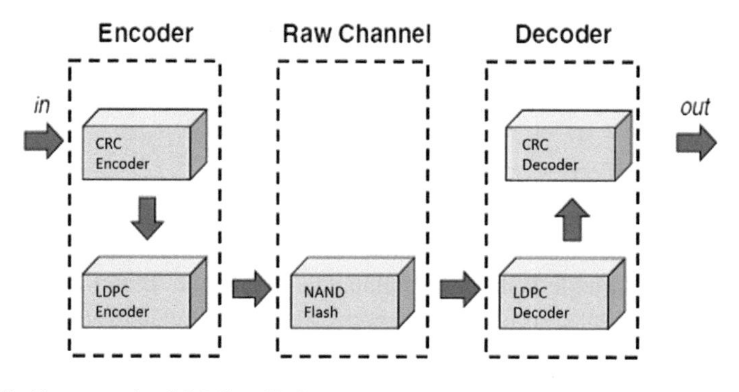

Fig. 11.30 Concatenation LDPC + CRC

References

1. C.E. Shannon, A mathematical theory of communication. Bell Syst. Tech. J. **27**(379–423), 623–656 (1948)
2. R.C. Bose, D.K. Ray-Chaudhuri, On a class of error-correcting binary group codes. Inform. Contr. **3**(1), 68–79 (1960)
3. A. Hocquenghem, Codes correcteurs d'erreurs. Chiffres **2** (1959)
4. I.S. Reed, G. Solomon, Polynomial codes over certain finite fields. J. SIAM **8**(2), 300–304 (1960)
5. C. Berrou, A. Glavieux, P. Thitimajshimima, Near Shannon limit error-correcting coding and decoding: turbo-codes, in *Proceedings of ICC'93* (Geneva, Switzerland, May 1993), pp. 1064–1070
6. R.G. Gallager, Low-density parity-check codes. IRE Trans. Inf. Theory IT **8**, 21–28 (1962)
7. D.J.C. MacKay, R.M. Neal, Near Shannon limit performance of low density parity check codes. Electron. Lett. **32**(18), 1645–1646 (1996)
8. R. Micheloni, A. Marelli, R. Ravasio, *Error Correction Codes for Non-volatile Memories* (Springer, Berlin, 2008)
9. S. Lin, D.J. Costello, *Error Control Coding* (Prentice Hall, Upper Saddle River, 2004)
10. T.K. Moon, *Error Correcting Coding—Mathematical Methods and Algorithms* (Wiley, Hoboken, 2005)
11. Y. Chen, K. Parthi, Small area parallel Chien search architecture for long BCH codes. IEEE Trans. Very Large Scale Integr. (VLSI) Syst. **12**(5), 545–549 (2004)
12. R. Micheloni et al., A 4 Gb 2b/cell NAND flash memory with embedded 5b BCH ECC for 36 MB/s system read throughput, in *ISCC Digest of Technical Papers*, San Francisco, Feb 2006
13. R. Micheloni, L. Crippa, A. Marelli, *Inside NAND Flash Memories* (Springer, Berlin, 2010)
14. E.R. Berlekamp, *Algebraic Coding Theory* (McGraw-Hill, New York, 1968)
15. H.O. Burton, Inversionless decoding of binary BCH codes. IEEE Trans. Inf. Theory **17**(4), 464–466 (1971)
16. I.S. Reed, M.T. Shih, T.K. Truong, VLSI design of inverse-free Berlekamp-Massey algorithm. IEEE Proc. **138**, 295–298 (1991)
17. S. Mizrachi, D. Stopler, Efficient method for fast decoding of BCH binary codes. US Patent 2003/0159103 A1, Aug 2003
18. W.W. Peterson, E.J. Weldon Jr., *Error-Correcting Codes*, 2nd edn. (MIT Press, Cambridge, 1972)
19. T. Kasami, T. Fujiwara, S. Lin, An approximation to the weight distribution of binary linear codes. IEEE Trans. Inf. Theory **31**(6), 769–780 (1985)
20. I. Krasikov, S. Litsyn, On spectra of BCH codes. IEEE Trans. Inf. Theory **41**, 786–788 (1995)
21. I. Krasikov, S. Litsyn, On the distance distribution of duals BCH codes. IEEE Trans. Inf. Theory **45**, 247–250 (2001)
22. F.J. MacWilliams, N.J.A. Sloane, *The Theory of Error-Correcting Codes*. North-Holland Mathematical Library, vol. 16 (North-Holland Publishing Company, Amsterdam, 1977)
23. O. Keren, S. Litsyn, More on the distance distribution of BCH Codes. IEEE Trans. Inf. Theory **1**, 251–155 (1999)
24. M. Sala, A. Tamponi, A linear programming estimate of the weight distribution of BCH(255, k). IEEE Trans. Inf. Theory **46**(6), 2235–2237 (2000)
25. M.G. Kim, J.H. Lee, Decoder error probability of binary linear block codes and its application to binary primitive BCH codes. IEICE Trans. Fundam. **E79-A**(4), 592–599 (1996)
26. Y. Lee, H. Yoo, I. Yoo, I.C. Park, 6.4 Gb/s multi-threaded BCH encoder and decoder for multi-channel SSD controllers, in *ISCC Digest of Technical Papers*, San Francisco, Feb 2012
27. A. Marelli, R. Micheloni, False Decoding Probability (Detection) of BCH and LDPC Codes, Flash Memory Summit 2016

28. M. Hagiwara, M.P.C. Fossorier, H. Imai, Fixed Initialization decoding of LDPC codes over binary simmetric channel, in *IEEE Transaction on Information Theory,* April 2012
29. S.M. Khatami, L. Danjean, D.V. Nguyen, B. Vasic, An Efficient Exhaustive Low-Weight Codeword Search for Structured LDPC Codes

Chapter 12
Low-Density Parity-Check (LDPC) Codes

E. Paolini

Abstract In this chapter, low-density parity-check (LDPC) codes, a class of powerful iteratively decodable error correcting codes, are introduced. The chapter first reviews some basic concepts and results in information theory such as Shannon's channel capacity and channel coding theorem. It then overviews the flash memory channel model. Next, it addresses binary LDPC codes describing both their structure and efficient implementation, and their belief propagation and reduced-complexity decoding algorithms. Non-binary LDPC codes and their belief propagation decoding algorithm are also addressed. Finally simulation results are provided.

12.1 Shannon Limit

12.1.1 Entropy and Mutual Information

Let X be a discrete random variable taking its values in a set $\mathcal{X}$, according to some probability mass function (pmf) $p(x) = \Pr\{X = x\}$. The entropy of X is defined as

$$H(X) = -\sum_{x} p(x) \log_2 p(x).$$

Intuitively, the entropy $H(X)$ may be thought as the uncertainty associated with the random variable. For example, a deterministic variable is characterized by a zero entropy while, for a given positive integer M, the random variable with the largest entropy among all discrete random variables whose support set $\mathcal{X}$ has cardinality M is the uniform one, i.e., $p(x) = 1/M$ for all $x \in \mathcal{X}$. In this latter case we obtain $H(X) = \log_2 M$.

E. Paolini (✉)
DEI, University of Bologna, Bologna, Italy
e-mail: e.paolini@unibo.it

R. Micheloni et al. (eds.), *Inside Solid State Drives (SSDs)*,
Springer Series in Advanced Microelectronics 37,
https://doi.org/10.1007/978-981-13-0599-3_12

Consider now a second discrete random variable $Y \in \mathcal{Y}$ characterized by a pmf $p(y)$. Let $p(y|x) = \Pr\{Y = y|X = x\}$ be the pmf of Y conditioned to the event $\{X = x\}$. The entropy of Y given the event $\{X = x\}$ is defined as

$$H(Y|X = x) = -\sum_{y} p(y|x) \log_2 p(y|x).$$

Next, the conditional entropy $H(Y|X)$ is defined as

$$\begin{aligned} H(Y|X) &= \sum_{x} p(x) H(Y|X = x) \\ &= -\sum_{x} \sum_{y} p(y|x) p(x) \log_2 p(y|x). \end{aligned}$$

Finally, the mutual information $I(X;Y)$ between X and Y is defined as

$$I(X;Y) = \sum_{x} \sum_{y} p(y|x) p(x) \log_2 \frac{p(y|x)p(x)}{p(x)p(y)}. \tag{12.1}$$

It can be shown that $I(X;Y) = H(Y) - H(Y|X) = H(X) - H(X|Y)$. As such, $I(X;Y)$ intuitively represents the reduction of uncertainty about X due to the fact that we can observe Y (equivalently, reduction of uncertainty about Y due to the fact that we can observe X). The mutual information is well-defined also for continuous random variables. In this case, $p(x)$, $p(y)$, and $p(y|x)$ are probability density functions (pdfs), and we have

$$I(X;Y) = \int p(y|x) p(x) \log_2 \frac{p(y|x)p(x)}{p(x)p(y)} dxdy. \tag{12.2}$$

Moreover, if X is a discrete random variable and Y is a continuous one, $I(X;Y)$ is defined as

$$I(X;Y) = \sum_{x} p(x) \int p(y|x) \log_2 \frac{p(y|x)p(x)}{p(x)p(y)} dy. \tag{12.3}$$

12.1.2 System Model and Channel Capacity

The fundamental limit of point-to-point digital communication over a noisy channel was established in 1948 by C. Shannon, who showed that a vanishing error probability can be attained at a finite information rate, provided this rate is smaller than the *capacity* of the noisy channel.

With reference to Fig. 12.1, a source S of information generates messages that must be delivered to a destination D through a noisy channel. The generic message,

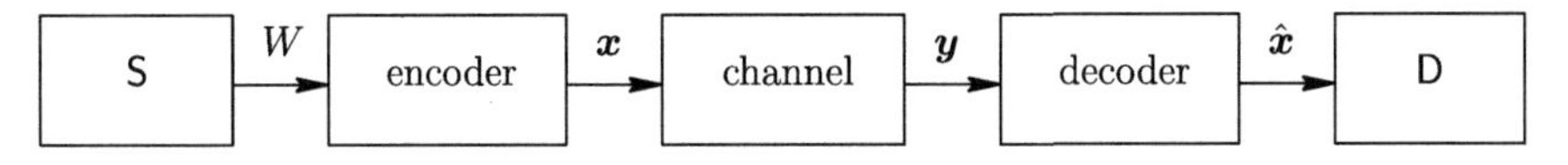

Fig. 12.1 Communication model

denoted by W, is drawn from a set of M possible messages $\{1, 2, \ldots, M\}$, where all messages are a priori equally likely. Prior to transmission over the channel, the message W is encoded through a *channel encoder*, that maps deterministically (and univocally) each message onto a *codeword* $x = [x_0, x_1, \ldots, x_{n-1}]$, i.e., an n-tuple of symbols belonging to some alphabet $\mathcal{X}$. The ratio

$$R = \frac{\log_2 M}{n}$$

is the code rate of the channel code and the code is named an $(n, 2^{nR})$ code. All n codeword symbols are then transmitted sequentially over the channel, resulting in a sequence $y = [y_0, y_1, \ldots, y_{n-1}]$ whose symbols belong to an alphabet $\mathcal{Y}$. A decoding algorithm is then performed by a *channel decoder* to decide which codeword, out of the set of M candidate codewords, had been transmitted over the channel, given the noisy observation y. The codeword $\hat{x}$ returned by the decoder is converted back to the corresponding message $\hat{W}$ that is finally delivered to the destination. As error occurs whenever $W \neq \hat{W}$, i.e., a wrong message is delivered.

A probability of error can be defined for each of the M transmitted messages as follows. The probability of error associated with the j-th message, $j \in \{1, 2, \ldots, M\}$, is denoted by $P_{e,j}$ and is defined as

$$P_{e,j} = \Pr\{\hat{W} \neq W | W = j\}.$$

Furthermore, the maximum probability of error is defined as

$$P_{e,max} = \max_{j \in \{1, 2, \ldots, M\}} P_{e,j} \tag{12.4}$$

and the average probability of error as

$$P_e = \frac{1}{M} \sum_{j=1}^{M} P_{e,j}. \tag{12.5}$$

The channel code along with its decoding algorithm shall be designed in order to make the maximum probability of error over the given channel as small as possible.

Assume that both the input alphabet $\mathcal{X}$ and the output alphabet $\mathcal{Y}$ are discrete. Let $X \in \mathcal{X}$ and $Y \in \mathcal{Y}$ be two discrete random variables, representing the input to the channel and the corresponding output. Moreover, assume that the channel is fully defined by the transition probabilities $p(y|x) = \Pr\{Y = y | X = x\}$. In this case, the channel is called a discrete memory-less channel (DMC). The capacity of a DMC is defined as

$$C = \max_{p(x)} I(X; Y) \tag{12.6}$$

i.e., as the maximum amount of uncertainty we can remove from the input symbol (which cannot be observed directly) by observing the output symbol, where the maximum is taken over all possible pmfs for the input symbol. The capacity is an intrinsic parameter of the channel, only depending on the cardinalities of $\mathcal{X}$ and $\mathcal{Y}$ and on the transition probabilities $p(y|x)$. It is expressed in terms of information bits (or Shannon) per channel use.

Example 12.1 The DMC depicted in Fig. 12.2 is characterized by $\mathcal{X} = \mathcal{Y} = \{+1, -1\}$ and by $\Pr\{Y = +1|X = +1\} = \Pr\{Y = -1|X = -1\} = 1 - p$, $\Pr\{Y = +1|X = -1\} = \Pr\{Y = -1|X = +1\} = p$. This channel is known as binary symmetric channel (BSC), and p is called the error (or crossover) probability. Every binary symbol input to the channel is received in error with probability p and is correctly received with probability $1 - p$. The capacity of the BSC is achieved for $\Pr\{X = +1\} = \Pr\{X = -1\} = 1/2$ and is given by[1]

$$C = 1 - [-p \log_2 p - (1 - p) \log_2(1 - p)]. \tag{12.7}$$

As we shall see later, the BSC is a possible channel model for SLC Flash memories. Assuming $p \leq 1/2$, its capacity is maximum for $p = 0$, where we have $C = 1$ (every binary symbol outcoming from the channel is reliable) and is minimum for $p = 1/2$, where we have $C = 0$ (no uncertainty is removed from X by observing Y).

The concept of capacity, so far introduced for a DMC, can be extended to time-discrete memory-less channels whose input symbol is either a discrete or a continuous random variable and whose output symbol is a continuous one. The capacity is still defined by (12.6), where the mutual information is now given by (12.2) if X is continuous, and by (12.3) if X is discrete. As opposed to the DMC case, however, additional constraints to the optimization problem may be introduced (for example, an upper bound on the average transmitted power). The reason is that the solution to the unconstrained optimization problem may correspond to an input variable X for which the channel is essentially noiseless.

Additive noise channels represent an important class of such channels. Here, the output symbol is obtained as $Y = X + Z$, where Z is a continuous random variable, namely, an additive noise. If Z is independent of X and is normally distributed with zero mean and variance σ^2,

---

[1]The capacity of the BSC only depends on the crossover probability and not on the values assumed by X and Y.

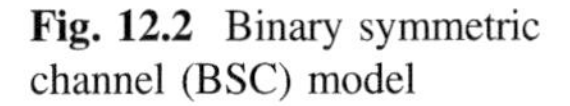
Fig. 12.2 Binary symmetric channel (BSC) model

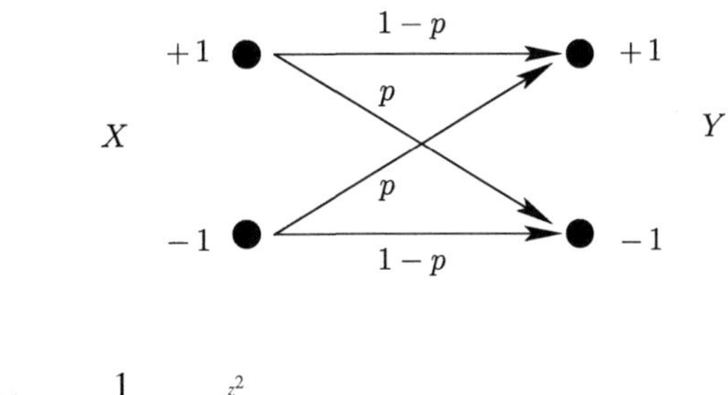

$$p(z) = \frac{1}{\sqrt{2\pi\sigma^2}} e^{-\frac{z^2}{2\sigma^2}},$$

then the corresponding channel is called an additive Gaussian channel.

Example 12.2 Consider the additive Gaussian channel depicted in Fig. 12.3, and assume that $\boldsymbol{X}$ is a Bernoulli (i.e., discrete with a binary alphabet) random variable. Without any further constraint, it is possible to achieve the capacity $C = 1$ (corresponding to a noiseless channel) *regardless* of σ^2 by letting $X \in \{-A, +A\}$, where $A > 0$ is a real, choosing $\Pr\{X = -A\} = \Pr\{X = +A\} = 1/2$, and letting $A \to \infty$. On the other hand, if the maximization problem is constrained to $(1/n)\sum_{i=0}^{n-1} x_i^2 \leq E_s$ for any transmitted codeword, then the maximum is attained for $X \in \{-\sqrt{E_s}, +\sqrt{E_s}\}$ and $\Pr\{X = -\sqrt{E_s}\} = \Pr\{X = +\sqrt{E_s}\} = 1/2$. In this case (12.3) yields

$$C = -\int p(y)\log_2\left(p(y)\sqrt{2\pi e\sigma^2}\right)dy,$$

where

$$p(y) = \frac{1}{\sqrt{8\pi\sigma^2}}\left(e^{-\frac{(y-\sqrt{E_s})^2}{2\sigma^2}} + e^{-\frac{(y+\sqrt{E_s})^2}{2\sigma^2}}\right).$$

and where the capacity, that does not admit a closed-form expression, must be computed via numerical integration. This channel model is known as the binary-input additive white Gaussian noise (Bi-AWGN) channel. It is possible to show that its capacity is a function of parameter E_s/N_0, where $N_0 = 2\sigma^2$. In general, the larger E_s/N_0 the higher C. Moreover, $C \to 1$ as $E_s/N_0 \to \infty$.

Example 12.3 Consider a channel $X \to Y' \to Y$ composed of the cascade of a Bi-AWGN channel and a one-bit quantizer, returning $Y = +1$ if $Y' > 0$ and $Y = -1$ otherwise (if $Y' = 0$, +1 or −1 is returned with equal probability). It is readily shown that this channel is equivalent to a BSC whose crossover probability p is

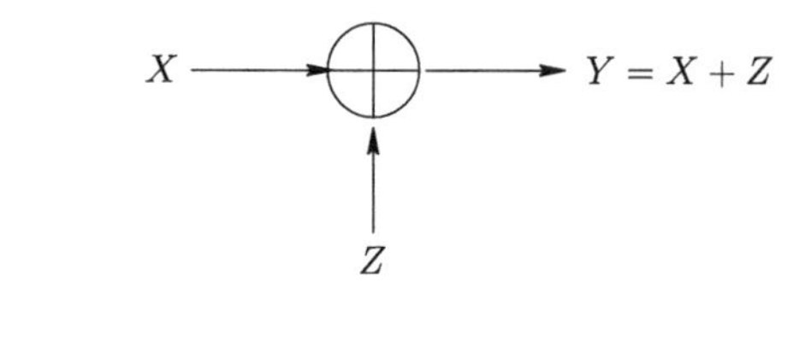

Fig. 12.3 Binary-input additive white Gaussian noise channel model

$$p = \frac{1}{2}\operatorname{erfc}\left(\sqrt{\frac{E_s}{N_0}}\right) \tag{12.8}$$

where

$$\operatorname{erfc}(x) = \frac{2}{\sqrt{\pi}}\int_x^{\infty} e^{-\theta^2}\,d\theta.$$

Again, the capacity is a monotonically increasing function of parameter E_s/N_0, and again $C \to 1$ as $E_s/N_0 \to \infty$. For the same value of E_s/N_0, the capacity of the output-quantized Bi-AWGN channel is always smaller than the capacity of the corresponding unquantized channel.

With reference to the last two examples, if Y is allowed to assume $q > 2$ different quantized values (which corresponds to adopting $\lceil \log_2 q \rceil$ quantization bits), the capacity of the obtained channel is upper bounded by that of the unquantized Bi-AWGN channel and is lower bounded by that of the one-bit quantized channel. (Note that the $q-1$ quantization thresholds shall be properly designed.) In general, the higher q the larger the capacity.

12.1.3 The Channel Coding Theorem

Adopting the formulation in [1], which makes use of the maximum error probability defined in (12.4), Shannon's channel coding theorem can be stated as follows. "For every rate $R < C$ there exists a sequence of $(n, 2^{nR})$ codes for which $\lim_{n\to\infty} P_{e,max}(n) = 0$. Conversely, if $\lim_{n\to\infty} P_{e,max}(n) = 0$ for a sequence of $(n, 2^{nR})$ codes, then $R \le C$." Note that $\lim_{n\to\infty} P_{e,max}(n) = 0$ implies $\lim_{n\to\infty} P_e(n) = 0$, where $P_e(n)$ is the average error probability defined in (12.5).

Essentially, Shannon's channel coding theorem states that communication over a noisy channel is possible with an arbitrarily small maximum error rate if and only if the code rate of the employed channel code does not exceed the channel capacity. On the other hand, from the proof of the converse, it is possible to show that, when $R > C$, the average probability of error probability is bounded away from zero. Specifically, we have

$$P_e(n) \geq 1 - \frac{C}{R} - \frac{1}{nR} \tag{12.9}$$

$$\to 1 - \frac{C}{R} \tag{12.10}$$

in the limit where $n \to \infty$. Inequality (12.9) defines a *non-achievable region* for the considered communication channel. No channel code of length n exists whose average probability of error over the considered channel is smaller than the right-hand side of (12.9). For $n \to \infty$, the non-achievable region is identified by (12.10). For a channel parametrized by some parameter γ (e.g., the crossover probability p for a BSC, or E_s/N_0 for the Bi-AWGN channel or its output-quantized version), the non-achievable region can be reported in the $P_e(n)$ versus γ plane for a specific code rate R, as illustrated in the following example.

Example 12.4 In Fig. 12.4 the non-achievable region is depicted for both the unquantized Bi-AWGN channel and its one-bit output-quantized version, for code rate $R = 9/10$ and infinite codeword length. Specifically, for fixed $R = 9/10$ the right-hand side of (12.10) is plotted as a function of E_b/N_0 (in logarithmic scale), where $E_b = RE_s$. If E_s is interpreted as the energy per transmitted binary symbol, E_b can be regarded as the energy per information bit. The dashed curve identifies a non-achievable region over the unquantized Bi-AWGN channel (i.e., no $(E_b/N_0, P_e)$ point inside the corresponding area is achievable), while the solid one a non-achievable region over its one-bit output-quantized version. That the unquantized non-achievable region is contained in the quantized one is coherent with the fact that the capacity of the Bi-AWGN channel is larger than the capacity of its output-quantized version, for the same value of E_s/N_0. In general, if $q > 2$

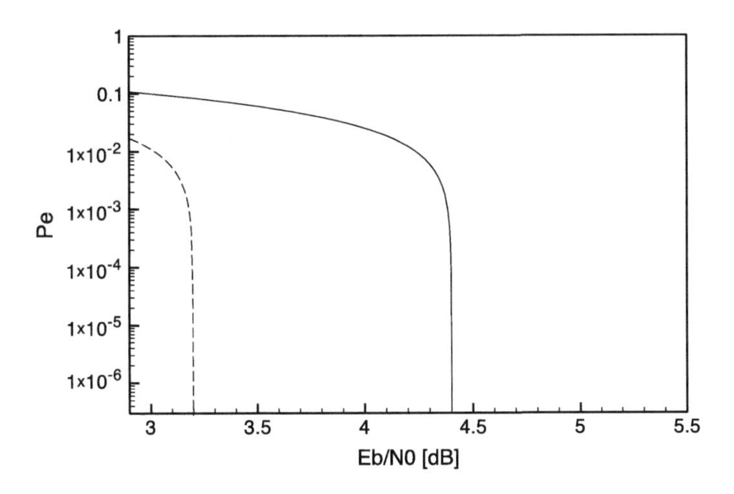

Fig. 12.4 Plot of the Shannon limit for code rate $R = 9/10$, over the Bi-AWGN channel and over the BSC obtained via one-bit quantization of the output of the Bi-AWGN channel

quantization levels are allowed, the corresponding non-achievable region is identified by a curve falling between the two plotted curves. This serves to illustrate how soft information at the decoder can be exploited to improve the system performance. The smallest value of E_b/N_0 for which communication is possible with a vanishing error probability at the given rate $R=9/10$ over the Bi-AWGN channel is about 3.198 dB. The corresponding value over the one-bit quantized Bi-AWGN channel is about 4.400 dB.

12.2 Maximum a Posteriori and Maximum Likelihood Decoding of Linear Block Codes

As from Sect. 12.1.2, decoding is essentially a decision problem. Given the observation $\boldsymbol{y}$ from the communication channel, the decoder has to decide which of the M codewords has been most likely transmitted, in order to minimize the maximum probability of error. Optimum decoding is based on *maximum a posteriori* (MAP) decision criterion, and consists of assuming as the transmitted codeword the one maximizing the a posteriori probability:

$$\hat{x} = \text{argmax}_x p(x|y).$$

When the codewords are a priori equally likely, then MAP decoding is equivalent to maximum likelihood (ML) decoding, that returns the codeword

$$\hat{x} = \text{argmax}_x p(y|x).$$

It is readily shown that, over a BSC, ML decoding is equivalent to returning the codeword exhibiting the minimum Hamming distance from the received word $\boldsymbol{y}$. (Recall that the Hamming distance between two sequences is the number of positions at which the corresponding symbols are different.) Moreover, over a Bi-AWGN channel, ML decoding consists of returning the codeword (whose symbols belong to the set $\{-\sqrt{E_s}, +\sqrt{E_s}\}$) exhibiting the minimum Euclidean distance from $\boldsymbol{y}$.

Optimum decoding is unfeasible for most codes (including linear codes), due to the need of computing M metrics, with M prohibitively large. Low-density parity-check codes, introduced in Sect. 12.4, are capable to perform close to the Shannon limit at a manageable complexity.

12.3 NAND Flash Memory Channel Model

In NAND flash memories, the generic memory cell is a floating gate transistor. Writing the cell consists of exploiting Fowler-Nordheim tunneling effect [2] to inject a certain amount of charges into the floating gate in order to program the threshold voltage V_{th} of the transistor. For an MLC memory with b bits per cell, there are 2^b nominal values for threshold voltage V_{th}, each bijectively associated with a word of b bits. (There are two nominal values for V_{th} in the particular case of an SLC memory.) The whole range of possible values of V_{th} is then partitioned into 2^b intervals, each corresponding to a nominal value of the threshold voltage.

Reading a cell is a decision problem consisting of picking one of the 2^b nominal values of V_{th} and forwarding the corresponding binary b-tuple. The value of V_{th}, however, cannot be observed directly. In order to read the cell, a word-line voltage must be applied and the corresponding transistor drain current measured. In this chapter, we refer to the word-line voltage simply as the "read voltage", denoting it by V_{READ}. If for some V_{READ} a sufficiently high drain current is detected then we conclude that $V_{READ} > V_{th}$, otherwise we conclude that $V_{READ} < V_{th}$. In this sense, the application of a specific read voltage value is capable to provide exactly one bit of information. Therefore, in order to read the full content of a cell in an MLC memory the drain current must be analyzed for a sufficiently large number of read voltage values. A single V_{READ} value is sufficient in the SLC case unless we wish to extract some soft information to improve the performance of the adopted error control coding scheme.

In ideal flash memories, after a cell is written the corresponding value of V_{th} is exactly equal to one of the 2^b nominal values. In real memories, however, the actual value of V_{th} may differ, even significantly, from its nominal value due to a number of possible physical impairments. For a thorough description of these impairments we refer the reader, for example, to [3, Chap. 4], [4]. As such, the actual value of V_{th} may fall into a voltage interval whose nominal voltage threshold is different from the one we attempted to set during the write operation. When this happens the forwarded binary b-tuple after a read operation differs from the one that was written into the cell. A bit error generated by an erroneous decision about the interval of voltage values V_{th} belongs to is called a *raw bit error*, and the probability of occurrence of raw bit errors is called the raw bit error probability.

The raw bit error probability may be analyzed by modeling the threshold voltage V_{th} of the generic cell as a continuous random variable whose pdf is here denoted by $p(V_{th})$. It must be pointed out that $p(V_{th})$ is not constant during the memory lifetime, as it is modified by subsequent write and read operations, leading to a progressive degradation of the channel in terms of increasing raw bit error probability. The threshold voltages for two different memory cells are typically assumed to be independent and identically distributed (i.i.d.) random variables. In the following two subsections, the channel model for SLC and MLC flash memories is addressed.

12.3.1 SLC Channel Model

The simplest channel model for an SLC flash memory consists of modeling the threshold voltage V_{th} of the generic cell as the weighted sum (with the same weights) of two independent Gaussian random variables with the same variance σ^2 neglecting that, in principle, Gaussian random variables assume their values over an infinite range. The mean values of the two Gaussian distributions are the two nominal values of the threshold voltage, namely, $V_{th,1}$ and $V_{th,2}$ where we assume $V_{th,1} < V_{th,2}$. Let $X \in \{0,1\}$ be a Bernoulli random variable with equiprobable values, representing the bit originally written into the memory cell. Moreover, let Y be the symbol read from the cell. Conditionally to X, the threshold voltage V_{th} is a Gaussian random variable with variance σ^2 and whose mean is $V_{th,1}$ if $X=1$ (erase state) and $V_{th,2}$ if $X=0$. This is depicted in Fig. 12.5. Overall, we have

$$\begin{aligned} p(V_{th}) &= \frac{1}{2} p(V_{th}|X=1) + \frac{1}{2} p(V_{th}|X=0) \\ &= \frac{1}{\sqrt{8\pi\sigma^2}} \left(e^{-\frac{(V_{th}-V_{th,1})^2}{2\sigma^2}} + e^{-\frac{(V_{th}-V_{th,2})^2}{2\sigma^2}} \right). \end{aligned}$$

If we apply only one read voltage $V_{th,1} < V_{READ,1} < V_{th,2}$ we get information about the actual value of V_{th} being larger or smaller than the applied read voltage value. Hence, if only one read voltage value is used, Y is a Bernoulli random

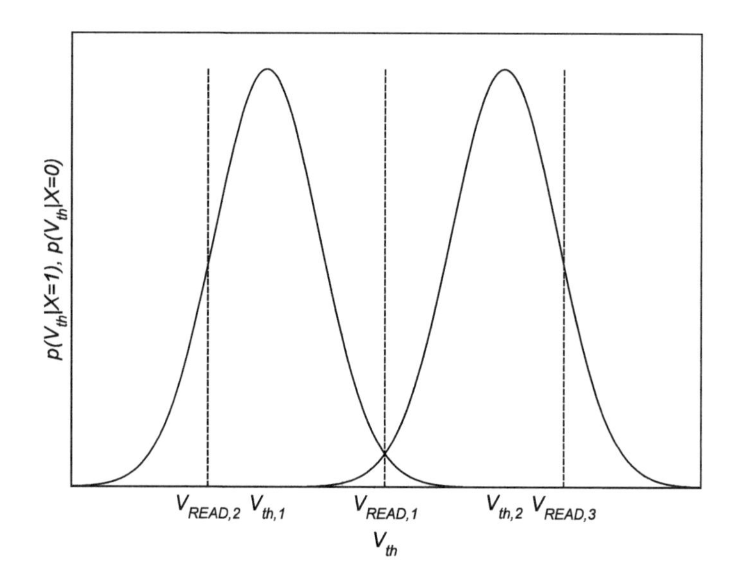

Fig. 12.5 Plot of $p(V_{th}|X=1)$ and $p(V_{th}|X=0)$ for an SLC flash memory where the threshold voltage V_{th} is modeled as the sum of two independent and identically distributed (i.i.d.) Gaussian random variables

variable as well as X. In particular, we have $Y = 1$ if $V_{th} < V_{READ}$ is detected, and $Y = 0$ otherwise. A raw bit error occurs any time $Y \neq X$, and the raw bit error probability is trivially minimized by setting $V_{READ,1} = (V_{th,1} + V_{th,2})/2$, as depicted in Fig. 12.5. In this situation, the channel is clearly equivalent to the cascade of a Bi-AWGN channel and a one-bit quantizer described in Example 12.2 (i.e., to a BSC), and the raw bit error probability is given by (12.8) where $E_s/N_0 = (V_{th,2} - V_{READ,1})^2/2\sigma^2$. At the beginning of the memory life, σ^2 is very small and the memory is almost ideal. Then, σ^2 increases with the memory use, increasing the raw error probability and degrading the channel. A typical value of the raw bit error probability towards the end of the memory life is 10^{-2}.

If an error correcting code is employed to protect the data stored in the flash memory, hard-decision decoding must be necessarily performed if only one V_{READ} value is used as no soft information is available at the decoder. As it will be shown in Sect. 0, however, the availability of soft information at the decoder input represents an essential feature to boost the performance of the coding scheme. In order to provide the decoder with soft information, and consequently to increase its coding gain, more read voltages must be applied sequentially. For example, with reference again to Fig. 12.5 we may employ three read voltage values $V_{READ,1}$, $V_{READ,2}$, and $V_{READ,3}$ and apply two of them for each cell read operation. Specifically, $V_{READ,1}$ is applied at first. if $V_{th} < V_{READ,1}$ then $V_{READ,2}$ is applied to discriminate between $V_{th} < V_{READ,2}$ and $V_{READ,2} < V_{th} < V_{READ,1}$. On the contrary, $V_{READ,3}$ is applied to discriminate between $V_{th} > V_{READ,3}$ and $V_{READ,1} < V_{th} < V_{READ,3}$. In this case the output symbol Y is a discrete random variable assuming the four possible values in the set $\{Y_1, Y_2, Y_3, Y_4\}$ and the channel may be represented as the DMC depicted in Fig. 12.6.

Each arrow in the depicted DMC is associated with a transition probability $p(y|x)$, where the transition probabilities depend on the choice of the read voltages $V_{READ,2}$ and $V_{READ,3}$. A "natural" approach to choose them consists of maximizing the mutual information between the random variables X and Y under the setting $\Pr(X = 0) = \Pr(X = 1) = 1/2$. This approach, proposed in [5], may be easily

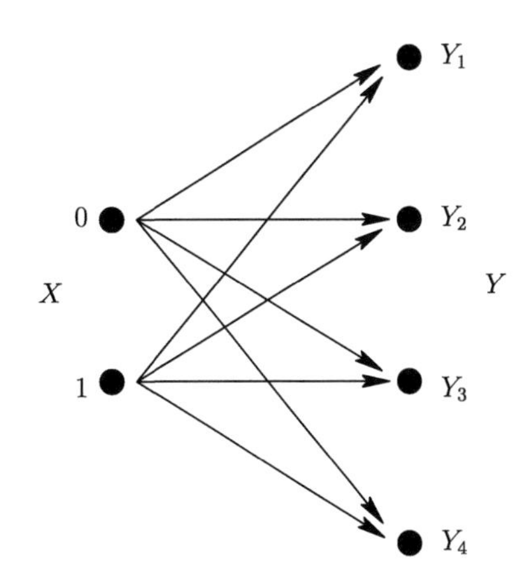

Fig. 12.6 Equivalent channel model for an SLC flash memory where the threshold voltage is modeled as the sum of two i.i.d. Gaussian random variables and where three read voltage values are employed. Each read operation involves two read voltages

extended to any number of read voltages. It may also be easily extended to different choices of the pdf $p(V_{th})$, and therefore to MLC Flash memories.

12.3.2 MLC Channel Model

While the channel model for SLC Flash memories is rather well-established, the development of an MLC channel model is still a subject of research and measurement campaigns, and several models may be found in the literature. These models typically assume the random variable V_{th} to be the weighted sum (with the same weights) of 2^b independent random variables, each corresponding to a nominal value of the threshold voltage. Among these models, the one described next has been adopted in several works [6]. Letting $\boldsymbol{X}$ denote the binary b-tuple that was written in the cell, the pdf $p(V_{th}|X_1 = 11\ldots1)$ associated with the lowest nominal threshold voltage value $V_{th,1}$ (erase state) is modeled as Gaussian with mean $V_{th,1}$ and variance σ_0^2, while the pdf $p(V_{th}|X_i)$ associated with any other nominal value $V_{th,i}$ $(X_i \neq 11\ldots1)$ is characterized by a uniform central region of size ΔV centered in the mean value $V_{th,i}$ and by two Gaussian tails of variance $\sigma^2 < \sigma_0^2$. Formally, for $i \in \{2, 3, \ldots, 2^b\}$ we have

$$p(V_{th}|X_i) = \begin{cases} \frac{1}{\sqrt{2\pi\sigma^2} + \Delta V} e^{-\frac{(V_{th} - V_{th,i} - \Delta V/2)^2}{2\sigma^2}} & V_{th} > V_{th,1} + \frac{\Delta V}{2} \\ \frac{1}{\sqrt{2\pi\sigma^2} + \Delta V} & V_{th,1} - \frac{\Delta V}{2} < V_{th} < V_{th,1} + \frac{\Delta V}{2} \\ \frac{1}{\sqrt{2\pi\sigma^2} + \Delta V} e^{-\frac{(V_{th} - V_{th,i} + \Delta V/2)^2}{2\sigma^2}} & V_{th} < V_{th,1} - \frac{\Delta V}{2} \end{cases}$$

and

$$p(V_{th}) = \frac{1}{2^b} \sum_{i=1}^{2^b} p(V_{th}|X_i).$$

A pictorial representation of the four conditional pdfs $p(V_{th}|X_i)$, $i \in \{1, 2, 3, 4\}$, for an MLC flash memory with $b = 2$ bits per cell and equally spaced threshold voltages is shown in Fig. 12.7.

In an analogous way as for the SLC case, a read is performed by applying sequentially a certain number of read voltages V_{READ} in order to identify the interval in which the actual value of the threshold voltage belongs. If $N \geq 2^b - 1$ different read voltages are employed, the equivalent communication channel is a DMC with 2^b equiprobable input symbols $\boldsymbol{X}$ and $N + 1$ output symbols $\boldsymbol{Y}$. Again, the larger the number of employed read voltages (i.e., the larger the number of intervals in which the range of possible V_{th} values is partitioned) the more accurate the soft information at the decoder input, the lower the bit error rate after decoding. Again, the values of the N read voltages must be properly designed, for instance, maximizing

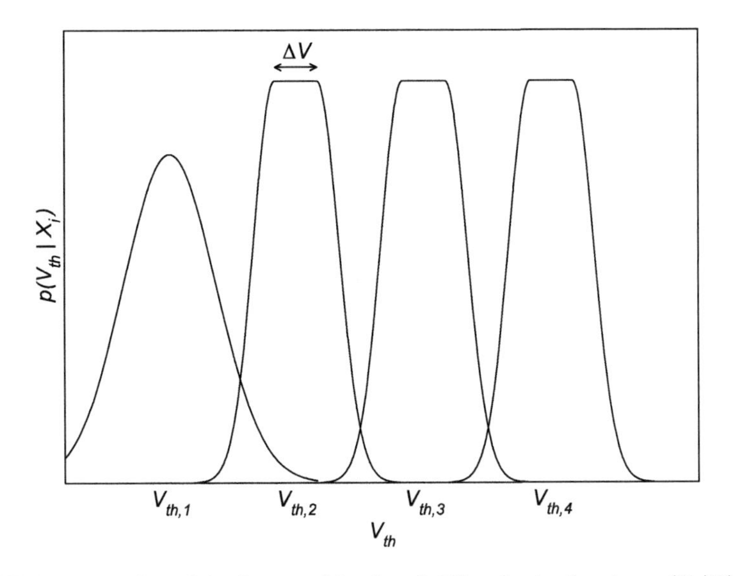

Fig. 12.7 Representation of the four conditional probability density functions $p(V_{th}|X_i)$ of the threshold voltage in an MLC flash memory with $b=2$ bits per cell

the mutual information $I(X;Y)$ under the assumption $\Pr(X=X_i)=2^{-b}$ for all $i\in\{1,2,\ldots,2^b\}$.

12.4 Low-Density Parity-Check Codes

Low-density parity-check (LDPC) codes were introduced by Gallager in [7] and have been almost forgotten for about 30 years. They gained a new interest only after the discovery of turbo codes [8], when it was shown that iterative decoding schemes can attain performances very close to the Shannon limit with a manageable complexity [9, 10].

A binary LDPC code is defined as a binary linear block code whose parity-check matrix $\boldsymbol{H}$ is characterized by a relatively small number of 1 entries, i.e., whose parity-check matrix is sparse. LDPC codes are often represented graphically through a bipartite graph $G=(\mathcal{V}\cup\mathcal{C},\mathcal{E})$ called the Tanner graph [11]. In the Tanner graph there are two different types of nodes, namely, the variable nodes (whose set is $\mathcal{V}$) and the check nodes (whose set is $\mathcal{C}$). The n variable nodes and the m check nodes are associated in a bijective way with the n encoded bits of the generic codeword and with the m parity-check equations, respectively. Each edge $e\in\mathcal{E}$ in the Tanner graph connects a variable node $V\in\mathcal{V}$ with a check node $C\in\mathcal{C}$ if and only if the bit corresponding to V is involved in the parity-check equation corresponding to C. Note that in general not all the m parity-check equations may be linearly independent, so that the actual code rate R of the LDPC code fulfills

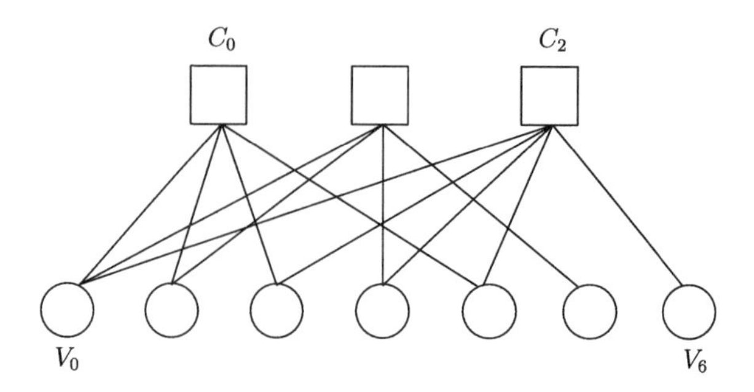

Fig. 12.8 Tanner graph of a $(7,4)$ Hamming code represented by $H=[1110100, 1101010, 10111001]$. There are seven variable nodes variable nodes $\{V_0, \ldots, V_6\}$, one for each encoded bit, and three check nodes, $\{C_0, \ldots, C_2\}$, one for each parity-check equation

$$R \geq \frac{n-m}{n}$$

where equality holds when all m equations are independent. In the Tanner graph of an LDPC code a cycle (or loop) is any closed path starting from a node and ending on the same node. The length of a cycle is the number of edges involved in the cycle. Moreover, the girth g of the Tanner graph is the length of its shortest loop. For reasons that will be clear in the next section, the Tanner graph of an LDPC code should exhibit a large girth. In the Tanner graph, the degree of a variable node or check node is the number of edges incident to it. An LDPC code is said to be regular if all of its variable nodes have the same degree and all of its check nodes have the same degree, and is said to be irregular otherwise.

The representation of LDPC codes in terms of their Tanner graphs is very convenient in order to describe their iterative decoding algorithm, known as belief propagation (BP). In fact, as it will be addressed in Sect. 12.5, BP decoding of LDPC codes may be interpreted as an iterative exchange of messages between the variable nodes and the check nodes along the edges of the Tanner graph. In principle, the Tanner graph can be drawn for any $\boldsymbol{H}$ matrix of any linear block code. As an example, in Fig. 12.8 the Tanner graph is depicted for the $(7,4)$ Hamming code represented by $H=[1110100, 1101010, 10111001]$.

In this section we provide a few details about binary LDPC code design, while LDPC decoding is discussed in the next section. One of the major issues in LDPC coding is represented by efficient encoding, i.e., the efficient computation of the encoded codeword of n bits from a message W represented by a binary k-tuple. Hence, we focus on the design of quasi-cyclic LDPC (QC-LDPC) codes based on circulant matrices, a class of LDPC codes characterized by low-complexity encoding and good performances [12]. In general, a linear block code is said to be quasi-cyclic when there exists some positive integer q such that a cyclic shift by q positions of any codeword results in another codeword. The encoder of

QC-LDPC codes may be implemented very efficiently in hardware using shift register-based circuits [13]. Efficient hardware implementations for the decoder are also available [14].

12.4.1 LDPC Code Ensembles

As opposed to classical algebraic codes, LDPC codes are typically analyzed in terms of average ensemble properties, where an LDPC code ensemble is formed by all LDPC codes having the same codeword length n and nominally the same rate R, and sharing common properties. This approach was introduced by Gallager to analyze his regular LDPC codes [7], and has been successfully adopted to design irregular LDPC codes performing very close to the Shannon limit [15, 16].

An example of LDPC code ensemble is the *unstructured* irregular one [15]. Let n and m be the numbers of variable and check nodes, respectively. Moreover, let Λ_i and P_i be the fractions of variable nodes and check nodes of degree i, respectively. Hence, in the Tanner graph there are $\Lambda_i n$ variable nodes with i sockets and $\mathrm{P}_i m$ check nodes with i sockets and the number of edges is $E = n\sum_{i=2}^{D} i\Lambda_i = m\sum_{i=2}^{H} i\mathrm{P}_i$ where D is the maximum variable node degree and H the maximum check node degree. For given Λ_i, $i = 2, \ldots, D$ and P_i, $i = 2, \ldots, H$,[2] the unstructured $\mathcal{C}(n, \Lambda, \mathrm{P})$ ensemble includes all LDPC codes corresponding to all possible $E!$ edge permutations between the variable node and the check node sockets, according to a uniform probability distribution.

Another example is the *protograph* ensemble [17] (see also the work [18] on LDPC codes from superposition). A protograph is defined as a small Tanner graph and represents the starting point to derive a larger Tanner graph via a "copy-and-permute" procedure. Specifically, the protograph is first copied Q times. Then, the edges of the individual replicas are permuted among the replicas, leading to a larger graph. The edge permutation is performed in such a way that, if an edge e connects a variable node V to a check node C in the protograph, then in the final graph any of the Q replicas of e may connect only a replica of V to a replica of C. Note that, while parallel edges between nodes are allowed in the protograph, they are avoided in the permutation phase. An example of this copy-and-permute procedure is depicted in Fig. 12.9. For a given protograph and a given Q the ensemble is composed of the LDPC codes corresponding to all possible edge permutations fulfilling the described constraints (again, the probability distribution over such permutations is uniform).

[2] For unstructured ensemble, the minimum variable and check nodes are usually set to. The reason for this choice is out of the scope of this chapter.

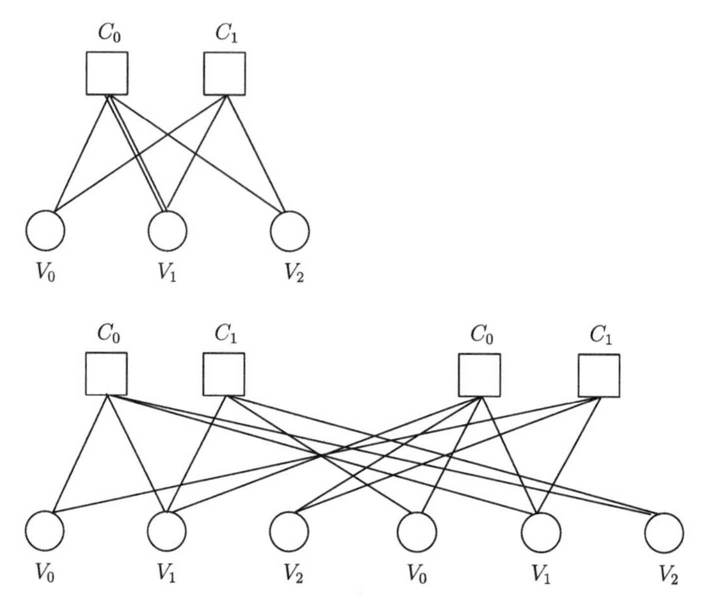

Fig. 12.9 Conceptual example of copy-and-permute protograph procedure

12.4.2 QC-LDPC Codes Construction

A very popular technique to design finite length LDPC codes consists of two subsequent steps. An ensemble of LDPC codes with desired properties is first designed and then a code from the ensemble is picked constructing its Tanner graph according to some graph-lifting algorithm. In the first design phase (ensemble optimization) *asymptotic ensembles* are considered, i.e., ensembles of LDPC codes whose codeword length tends to infinity (examples are the unstructured $\mathcal{C}(\infty, \Lambda, \mathrm{P})$ ensemble and the protograph ensemble defined by a specific finite-length protograph in the limit where $Q \to \infty$). The main parameter characterizing an asymptotic ensemble of LDPC codes under iterative decoding is the *asymptotic decoding threshold* [19, 15]. Letting ℓ be the iteration index and assuming that the communication channel is parameterized by some real parameter θ such that $\theta_1 < \theta_2$ means that the channel corresponding to θ_2 is a degraded version of the channel corresponding to θ_1, the asymptotic threshold θ^* is defined as

$$\theta^* = \sup\{\theta \,\text{s.t.}\, P_{e,\ell}^{\infty} \to 0 \text{ as } \ell \to \infty\}$$

where $P_{e,\ell}^{\infty}$ is the average error probability under iterative decoding over the asymptotic ensemble (i.e., the expected probability of error for an LDPC code randomly picked in the asymptotic ensemble). For example, over a BSC the parameter θ is the crossover probability p, while over a Bi-AWGN channel it is the noise power σ^2 for given E_s (therefore over the Bi-AWGN channel the threshold may be expressed as $(E_b/N_0)^*$ where $E_b = RE_s$ and R is the nominal ensemble rate). Note that for the same ensemble, the threshold is different for different message passing decoders. For unstructured ensembles the threshold may be calculated exactly via a procedure called *density evolution* [15] or approximately via a tool known as EXIT chart [20]. For protograph ensembles it may be calculated with good approximation via multi-dimensional EXIT analysis [21]. In Sect. 12.6.2 density evolution is reviewed for unstructured regular LDPC ensembles and for a very simple decoder called the Gallager B decoder.

Once a protograph ensemble with a satisfying threshold over the channel of interest has been designed, a QC-LDPC code can be constructed from the protograph. This step is usually performed by first representing the protograph as a *base matrix* $\boldsymbol{B}$. The number of rows and columns in the base matrix equal the number of check and variable nodes in the protograph, respectively. Moreover, the (j, i)th entry of $\boldsymbol{B}$ is equal to the number of connection between check node C_j and variable node V_i in the protograph. For example, the base matrix corresponding to the protograph depicted in Fig. 12.9 is

$$B = \begin{bmatrix} 1 & 2 & 1 \\ 1 & 1 & 1 \end{bmatrix}.$$

In order to construct the parity-check matrix $\boldsymbol{H}$ of a QC-LDPC code from $\boldsymbol{B}$, each entry in the base matrix is replaced with a $Q \times Q$ circulant matrix, where a circulant matrix is any square matrix such that every row is obtained from the previous row by a cyclic shift to the right by one position. An entry in $\boldsymbol{B}$ equal to t is replaced by a circulant matrix whose rows and columns all have Hamming weight t. (Null entries in $\boldsymbol{B}$ are replaced by zero $Q \times Q$ square matrices.) If the number of variable nodes in the protograph is n_p then the final LDPC code has length Qn_p. Moreover, it is a QC-LDPC code as the cyclic shift of any codeword by n_p positions results in another codeword. The specific circulant matrices used to replace the entries of the base matrix are chosen according to algorithms aimed at increasing the girth g of the graph, making it suitable to iterative message-passing decoding. It is pointed out that sometimes the parity-check matrix $\boldsymbol{H}$ is obtained by lifting the base matrix in several steps. For example, instead of replacing each entry of $\boldsymbol{B}$ by a $Q \times Q$ matrix (for large Q), $\tilde{Q} \times \tilde{Q}$ circulant matrices may be used at first, with Q being a multiple of $\tilde{Q}$, and then circulant permutation matrices of size $Q/\tilde{Q}$ may replace each entry in the "intermediate" matrix.[3]

[3]The described protograph-based technique is not the only one to construct good QC-LDPC codes. Another possible approach is based on Euclidean and projective finite geometries [22, 23].

12.4.3 Error Floor

Finite length LDPC codes are affected by a phenomenon known as the "error floor" [24, 25]. Considering again a communication channel parameterized by a real parameter θ indicating the level of channel noise, the error floor consists of a sudden reduction in the slope of the LDPC code performance curve when θ becomes lower than some value. For example, over the BSC the error floor appears at sufficiently low values of the error probability p, while over the Bi-AWGN channel it appears at sufficiently high values of E_b/N_0. An example performance curve in term of bit error rate (BER) versus E_b/N_0 exhibiting an error floor is depicted in Fig. 12.10. In NAND Flash memories applications, very pressing requirements are usually imposed on the error floor. More specifically, it is often required that the error floor must not appear above page error rate (i.e., codeword error rate) 10^{-15}.

The error floor of LDPC codes under belief propagation decoding is mainly due to graphical structures in the Tanner graph called *trapping sets* [25]. Given a subset $\mathcal{W}$ of the variable nodes, the subgraph induced by $\mathcal{W}$ is the bipartite graph composed of $\mathcal{W}$, of the subset $\mathcal{U}$ of check nodes connected to $\mathcal{W}$ and of the corresponding edges. By definition, an (a, b) trapping set is any size-a subset $\mathcal{W}$ of the variable nodes, such that there are exactly b check nodes of odd degree (an arbitrary number of check nodes of even degree) in the corresponding induced subgraph. The parameter a is called the size of the trapping set. If there are only degree-1 and degree-2 check nodes in the induced subgraph, then the trapping set is said to be *elementary*. Elementary trapping sets of small size are a major cause of error floor

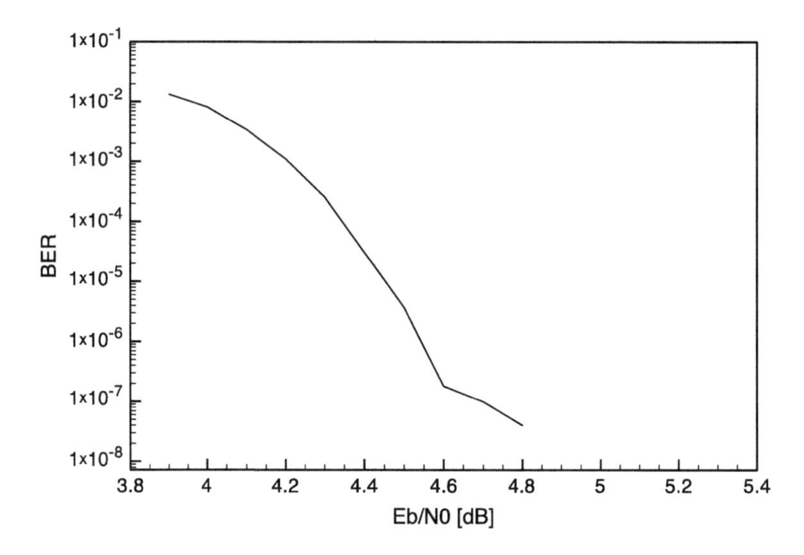

Fig. 12.10 Performance curve (in terms of BER vs. E_b/N_0) exhibiting an error floor at BER $\approx 10^{-7}$ ($E_b/N_0 > 4.6$ dB)

for iteratively decoded LDPC codes. We point out that small weight codewords may also contribute to the error floor together with trapping sets.

The need to construct LDPC codes characterized by very low error floors imposes some modifications to the QC-LDPC code design procedure described in the previous subsection, which becomes more involved. The asymptotic decoding threshold is not the only metric to be taken into account during the ensemble optimization phase, as other asymptotic parameters such as the typical relative minimum distance or smallest trapping set size must be considered [26, 27]. We also point out that reliable error floor analysis at very low error rates of LDPC codes for storage applications still represents an open issue. In fact, Monte Carlo software simulation is not feasible at very low error rates because of prohibitively long simulation times. Approaches proposed in the literature are hardware simulation, importance sampling [6, 28], and estimation techniques [29].

12.5 Belief Propagation (BP) Decoding of LDPC Codes

12.5.1 Introduction

As opposed to MAP and ML decoding algorithms (Sect. 12.2), that are block-wise algorithms, BP is a *bit-wise* decoding algorithm, working iteratively. More specifically, at the end of each decoding iteration a separate decision is taken about each bit in the codeword, and then it is checked whether the currently decoded hard-decision sequence is a codeword or it is not. Letting $y = [y_0, y_1, \ldots, y_{n-1}]$ denote the sequence outcoming from the communication channel, the decision about encoded bit c_i, $i = 0, \ldots, n-1$, is taken according to its a posteriori likelihood ratio (LR), namely,

$$L(c_i|y) = \frac{\Pr(c_i = 0|y)}{\Pr(c_i = 1|y)} \underset{\hat{c}_i = 1}{\overset{\hat{c}_i = 0}{\gtrless}} 1.$$

Unfortunately, the only information available at variable node i at the beginning of the decoding process is the a priori LR

$$L(c_i|y_i) = \frac{\Pr(c_i = 0|y_i)}{\Pr(c_i = 1|y_i)}$$

i.e., the LR conditioned only to the local observation, not the a posteriori LR $L(c_i|y)$ as required. Indeed, the task of the BP decoder consists of calculating the a posteriori LR for each variable node, starting from the individual a priori LRs, exploiting an iterative exchange of information among the nodes of the bipartite graph. In the following description of the BP decoder, we will not make any assumption on the communication channel, but that the channel is memory-less with binary input and equally likely input values.

12.5.2 Preliminaries

We start with some preliminary material that will be useful to properly describe BP decoding of LDPC codes.

Let us consider a Bernoulli random variable B taking the values 0 and 1 with equal probabilities. As depicted in Fig. 12.11, assume that N random experiments are performed to get information about the value assumed by B and that all these experiments are independent. The outcome of the n-th experiment (n-th observation) is denoted by ω_n, while the vector of N observables by $\boldsymbol{\omega} = [\omega_1, \omega_2, \ldots, \omega_N]$. We define the likelihood ratio (LR) of B conditioned to the observation ω_n as

$$L(B|\omega_n) = \frac{\Pr(B=0|\omega_n)}{\Pr(B=1|\omega_n)} \tag{12.11}$$

and the a posteriori likelihood ratio of B (i.e., conditioned to the whole set of N independent observations), as

$$L(B|\boldsymbol{\omega}) = \frac{\Pr(B=0|\boldsymbol{\omega})}{\Pr(B=1|\boldsymbol{\omega})}. \tag{12.12}$$

We now seek for an expression of the a posteriori LR, $L(B|\boldsymbol{\omega})$, as a function of the individual LRs, each conditioned to a specific observation. By Bayes rule we have

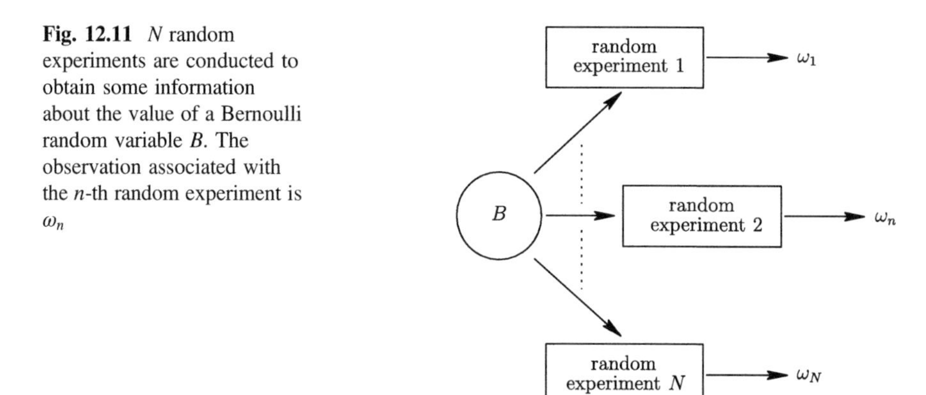

Fig. 12.11 N random experiments are conducted to obtain some information about the value of a Bernoulli random variable B. The observation associated with the n-th random experiment is ω_n

$$
\begin{aligned}
L(B|\boldsymbol{\omega}) &= \frac{p(\boldsymbol{\omega}|B=0)}{p(\boldsymbol{\omega}|B=1)} \\
&= \prod_{n=1}^{N} \frac{p(\omega_n|B=0)}{p(\omega_n|B=1)} \\
&= \prod_{n=1}^{N} L(B|\omega_n),
\end{aligned} \tag{12.13}
$$

where the second equality follows from independence of the random experiments.

We also observe that, through (12.12) and the relationship $\Pr(B=0|\boldsymbol{\omega}) + \Pr(B=1|\boldsymbol{\omega}) = 1$, the probabilities $\Pr(B=0|\boldsymbol{\omega})$ and $\Pr(B=1|\boldsymbol{\omega})$ may be expressed as functions of the a posteriori LR as follows:

$$
\Pr(B=0|\boldsymbol{\omega}) = \frac{L(B|\boldsymbol{\omega})}{1+L(B|\boldsymbol{\omega})}, \tag{12.14}
$$

$$
Pr(B=1|\boldsymbol{\omega}) = \frac{1}{1+L(B|\boldsymbol{\omega})}. \tag{12.15}
$$

This is sometimes referred to as *soft bit*. Analogous relationships may be derived for $\Pr(B=0|\omega_n)$ and $\Pr(B=1|\omega_n)$.

Next, consider n statistically independent Bernoulli random variables $B_1, B_2, \ldots, B_n$ each taking its value in $\{0, 1\}$. We allow $\Pr(B_k=1) \neq \Pr(B_l=1)$ if $k \neq l$. We ask what is the probability that the n variables sum to 0 (in binary algebra), i.e., the probability that an even number of such random variables take value 1. This problem was solved in [7], where it was shown that

$$
\Pr(B_1+B_2+\cdots+B_n=0) = \frac{1+\prod_{\mathrm{k}=1}^{n}(1-2\Pr(B_k=1))}{2}. \tag{12.16}
$$

Consider now n Bernoulli random variables $B_1, B_2, \ldots, B_n$ fulfilling a parity constraint $B_1+B_2+\cdots+B_n=0$. Moreover, assume that some reliability information is known about variables $B_1, \ldots, B_{i-1}, B_{i+1}, \ldots, B_n$, in terms of LRs $L(B_k)$, $k \in \{1, \ldots, i-1, i+1, \ldots, n\}$ and that $B_1, \ldots, B_{i-1}, B_{i+1}, \ldots, B_n$ are statistically independent. We seek for an expression of the LR $L(B_i)$, conditional on all available information about the other $n-1$ variables. Since $\Pr(B_i=0) = \Pr(B_1+\cdots B_{i-1}+B_{i+1}+\cdots+B_n=0)$, through (12.16) we obtain

$$
\begin{aligned}
&\Pr(B_i=0|L(B_1), \ldots, L(B_{i-1}), L(B_{i+1}), \ldots, L(B_n)) \\
&= \frac{1+\prod_{\mathrm{k}\neq\mathrm{i}}(1-2\Pr(B_k=1))}{2}
\end{aligned}
$$

and, consequently,

$$\Pr(B_i = 1 | L(B_1), \ldots, L(B_{i-1}), L(B_{i+1}), \ldots, L(B_n))$$
$$= \frac{1 - \prod_{\mathrm{k} \neq \mathrm{i}} (1 - 2\Pr(B_k = 1))}{2}.$$

Note that each term $\Pr(B_k = 1)$ involved in the multiplication may be expressed in terms of the corresponding $L(B_k)$ through (12.15). From the term-by-term ratio between these two latter equations, we obtain

$$\mathrm{L}(B_i | L(B_1), \ldots, L(B_{i-1}), L(B_{i+1}), \ldots, L(B_n)) = \frac{1 + \prod_{\mathrm{k} \neq \mathrm{i}} (1 - 2\Pr(B_k = 1))}{1 - \prod_{\mathrm{k} \neq \mathrm{i}} (1 - 2\Pr(B_k = 1))}.$$

Through (12.15), after a few calculations this leads to

$$\mathrm{L}(B_i | L(B_1), \ldots, L(B_{i-1}), L(B_{i+1}), \ldots, L(B_n)) = \frac{\prod_{k \neq i} \frac{L(B_k) + 1}{L(B_k) - 1} + 1}{\prod_{k \neq i} \frac{L(B_k) + 1}{L(B_k) - 1} - 1}. \quad (12.17)$$

12.5.3 *Algorithm Description*

12.5.3.1 Overview

For ease of presentation, in the description of the algorithm we omit the decoding iteration index. We denote by r_i^j the message sent by variable node V_i, $i = 0, \ldots, n-1$, to check node C_j, $j = 0, \ldots, m-1$ during the current iteration, and by m_j^i the message sent back by check node C_j, to variable node V_i, during the same iteration. For $i = 0, \ldots, n-1$, we also denote by w_i the a priori LR for variable node V_i, i.e.,

$$w_i = \frac{\Pr(c_i = 0 | y_i)}{\Pr(c_i = 1 | y_i)}.$$

This is illustrated in Fig. 12.12.

Belief-propagation decoding is composed of four steps, namely[4]:

- initialization;
- horizontal step;

[4]The words "horizontal" and "vertical" remind us that the check nodes and the variable nodes are associated with the rows and the columns of the parity-check matrix, respectively.

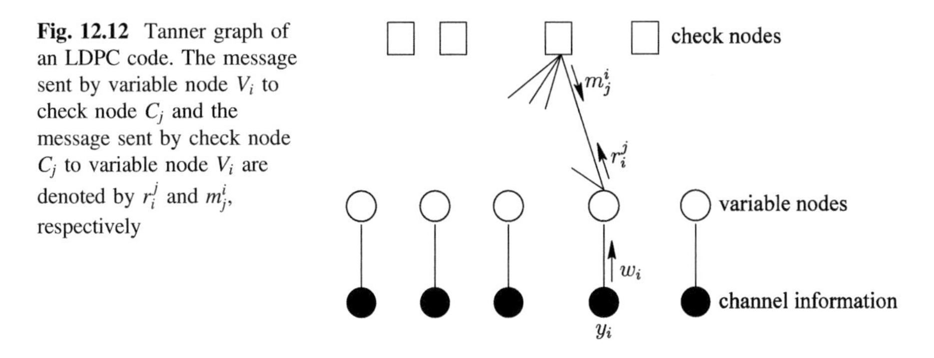

Fig. 12.12 Tanner graph of an LDPC code. The message sent by variable node V_i to check node C_j and the message sent by check node C_j to variable node V_i are denoted by r_i^j and m_j^i, respectively

- vertical step;
- hard decision and stopping criterion step.

Out of them, the initialization step is executed only once, at the beginning of decoding. The other three steps are executed iteratively, until a termination condition is verified or a maximum number of iterations, denoted by I_{max}, is reached. Each decoding iteration is split into two half-iterations. During the first half-iteration (horizontal step), check nodes process messages incoming from their neighboring variable nodes. Then, each check node sends one message along every edge incident on it. Thus, every check node sends one message per iteration to each of its neighboring variable nodes. During the second half-iteration (vertical step) variable nodes process messages incoming from their neighboring check nodes. Similar to the previous half-iteration, at the end of this processing each variable node sends one message along each edge incident on it. Thus, every variable node sends one message per iteration to each of its neighboring check nodes. At the end of the two half-iterations, a hard decision is taken in each variable node, about the value of the corresponding encoded bit.

The message transmitted by check node $C_j, j=0, \ldots, m-1$, to variable node V_i, $i=0, \ldots, n-1$, where V_i belongs to the neighborhood of C_j, may be interpreted as the best estimate C_j has about the value of V_i up to the current iteration. This is the estimate of the value of V_i given all information about V_i the check node has got from the variable nodes connected to it *other than* V_i. This is known as *extrinsic information*. Analogously, the message sent back by variable node V_i to check node C_j may be interpreted as the best estimate V_i has about itself up to the current iteration. This is the estimate of its value given all information the variable node has got from the communication channel and from the check nodes connected to it *other than* C_j (extrinsic information). All messages exchanged between variable nodes and check nodes are LRs or, equivalently, soft bits.

At the end of the vertical step, each variable node takes a hard decision about the value of its associated bit, based on the a priori information incoming from the channel and on all estimates incoming from the check nodes connected to it. If the obtained hard-decision binary sequence $\hat{c}$ is a codeword of the LDPC code, i.e., if every check node is connected to an even number of variable nodes whose current estimate is 1, then a decoding success is declared, decoding is terminated, and $\hat{c}$ is returned as the decoded codeword. Otherwise, a new iteration is started, unless the maximum number of iterations has been reached. In this latter case, no codeword has been found and a decoding failure is declared. LDPC codes decoded via belief propagation are then characterized by two different error events: detected errors and undetected errors. A detected error takes place whenever no codeword is found up to the maximum number of iterations. An undetected error takes place whenever, at some iteration, the hard-decision sequence $\hat{c}$ is a codeword but not the transmitted one. Undetected errors may be extremely dangerous is some contexts, including NAND Flash memories.

12.5.3.2 Initialization

At the beginning, each variable node broadcasts to all its neighboring check nodes the a priori LR received from the communication channel. Hence, we have

$$r_i^j = w_i$$

for all $j \in N(i)$, where $N(i)$ is the set of indexes of check nodes connected to V_i. The expression of w_i depends on the nature of the channel. For example, it is easy to check that over a BSC with error probability p and antipodal mapping $x_i = 1 - 2c_i \in \{-1, +1\}$, we have

$$w_i = \begin{cases} \frac{1-p}{p} & \text{if } y_i = +1 \\ \frac{p}{1-p} & \text{if } y_i = -1. \end{cases} \tag{12.18}$$

As another example, over a Bi-AWGN channel and again antipodal mapping $x_i = 1 - 2c_i$, (meaning E_s normalized to 1) we have

$$w_i = e^{(2/\sigma^2) y_i}. \tag{12.19}$$

Importantly, the initialization step requires a knowledge of the channel. For instance, in the case of a BSC the error probability p must be known, as well as the noise power σ^2 in the Bi-AWGN case.

12.5.3.3 Horizontal Step

For $j = 0, \ldots, m-1$, check node C_j, of degree h_j, sends to each of the h_j variable nodes connected to it its current estimate of the corresponding bit. If variable node V_i is connected to C_j, the message from C_j to V_i is the LR of bit c_i, conditional on the information available at C_j incoming from all its neighboring variable nodes, except the information incoming from V_i. A pictorial representation of this process

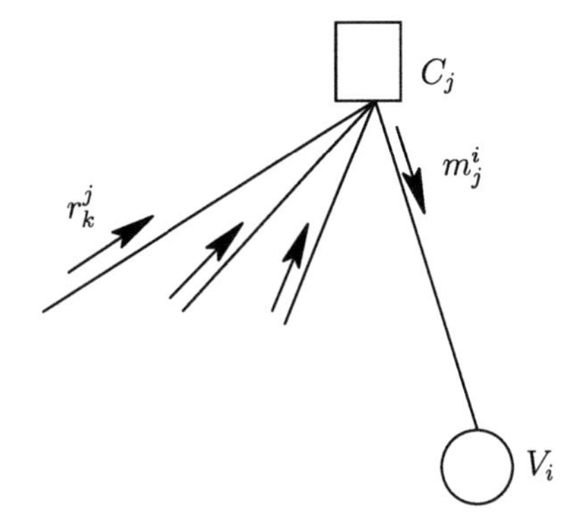

Fig. 12.13 Check node processing of incoming messages during the horizontal step

is provided in Fig. 12.13. Note that two different variable nodes connected to C_j will receive, in general, different messages.

The message m_j^i from C_j to V_i can be calculated exploiting one of the results introduced in Sect. 12.5.2. In fact, each of the h_j incoming messages is the LR of a specific bit on which the check node imposes a parity constraint. Hence, under independence hypothesis, denoting by $N(j)\backslash\{i\}$ the set of indexes of variable nodes connected to C_j except V_i, from (12.17) we immediately obtain

$$m_j^i = \frac{\prod_{k \in N(j)\backslash\{i\}} \frac{r_k^j + 1}{r_k^j - 1} + 1}{\prod_{k \in N(j)\backslash\{i\}} \frac{r_k^j + 1}{r_k^j - 1} - 1}. \tag{12.20}$$

Note that the independence hypothesis is fulfilled only during the first $g/2$ decoding iterations, where g is the girth of the Tanner graph. On the other hand, it represents an approximation during all subsequent iterations.

12.5.3.4 Vertical Step

For $i = 0, \ldots, n-1$, variable node V_i, of degree d_i, sends to each of its d_i neighboring check nodes its current estimate of the associated bit. With reference to Fig. 12.14, the message r_i^j sent to check node C_j is the LR about bit c_i, conditional on the a priori information available from the communication channel and on the information incoming from all check nodes connected to it, except C_j. Again, two different check nodes connected to V_i will receive, in general, different messages.

The message r_i^j that variable node V_i sends to check node C_j connected to it can be easily computed based on the result in Sect. 12.5.2. In fact, each of the d_i messages incoming towards the variable node (including the message w_i incoming

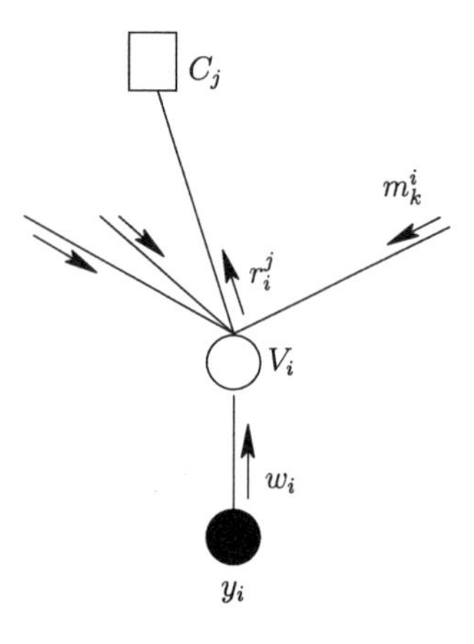

Fig. 12.14 Variable node processing of incoming messages during the vertical step

from the channel), represents the LR of c_i conditioned to some observation. Under the hypothesis of independence for the d_i observations, denoting by $N(i)\backslash\{j\}$ the set of indexes check nodes connected to V_i except check node of index j, we have

$$r_i^j = w_i \prod_{k \in N(i)\backslash\{j\}} m_k^i. \tag{12.21}$$

(Again, the independence hypothesis is valid rigorously only during the first $g/2$ decoding iterations.)

12.5.3.5 Hard Decision and Stopping Criterion

At the last step of each iteration, every variable node takes a decision about its associated encoded bit. This decision is based on all currently available information about the bit, i.e., on the a priori information from the communication channel and on *all* messages incoming from the check nodes. Let m^i denote the list of all messages incoming towards the variable node V_i. Applying again the result developed in Sect. 12.5.2 under the hypothesis of independence of the incoming messages, we may write

$$L(c_i | w_i, m^i) = w_i \prod_{k \in N(i)} m_k^i. \tag{12.22}$$

(Again, the independence hypothesis is fulfilled rigorously only during the first $g/2$ decoding iterations.) The decision about encoded bit c_i at the end of the generic iteration is then

$$L(c_i|w_i, m^i) \underset{\hat{c}_i = 1}{\overset{\hat{c}_i = 0}{\gtrless}} 1.$$

If the current hard-decision sequence $\hat{c}$ is a codeword ($\hat{c}H^T = 0$, where $\boldsymbol{H}$ is any parity-check matrix of the code) then the algorithm is terminated and $\hat{c}$ is returned as the decoded codeword. Else, if $\hat{c}$ is not a codeword and the maximum number of iterations I_{max} has been reached, the algorithm is terminated and a failure is reported. Else, a new iteration is started jumping to the horizontal step. Belief propagation decoding of LDPC codes may be summarized as follows.

Belief-Propagation Decoding of LDPC Codes

1: set $I = 1$. For $i = 0, \ldots, n-1$, for $j \in N(i)$, set $r_i^j = w_i$;
2: for $j = 0, \ldots, m-1$
    for $i \in N(j)$ calculate m_j^i according to (11.20);
3: for $i = 0, \ldots, n-1$
    for $j \in N(i)$ calculate r_i^j according to (11.21);
4: for $i = 0, \ldots, n-1$ {
    calculate $L(c_i|w_i, \boldsymbol{m}^i)$ according to (11.22);
    if $L(c_i|w_i, \boldsymbol{m}^i) \geq 1$ then set $\hat{c}_i = 0$;
    else set $\hat{c}_i = 1$;
  }
  if $\hat{\boldsymbol{c}}\boldsymbol{H}^T = \boldsymbol{0}$ then return $\hat{\boldsymbol{c}}$;
  else {
    if $I = I_{max}$ exit;
    else {
      $I = I + 1$;
      goto 2;
      }
  }

12.5.4 Log-Domain BP Decoder

The main issue when implementing BP decoding described in Sect. 12.5.3 is represented by the need to handle and combine, through multiplications and divisions, likelihood ratios whose values may differ by several orders of magnitude.

For this reason, a log-domain implementation is usually preferred from an implementation viewpoint. In the log-domain version of BP decoding, log-likelihood ratios (LLRs) of the encoded bits are exchanged between variable and check nodes. Next, we discuss how the above-described BP decoding shall be modified in the log-domain. All logarithms are assumed to be natural logarithms. Moreover, $\mathrm{sgn}(x)$ will denote the sign function, i.e., $\mathrm{sgn}(x) = +1$ if $x \geq 0$ and $\mathrm{sgn}(x) = -1$ otherwise.

The initialization step remains the same, the only difference being that the first message each variable node sends to all its neighboring check nodes is the a priori LLR of the corresponding encoded bit. Neglecting again the iteration index and denoting by R_i^j the message sent from variable node $i \in \{0, \ldots, n-1\}$ to check node $j \in N(i)$, we have

$$R_i^j = W_i,$$

where $W_i = \log w_i$. For instance, assuming antipodal mapping $x_i = 1 - 2c_i$, over a BSC with error probability p we have

$$W_i = \begin{cases} \log \frac{1-p}{p} \, if y_i = +1 \\ \log \frac{p}{1-p} \, if y_i = -1 \end{cases} \tag{12.23}$$

while, over a Bi-AWGN channel,

$$W_i = \frac{2}{\sigma^2} y_i. \tag{12.24}$$

The development of check node message processing (horizontal step) in the log domain is more involved. Denoting $R_i^j = \log r_i^j$ and $M_j^i = \log m_j^i$, from (12.20) we may write

$$\begin{aligned} M_j^i &= \log \frac{\prod_{k \in N(j) \setminus \{i\}} \frac{e^{R_k^j} + 1}{e^{R_k^j} - 1} + 1}{\prod_{k \in N(j) \setminus \{i\}} \frac{e^{R_k^j} + 1}{e^{R_k^j} - 1} - 1} \\ &= \log \frac{\prod_{k \in N(j) \setminus \{i\}} \mathrm{sgn}\left(R_k^j\right) \cdot \prod_{k \in N(j) \setminus \{i\}} \frac{e^{|R_k^j|} + 1}{e^{|R_k^j|} - 1} + 1}{\prod_{k \in N(j) \setminus \{i\}} \mathrm{sgn}\left(R_k^j\right) \cdot \prod_{k \in N(j) \setminus \{i\}} \frac{e^{|R_k^j|} + 1}{e^{|R_k^j|} - 1} - 1}, \end{aligned}$$

where we have exploited the fact that any odd function fulfills $f(x) = \mathrm{sgn}(x) f(|x|)$ and the fact that $f(x) = (e^x + 1)/(e^x - 1)$ is odd. The obtained expression of M_j^i can be further developed through the identity $\log((x+1)/(x-1)) = \mathrm{sgn}(x) \cdot \log((|x|+1)/(|x|-1))$ and through the fact that $e^{|R|} \geq 1$. This yields

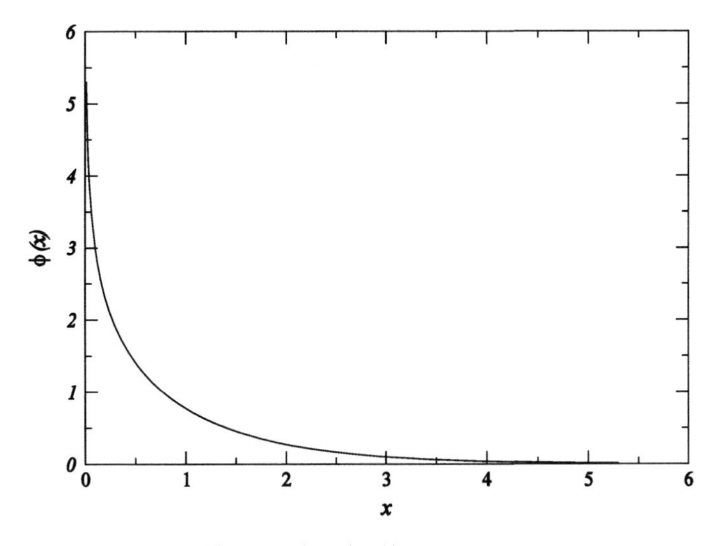

Fig. 12.15 Plot of function $\varphi(x) = -\log(\tanh(x/2))$.

$$
\begin{aligned}
M_j^i &= \prod_{k \in N(j)\setminus\{i\}} \operatorname{sgn}\left(R_k^j\right) \cdot \log \frac{\prod_{k \in N(j)\setminus\{i\}} \frac{e^{|R_k^j|}+1}{e^{|R_k^j|}-1} + 1}{\prod_{k \in N(j)\setminus\{i\}} \frac{e^{|R_k^j|}+1}{e^{|R_k^j|}-1} - 1} \\
&= \prod_{k \in N(j)\setminus\{i\}} \operatorname{sgn}\left(R_k^j\right) \cdot \log \frac{e^{\sum_{k \in N(j)\setminus\{i\}} \log \frac{e^{|R_k^j|}+1}{e^{|R_k^j|}-1}} + 1}{e^{\sum_{k \in N(j)\setminus\{i\}} \log \frac{e^{|R_k^j|}+1}{e^{|R_k^j|}-1}} - 1} \\
&= \prod_{k \in N(j)\setminus\{i\}} \operatorname{sgn}\left(R_k^j\right) \cdot \varphi\left(\sum_{k \in N(j)\setminus\{i\}} \varphi(|R_k^j|) \right)
\end{aligned}
\tag{12.25}
$$

where, for $x > 0$, we have introduced the nonlinear function

$$
\varphi(x) = \log \frac{e^x + 1}{e^x - 1} = -\log(\tanh(x/2)).
$$

A plot of this function is depicted in Fig. 12.15. Note that the function coincides with its inverse, i.e., $\varphi(\varphi(x)) = x$.

The transposition of the variable node processing (vertical step) to the logarithmic domain is much simpler. In fact, from (12.21) we immediately obtain

$$
R_i^j = W_i + \sum_{k \in N(i)\setminus\{j\}} M_k^i. \tag{12.26}
$$

Analogously, (12.22) shall be updated as

$$\log L\left(c_i \middle| W_i, \boldsymbol{M}^i\right) = W_i + \sum_{k \in N(i)} M_k^i. \tag{12.27}$$

The algorithm may be then summarized as follows.

Log-Domain Belief-Propagation Decoding of LDPC Codes

1: set $I = 1$. For $i = 0, \dots, n-1$, for $j \in N(i)$, set $R_i^j = W_i$;
2: for $j = 0, \dots, m-1$
 for $i \in N(j)$ calculate M_j^i according to (11.25);
3: for $i = 0, \dots, n-1$
 for $j \in N(i)$ calculate R_i^j according to (11.26);
4: for $i = 0, \dots, n-1$ {
 calculate $\log L(c_i | W_i, \boldsymbol{M}^i)$ according to (11.27);
 if $\log L(c_i | W_i, \boldsymbol{M}^i) \geq 0$ then set $\hat{c}_i = 0$;
 else set $\hat{c}_i = 1$;
}
if $\hat{\boldsymbol{c}}\boldsymbol{H}^T = \boldsymbol{0}$ then return $\hat{\boldsymbol{c}}$;
else {
 if $I = I_{max}$ exit;
 else {
  $I = I + 1$;
  goto 2;
  }
}

Although an enhanced numerical stability is achieved operating on log-likelihood ratios, as well as a lower complexity (as, for instance, products in (12.21) and (12.22) are transformed in sums in (12.25) and (12.26), respectively), check node processing in the log-domain imposes the evaluation of the nonlinear function φ. For a single check node C_j of degree h_j, this function should in principle be evaluated $(h_j)^2$ times per iteration (even if techniques to limit the number of φ evaluations exist). The calculation of function φ is typically performed by means of lookup-tables. Note that, however, for small x the graph of $\varphi(x)$ is very steep, thus requiring a very fine (in general, nonuniform) discretization of the corresponding region of the function domain, and that the implementation of $\varphi(x)$ through a lookup table may be quite inconvenient in hardware implementation. For these reasons, extensive work has been carried out to develop either approximations of the log-domain BP decoder or other reduced-complexity decoding schemes.

All of these decoders offer a reduced error correction capability than actual BP. However, they also exhibit a lower decoding complexity and, hence, a higher decoding speed.

12.6 Reduced-Complexity Decoders

So far we have focused on the BP decoder (both in probability domain and log-domain) originally developed by Gallager. Next, we present a few reduced-complexity, implementation-friendly decoders for LDPC codes. It must be pointed out that a large amount of reduced-complexity decoding schemes for LDPC codes have been developed in the last decade [30]. Most of these decoding schemes may be seen as approximations of the BP decoder, in the sense that they are characterized by approximations of the most complex step of BP decoding, namely, the horizontal step (consisting of the calculation of extrinsic messages from the check nodes to the variable nodes). As such, these approximate BP decoding algorithms can be formalized via the same pseudo-code we have adopted for the log-domain BP decoder, with a difference in step 2.

We only present the most famous approximation of the BP decoder, called the Min-Sum (MS) decoder. We then move to describe decoders exhibiting an even lower complexities. More specifically, we present a binary message-passing algorithm known as "Gallager B" (and originally proposed in [7]) and a class of non-message-passing decoders named "flipping algorithms" (the idea of bit flipping appears again in [7]). These very low complexity decoding algorithms (along with some of their modifications, not addressed in this chapter) are of interest in NAND Flash memories at the beginning of the memory life, when the raw bit error probability is extremely low.

12.6.1 Min-Sum Decoder

The MS decoder can be directly developed from the log-domain BP decoder as follows. From Fig. 12.15 observe that the graph of function $\varphi(x)$ is very steep for small values of x. Then, when x assumes small values, a small perturbation in terms of x determines a large deviation in terms of $\varphi(x)$. For this reason, if at least one of the magnitudes $|R_k^j|$ in the summation appearing in (12.25) is sufficiently small, the corresponding value of $\varphi(|R_k^j|)$ dominates the other summands. Hence, we can write

$$
\begin{aligned}
M_j^i &= \prod_{k \in N(j)\setminus\{i\}} \mathrm{sgn}\left(R_k^j\right) \cdot \varphi\left(\sum_{k \in N(j)\setminus\{i\}} \varphi(|R_k^j|)\right) \\
&\approx \prod_{k \in N(j)\setminus\{i\}} \mathrm{sgn}\left(R_k^j\right) \cdot \varphi\left(\max_{k \in N(j)\setminus\{i\}} \varphi(|R_k^j|)\right) \\
&= \prod_{k \in N(j)\setminus\{i\}} \mathrm{sgn}\left(R_k^j\right) \cdot \min_{k \in N(j)\setminus\{i\}} |R_k^j|
\end{aligned}
\tag{12.28}
$$

where the last equality follows from $\varphi(x)$ being self-invertible (i.e., $\varphi(\varphi(x))=x$) and monotonically decreasing. The MS decoding algorithm is summarized next.

Min-Sum Decoding of LDPC Codes

1: set $I = 1$. For $i = 0, \ldots, n-1$, for $j \in N(i)$, set $R_i^j = W_i$;
2: for $j = 0, \ldots, m-1$
  for $i \in N(j)$ calculate M_j^i according to (11.28);
3: for $i = 0, \ldots, n-1$
  for $j \in N(i)$ calculate R_i^j according to (11.26);
4: for $i = 0, \ldots, n-1$ {
  calculate $\log L(c_i | W_i, \boldsymbol{M}^i)$ according to (11.27);
  if $\log L(c_i | W_i, \boldsymbol{M}^i) \geq 0$ then set $\hat{c}_i = 0$;
  else set $\hat{c}_i = 1$;
 }
 if $\hat{\boldsymbol{c}}\boldsymbol{H}^T = \boldsymbol{0}$ then return $\hat{\boldsymbol{c}}$;
 else {
  if $I = I_{max}$ exit;
  else {
   $I = I + 1$;
   goto 2;
   }
 }

Several improvements to the MS decoder have been proposed in the literature, to reduce the gap between its performance and that of BP decoding, at the expense of a small increase in terms of computational cost. These refinements are out of the scope of this book. Interested readers may refer, for example, to [31, 32].

12.6.2 Gallager B Decoder

The BP and MS decoders are characterized by real-valued (properly quantized, in hardware implementation) messages exchanged between the variable nodes and the check nodes. Moreover, as previously emphasized, both algorithms remain unchanged over a wide range of communication channels. In contrast, Gallager B decoder, first proposed in [7], is a message-passing decoding algorithm for LDPC codes characterized by binary-valued messages and is specifically tailored for the BSC (i.e., no soft information is available at the decoder input). Although its performance is poor compared with that of BP and MS algorithms over the BSC, it has been proved that it represents the optimum LDPC decoder over the BSC when the extrinsic messages are constrained to be binary.

The algorithm works as follows. Assuming transmission over a BSC with error probability p and input and output alphabets $\mathcal{X}=\mathcal{Y}=\{0,1\}$, for $i=0,\ldots,n-1$ variable node V_i is fed with the corresponding binary symbol $y_i \in \{0,1\}$ received from the channel. (In contrast, to perform BP decoding over the BSC variable node i is initialized according to (12.18) or to its logarithmic version (12.23).) The symbol y_i is broadcasted by variable node V_i to each of its neighboring check nodes. The algorithm is then structured in a similar way as BP or MS, where the horizontal, vertical, and stopping criterion steps are specified as follows.

During the horizontal step, for $j=0,\ldots,m-1$ the message propagating from check node C_j to variable node V_i, $i \in N(j)$, is simply the modulo-2 summation of all binary messages incoming from variable nodes connected to C_j but the message incoming from V_i. Hence, we can write

$$m_j^i = \sum_{k \in N(j)\setminus\{i\}} r_k^j \tag{12.29}$$

where the summation is modulo-2. (Note that $r_i^j = y_i$ for all $i=0,\ldots,n-1$ at the first iteration.) During the vertical step, for $i=0,\ldots,n-1$ the message from variable node V_i to check node $C_j, j \in N(i)$, is equal to the modulo-2 complement of y_i if the number of incoming extrinsic messages different from y_i is above some threshold, and is equal to y_i otherwise. Letting

$$X_j^i = \left|\left\{m_k^i \neq y_i \ s.t. \ k \in N(i)\setminus\{j\}\right\}\right|$$

and $T^{(i)}$ be the number of such extrinsic messages and the threshold at the current iteration, respectively, and letting $C(y_i)$ be the modulo-2 complement of y_i, we have

$$r_i^j = \begin{cases} C(y_i) & \text{if } X_j^i \geq T^{(i)} \\ y_i & \text{otherwise.} \end{cases} \tag{12.30}$$

At the end of each decoding iteration, for each variable node V_i the decision about the current value of the local bit $\hat{c}_i$ is taken according to a majority policy.

More specifically, if the variable node degree d_i is even, then $\hat{c}_i$ is set equal to the value assumed by the majority of the incoming messages m_j^i and of y_i. On the other hand, if the variable node degree is odd, then $\hat{c}_i$ is set equal to the value assumed by the majority of the incoming messages m_j^i (y_i is not considered).

Gallager B Decoding of LDPC Codes

1: set $I = 1$. For $i = 0, \ldots, n-1$, for $j \in N(i)$, set $r_i^j = y_i$;
2: for $j = 0, \ldots, m-1$
  for $i \in N(j)$ calculate m_j^i according to (11.29);
3: for $i = 0, \ldots, n-1$
  for $j \in N(i)$ calculate r_i^j according to (11.30);
4: for $i = 0, \ldots, n-1$ {
  if $d_i \bmod 2 = 0$ then set $\hat{c}_i$ to the value assumed by the majority of the incoming messages m_j^i and of y_i ;
  else set $\hat{c}_i$ to the value assumed by the majority of the incoming messages m_j^i ;
 }
 if $\hat{\boldsymbol{c}}\boldsymbol{H}^T = \boldsymbol{0}$ then return $\hat{\boldsymbol{c}}$;
 else {
  if $I = I_{max}$ exit;
  else {
   $I = I + 1$;
   goto 2;
   }
 }

Appropriate values for the threshold $T^{(i)}$ range between $\lfloor (d_i - 1)/2 \rfloor$ and d_i, as the number of incoming extrinsic messages enforcing an outgoing message different from $\tilde{y}_i$ must be sufficiently high. Note that in principle, for irregular codes the value of the threshold may be different for two different variable nodes, even during the same iteration. Also note that, for the same variable node, the value of the threshold may not remain constant with the iteration index, as it may be adjusted dynamically. In [7] it was shown that for a regular (d, h) LDPC code, the optimum value of the threshold (the same for all variable nodes at the same iteration) is the smallest integer T for which the inequality

$$\frac{1-p}{p} \leq \left(\frac{1 + (1 - 2\varepsilon)^{h-1}}{1 - (1 - 2\varepsilon)^{h-1}} \right)^{2T - d + 1} \tag{12.31}$$

is fulfilled, where p is the BSC error probability and ε is the extrinsic error probability. This latter parameter represents the average probability that an edge in the Tanner graph carries an error message from the variable node set to the check node set at the considered iteration, and varies over iterations. In the asymptotic setting where the Tanner graph is assumed to be cycle-free, the update equation for ε for regular LDPC codes is [7]

$$\begin{aligned}\varepsilon_{\ell+1} = p - p \sum_{z=T_\ell}^{d-1} \binom{d-1}{z} \left[\frac{1+(1-2\varepsilon_\ell)^{h-1}}{2}\right]^z \left[\frac{1-(1-2\varepsilon_\ell)^{h-1}}{2}\right]^{d-1-z} \\ + (1-p) \sum_{z=T_\ell}^{d-1} \binom{d-1}{z} \left[\frac{1-(1-2\varepsilon_\ell)^{h-1}}{2}\right]^z \left[\frac{1+(1-2\varepsilon_\ell)^{h-1}}{2}\right]^{d-1-z}\end{aligned} \quad (12.32)$$

where $\ell \geq 0$ is the iteration index and where $\varepsilon_0 = p$.

Example 12.5 Equation (12.32) represents density evolution recursion for Gallager B decoding of regular unstructured (d, h) LDPC code ensembles. The asymptotic decoding threshold p^* for this ensemble under Gallager B decoding is then the sup of the set of all $p > 0$ such that $\lim_{\ell \to \infty} \varepsilon_\ell = 0$. For given d and h, whether or not some p is above or below threshold can be easily checked by running the recursion (with starting point $\varepsilon_0 = p$), adapting the value of T_ℓ at each iteration according to (12.31) for the current value of ε_ℓ. For example, for $d = 4$ and $h = 40$ (which corresponds to a rate $R = 9/10$ ensemble) we obtain a threshold $p^* = 0.0041$. Through (12.8) and $E_s = RE_b$, this corresponds to a threshold $(E_b/N_o)^* = 5.892$ dB, about 1.5 dB away from the Shannon limit relevant to the one-bit quantized Bi-AWGN channel.

12.6.3 Flipping Algorithms

Flipping algorithms are a class of low-complexity, iterative decoding algorithms for LDPC codes over the BSC different from message-passing ones. The decoding strategy consists of flipping, at the end of each decoding iteration, the current value of a subset of variable nodes for which a certain flipping condition is fulfilled. If the obtained binary sequence is a codeword, decoding is stopped and the codeword is returned. Otherwise, a new iteration is started. The process continues until a codeword is found or a maximum number of iterations is reached. Different flipping algorithms are characterized by different criteria to identify the variable nodes to be flipped.

A popular flipping algorithm, hereafter referred to simply as bit-flipping (BF) algorithm, consists of flipping at each iteration those variable nodes for which the number u of unsatisfied check nodes is maximum. A BSC with input and output alphabets $\mathcal{X} = \mathcal{Y} = \{0, 1\}$ is assumed.

Bit-Flipping Decoding of LDPC Codes

1: set $I = 1$. For $i = \{0, \ldots, n-1\}$, set $\hat{c}_i = y_i$;
2: if $\hat{\boldsymbol{c}}\boldsymbol{H}^T = 0$ then return $\hat{\boldsymbol{c}}$;
3: for $i = 0, \ldots, n-1$
    calculate u_i ;
4: calculate u_{max} ;
5: for each i such that $u_i = u_{max}$ set $\hat{c}_i = \left(\hat{c}_i + 1\right) mod2$;
5: if $\hat{\boldsymbol{c}}\boldsymbol{H}^T = 0$ then return $\hat{\boldsymbol{c}}$;
  else {
    if $I = I_{max}$ exit;
    else {
      $I = I + 1$;
      goto 2;
      }
  }

12.7 Non-binary LDPC Codes

The so far introduced LDPC codes are binary, in that the code represents an Rn-dimensional subspace of the vector space $\mathrm{GF}(2)^n$, where R is the code rate, n is the codeword length, and GF(2) is the Galois field of order 2. More specifically, the LDPC code is the (Rn-dimensional) null space of an $m \times n$ sparse parity-check matrix $\boldsymbol{H}$. All n encoded vectors belong to GF(2) as well as all of the elements of $\boldsymbol{H}$. If row vector $\boldsymbol{a}$ belongs to $\mathrm{GF}(2)^n$, then the syndrome of $\boldsymbol{a}$ is $s = aH^T \in \mathrm{GF}(2)^m$, where all operations are performed in GF(2). Vector $\boldsymbol{a}$ is a codeword if and only if its syndrome is null.

Like other classes of linear block codes, also LDPC codes may be constructed on Galois fields of order $q > 2$ [33]. In this case the code can be represented by a sparse parity-check matrix $\boldsymbol{H}$ on GF(q) i.e., a matrix whose elements $h_{j,i}$, $j \in \{0, 1, \ldots, m-1\}$ and $i \in \{0, 1, \ldots, n-1\}$, belong to GF($q$) and with a relatively small number of nonzero elements. The code is an Rn-dimensional subspace of the vector space $\mathrm{GF}(q)^n$, where R is still the code rate and the codeword length n is expressed in Galois field symbols. Letting row vector $\boldsymbol{a}$ belong to $\mathrm{GF}(q)^n$, the syndrome of $\boldsymbol{a}$ is still $s = aH^T \in \mathrm{GF}(q)^m$, where now all operations are performed in

GF(q). Still, $\boldsymbol{a}$ is a codeword if and only if its syndrome is null. Hereafter we focus on LDPC codes constructed on extension fields GF(q) with $q=2^p$ for integer $p>2$. We denote by α a primitive element of GF(q). We use the terminology non-binary LDPC (NB-LDPC) code to refer to an LDPC code constructed on the Galois field GF(q).

12.7.1 NB-LDPC Code Ensembles

As binary LDPC codes, also NB-LDPC ones admit a graphical representation through a Tanner graph $G=(\mathcal{V}\cup\mathcal{C},\mathcal{E})$. Again, $\mathcal{V}=\{V_0,V_1,\ldots,V_{n-1}\}$ is the set of variable nodes, $\mathcal{C}=\{C_0,C_1,\ldots,C_{m-1}\}$ is the set of check nodes, and $\mathcal{E}$ is the set of edges. The number of edges, equal to the number of non-zero entries of $\boldsymbol{H}$, is still denoted by E. The n variable nodes and the m check nodes are still bijectively associated with the n codeword symbols and with the m parity-check equations, respectively; each encoded symbol now belongs to GF(q) and each parity-check equation is a linear equation in GF(q). In the Tanner graph, variable node $V_i\in\mathcal{V}$ is connected to check node $C_j\in\mathcal{C}$ by an edge if and only if $h_{j,i}\in\mathrm{GF}(q)\backslash\{0\}$, i.e., if and only if the element of $\boldsymbol{H}$ in row $j\in\{0,1,\ldots,m-1\}$ and column $i\in\{0,1,\ldots,n-1\}$ is non-zero. Equivalently, V_i is connected to C_j if and only if the non-binary codeword symbol c_i, associated with V_i, is involved in the parity-check equation corresponding to C_j. Edge labeling represents the main difference between the Tanner graphs of binary and non-binary LDPC codes. As opposed to the Tanner graph of a binary LDPC code, in fact, in the Tanner graph of a NB-LDPC code the edge connecting variable node V_i to check node C_j is labeled by the corresponding non-zero element $h_{j,i}$ of the parity-check matrix.

As an example, the Tanner graph of a linear block code with codeword length $n=5$ and dimension $k=3$ (where both n and k are measured in field symbols) is shown in Fig. 12.16. The Tanner graph has two check nodes, each imposing a linear constraint on the variable nodes connected to it, and five variable nodes, each

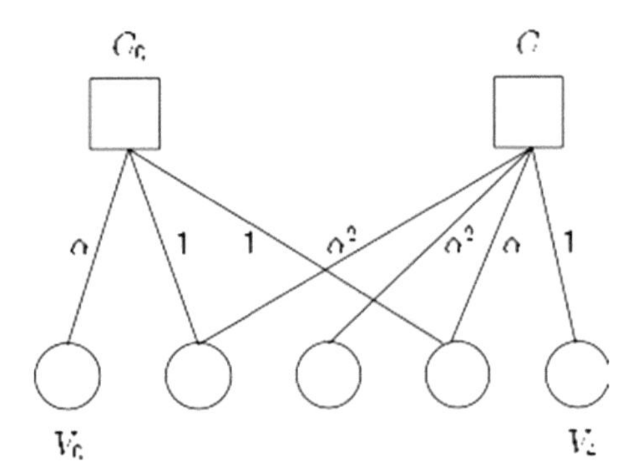

Fig. 12.16 Tanner graph of a non-binary linear block code over GF(4) with codeword length 5 (field symbols) and code rate 3/5. Each edge in the Tanner graph is labeled with a non-zero element of GF(4)

representing a codeword symbol. The parity-check matrix of the corresponding linear block code is

$$H = \begin{bmatrix} \alpha & 1 & 0 & 1 & 0 \\ 0 & \alpha^2 & \alpha^2 & \alpha & 1 \end{bmatrix}.$$

Similarly to their binary counterparts, NB-LDPC codes are usually analyzed in terms of ensemble average. Ensembles of NB-LDPC codes are defined similarly to ensembles of binary LDPC codes, with the difference that edge labeling is also considered in the ensemble definition. For example, the unstructured ensemble of NB-LDPC codes over GF(q) of length n and degree distribution (Λ, P), denoted by $\mathcal{C}_q(n, \Lambda, \mathrm{P})$, includes the LDPC codes constructed GF(q) corresponding to all possible $E!$ edge permutations between the variable node and the check node sockets, according to a uniform probability distribution and, for each such permutation, all possible edges labelings with non-zero elements of GF(q) again according to a uniform probability measure. Ensembles of ultra-sparse NB-LDPC codes (where all variable nodes have degree 2) have attracted an increasing interest in the past decade [34, 35]. Ensembles of protograph-based NB-LDPC codes may also be defined similarly to their binary counterparts, by including edge labeling in the ensemble definition [36, 37].

12.7.2 *Iterative Decoding of NB-LDPC Codes*

Similarly to binary LDPC codes, NB-LDPC codes may be decoded iteratively via BP decoding. The BP decoder for NB-LDPC codes may be regarded as a generalization of the above-described BP decoder for binary LDPC codes. Hereafter, we provide a description of such a decoder, focusing on its probability-domain implementation. We assume an extension field of order $q = 2^p$ for integer $p > 2$. For the sake of clarity, we divide the algorithm into six steps called initialization, message permutation, horizontal step, message de-permutation, vertical step and hard decision and stopping criterion. Out of these six steps, the first one (initialization) is performed only once, at the beginning of the algorithm, while the others are performed iteratively until a stopping rule is verified.

In the non-binary BP decoder, each message still represents extrinsic information. As opposed to the binary case, in which each message exchanged between a variable node V_i and a check node C_j is (in the log-domain implementation) a scalar value representing a likelihood ratio or log-likelihood ratio, in the non-binary setting each message is a vector of length $q = 2^p$ representing a pmf for the non-binary symbol associated with V_i. For example, for an LDPC code constructed over the Galois field GF(4), the message m_j^i from check node C_j to variable node V_i is a vector with four elements having the form $m_j^i = \left(m_j^i(0), m_j^i(1), m_j^i(\alpha), m_j^i(\alpha^2)\right)$

where $m_j^i(0) = \Pr(V_i = 0)$, $m_j^i(1) = \Pr\{V_i = 1\}$, $m_j^i(\alpha) = \Pr\{V_i = \alpha\}$, and $m_j^i(\alpha^2) = \Pr\{V_i = \alpha^2\}$, each probability being conditioned to the extrinsic information received by the check node along all of its edges but the one towards V_i.

12.7.2.1 Initialization

In the initialization step, each variable node receives a priori information from the channel and simply broadcasts it along all of its edges, towards the check nodes that are connected to it. Hereafter we denote by r^i a priori information for variable node V_i. As well as messages exchanged between variable nodes and check nodes, r_i is a pmf for the non-binary symbol $c_i \in \mathrm{GF}(2^p)$ associated with V_i. The way a priori information r_i is computed depends on the channel.

For example, let us consider transmission of a NB-LDPC code constructed on $\mathrm{GF}(2^p)$ over the Bi-AWGN channel depicted in Fig. 12.16. Let $c = (c_0, c_1, \ldots, c_{n-1})$ be the NB-LDPC codeword, $c_i \in \mathrm{GF}(2^p)$ for $i \in \{0, 1, \ldots, n-1\}$. In this case the generic non-binary codeword symbol c_i is first converted to its binary representation $c_i = (c_{i,0}, c_{i,1}, \ldots, c_{i,p-1})$, $c_{t,j} \in \mathrm{GF}(2)$ for $j \in \{0, 1, \ldots, p-1\}$. Then, the binary representation is mapped onto a word of p antipodal symbols $x_i = 1 - 2c_i$ (meaning E_s normalized to 1), yielding a sequence $x = (x_0, x_1, \ldots, x_{n-1})$ of np channel symbols that are transmitted sequentially over the channel. Letting $y = (y_0, y_1, \ldots, y_{n-1})$ be the corresponding Bi-AWGN channel output, it is easy to verify that, for all $i \in \{0, 1, \ldots, n-1\}$ and for each $\beta \in \mathrm{GF}(2^p)$, we have

$$\Pr(c_i = \beta | y_i) \, \alpha \, (2\pi\sigma^2)^{-\frac{p}{2}} \exp\left(- \frac{||\mathbf{x_i}(\beta)||^2 + ||\mathbf{y_i}||^2}{2\sigma^2} \right) \exp\left(\frac{<\mathbf{x_i}(\beta), \mathbf{y_i}>}{\sigma^2} \right) \tag{12.33}$$

where $x_i(\beta)$ is the antipodal version of the binary representation of β, where σ^2 is the variance of each noise sample, and where $\langle x_i(\beta), y_i \rangle$ is the inner product between $x_i(\beta)$ and y_i. In this example, a priori information for V_i (coinciding with the message V_i sends to all of its neighboring check nodes during the initialization step) is therefore $r_i = (\Pr\{c_i = 0\}, \Pr\{c_i = 1\}, \Pr\{c_i = \alpha\}, \ldots, \Pr\{c_i = \alpha^{q-2}\})$ where each element of the pmf r^i is computed according to (12.33). Over an SLC or MLC channel model, a priori information shall be appropriately computed, usually based again on the binary representation of each non-binary codeword symbol.

All subsequent steps of the BP decoder for NB-LDPC codes, described next, remain the same regardless of the specific channel model and therefore irrespective of how a priori information is computed.

12.7.2.2 Message Permutation

As previously described, each edge in the Tanner graph of a NB-LDPC code is labeled by the corresponding non-zero element of the parity-check matrix. Considering check node C_j and letting $c_i \in \mathrm{GF}(2^p)$ be the jth codeword symbol, the check node imposes the constraint

$$\sum_{k \in N(j)} h_{j,k} c_k = 0. \tag{12.34}$$

where $h_{j,k} \in \mathrm{GF}(2^p) \backslash \{0\}$ is the element of $\boldsymbol{H}$ in position (j,k). This means that the value of each variable node V_i, connected to C_j, is first multiplied by the corresponding edge label and then is checked by the check node through (12.34). In terms of BP decoding, where message r_i^j is a pmf for symbol $c_i \in \mathrm{GF}(2^p)$, multiplication of by the non-zero edge label simply entails a permutation of the elements of r_i^j. To make a distinction between the message sent by V_i and the message received by C_j (after the permutation), hereafter we denote the former by r_i^j and the latter by $\Pi(r_i^j)$. An example is provided next.

Example 12.6 Let the Galois field order be $q = 4$. Let the edge connecting variable node V_i and check node C_j be labeled by $\alpha \in \mathrm{GF}(4)$, and the message sent by V_i be $r_i^j = (0.4, 0.3, 0.2, 0.1)$. Since in GF(4) we have $0 \cdot \alpha = 0$, $1 \cdot \alpha = \alpha$, $\alpha \cdot \alpha = \alpha^2$, and $\alpha^2 \cdot \alpha = 1$, the effect of the edge label α on the message is a permutation of its elements, leading to the message $\Pi(r_i^j) = (0.4, 0.1, 0.3, 0.2)$ received by C_j. Each non-zero edge label induces a specific permutation.

12.7.2.3 Horizontal Step

Check node C_j, $j \in \{0, 1, \ldots, m-1\}$, receives one message $\Pi(r_i^j)$, $i \in N(j)$, from each of its neighboring variable nodes and sends back one message m_j^i, $i \in N(j)$, to each of them. To understand how extrinsic information shall be generated at the check node and forwarded to the relevant variable node, we can look at (12.34) that we recast in the form $\sum_{k \in N(j)} z_k = 0$ by defining $z_k = h_{j,k} c_k$. For some $i \in N(j)$, the constraint imposed by the check node is $z_i = -\sum_{k \in N(j) \backslash \{i\}} z_k = \sum_{k \in N(j) \backslash \{i\}} z_k$ where the "$-$" sign can be omitted owing to the fact that $q = 2^p$. We may regard each summand z_k as a random variable taking values in $\mathrm{GF}(q)$ and with pmf equal to that of the incoming message $\Pi(r_k^j)$. Under the assumption that all z_k are independent, the pmf of their sum (hence the pmf of z_i) is the convolution of their pmfs. That is, we may write

$$m_j^i = \circledast_{k \in N(j) \backslash \{i\}} \Pi\left(r_k^j\right) \tag{12.35}$$

where $\circledast$ denotes convolution between pmfs.

Since the complexity of convolution scales quadratically with the vector size, a naïve implementation of the horizontal step based on (12.35) leads to a complexity scaling as $\mathcal{O}(q^2)$, such a complexity dominating the overall decoding complexity and becoming problematic even for moderate q. A reduced-complexity but equivalent implementation of the horizontal step is based on applying fast Hadamard transform to both sides of (12.35). Hadamard transform turns vector convolution into element-wise multiplication of the transformed vectors; therefore, letting $\mathcal{H}$ denote the Hadamard transform and recalling that Hadamard transform coincides with its inverse, m_j^i in (12.35) may equivalently be calculated as

$$m_j^i = \mathcal{H}\left(\otimes_{k \in N(j) \backslash \{i\}} \mathcal{H}\left(\Pi\left(r_k^j\right)\right)\right) \tag{12.36}$$

where $\otimes$ denotes element-wise product between two vectors. Using fast Hadamard transform reduces the horizontal step complexity (and more in general the complexity of the whole decoder) to $\mathcal{O}(q \log q)$.

12.7.2.4 Message De-permutation

In the previously described message permutation step, the elements of message r_i^j, sent by variable node V_i to check node C_j, are permuted according to the permutation established by the edge label $h_{j,i}$. The message m_j^i, sent by C_j towards V_i, must undergo the inverse permutation (equivalently, the permutation established by the inverse label $h_{j,k}^{-1}$) before reaching V_i. For the sake of clarity, in order to distinguish the message sent by C_j from the message received by V_i after de-permutation, we keep denoting by m_j^i the former and by $\Pi^{-1}\left(m_j^i\right)$ the latter.

12.7.2.5 Vertical Step

Variable node V_i, $i \in \{0, 1, \ldots, n-1\}$, receives the d_i messages $\Pi^{-1}\left(m_j^i\right)$, $j \in N(i)$, and generates d_i messages r_i^j, $j \in N(i)$, each of which is sent towards a specific edge to the corresponding check node. Each message r_i^j is computed based on a priori information r_i available from the channel and on extrinsic information $\Pi^{-1}\left(m_k^i\right)$, $k \in N(i) \backslash \{j\}$. Specifically, assuming independence between all of the incoming messages (including a priori information), r_i^j is computed as

$$r_i^j = \gamma_j \cdot r_i \otimes \left(\otimes_{k \in N(i) \setminus \{j\}} \Pi^{-1} \left(m_k^i \right) \right) \tag{12.37}$$

where again $\otimes$ denotes element-wise product between two vectors and where the scalar γ_j is a scaling factor whose value makes the sum of the elements of r_i^j equal to 1.

12.7.2.6 Hard Decision and Stopping Criterion

At the end of each BP decoding iteration, a hard decision is made about the value taken by each variable node; this hard decision exploits a posteriori information for the variable node. In probability-domain BP decoding of NB-LDPC codes, a posteriori information is represented by the pmf of the Galois field symbol c_i associated with the variable node given all incoming messages and a priori information. Under independence assumption this is given by

$$r_i^{\mathrm{APP}} = \gamma \cdot r_i \otimes \left[\otimes_{j \in N(i)} \left(\Pi^{-1} \left(m_j^i \right) \right) \right] \tag{12.38}$$

where again γ is a normalization factor. Let $\mathbf{\Pi}^{-1}(m^i)$ be the ordered list of messages received by variable node V_i. Once the a posteriori pmf r_i^{APP} for symbol c_i has been computed, a symbol-wise MAP decision is made, namely,

$$\hat{c}_i = \mathrm{argmax}_{c \in \mathrm{GF}(q)} \Pr\left\{ c | r^i, \mathbf{\Pi}^{-1}(m^i) \right\}. \tag{12.39}$$

In other words, $\hat{c}_i$ is the element of GF(q) that corresponds to the largest element of the pmf r_i^{APP}. Note that, as the scaling factor γ is the same for all elements of r_i^{APP}, it does not affect the final decision and therefore it can be set to 1 for all $i \in \{0, 1, \ldots, n-1\}$.

Similarly to the binary case, if the current hard-decision sequence $\hat{c} = (\hat{c}_0, \hat{c}_1, \ldots, \hat{c}_{n-1})$ fulfills $\hat{c}H^T = \mathbf{0}$, then the algorithm terminates and $\hat{c}$ is returned as the detected codeword. Else, if $\hat{c}$ is not a codeword and the maximum number of iterations I_{max} has been reached, the algorithm is terminated and a failure is reported. Else, a new iteration is started jumping to the message permutation step.

Belief propagation decoding of LDPC codes over GF(q), $q = 2^p$, may be summarized as follows.

Belief-Propagation Decoding of NB-LDPC Codes over GF(2^p)

```
1: set I = 1. For i = 0,...,n − 1, for j ∈ N(i), set r_i^j = r_i;
2: for i = 0,...,n − 1
      for j ∈ N(i) permute the elements of r_i^j based on h_{j,i};
3: for j = 0,...,m − 1
      for i ∈ N(j) calculate m_j^i according to (11.36);
4: for i = 0,...,n − 1
      for j ∈ N(i) permute the elements of m_j^i based on h_{j,i}^{-1};
5: for i = 0,...,n − 1
      for j ∈ N(i) calculate r_i^j according to (11.37);
6: for i = 0,...,n − 1 {
      calculate r_i^APP according to (11.38);
      set the value of ĉ_i according to (11.39);
   }
   if ĉH^T = 0 then return ĉ;
   else {
      if I = I_max exit;
      else {
         I = I + 1;
         goto 2;
         }
   }
```

12.8 Numerical Example

In this section, we present some numerical results aimed at comparing the performance of binary LDPC and BCH codes, with the purpose to highlight the potential of LDPC codes in Flash memories applications. We assume an SLC memory as the reference channel model. We compare the performance of a regular QC-LDPC code, under several decoding algorithms offering different tradeoffs between performance and complexity, with the performance of a narrowsense binary BCH code with similar parameters, decoded via bounded distance decoding.

The LDPC code is characterized by a length $n_{\mathrm{LDPC}} = 8200$ and a dimension $k_{\mathrm{LDPC}} = 7379$ bits, and therefore by a code rate R very close to 9/10. Its minimum distance, estimated with the impulse method proposed in [38], is equal to $d_{\mathrm{LDPC}} = 114$. All variable nodes of the LDPC code have degree 4, and all of its check nodes have degree 40. Its 820×8200 parity-check matrix is in block circulant form, where the generic block is a 205×205 circulant permutation matrix, and has been constructed according to a block circulant version of the progressive edge-growth (PEG) algorithm. The performance of this code has been evaluated via Monte Carlo software simulation, under BP, MS, and BF decoding algorithms. The performance curves under both BP and MS decoding have been obtained under two different settings, namely, soft-decision and hard-decision decoding. These two settings correspond to assuming the Bi-AWGN channel with unquantized output (Example 12.2) and with one-bit quantized output (Example 12.3), respectively, as the channel model. The first setting is equivalent to assuming an SLC memory with an infinite number of reads per bit, while the second one to assuming an SLC memory with one read per bit. The variable nodes are initialized according to (12.19) in the unquantized case and according to (12.18) in the quantized one. In the quantized case, the raw bit error rate of the channel can be obtained from E_b/N_0 according to (12.8), where $E_s/N_0 = RE_b/N_0$. For instance, $E_b/N_0 = 5\,\text{dB}$ corresponds to a raw bit error rate $p = 8.5 \cdot 10^{-3}$. The Shannon limit for the unquantized case and for the one-bit quantized case are also evaluated, for benchmarking purposes.

The competitor BCH code has nominal parameters $n_{\text{BCH}} = 8191$, $k_{\text{BCH}} = 7372$, $t = 63$ (error correction capability), and minimum distance $d_{\text{BCH}} = 127$. Its code rate is approximately equal to $9/10$, similar to the code rate of the QC-LDPC code. The codeword error rate (CER) and the bit error rate (BER) of the BCH code under hard decision bounded distance decoding have been evaluated analytically according to the relationships

$$P_e = \sum_{r=t+1}^{n_{\text{BCH}}} \binom{n_{\text{BCH}}}{r} p^r (1-p)^{n_{\text{BCH}}-r} \tag{12.40}$$

and

$$P_b \approx \frac{d_{\text{BCH}}}{k} \cdot P_e \tag{12.41}$$

respectively.

With reference to Fig. 12.17, we see that over the hard-decision channel (SLC with one read) the BCH code exhibits nearly the same performance as the QC-LDPC code decoded via BP and that its performance is even slightly better at low error rates. This is not surprising, as BCH codes are well known to offer very good performances over hard-decision channels, especially at high code rates. As opposed to BCH codes, however, LDPC codes can handle in a very natural way soft information incoming from the communication channel, which allows to attain substantial performance improvements over the error correction capabilities achievable with hard-decision decoding. In our example, when the LDPC decoder is fed with unquantized soft information, its coding gain with respect to that achieved under hard-decision decoding is improved by about 1.6 dB under both BP

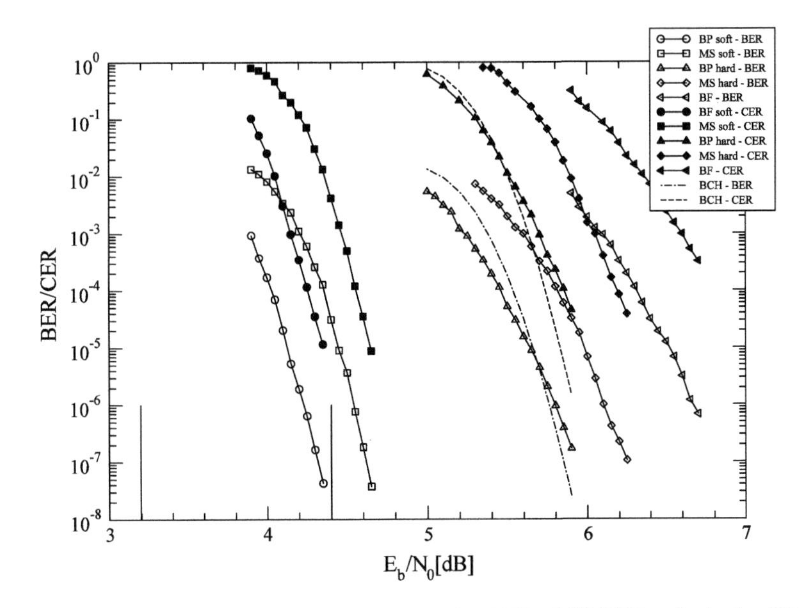

Fig. 12.17 Bit and codeword error rates for an (8191, 7372) QC-LDPC code (under different decoding algorithms) and an (8191, 7372), t = 63 narrowsense binary BCH code under bounded distance decoding, over an SLC flash memory channel. Curves corresponding to filled and empty symbols illustrate the codeword error rates and the bit error rates of the LDPC code, respectively. The dashed and dot-dashed lines illustrate the codeword error rate and the bit error rate of the BCH code, respectively. The two straight solid lines are the Shannon limits for rate R = 9/10 under soft-decision and hard-decision decoding, respectively

and MS decoding algorithms at $\text{CER} = 10^{-4}$. Moreover, again at $\text{CER} = 10^{-4}$, the LDPC code under unquantized BP decoding performs only 0.8 dB away from the corresponding Shannon limit, in terms of BER.

For the same decoding algorithm (BP or MS), the performance curves of the LDPC code labeled as "soft" and "hard" represent the two extreme cases in which unconstrained soft information is available at the decoder, and no soft information is available. In general, when a finite number of cell reads is performed with different read voltage values, the corresponding performance curve will lie between the two extreme curves: The larger the number of cell reads, the closer the performance curve to the "soft" one. Therefore, LDPC codes can largely outperform BCH codes in Flash memory applications, provided a sufficient amount of soft information is available at the decoder. It is also pointed out that the design of appropriate QC irregular LDPC codes can favor an even larger coding gain with respect to BCH codes.

We also highlight how very simple decoding algorithms of LDPC codes such as BF (or Gallager B) decoding, can be of interest at the beginning of the memory life, i.e., when the raw bit error rate is very small. For example, as from Fig. 12.17, BF decoding could become of interest for values of E_b/N_0 larger of 7.0 dB, corresponding to a raw bit error rate smaller than $1.3 \cdot 10^{-3}$.

Acknowledgements The author wishes to thank R. Micheloni and A. Marelli for their careful proofcheck of this chapter.

References

1. T.M. Cover, J.A. Thomas, *Elements of Information Theory* (Wiley, 1991)
2. R.D. Fowler, L. Nordheim, Electron emission in intense electric fields. Proc. R. S. Lond. **119**, 173–181 (1928)
3. R. Micheloni, L. Crippa, A. Marelli (eds.), *Inside NAND Flash Memories* (Springer, 2010)
4. N. Mielke et al., Bit error rate in NAND Flash memories, in *Proceedings of the 2008 IEEE International Symposium on Reliability Physics*, Phoenix, AZ, USA, April/May 2008, pp. 9–19
5. J. Wang, T. Courtade, H. Shankar, R. Wesel, Soft information for LDPC decoding in flash: mutual-information optimized quantization, in *Proceedings of the 2011 IEEE Global Telecommunication Conference*, Houston, TX, USA, Dec 2011
6. S. Li, T. Zhang, Improving multi-level NAND flash memory storage reliability using concatenated BCH-TCM coding. IEEE Trans. VLSI **18**, 1412–1420 (2010)
7. R.G. Gallager, *Low-Density Parity-Check Codes* (MIT Press, Cambridge, Massachusetts, 1963)
8. C. Berrou, A. Glavieux, P. Thitimajshima, Near Shannon limit error-correcting coding and decoding: turbo-codes, in *Proceedings of the 2003 International Symposium on Communication*, vol. 2, May 1993, pp. 1064–1070
9. T. Richardson, R. Urbanke, The renaissance of Gallager's low-density parity-check codes. IEEE Commun. Mag. **41**, 126–131 (2003)
10. N. Bonello, S. Chen, L. Hanzo, Low-density parity-check codes and their rateless relatives. IEEE Commun. Surv. Tutor. **13**, 3–26 (2011)
11. M. Tanner, A recursive approach to low complexity codes. IEEE Trans. Inf. Theory **27**, 533–547 (1981)
12. M. Fossorier, Quasi-cyclic low-density parity-check codes from circulant permutation matrices. IEEE Trans. Inf. Theory **50**, 1788–1793 (2004)
13. Z. Li, L. Chen, L. Zeng, S. Lin, W. Fong, Efficient encoding of low-density parity-check codes. IEEE Trans. Commun. **54**, 71–81 (2006)
14. M. Mansour, High-performance decoders for regular and irregular repeat-accumulate codes, in *Proceedings of the IEEE 2004 IEEE Global Telecommunications Conference*, Nov/Dec 2004, pp. 2583–2588
15. T. Richardson, M. Shokrollahi, R. Urbanke, Design of capacity-approaching irregular low-density parity-check codes. IEEE Trans. Inf. Theory **47**, 619–637 (2001)
16. S.-Y. Chung, G.D. Forney Jr., T. Richardson, R. Urbanke, On the design of low-density parity-check codes within 0.0045 dB of the Shannon limit. IEEE Commun. Lett. **5**, 58–60 (2001)
17. J. Thorpe, Low-density parity-check (LDPC) codes constructed from protographs, JPL INP, Technical Report, Aug 2003, pp. 42–154
18. J. Xu, L. Chen, L. Zeng, L. Lan, S. Lin, Construction of low-density parity-check codes by superposition. IEEE Trans. Commun. **53**, 243–251 (2005)
19. T. Richardson, R. Urbanke, The capacity of low-density parity-check codes under message-passing decoding. IEEE Trans. Inf. Theory **47**, 599–618 (2001)
20. S. ten Brink, Convergence behavior of iteratively decoded parallel concatenated codes. IEEE Trans. Commun. **49**, 1727–1737 (2001)
21. G. Liva, M. Chiani, Protograph LDPC codes design based on EXIT analysis, in *Proceedings of the 2007 IEEE Global Telecommunications Conference*, Washington, DC, USA, Nov 2007, pp. 3250–3254

22. L. Chen, J. Xu, I. Djurdjevic, S. Lin, Near Shannon limit quasi cyclic low-density parity-check codes. IEEE Trans. Commun. **52**, 1038–1042 (2004)
23. H. Tang, J. Xu, Y. Kou, S. Lin, K. Abdel-Ghaffar, On algebraic construction of Gallager and circulant low density parity-check codes. IEEE Trans. Inf. Theory **50**, 1269–1279 (2004)
24. M. Chiani, A. Ventura, Design and performance evaluation of some high-rate irregular low-density parity-check codes, in *Proceedings of the 2001 Global Telecommunication Conference*, San Antonio, TX, USA, Nov 2001, pp. 990–994
25. T. Richardson, Error floors of LDPC codes, in *Proceedings of the 41st Annual Allerton Conference on Communication, Control and Computing* (2003)
26. S. Abu-Surra, D. Divsalar, W.E. Ryan, Enumerators for protograph-based ensembles of LDPC and generalized LDPC codes. IEEE Trans. Inf. Theory **57**, 858–886 (2011)
27. M. Flanagan, E. Paolini, M. Chiani, M. Fossorier, On the growth rate of the weight distribution of irregular doubly-generalized LDPC codes. IEEE Trans. Inf. Theory **57**, 3721–3737 (2011)
28. D. Cavus, C. Haymes, Low BER performance estimation of LDPC codes via application of importance sampling to trapping sets. IEEE Trans. Commun. **57**, 1886–1888 (2009)
29. L. Dolecek et al., Predicting error floors of structured LDPC codes: Deterministic bounds and estimates. IEEE J. Sel. Areas Commun. **27**, 908–917 (2009)
30. J. Chen, A. Dholakia, E. Eleftheriou, M. Fossorier, X.-Y. Hu, Reduced-complexity decoding of LDPC codes. IEEE Trans. Commun. **53**, 1288–1299 (2005)
31. J. Zhao, F. Zarkeshvari, A. Banihashemi, On implementation of min-sum algorithm and its modifications for decoding low-density parity-check (LDPC) codes. IEEE Trans. Commun. **53**, 549–554 (2005)
32. J. Chen, M. Tanner, C. Jones, Y. Li, Improved min-sum decoding algorithms for irregular LDPC codes, in *Proceedings of the 2005 IEEE International Symposium on Information Theory*, Sept 2005, pp. 449–453
33. M. Davey, D. MacKay, Low-density parity check codes over GF(q). IEEE Commun. Lett. **2** (6), 165–167 (1998)
34. C. Poulliat, M. Fossorier, D. Declercq, Design of regular (2, d_c) -LDPC codes over GF(q) using their binary images. IEEE Trans. Commun. **56**(10), 1626–1635 (2008)
35. G. Liva, E. Paolini, B. Matuz, S. Scalise, M. Chiani, Short turbo codes over high order fields. IEEE Trans. Commun. **61**(6), 2201–2211 (2013)
36. L. Dolecek, D. Divsalar, Y. Sun, B. Amiri, Non-binary protograph-based LDPC codes: enumerators, analysis, and designs. IEEE Trans. Inf. Theory **60**(7), 3913–3941 (2014)
37. E. Paolini, M. Flanagan, Efficient and exact evaluation of the weight spectral shape and typical minimum distance of protograph LDPC Codes. IEEE Commun. Lett. **20**(11), 2141–2144 (2016)
38. X.-Y. Hu, M. Fossorier, E. Eleftheriou, On the computation of the minimum distance of low-density parity-check codes, in *Proceedings of the 2004 International Conference on Communication*, June 2004, pp. 767–771

Chapter 13
Protecting SSD Data Against Attacks

Alessia Marelli and Rino Micheloni

Abstract When a drive is broken and we have to throw it away, we want to be sure that no hackers can recover the data stored in that disk, especially in the enterprise environment where sensitive date are stored on the drive, such as financial transactions or military applications. As the SSD market is growing, the security issue must be carefully considered. Some methods used with HDDs, such as degaussian, are not applicable to SSDs, due to the different storage technique. Recent studies indicate that encryption is the necessary step to protect data stored in SSD against hackers attacks. This chapter describes the SSD security approach in comparison to HDD, then it walks the reader through the encryption world: how a cryptosystem is built, how a cryptosystem is broken, different encryption applications, and then the AES cryptosystem as it is the most used in SSDs; finally, it addresses the security applications in SSDs.

13.1 Challenges of SSD Security Versus HDD

Hard Disk Drives as well as Solid State Disks contain a number of sensible data that must be kept secret. When a disk is thrown away or stolen, it is very important that nobody can access these data.

The purpose of HDD is to store data and protect them from corruption or accidental erase. In this latter case, procedures like folder or un-erase are used. In addition, data erasure is unlikely to occur because it takes a lot of time, hence reducing performances. The drawback is that user data are vulnerable to recovery by unauthorized person. Increased storage of sensitive data, combined with rapid technological change and the shorter lifespan of IT assets, has driven the need for permanent data erasure of electronic devices as they are retired.

A. Marelli (✉) · R. Micheloni
Storage Solutions, Microsemi Corporation, Vimercate, MB, Italy
e-mail: alessiamarelli@gmail.com

R. Micheloni
e-mail: rino.micheloni@ieee.org

R. Micheloni et al. (eds.), *Inside Solid State Drives (SSDs)*,
Springer Series in Advanced Microelectronics 37,
https://doi.org/10.1007/978-981-13-0599-3_13

If data erasure does not occur when a disk is retired or lost, an organization or a user faces the possibility that data will be stolen and compromised, leading to identity theft, loss of corporate reputation, threats to regulatory compliance and financial impacts. There are well-known cases of sensible data loss such as CardSystems Solutions where Credit card breach exposed 40 million accounts in 2005. In addition, government laws oblige disk makers to have a method to secure data. Nowadays, there are four methods to secure data:

- physical drive destruction;
- degaussian;
- secure erase;
- encryption.

In the following we will see what these methods are and how they are applied to SSD and HDD.

To prevent data from recovery, disks can be broken up to microscopic pieces. However, such physical destruction is not absolute if any remaining disk pieces are larger than a single 512-byte block. In case of HDD this is not easy and a magnetic microscopy is able to recover the data. In case of SSD it is easier to destroy the physical component but this method is old and not used.

Degaussian uses magnetic field to erase data stored on HDD. Degaussers create high intensity magnetic fields that erase all the magnetic recordings in a hard disk drive, including the sector header information on drive data tracks. Like physical destruction, once this procedure is applied, the disk is no longer usable. However, as the storage density increases, higher magnetic fields are required, so that old degaussers cannot be reused in modern HDD. In addition, new perpendicular recording drives my not be erasable by present degaussers designed for past longitudinal recording drives. Due to the different physical media, degaussian procedure is not applicable to SSD. However, there are companies [1] that build a self-destructive SSD by applying an over-current to the NAND Flash memories.

As regards erase, four security levels are defined: weak erase (deleting files), block erase (overwrite by external software), secure erase and fast secure erase. There is a big difference in terms of security achieved and time required by these four levels as depicted in Fig. 13.1.

Sanitation of HDD through erase is not easy, because when we delete a file, we just remove its name from the directory structure. The user data remain on the drive where they can be retrieved until the sectors are overwritten by new data. Even reformatting the drive only file directories and links among sectors are cleared, but the user data remain and can be recovered. Moreover, software utilities that overwrite files are susceptible to error or malicious virus attack and require constant update.

Secure Erase (SE) is the name given to a set of commands available in PATA and SATA hard drives. The Secure Erase commands are used as data sanitization method to completely overwrite all the data on a hard drive. The method is very simple: it writes a binary one or zero in all the locations.

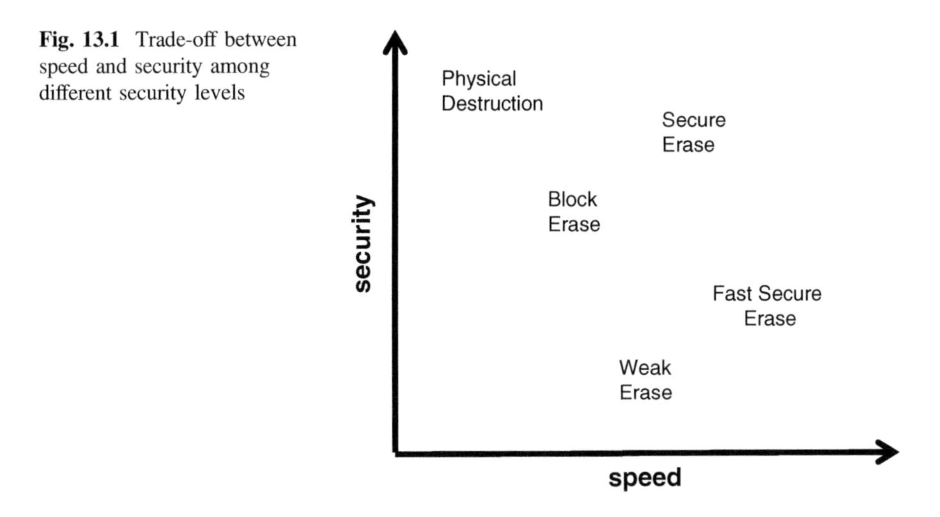

Fig. 13.1 Trade-off between speed and security among different security levels

After SE file recovery programs will not be able to extract data from the drive. Secure Erase is a simple addition to the existing "format drive" command and adds no cost to hard disk drives. Usually, HDDs ask for multiple SE operations; in the SSD case a single erase should be enough because data are erased in blocks. The bad news is that the operating system is not aware of where data are physically stored; only the Flash controller inside the SSD knows the logical-to-physical mapping (Chap. 2). Recently it has been published [2] a study on limitations about secure erase applied to SSDs.

- First, ATA and SCSI built-in commands are effective, but manufacturers sometimes implement them incorrectly. Moreover, sometimes they are not implemented in SSDs.
- Second, overwriting the entire visible address space of an SSD twice is usually, but not always, sufficient to sanitize the drive. In addition, due to the Firmware Transaction Layer (FTL) (Chap. 2) the procedure is more complex and time consuming compared to HDD.
- Third, none of the existing hard drive techniques for individual file sanitization are effective on SSDs.

Even if there is a lot of effort on developing a stronger secure erase for SSDs, nowadays encryption is the preferred method. Encryption should be used on the drive since beginning of life: when we want to destroy data, it is enough to delete all the keys in order to be sure that all the data are un-recoverable.

The next section walks the reader through the encryption world before discussing encryption applied to SSDs.

13.2 Introduction to Cryptography

The fascinating art of cryptography was born as soon as the civilized man began to communicate information to another man. In fact, quite at the beginning he felt the need of secrecy or privacy, so that if Alessia wants to send a message to Rino, she doesn't want Kam, who heard the message, to understand its meaning.

There are evidence of cryptographic schemes in the ancient Jew population and their atbash schemes, the Spartans with their scytale (Fig. 13.2) but the first "published" encryption scheme is the Caesar ciphrary invented by emperor Caius Julius Caesar. From then a number of different schemes were used during the ages, till the popular Enigma during the Second World War, used by the Germans to send encrypted messages to U-boots (Fig. 13.3).

Together with encryption methods, more and more efforts were put on the opposite side of the story: the codebreakers that invented the science of crypt-analysis. The most famous were the scientists of Bletchey park (Alan Touring was one of them) that were able to decrypt the messages sent with Enigma. This was a key point in the defeat of Germany in the Second World War.

Modern encryption science was born in 1949 with Shannon [3], the father of Information theory with the paper "Communication Theory of Secrecy Systems". After that, the encryption science was pushed by military industry and then applied to telephone lines, computer networks, financial transactions and so on.

More and more complex schemes were discovered and then analyzed to find their weakness. The next sub-sections introduce the basic concept of a crypto-graphic system and how it is possible to find if it is secure or not. Last sub-section describes encryption applied to MAC (Message Authentication Code) and digital signatures.

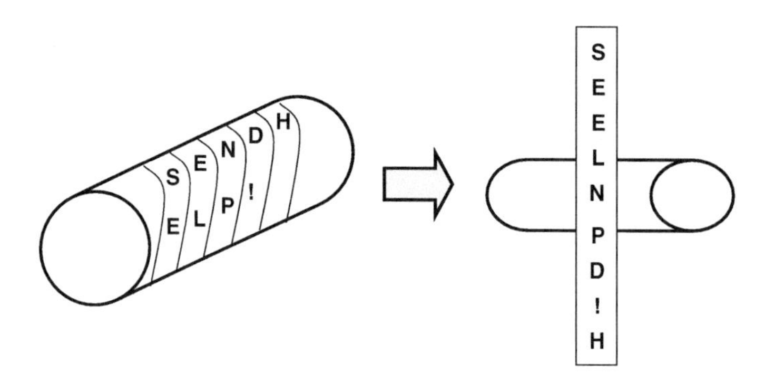

Fig. 13.2 The scytale used by spartans to encrypt codes: it was a wooden stick used to roll the message to be encrypted

Fig. 13.3 The enigma machine

13.2.1 Basic Concepts

As the name cryptography suggests (from the greek kryptos = hidden and graphia = written language) the purpose of this science is to hide an information under an apparent random message.

Let's say Alessia wants to send a message to Rino and be sure that listener Kam doesn't understand the message (Fig. 13.4). The message Alessia wants to send is called *plaintext*. She applies an *encryption* function, that generally involves a *key*, to the plaintext in order to get a *ciphertext* to be sent to Rino. On the other side, Rino receives the ciphertext and applies his *decryption* function, that generally involves another key, in order to recover the original plaintext. If Kam hears the ciphertext, he is unable to recover the plaintext because he hasn't the key.

A basic example of an encryption scheme is based on letter substitution. Figure 13.5 shows the Caesar code (Fig. 13.5). The key is the width of the rotation, 3 in this example.

Alessia wants to send the plaintext "Caesar" to Rino. She uses her key (rotation of 3 positions) to obtain the cyphertext "Zxbpxo". Rino receives the message and rotates back of 3 positions to read the original message.

This is a very simple example where Rino and Alessia have the same key. Of course, a number of modifications have been introduced in order to have a different

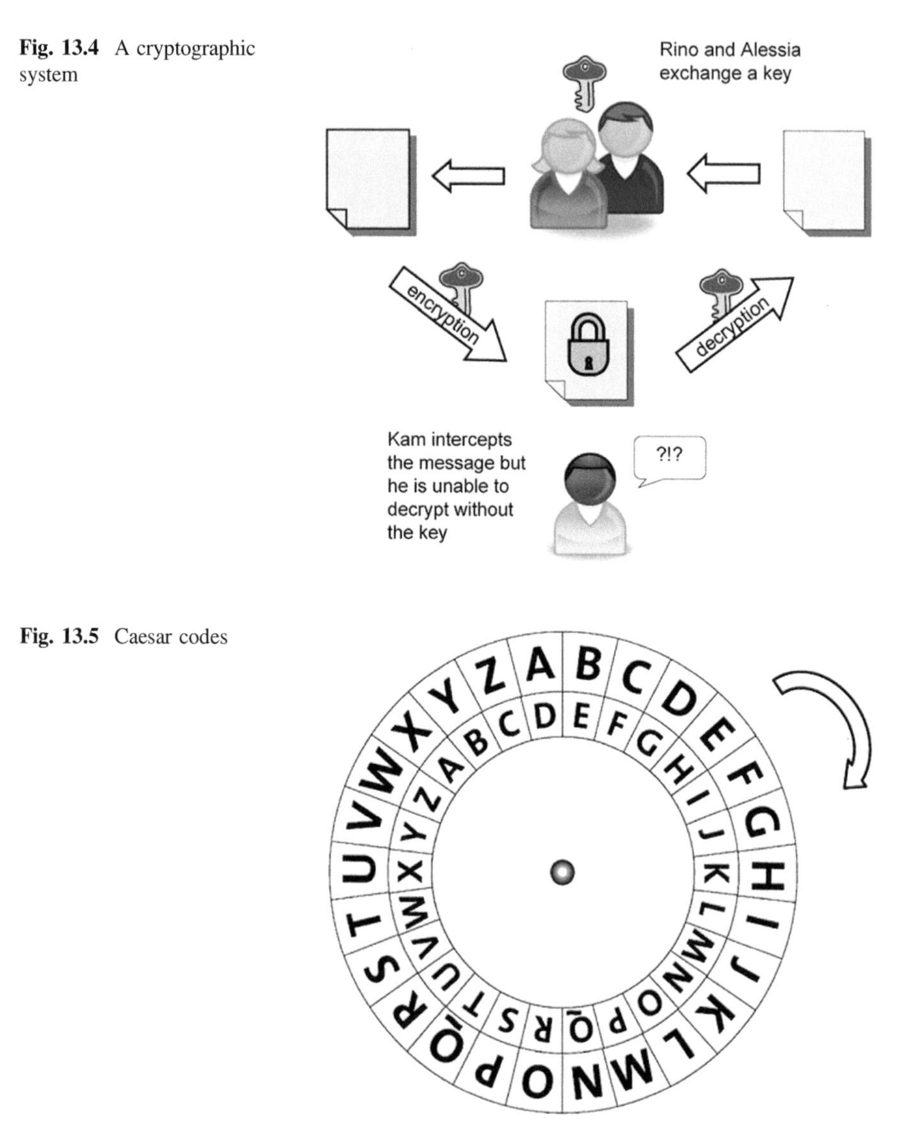

Fig. 13.4 A cryptographic system

Fig. 13.5 Caesar codes

number of rotations for each letter of the message (Vigenère codes) or different keys for Alessia and Rino, or different encryption methods.

However, it is necessary to have a "metric" to evaluate the security of a cryptosystem.

Shannon was able to give a mathematical structure to the encryption science, first of all by evaluating the secrecy of a system. In fact there are different levels of security of a cryptosystem as shown in Fig. 13.6.

The low level of security is the *computational security*. This is a measure of the computational effort required to break a cryptosystem; in other words a system is

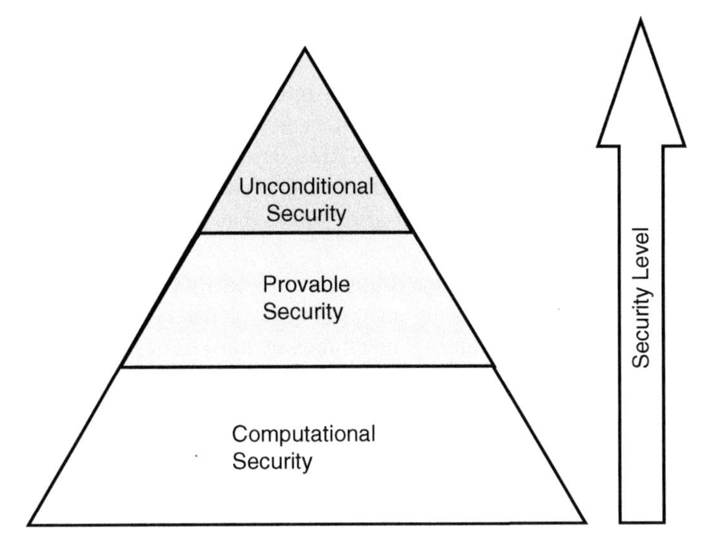

Fig. 13.6 The pyramid of security

considered computational secure if it requires at least N operations. However, given the speed of the technology evolution, what is secure today will unlikely be secure tomorrow. Moreover, there aren't any practical secure cryptosystems based on this definition. The problem is that people study the computational security of a system under a specific attack, but this does not guarantee its security under another attack.

The second level of security is the *provable security*. A cryptosystem is said to be provable secure if its construction is based on a very difficult mathematical problem, not yet theoretically solved. For example, as it will be discussed later, RSA system is based on integer factorization. Until now, there aren't any methods that can easily factorize an integer. If some day a method will be found, RSA will be easily broken, but until then it is provable secure.

The highest level of security is the *unconditional security*. In this case there are no bounds on the computational effort that Kam can use: the cryptosystem can't be broken even with infinite computational resources.

We won't go through all the mathematical description of this analysis, but we report here only an interesting result: the Vigenère cipher is unconditional secure if the keyword has the same length of the plaintext. It is even more secure if the key is used only once.

Nowadays, there is only one cryptosystem known as unconditional secure: the One-Time Pad. Historically, this encryption method was used by KGB agents. The system was so secure that some messages have been decrypted only when agents re-used the same key more than once or some spies have been arrested and revealed the keys. We explain this method with an example.

Alessia wants to send the message "hello" to Rino. They have the same pads of keys to be used only one time and they decided for "xmckl" (same length of the message). The encryption method follows Fig. 13.7. Based on the alphabet, letters are translated in numbers, and the sum of message and key gives the ciphertext.

The sum is performed mod(26). Alessia immediately destroys the key. Rino receives the message “eqnvz”, translates it in numbers, and subtracts the key to obtain the original message. At this point Rino destroys the key.

If Kam hears the cipthertext and tries to decrypt it with infinite computing power, he fill find “xmckl” as key but also “tquri” that gives the word “later” with same probability.

This is a very simple and fast encryption method, easily performed by xoring the key with either the plaintext (during encryption) or ciphertext (during decryption).

Difficulties arise in the key management: the key must be as long as the message, it must be random, it must be used only once and destroyed immediately after use. In addition, it is very difficult to distribute keys among multiple users. Especially the requirement on the key length is so difficult to achieve that different encryption methods are preferred, such as AES (Sect. 13.3) even if not unconditional secure.

This discussion leads us to the problem of the key. In fact, till few years ago all those methods were based on a symmetric encryption [4, 5]. In other words it is very easy to understand the key that Rino has, given Alessia’s key. In particular most of the time the key is the same. It follows that this key must be secret otherwise all the messages will be decrypted by Kam.

This leads us to some kind of paradox: we want to send secret messages but we must exchange a secure key over a secure channel. This is what happens in internet, when we are accessing a secure channel (e.g. home banking, credit card payment, etc.): we are exchanging a secure key to encrypt and decrypt messages. In financial transactions, however, we have the logistic problem of keys distribution. In other words, a bank must provide a different key to each user: handling of all these keys is translated in time and cost. It’s not the purpose of this chapter to address the problem of keys distribution. One way to solve it is the use of the Diffie-Hellman algorithm, i.e. an asymmetric encryption. The interested reader can refer to [5–8].

In order to overcome the problem of the key exchange, the *public-key cryptosystem* has been developed. The idea behind is that it might be unfeasible to find out Alessia’s key *d*, given Rino’s key *k*. It follows that Rino can publish his key and Alessia uses it to encrypt the message (Fig. 13.8). The sent message is received by Rino, that now uses his private key to decrypt the message.

```
               H          E          L          L          O  message
   7 (H)     4 (E)    11 (L)    11 (L)    14 (O)  message
+ 23 (X)    12 (M)     2 (C)    10 (K)    11 (L)  key
= 30        16        13        21        25      message + key
=  4 (E)    16 (Q)    13 (N)    21 (V)    25 (Z)  message + key
(mod 26)
       E         Q         N         V         Z  → ciphertext
```

Fig. 13.7 Example of one-time pad encryption

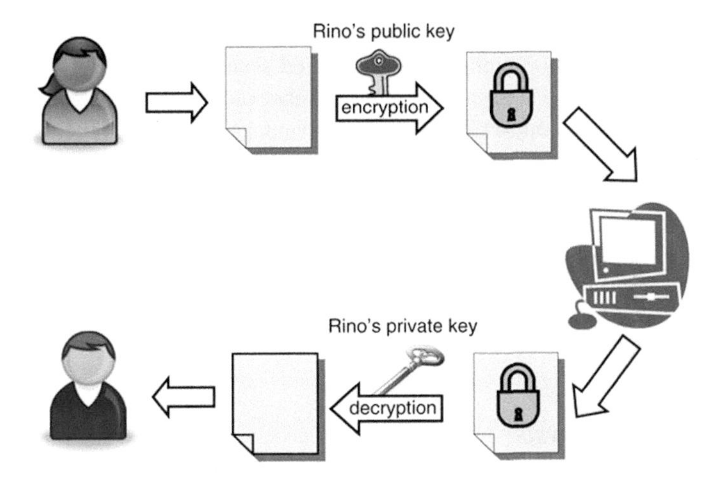

Fig. 13.8 The asymmetric cryptosystem

Observe that this method can also be reversed, that is Rino can use his private key to encrypt the message and sends it to Alessia. Alessia uses Rino's public key to decrypt the message. In this case everyone can decrypt the message, since Rino's key is public, but we are sure of the authenticity of the message, because it was encrypted using Rino's private key (Sect. 13.2.3).

The advantage of the public-key cryptosystems is that Alessia can send messages using the public key without any secret prior exchange of keys and be sure that only Rino is able to decrypt the message.

The public-key cryptosystem was first discussed by Rivest, Shamir and Adleman in 1978 with the very famous system called RSA [9]. Several systems have then be proposed, but their security remains computational. In fact, asymmetric encryption could never provide unconditional security. When Kam intercepts the ciphertext y, he can encrypt each possible plaintext using the public encryption rule until he finds the unique solution so that $y = e(x)$. This x is the decryption of y.

Public-key cryptosystem is based on one-way functions which are very easy to compute but very difficult to be inverted. There are a lot of functions that are believed to be one-way but never proven.

An example of such function is the factorization of an integer into two prime numbers, used in RSA. This cryptosystem can be summarized as follows:

- Rino picks up two large prime numbers p and q;
- Rino sends the number $n = p \times q$ to Alessia. Everyone can see it;
- Alessia uses n to encrypt the message;
- Alessia sends the cipthertext to Rino. Everyone can see it but nobody can decrypt it;
- Rino receives the message and, knowing p and q, is able to decrypt it.

The difficulty of this algorithm is the primality test of large integers. Today only numbers with al least 300 ciphers are considered secure [10, 11].

Asymmetric cryptosystems are used in a number of different protocols like SSH, Internet Key Exchange and PGP. The main advantage is that the generation of the key pair solves the logistic problem of key distribution and the problem of authentication (Sect. 13.2.3).

These systems are not broadly used because they are too slow and can limit performances in most of the cases, like in SSDs. A solution that sometimes is adopted is to transmit the keys with a public-key cryptosystems and then switch to a symmetric cryptosystem.

13.2.2 Cryptanalysis

Let's analyze the cryptosystem from Kam's side. Kam is not the bad guy of the story; of course, he could be a hacker that wants to intercept our credit card but he could also be a secret agent that needs to intercept a terroristic attack. This is the reason why the government puts a lot of effort and money in finding a good code but also in breaking codes.

Cryptoanalysis science, as the name suggests (from Greek cryptos = "hidden" and analyein = "to untie") has the purpose to break codes.

Generally speaking, we suppose that Kam knows the cryptosystem in use: this is known as the Kerckhoffs' principle. Hence, given a ciphertext, Kam's goal is to understand the key of the system.

Different attacks are based on the amount of information that Kam has.

- *Ciphertext only* attack: Kam knows a ciphertext or a part of it.
- *Known plaintext* attack: Kam knows plaintexts and their corresponding ciphertexts.
- *Chosen plaintext* attack: Kam can choose a set of plaintexts and encrypt them.
- *Chosen ciphertext* attack: Kam can choose a set of ciphertexts and decrypt them.

The attacks are based on available resources, i.e. computing power, storage memory, and time.

At this point we need to clarify what "break the code" means. Generally speaking, Kam wants to know the key, but if he is unable to recover the key, he could attempt a partial break of the code.

In the pyramid of Fig. 13.9 the highest level is the *total break* where Kam understands the key. The second level is the *global deduction*: Kam does not know the key but he discovers a functionally equivalent encryption and decryption method. Then we have *instance deduction*: Kam produces additional plaintexts or ciphertexts. Finally, we have *distinguishing algorithm*: Kam is able to distinguish a ciphertext from a random permutation.

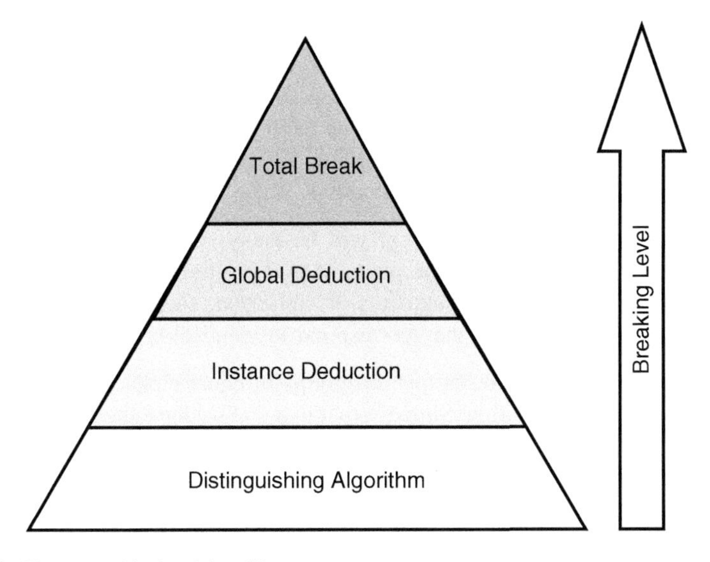

Fig. 13.9 The pyramid of codebreaking

For example, if we want to discover a key of a ciphertext obtained with a Caesar code (and all the substitution cryptosystems) we can use an attack called *Frequency Analysis*. This attack is based on the analysis of the frequency of letters or group of letters in a particular language. Typical distribution of letters in English language is shown in Fig. 13.10.

When Kam intercepts a message, he can easily find out the most frequent letter and decrypts it as either E or A or T, but unlikely as Z. By analyzing letter's frequency, and group of letters together, he can recover the plaintext.

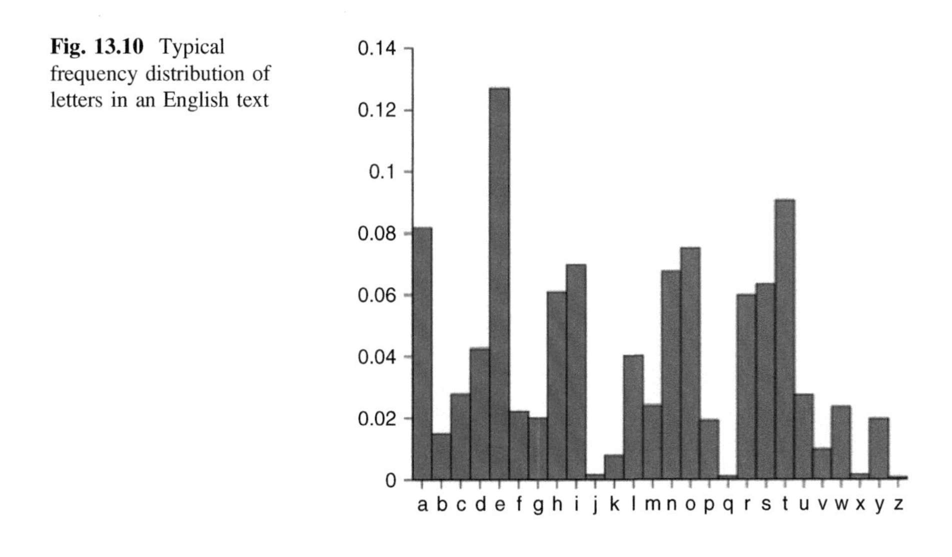

Fig. 13.10 Typical frequency distribution of letters in an English text

An evolution of this attack, used in more complex cryptosystems, as Vigenère codes, is called *Kasiski method*. The purpose of this attack is to understand the length of the key and then reduce the ciphertext to a cipher substitution that can be analyzed with frequency analysis attack. The method was discovered by Kasiski in 1863 and independently by Babbage in 1846.

The method is based on these observations:

1. two identical segments of plaintext will be encrypted to the same ciphertext whenever their occurrence in the plaintext is d position apart;
2. if we observe two identical segments of ciphertext, each of length at least 3, there is a good chance that they correspond to identical segment of plaintext.

Hence, the first thing to do is to find groups of equal characters, of at least 3 letters, and record their position. Suppose that in a text we have the same group of 3 letters separated of 165, 235, 275 and 285 positions. The greatest common divisor is 5 and it is very likely to be the keyword length. Now that the length is known, every group of 5 letters can be broken via the frequency analysis attack.

These two attacks (Frequency analysis and Kasiski method) are based on linguistic statistics, but as the cryptosystems complexity increases, more mathematics and computational power are required.

There are cases where codes are broken not because of the weakness of the code itself, but because of an erroneous or insecure usage. For example, encrypting two messages with the same key is an insecure process, the messages are said to be *in depth*: Kam gains a lot of information by analyzing more than one ciphertext encrypted with the same key.

Another weakness that historically helped breaking a code is the *indicator* transmission with the Enigma machine.

The key was kept constant for a period of time, generally a day. However, a different rotor position (Fig. 13.11) was used for each message, a sort of initialization message.

The starting position of these rotors was transmitted just before the ciphertext. It was design weakness and operator sloppiness in this indicator procedure that broke Enigma. The procedure works as follows: the operator sets the rotor as indicated by his list to the initial setting, i.e. to some specific combination of letters (e.g. RDKP) visible in the rotor window. Then the operator chooses a starting position for his message which becomes the indicator to be sent with the message (e.g. ABGY). He then types ABGY two times in the machine so that the message is encoded twice, for example in SWTHNQLM. He transmits this string and then the encrypted plaintext.

At the receiver side, the operator sets the rotor in the initial settings and then types SWTHNQLM. Immediately RDKP pops up, the receiver sets the rotors in that position and starts typing the ciphertext to obtain the plaintext.

The weakness of this scheme is that it is used as a worldwide setting. Moreover, the repetition of this value causes a security flaw.

Fig. 13.11 An example of a rotor position to send an indicator for the Enigma machine

The attacks used for symmetric-type cryptosystems are based on difficult mathematical problems. The most obvious way to attack this system is solving those mathematical problems. In case of RSA cryptosystem, 3 algorithms seem to be the most effective to factorize integers: quadratic sieve, elliptic-curve factorization and number field sieve [12–15].

Today Cryptanalysis tries to break RSA encryption by using a huge computational power. In 1980 10^{12} CPU operations were required to factor a number of 50 digits. The same number of operations was required to factor a number of 75 digits in 1984. Nowadays, it is possible to factor a number of 150 digits. Given the speed trend of CPUs, more and more digits are required to secure RSA cryptosystems.

13.2.3 Hash Functions

The previous sections described the use of encryption for the general case where Alessia wants to send a message to Rino and doesn't want Kam, who hears the message, to understand. However, encryption is used to solve also a number of other issues in telecommunication world. In this section we address these issues and how they are solved.

The hash function is any algorithm that maps a large bunch of data of variable length to a smaller set of data of fixed length [16, 17]. A cryptographic hash function is used to provide data integrity: in some way, it builds a fingerprint of data, so that when data change, the fingerprint is not valid anymore. It is also used when data are stored in an insecure location: fingerprints are re-computed from time

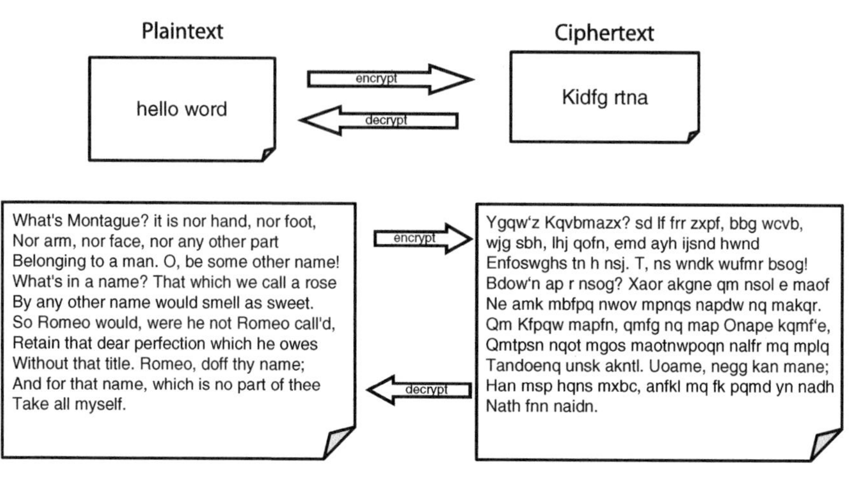

Fig. 13.12 General properties of encryption

to time to verify that they have not changed. This fingerprint is usually called *digest*. With a good hash function it is easy to compute a digest given a message, but it is unfeasible to find the message given the hash; in other words it is an unidirectional function. This is very different from encryption where we encrypt and decrypt, and the ciphertext has the same length of plaintext (Fig. 13.12). On the contrary, hash functions are unidirectional and the length of the digest is fixed despite the length of the message (Fig. 13.13).

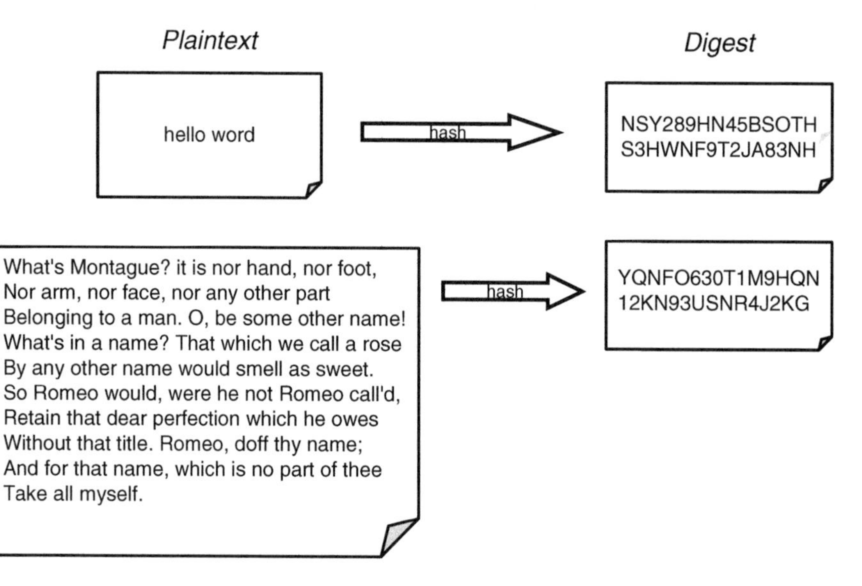

Fig. 13.13 General properties of hash function

Being unidirectional is a basic requirement for security. If a hash function hasn't this property, we say that it has *preimage resistance*. Another bad property is called *collision resistance*: given a message and its digest, there is another message with the same digest.

These bad properties imply that somebody can change a message without changing its digest. As discussed, hash functions are used to verify data integrity. For example, when we download a file from the web, our PC computes the hash function and compares it with the one published on the website as data integrity check [18, 19]. Please note that the digest is not visible on the screen but embedded in the properties of the file.

Another application is the password storage (Fig. 13.14).

PCs do not store cleartext password, because it would be too dangerous if the personal computer is stolen or somebody has access to its storage area. Therefore, the hash function of the password is stored, since it is unfeasible to recover the cleartext password from the hash. On the following login, the system re-computes the hash for the cleartext password and compares it with the stored one. Since it is impossible to have two messages with the same hash, the user must have typed the correct password to login in the system.

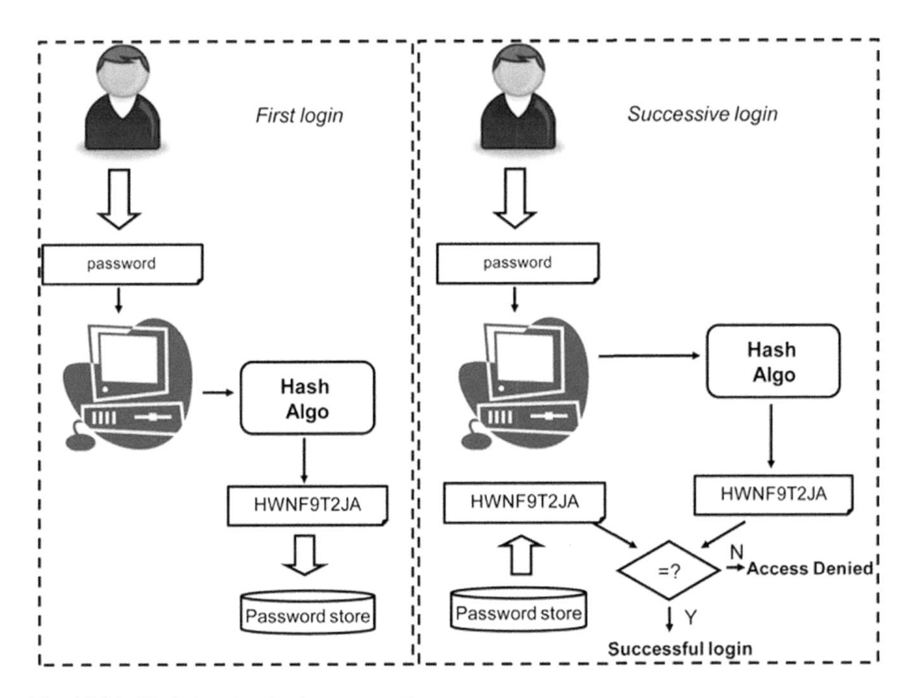

Fig. 13.14 Hash function in the password storage

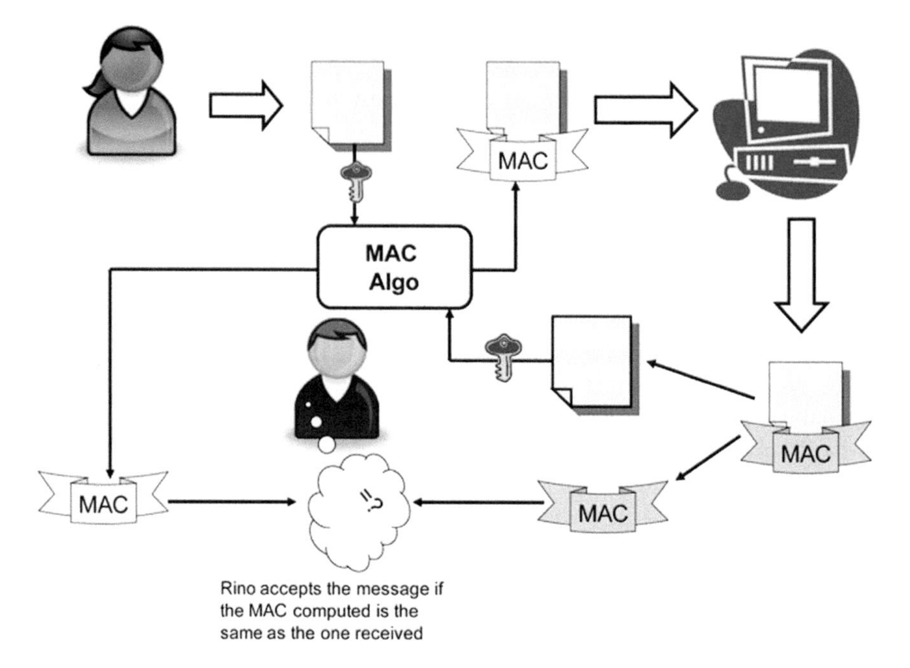

Fig. 13.15 Block scheme of the MAC usage

A special application is the *Message Authentication Code* (MAC). In this case keyed cryptographic functions are used [20–22]. These functions have stringent security requirements: specifically, even if the attacker is able to generate MACs for some messages, the attacker cannot guess the MAC for other messages without performing unfeasible amounts of computations.

The power of the MAC is that it guarantees both data integrity and authenticity of the message (Fig. 13.15). Moreover, since MACs require the same key for both receiver and sender, MAC functions are similar to symmetric encryption functions.

Another important usage of the hash function is the *digital signature*. There are three reasons to use a digital signature.

1. *Authentication*: this is the same reason why we sign documents. We want to authenticate the source of the messages. This is especially true in financial transactions.
2. *Integrity*: sender and receiver want to be sure that the message has not been corrupted during transition, even if it has been encrypted. Since there is no valid way to change a message without changing its signature, a non-valid signature detects a corrupted message.
3. *Non-repudiation*: once we have signed a document, we can't later deny it.

Although the discussion is very complex about how to digitally sign a document, high level blocks are sketched in Fig. 13.16.

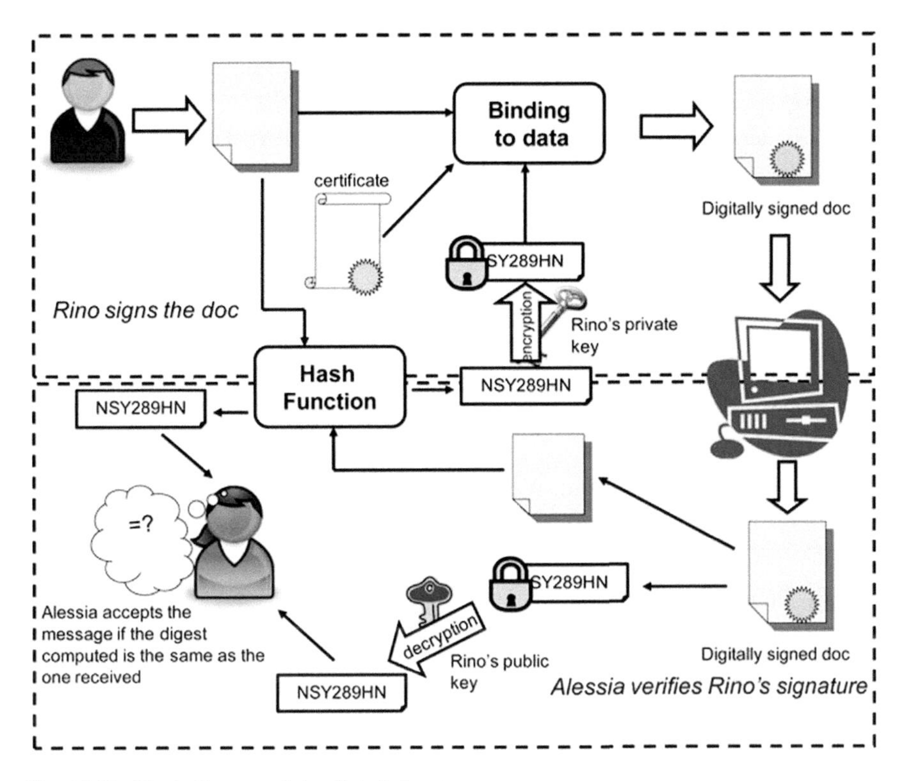

Fig. 13.16 Block diagram of the digital signature

First of all, the asymmetric encryption is used. This is because everybody should be able to decrypt, but nobody could modify the signed document. The hash function is computed on data. At this point the resulting digest is encrypted using Rino's private key, in order to produce the signature. Finally, there is a certificate that binds the signature to the document so that they can't be split.

On the receiver side, Alessia reads the signed message and decrypts the digest using Rino's public key to obtain the received cleartext digest.

She computes the hash function on the received document and compares the result with the obtained cleartext digest. If they are equal, she accepts the message from Rino, otherwise she repudiates it.

13.3 AES

Advanced Encryption Standard (AES), or its variant XTS-AES, is the encryption system generally used in Solid State Disks. It is an iterative symmetric encryption method, it supports 128, 192 and 256 bits as key length, and is available worldwide on a royalty-free basis. The algorithm was originally proposed by Daemen and Rijmen (called Rijndael) and it was published in the Federal Register on December 4, 2001 [23].

AES is iterative and the number of iterations (rounds) Nr depends on the key length: $Nr = 10$ if the key length is 128, $Nr = 12$ if the key length is 192 and $Nr = 14$ if the key length is 256.

AES works on a basic unit called *state*. Each state consists of a matrix: 4×4 bytes in the 128 case, and 4×8 bytes in the 256 case. We can split the algorithm in two parts: key generation and core algorithm [24–27].

13.3.1 Key Generator

Every key is split in 32-bit word. We have 8 words in the 256 case. At iteration i we have 32-bit word as input and 32-bit word as output. The algorithm proceeds as follows:

1. copy the input over the output;
2. *rotate operation* to rotate 8 bits to the left;
3. apply *S-box* to the 4 bytes individually;
4. on the first (leftmost) byte of the output word, XOR the byte with $2\ (i - 1)$. In other words, perform the *rcon operation* with i as the input, and XOR the *rcon* output with the first byte of the output word.

As the name may suggest, the rotate operation cyclically shifts bytes to the left:

$$\text{rotate}\ (B_0,\ B_1,\ B_2,\ B_3) = (B_1,\ B_2,\ B_3,\ B_0).$$

The *rcon* operation is equal to

$$rcon(i) = x^{i-1} \text{ in GF}(2^8) \text{ or } rcon(i) = x^{i-1} \bmod x^8 + x^4 + x^3 + x + 1 \text{ in GF(2)}.$$

For example rcon(1) = 1, rcon(4) = 3 and rcon(9) = 27.

Finally, we define the S-box in Fig. 13.17: it indicates a substitution to be made for each byte combination.

For example: S-box(9c) = de or S-box(f2) = 89.

13.3.2 AES Algorithm Core

Once we have defined how the keys and sub-keys are computed we now describe the AES algorithm.

	x0	x1	x2	x3	x4	x5	x6	x7	x8	x9	xa	xb	xc	xd	xe	xf
0x	63	7c	77	7b	f2	6b	6f	c5	30	01	67	2b	fe	d7	ab	76
1x	ca	82	c9	7d	fa	59	47	f0	ad	d4	a2	af	9c	a4	72	c0
2x	b7	fd	93	26	36	3f	f7	cc	34	a5	e5	f1	71	d8	31	15
3x	04	c7	23	c3	18	96	05	9a	07	12	80	e2	eb	27	b2	75
4x	09	83	2c	1a	1b	6e	5a	a0	52	3b	d6	b3	29	e3	2f	84
5x	53	d1	00	ed	20	fc	b1	5b	6a	cb	be	39	4a	4c	58	cf
6x	d0	ef	aa	fb	43	4d	33	85	45	f9	02	7f	50	3c	9f	a8
7x	51	a3	40	8f	92	9d	38	f5	bc	b6	da	21	10	ff	f3	d2
8x	cd	0c	13	ec	5f	97	44	17	c4	a7	7e	3d	64	5d	19	73
9x	60	81	4f	dc	22	2a	90	88	46	ee	b8	14	de	5e	0b	db
ax	e0	32	3a	0a	49	06	24	5c	c2	d3	ac	62	91	95	e4	79
bx	e7	c8	37	6d	8d	d5	4e	a9	6c	56	f4	ea	65	7a	ae	08
cx	ba	78	25	2e	1c	a6	b4	c6	e8	dd	74	1f	4b	bd	8b	8a
dx	70	e3	b5	66	48	03	f6	0e	61	35	57	b9	86	c1	1d	9e
ex	e1	f8	98	11	69	d9	8e	94	9b	1e	87	e9	ce	55	28	df
fx	8c	a1	89	0d	bf	e6	42	68	41	99	2d	0f	b0	54	bb	16

Fig. 13.17 S-box for AES

- State = plaintext. Perform the *AddRoundKey* operation between the state and the key.
- For each iteration:
  - Execute *SubBytes*
  - Execute *ShiftRows*
  - Execute *MixColumns*
  - Execute *AddRoundKey*
  - Execute *SubBytes*
  - Execute *ShiftRows*
  - Execute *AddRoundKey*
- The resulting ciphertext = State.

The AddRoundKey operation is simply the XOR (Fig. 13.18) between the State and the subkey obtained at that point using the key generator.

In the SubBytes step, each byte in the state matrix is replaced with a SubByte using an 8-bit substitution box, the S-box (Fig. 13.19). This operation provides the non-linearity in the cipher.

The ShiftRow operation operates on the rows of the State. Each byte of the row is cyclically shifted to the left by some locations. The first row does not shift, the second row shifts by one location, the third row by two locations and so on and so forth (Fig. 13.20).

In the MixColumn operation four bytes of each column of the state are combined using an invertible linear transformation. Together with the ShiftRow operation it provides *diffusion*, i.e. non-uniformity of the ciphertext.

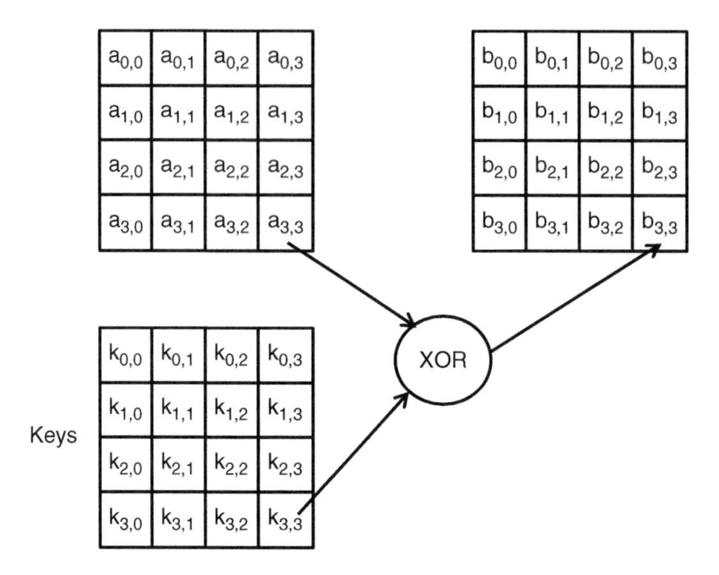

Fig. 13.18 Representation of the AddRoundKey operation

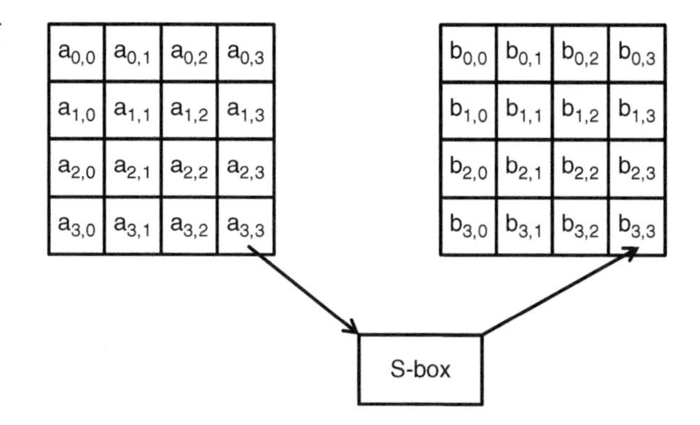

Fig. 13.19 Representation of the SubBytes operation

Each column is multiplied by a known matrix which is

$$\begin{pmatrix} 2 & 3 & 1 & 1 \\ 1 & 2 & 3 & 1 \\ 1 & 1 & 2 & 3 \\ 3 & 1 & 1 & 2 \end{pmatrix}$$

in the 128 case. Multiplication by 1 means no change, multiplication by 2 means shifting to the left, and multiplication by 3 means shifting to the left and then performing XOR with the initial unshifted value. After shifting, a conditional XOR with 0x1B should be performed if the shifted value is larger than 0xFF. This operation is represented in Fig. 13.21.

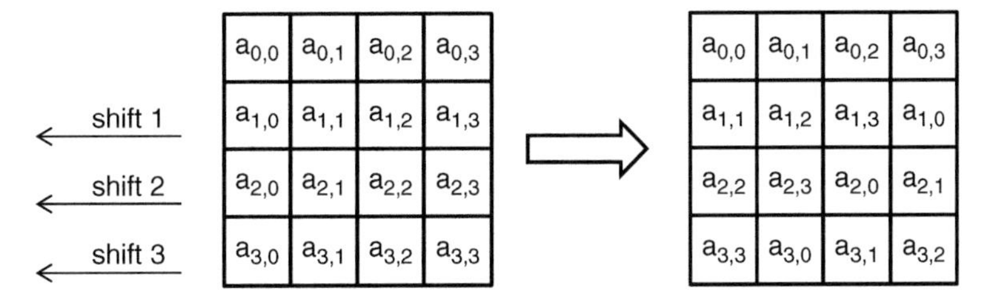

Fig. 13.20 Representation of the ShiftRow operation

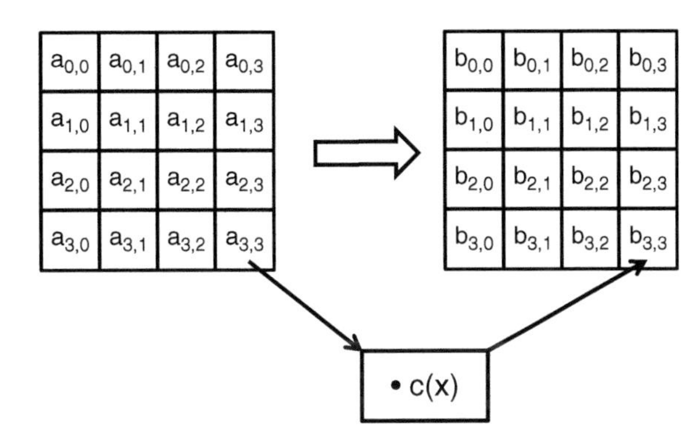

Fig. 13.21 Representation of the MixColumn operation

So far there aren't any known successful attacks to AES. Especially AES-256 is considered very secure, because all the operations are studied to mix data and avoid any linearity or uniformity.

13.4 SSD Security and Applications

As described in Sect. 13.1, SSDs are gaining popularity, but security is a hard matter. More and more companies build military-grade SSDs, protecting sensitive data from environmental and human threats. In fact, this is a very important issue in defense applications or financial applications where sensitive data are treated.

SSD security is so difficult because they are based on industry-standard NAND Flash chips that were designed for cameras and MP3 players: these memories have no physical security hooks that prevent them from being removed from enclosures. A hacker could easily unsolder NAND chips and read data using a standard Flash programmer. Once raw data are read, corresponding files could be reassembled using data recovery software.

When the SSD is broken, we want data to be erased or unreadable before throwing away the SSD. Secure Erase command exists but it has its own drawbacks. First of all, if the SSD is broken, it could be possible that some blocks become un-erasable, but a hacker can read back data from those blocks.

In addition there isn't a mechanism to erase single files, but the entire SSD must be erased.

The logical-to-physical mapping of SSDs makes files even harder to be completely erased. In fact, the erase operation is a slow operation in NAND Flash, so it happens that files are not really erased but just "marked" as erased to avoid a drop in performance. The problem is that the file-system does not know the real blocks where data are stored. Logical-to-physical mapping is managed by the Flash controller inside the SSD. In other words, it's like saying that the file-system hasn't a full control on the block locations. In this context, the most common way to increase security is encryption, and it must be done within the SSD itself.

Here is what happens. Data are input by the host, encrypted by the Flash controller, and then stored in NAND. During read operation, data are read from NAND, decrypted and output to the host. Encryption and key generation are completely transparent to the host. In this way, when we want to make data unreadable, it is enough to erase the locations where keys are stored. This location can be a NAND block or a RAM block in the Flash controller.

As already pointed out, AES-256 or the XTS-AES-256 are generally used in SSDs. The firmware running on the Flash controller sets the first key; following keys are computed by the key generator described in Sect. 13.3. All the keys are stored in specific NAND blocks.

Finally, we can state that encryption is the first step to secure data on SSDs, and the sooner we use it the more secure system we have. While it is easy to encrypt data already stored on a HDD, because we can re-write encrypted data in the same locations, this is not so easy with SSDs. NAND storage doesn't allow to re-write data on the same locations: actually, encrypted data are stored in different locations (logical-to-physical mapping). At the end of the day, encryption must be activated when the device is fresh in order to secure data from external attacks.

References

1. www.runcore.com
2. M. Wei, L.M. Grupp, F.E. Spada, S. Swanson, Reliably erasing data from flash-based solid state drives, in *Usenix FAST 11 Conference* (San Jose, 2011)
3. C. Shannon, Communication theory of secrecy systems. Bell Syst. Tech. J. **27**, 379–423 (1949)
4. O. Goldreich, *Foundations of Criptography: Basic Tools* (Cambridge University Press, Cambridge, 2001)
5. D.R. Stinson, *Cryptography: Theory and Practice* (Chapman & Hall/CRC, London, 2006)
6. W. Diffie, M.E. Hellman, Multiuser cryptographic techniques. Fed. Inf. Process. Stand. Conf. Proc. **45**, 109–112 (1979)
7. U. Maurer, S. Wolf, The Diffie-Hellman protocol. Des. Codes Cryptogr. **19**, 147–171 (2000)
8. B. Schneier, *Secrets and Lies: Digital Security in a Networked World* (Wiley, New York, 2000)

9. R.L. Rivest, A. Shamir, L. Adleman, A method for obtaining digital signatures and public key cryptosystems. Commun. ACM **21**, 120–126 (1978)
10. A.K. Lenstra, E.R. Verheaul, Selecting cryptographic key sizes. J. Cryptolo. **14**, 255–293 (2001)
11. M.O. Rabin, Probabilistic algorithms for testing primality. J. Number Theory **12**, 128–138 (1980)
12. M.J. Wiener, Cryptoanalysis of short RSA secret exponents. IEEE Trans. Inf. Theory **36**, 553–558 (1990)
13. A.K. Lenstra, Integer factoring. Des. Codes Cryptogr. **19**, 101–128 (2000)
14. D. Boneh, G. Durfee, Cryptoanalysis of RSA with private key d less than $N^{0.292}$. IEEE Trans. Inf. Theory **46**, 1339–1349 (2000)
15. D. Boneh, Twenty years of attacks on the RSA cryptosystem. Not. Am. Math. Soc. **46**, 203–213 (1999)
16. N. Ferguson, B. Schneier, *Practical Cryptography* (Wiley, New York, 2003)
17. H. Delfs, H. Knebl, *Introduction to Cryptography: Principles and Applications* (Springer, New York, Berlin, 2002)
18. R. Churchhouse, *Codes and Ciphers: Julius Caesar, the Enigma and the Internet* (Cambridge University Press, Cambridge, 2002)
19. M. Bellare, R. Canetti, H. Krawczyk, Keying hash function for message authentication. Lect. Notes Comput. Sci. **1109**, 1–15 (1996)
20. P. Preneel, P.C. Van Oorschot, On the security of iterated message authentication codes. IEEE Trans. Inf. Theory **45**, 188–199 (1999)
21. D. Pointcheval, J. Stern, Security arguments for signature schemes and blind signatures. J. Cryptol. **13**, 361–396 (2000)
22. T.P. Pedersen, Signing contracts and paying electronically. Lect. Notes Comput. Sci. **1561**, 134–157 (1999)
23. Advanced Encryption Standard in Federal Information Processing Standard (FIPS) Publication 197 (2001)
24. J. Nechvatal, E. Barker, L. Bassham, W. Burr, M. Dworkin, J. Foti, E. Roback, Report on the development of the advanced encryption standard (AES), 2 Oct 2000
25. S. Murphy, M.J.B. Robshaw, Essential algebraic structure within AES. Lect. Notes Comput. Sci. **2442**, 1–16 (2002)
26. S. Landau, Polynomials in the nation's service: using algebra to design the advanced encryption standard. Am. Math. Mon. **111**, 89–117 (2004)
27. S. Landau, Standing the test of time: the data encryption standard. Not. Am. Math. Soc. **47**, 341–349 (2000)

Index

A
Access time, 7, 20, 49, 193, 325
AC characteristics, 149
Adaptive memory, 48, 49
AddRoundKey, 473, 474
Adleman, L., 463
Advanced Encryption Standard (AES), 455, 471–473, 475, 476
AES, *see* Advanced Encryption Standard (AES)
All Bit Line (ABL), 85, 160–163
ALU, *see* Arithmetic Logic Unit (ALU)
Application classes, 222
Arithmetic Logic Unit (ALU), 141
Asymmetric coding, 153, 462, 471
Asynchronous interface, 6–8, 139, 144
ATA, 20, 33, 34, 36, 38, 457
Authentication, 458, 464, 470

B
Babbage, C., 466
Backwards-compatibility, 34
Bad block, 1, 10, 12, 56, 190, 245–247, 285
Bad Block Management (BBM), 12
BAR, *see* Base Address Register (BAR)
BBM, *see* Bad Block Management (BBM)
BCH, 12, 184–186, 189, 212–214, 369, 370, 374–377, 379–385, 388–398, 401, 402, 449–451
Belief propagation decoding, 402, 403, 407, 424, 433, 448
BER, *see* Bit Error Rate (BER)
Berlekamp, 184, 293, 377, 378, 380, 399–401
BET, *see* Block Erasing Table (BET)
BGA package, 2
Binary-input AWGN channel, 411, 413, 417, 422, 424, 430, 445, 450
Binary Symmetric Channel (BSC), 382, 410, 411, 413, 414, 417, 422, 424, 430, 434, 439, 441, 442
Bit Error Rate (BER) estimation, 18, 424, 450
Bit flipping decoding, 437
Bitline
  capacitance, 156, 158, 159
  parasitic, 156, 158, 160
  pitch, 80, 84, 91, 94
  precharge, 158
BLe, *see* Even page (BLe)
Block Erasing Table (BET), 346–351
BLO, *see* Odd page (BLO)
Block protection, 353–355
Blockwearing information table, 362, 363
Boot time, 31, 33
Bose, R.C., 370
Bose–Chaudhuri–Hocquenghem (BCH), 245, 289–291, 293, 302, 303, 305, 306
BSC, *see* Binary Symmetric Channel (BSC)
Bulk, 65, 75, 113, 167, 221
BWI-table merge, 364

C
Cached file selection algorithm, 50
Caching, 3, 44, 50, 193, 194
Caesar, J.C., 458
Capacitive coupling model, 64–66
Carlitz-Uchiyama inequality, 375, 389
CDM, *see* Cold-Data Migration (CDM)
CD variations, 93
Cell-to-cell variability, 93
Channel capacity, 407, 408, 412

R. Micheloni et al. (eds.), *Inside Solid State Drives (SSDs)*,
Springer Series in Advanced Microelectronics 37,
https://doi.org/10.1007/978-981-13-0599-3

Channel coding theorem, 407, 412
Channel encoder, 409
Charge pump, 171–175
Charge retention, 3, 64, 66, 76–78, 87
Charge trapping cells, 100
Charge Trapping Layer (CTL), 94–97, 99
Check nodes, 295, 296, 300, 419–421, 424, 428–432, 434, 437, 439, 442, 443, 445, 449
Chemical-Mechanical Planarization (CMP), 92
Chiang, M.L., 345
Chien, E.K., 399
Ciphertext, 459, 462–466, 468, 473
Client, 1, 26, 33, 219, 220, 222, 223, 225–227, 229
CMP, *see* Chemical-Mechanical Planarization (CMP)
Codebreakers, 458
Code ensemble, 421
Code Rate (CR), 371
Co-design, 196, 198
Cold data, 52, 285, 288, 343, 346–349, 351, 353–357, 359
Cold-Data Migration (CDM), 354, 356
Cold pool, 343, 346, 354–357, 359
Cold-Pool Adjustment (CPA), 356, 357
Collision resistance, 469
Computational security, 460, 461
Concatenation, 373, 404
Configuration read messages, 17
Consumer SSD, 31, 33
Control Interface (CI), 136–139
Correction capability, 184, 185, 243, 248, 259, 271, 274, 276, 279, 280, 282, 283, 293, 304, 306, 307, 369, 372, 374, 380, 381, 383, 386, 389, 391, 398, 400, 437, 450
CPA, *see* Cold-Pool Adjustment (CPA)
CR, *see* Code Rate (CR)
CRC, *see* Cyclic Redundancy Check (CRC)
Cryptoanalysis, 464
Cryptosystem, 455, 460, 461, 463, 464, 467
CTL, *see* Charge Trapping Layer (CTL)
Current consumption, 157
Cyclic code, 370, 374
Cyclic Redundancy Check (CRC), 18, 19, 32, 219, 220, 404

D
Daemen, J., 472
Data buffer, 190
Data center, 1, 25, 29–31, 33, 37, 40, 41, 240, 267
Data eviction, 194
Data fragmentation, 91, 94, 100
Data integrity, 8, 32, 467, 469, 470
Data link layer, 17–19
Datapath, 137–140
Data sensor, 52
Data strobe signal (DQS), 8, 144, 145
DC-DC converter, 2, 174, 175
DCO, *see* Digital Controlled Oscillator (DCO)
Decoder, 135, 136, 141, 176, 178, 184–188, 276, 291–293, 295–300, 304, 325, 375, 381, 398, 401–404, 409, 414, 417, 418, 421, 423, 425, 433, 436–439, 444, 445, 447, 450, 451
Degaussian, 455, 456
Density evolution, 423, 441
Detection, 12, 35, 152, 244, 269, 325, 369, 374, 381, 396, 398, 401
Detection capability, 212, 371
Diffusion, 108, 129, 221, 473
Digest, 468, 469, 471
Digital signatures, 458
DIMM, 47
Discrete capacitors, 220, 221
Distinguishing algorithm, 464
Disturbs, 11, 74, 118, 128, 152, 167, 183, 184, 211, 283
Double Data Rate (DDR) interface, 144
Double Data Rate (DDR) memory, 144
Double patterning, 91, 92
Double-supply voltage regulator, 175, 177
Downstream (DP), 15–17, 139, 141
DP, *see* Downstream (DP)
DQS, *see* Data Strobe Signal (DQS)
Drain, 65, 66, 89, 114, 118, 119, 121, 123, 128, 176, 212, 254, 256, 310, 415
DRAM, 20, 43, 44, 47, 91, 144, 190, 193–195, 197, 198, 206, 218, 219, 236, 239, 244, 245, 314–327
Dual-pool, 343, 346, 352–354, 356–359, 362–366
Dual port, 34, 35
Dynamic wear leveling, 11, 215

E
ECCs, *see* Error Correcting Codes (ECCs)
Electrical IPD thickness (EOT), 70, 81, 83, 84, 87, 97
Electrical physical layer, 18
Elias, P., 370
Embedded microcontroller, 135
Encoder, 184, 185, 376, 398, 420
Encryption, 455–464, 467, 468, 471, 476
EndPoints, 15–17

Endurance, 11, 31–33, 39, 41, 52, 53, 79, 98, 105, 183, 184, 187, 188, 193, 206, 209–211, 213, 215, 221–227, 229, 230, 248, 250, 260, 275, 283, 288, 313, 343, 344, 346, 360, 365, 366
Endurance stress, 210, 223, 225–227, 229, 230
Enigma, 458, 459, 466, 467
Enterprise, 1, 15, 19–23, 26, 29–37, 39–41, 52, 183, 195, 197, 207, 219–223, 225, 240, 401, 455
Enterprise SSD, 31, 33, 34, 229
Entropy of random variables, 407
Enumeration process, 16, 17
EOT, *see* Electrical IPD thickness (EOT)
EPROM, *see* Erasable Programmable Read Only Memory (EPROM)
Erasable Programmable Read Only Memory (EPROM), 3
Erase-history table, 363
Erase saturation, 71, 81, 82, 95, 99
Error Correcting Codes (ECCs), 184, 205, 344
Error floor, 424, 425
Error locator polynomial, 184, 375, 377–380, 399
Error prediction LDPC, 402, 403
Error reduction pulse, 183
Evenness-aware, 343, 346, 350–352, 365
Even page (BLe), 6
Extension, 25, 373, 382, 383, 388, 389, 391, 443, 444
External NAND, 45

F
Factorization, 461, 463, 467
FAT, *see* File Allocation Table (FAT)
FeRAM, 46
FFR, *see* Functional Failure Requirement (FFR)
FFS, *see* Flash File System (FFS)
FG, *see* Floating Gate (FG)
Fiber Channel (FC)
File Allocation Table (FAT), 10, 360
Filesystem, 44
Firmware Transaction Layer (FTL), 457
Flash cache, 22
Flash channel, 13, 145, 146
Flash controller, 1, 9, 23, 24, 55, 135, 200, 457, 476
Flash File System (FFS), 10
Flash management, 24, 31, 33, 39, 345–347, 352
Flash Transaction Layer (FTL), 10, 182, 183, 190, 196–201, 215, 216, 219, 346, 360, 361, 457
Floating Gate (FG), 3, 62–66, 68–72, 74–85, 87–89, 93–96, 98–100, 106, 115, 117–121, 123–127, 207, 208, 210–212, 233, 234, 251, 254, 255, 264, 309, 310
Floating gate NAND, 61–64, 68, 76, 79, 81, 82, 88, 90, 94, 117
Floorplan, 125, 135, 136
Form factor, 20, 21, 37, 234
Fowler-Nordheim tunneling, 67, 68, 86, 95, 168, 415
Frequency analysis, 465, 466
FTL, *see* Firmware Transaction Layer (FTL)
Functional Failure Requirement (FFR), 222–224, 229, 230

G
Gallager, R.G., 370, 419, 437
Gallager B decoding, 441
Garbage collection, 1, 10–12, 44, 56, 190, 196, 197, 241–243, 249, 250, 285, 289, 344, 346–349, 352, 353, 355, 356, 359–361, 365
Generator matrix, 290, 295, 300, 371
Global deduction, 464
Global Wordlines (GWLs), 176
GWLs, *see* Global wordlines (GWLs)

H
Hamming, R., 369
Hamming distance, 252, 290, 371, 402, 414
Hard-decision decoding, 213, 417, 450, 451
Hard Disk Drive (HDD), 38, 43–45, 47, 48, 50, 51, 196, 205, 206, 223, 455–457, 476
Hash function, 467–471
HDD, *see* Hard Disk Drive (HDD)
HDD reliability, 206
High speed, 144, 150
High voltage management, 135, 171
High voltage PMOS, 176
Hocquenghem, A., 370
Host data transfer, 31
Hot-cold swapping, 345
Hot data, 52, 53, 284, 343, 346, 347, 350, 351, 353, 356
Hot plug, 35, 36
Hot pool, 343, 346, 354–357, 359, 363
Hot Pool Adjustment (HPA), 356, 357
HPA, *see* Hot Pool Adjustment (HPA)
HPC, *see* High Performance Computing (HPC)
Hybrid SSD, 51, 53, 55

I

I/O, 13, 21, 23, 25, 45, 50, 68, 95–97, 137, 139, 141, 144–147, 150, 152, 153, 155, 192, 194, 197–200, 234, 238–240, 242
I/O operations per second (IOPS), 3, 6, 8, 11–13, 19, 21, 23, 25, 32, 38, 39, 45, 50, 72, 92, 94, 108, 125, 137, 139, 141, 144–147, 150, 152–155, 159, 163, 171, 174, 176, 183, 184, 186, 188–194, 197–200, 206, 207, 210–213, 217, 220, 343, 344, 346–348, 350–352, 356, 359–361, 365, 369, 371–374, 376–383, 385, 386, 389, 407, 409–411, 413, 415–420, 422, 424–429, 432, 434, 435, 438, 439, 442, 443, 451, 462, 464, 466, 472, 473
Incremental Step Pulse Programming (ISPP) algorithm, 67–70, 72, 74, 81, 84, 95, 99, 165, 166, 170, 255
Indicator, 466, 467
Information deduction, 464
Instance deduction, 464
Integrity, 32, 144, 151, 153, 470
Interleaved architecture, 6, 14, 160
Interleaving, 6, 13, 52, 158–160, 255
IPD layer, 68, 69, 81
ISPP, *see* Incremental step pulse programming (ISPP) algorithm

K

Kang, D., 62
Kasiski method, 466
Key, 34, 41, 47, 65, 120, 124, 151, 183, 199, 243, 268, 272, 273, 275, 277, 280, 284, 303, 318, 326, 345, 365, 374, 398, 401, 458, 459, 461–466, 470–473, 476
Key generator, 472, 473, 476
Kim, H.J., 345
Krawtchouk polynomial, 386

L

Lane, 15, 16, 18, 21
LBF, *see* Line-by-fill (LBF)
LBS, *see* Line-by-spacer (LBS)
LDPC, *see* Low-density Parity-check (LDPC) codes
Leakage current, 77, 78, 85, 129, 256
Least Significant Bit (LSB), 6, 169, 170, 252
Legacy flash memory, 23
Legacy NAND, 144
Likelihood Ratio (LR), 425–428, 430–432
Line-by-fill (LBF), 92–94
Line-by-spacer (LBS), 92–94
Logical block, 6, 10, 194, 344, 360, 361, 364
Logical page, 6, 302
Logical physical layer, 18
Logical segment, 360, 362
Low-Density Parity-Check (LDPC) codes, 13, 184–187, 212–214, 289, 293, 369, 370, 401–404, 407, 414, 419–426, 429, 430, 433, 437, 439–446, 448–451
LR, *see* Likelihood Ratio (LR)
LRU, 51, 53, 193, 194
LSB, *see* Least Significant Bit (LSB)

M

MacWilliams equality, 372
MacWilliams, F.J., 372
Matrix ip-well, 167, 168
Mechanical reliability, 39, 66
Memory controller, 8, 9, 11–13, 16, 47, 143, 183, 233, 321, 325
Message Authentication Code (MAC), 458, 470
Miscorrection probability, 381, 404
MixColumns, 473
MLC, *see* Multi-level cells (MLC)
MLC channel model, 415, 418, 445
Most-Recently-Used (MRU) algorithm, 355
Most Significant Bit (MSB), 6, 170, 252
MRAM, 43, 197, 326
MSB, *see* Most Significant Bit (MSB)
Multi-channel, 13, 21, 369, 398, 401
Multi-core system, 25
Multi-Level Cells (MLC), 4, 6, 11, 43, 51–53, 71, 72, 75, 84, 85, 90, 95–97, 129, 156, 169, 170, 183, 208, 210, 215, 217, 220, 233, 234, 251, 253, 295, 343, 344, 350–352, 415, 418, 419, 445
Multi-thread, 32
Mutual information, 304, 407, 408, 410, 417, 419

N

NAND channel, 2, 3, 5, 7, 13, 72, 74, 89, 112, 121, 144, 167, 198, 424
NAND flash, 3, 4, 6–8, 12, 21, 39, 41, 43, 47, 48, 51, 52, 55, 61, 62, 71, 75, 76, 79, 83, 85, 87, 88, 90, 91, 96, 98, 100, 105, 106, 111, 113, 117–122, 125, 126, 129, 135, 136, 166, 177, 178, 181–188, 190–193, 195–201, 205–221, 233–237, 239, 241, 245, 251, 252, 254, 255, 257–261, 263, 264, 266, 267, 270, 272, 275, 278, 280, 283, 286, 288, 291, 297, 300, 308–315, 319, 327, 328, 344, 360, 363, 369, 415, 424, 430, 437, 456, 475, 476
NAND interface, 6, 144

NAND on motherboard, 47
NMOS transistors, 152, 172
Non-repudiation, 470
Non-transparent bridging (NTB), 17, 18
Non Volatile Memory (NVM), 3, 24
Non-volatile RAM, 2, 197
NOR memories, 3, 61, 95
NTB, *see* Non-Transparent Bridging (NTB)
NVM, *see* Non Volatile Memory (NVM)

O
OCD, *see* Off Chip Driver (OCD)
Odd page (BLO), 6
Off Chip Driver (OCD), 146, 150, 150–152
One-time pad, 461, 462
One-way function, 463
ONFI, *see* Open NAND flash interface (ONFI)
ONFI interface, 47, 144
ONO layer, 68, 70
Open-drain, 151
Open NAND Flash Interface (ONFI), 7, 8, 144, 145
Operating System (OS), 24, 48, 55, 196, 199–201
Optical litho gap, 91
OS, *see* Operating system (OS)
Over-programming, 208
Overprovisioning, 248, 250, 285–288

P
Paired storage computer, 47
Parallel BCH, 15, 34, 155, 184, 369, 398
Parity, 13, 184, 212, 217–220, 237, 244–247, 250, 290–292, 294–296, 300, 302–305, 370, 371, 373–375, 404, 419, 420, 427, 431, 443
Parity-check matrix, 419, 423, 428, 433, 442–444, 446, 449
Pass disturb, 73–76, 86, 167
Passtransistor (PT), 176, 177, 371
Pass window, 75
Pattern-based PDC (PB-PDC), 50
p-BICS, *see* Pipe-shaped bit cost scalable (p-BICS)
PB-PDC, *see* Pattern-based PDC (PB-PDC)
PCB, 2, 152
PCI, 1, 2, 14–16, 183, 187, 188, 198, 199, 201, 398
PCI express (PCIe), 1, 9, 15, 16, 18, 21, 23, 25, 26, 195
  switch, 16
PCI-PCI Bridges, 16
PDC, *see* Popular Data Concentration (PDC)
Peterson estimation, 383, 396, 397
Physical destruction, 456
Physical segment, 360–362, 364
Pinheiro, E., 50
Plaintext, 459, 461–466, 468, 473
Popular Data Concentration (PDC), 50
Power consumption, 14, 19, 31, 43, 49, 106, 115, 145, 159, 167, 174, 183, 184, 196, 317, 344
Power failure, 220
Power failure protection, 220
Preconditioning, 39
Preimage resistance, 469
Program counter, 141
Program disturb, 48, 73, 115, 128, 129, 167, 170, 211
Program enable signal, 6, 147, 177
Program inhibit, 72, 74, 86, 167
Program pulse, 67, 68, 72, 115, 170, 175
Program saturation, 69–71, 94–96, 100
Provable security, 461
Public-key cryptosystem, 462, 463
Pushpull, 147
PVG, *see* Program Voltage Generator (PVG)

Q
Quadruple Patterning (QP), 94
Quasi-Cyclic LDPC (QCLDPC), 420
Queue-Head table (QH table), 363

R
Random Access Memories (RAM), 3, 16, 24, 141, 142, 327, 345, 347, 348, 360, 362–366, 476
Random Telegraph Noise (RTN), 88–90, 100, 105, 210
Raw Bit Error Rate (RBER), 183, 186–188, 206, 209–213, 215, 216, 228, 245, 287, 305
Ray-Chaudhuri, D.K., 472
Rcon operation, 472
Read disturb, 48, 76, 170, 190, 193, 209, 211, 215, 257, 261, 266, 267, 274, 275, 277, 282, 283, 288, 289, 307, 308, 313, 314, 319–321, 326, 327
Read Only Memories (ROM), 3, 141, 142
Readyboost, 45
Ready/busy signal, 7, 84, 144, 200, 355, 364
Readydrive, 49
Recent erase-cycle count, 354, 356–359, 362, 363
Redundancy, 33, 136, 137

Redundant Array of Independent Disks (RAID), 23, 206, 207, 217, 218
Reed, I.S., 12, 370
Replacement algorithm, 50
ReRAM, 43
Resetting interval, 347–352
Retention, 8, 48, 52, 72, 77, 79, 96, 97, 105, 110, 115, 121, 183, 206, 209, 210, 221, 225, 227–229
  stress, 223
  time, 223
Rijmen, V., 472
Rijndael, 472
Rivest, R.L., 463
Root complex, 15–17
Rotate operation, 472
Rotational latency, 20, 38
RSA, 461, 463, 467

S
Sample Size (SS), 223, 224
Sanitation, 456
SAS, 9, 20–24, 29, 33–41, 52, 183, 195
SAS expander, 34
SATA, 9, 20–24, 29, 33–41, 52, 182, 183, 195, 398, 456
S-box, 472, 473
SCSI, 20, 34, 36, 38
SCSI express, 33, 34, 38, 457
Second preimage resistance, 469
Sector size, 47, 75, 206, 217, 218, 225, 361, 456
Secure erase, 456, 457, 475
Seek latency, 38
Self-Aligned Double Patterning (SADP), 91–94
Self-Aligned STI (SASTI), 93
Self boost, 167
Self-Boosted Program Inhibit (SBPI), 72–74
Sense amplifier, 84, 85, 156, 160, 161, 254, 311
Sequential read, 20
Serial ATA Tunneled Protocol (SATP), 36
Shannon, C., 408
Shannon, E.C., 369
Shannon limit, 13, 293, 407, 413, 414, 419, 421, 441, 450, 451
ShiftRows, 473
Shortening, 373
SILC, *see* Stress Induced Leakage Current (SILC)
Simultaneous Switching Noise (SSN), 147, 149, 150
Single Data Rate (SDR), 145, 150
Single-Level Cells (SLC), 4, 6, 11, 43, 51–53, 71, 72, 76, 156, 169, 182, 184, 208, 234, 251, 344, 410, 415–418, 445, 449–451
SLC channel model, 410, 415, 418
Slew rate, 148, 150–152
Soft decoding, 185, 205, 282, 295, 300, 303–306, 402
Solomon, G., 12, 370
Source, 4, 61, 64–66, 74, 75, 90, 106, 108, 110, 112, 114, 118, 119, 121, 123, 125, 153, 155, 158, 159, 220, 236, 251, 256, 263, 267, 270, 281, 289, 310, 313, 314, 408, 470
Source line, 4, 61, 75, 106, 110, 125, 159
  capacitance, 65, 74, 83, 119, 145, 146, 172, 221
  program, 75, 113
SRAM, 47, 91, 144
SSD form factor, 1, 20, 33, 37
SSD interface, 20
SSD performance(s), 21, 39, 43, 194, 195, 198
SSD power consumption, 14
Static wear leveling, 11, 215, 345, 346, 348–352
Storage Class Memories (SCM), 43
Storage management, 47, 51
Stress Induced Leakage Current (SILC), 77, 78, 210, 211, 264, 265
Stress methods, 223
SubBytes, 473, 474
Supercapacitors, 197, 221
SWL-BETUpdate, 348–350
SW Leveler, 346–348
SWL-Procedure, 348–351
Symmetric encryption, 462, 470, 471
Syndrome, 184, 291–293, 376–378, 399, 400, 442, 443
Systematic code, 371
Sze, S.M., 62

T
Tanner graph, 295, 296, 419–422, 424, 429, 431, 441, 443, 446
TANOS, 95–98
Temp sensor, 3
Terabit Cell Array Transistor (TCAT) technology, 112
Testing, 12, 31, 137, 141, 206, 221–223, 225, 226, 246, 258, 282, 324, 381
Test Interface (TI), 137–139
Testmode, 136, 138–140
Test of primality, 464

The NVM express (NVMe), 1, 24–26, 33
The sense amplifier, 153, 157, 254, 311
3D memory cell integration, 125
Threshold voltage, 3, 63, 66–71, 74, 76, 78, 83, 84, 86–88, 90, 153, 156, 157, 164, 165, 177, 184, 185, 207–211, 217, 234, 251–262, 264–267, 272, 273, 275, 276, 278, 279, 281, 282, 285, 288, 297, 298, 304, 307–310, 313, 415–419
Through Silicon Vias (TSV), 146
Toggle mode, 144
Total break, 464
TOX field, 66, 68, 100
Transaction Layer Package (TLP), 17–19
Trap Assisted Tunneling (TAT), 210, 211, 264, 265
Trapping set, 424, 425
TSOP package, 7
TSV-integrated SSD, 205
Tunnel dielectric, 66, 67, 95
Tunnel Oxide (TOX), 64–70, 77–79, 81, 83, 84, 87–89, 95, 98–100

U
Unconditional security, 461, 463
Uncorrectable Bit Error Rate (UBER), 206, 207, 212, 222–225, 229, 230, 305
Unevenness level, 348, 349, 351
Upstream (UP), 15–17, 139, 141
Usermode, 138, 139
Usertestmode, 138, 139
Utilization limiter, 52

V
Vertical-stacked-array-transistor (VSAT), 112
Video-on-demand (VOD), 171–174
Voltage doubler, 171–174
Voltage regulator(s), 135, 136, 164, 174, 175
VTH window, 79, 170

W
Wear-leveling, 39, 56, 345, 354, 364, 366
Weight distribution, 372, 373, 382, 383, 385, 388
Weighted probability, 381
Wordline, 105, 113, 115, 129, 135, 136, 158, 159, 166–168, 176, 178, 220, 252, 254, 255, 261–263, 266, 271–273, 275, 276, 288, 325
Wordline decoder, 176
Write Amplification Factor (WAF), 43, 56, 190

X
XLC, 85, 87, 169, 170
XTS-AES, 471, 476